Independent Groups t Groups (Continued)

[10.5] $\hat{s}_{\overline{X}_1 - \overline{X}_2} = \sqrt{\left[\dfrac{(n_1 - 1)\hat{s}_1^2 + (n_2 - 1)\hat{s}_2^2}{n_1 + n_2 - 2}\right]\left(\dfrac{1}{n_1} + \dfrac{1}{n_2}\right)}$

[10.6] $\hat{s}_{\overline{X}_1 - \overline{X}_2} = \sqrt{\left(\dfrac{SS_1 + SS_2}{n_1 + n_2 - 2}\right)\left(\dfrac{1}{n_1} + \dfrac{1}{n_2}\right)}$

[10.7] $t = \dfrac{(\overline{X}_1 - \overline{X}_2) - (\mu_1 - \mu_2)}{\hat{s}_{\overline{X}_1 - \overline{X}_2}}$ [10.10] $eta^2 = \dfrac{SS_{\text{EXPLAINED}}}{SS_{\text{TOTAL}}}$ [10.11] $eta^2 = \dfrac{t^2}{t^2 + df}$

Correlated Groups t Test

[11.1] $\hat{s}_{\overline{D}} = \dfrac{\hat{s}_D}{\sqrt{N}}$ [11.2] $t = \dfrac{\overline{D} - \mu_D}{\hat{s}_{\overline{D}}}$ [11.4] $eta^2 = \dfrac{t^2}{t^2 + df}$

One-Way Between-Subjects Analysis of Variance

[12.10] $F = \dfrac{MS_{\text{BETWEEN}}}{MS_{\text{WITHIN}}}$ [12.15] $eta^2 = \dfrac{SS_{\text{BETWEEN}}}{SS_{\text{TOTAL}}}$

[12.16] $eta^2 = \dfrac{(df_{\text{BETWEEN}})F}{(df_{\text{BETWEEN}})F + df_{\text{WITHIN}}}$ [12.17] $CD = q\sqrt{\dfrac{MS_{\text{WITHIN}}}{n}}$

One-Way Repeated Measures Analysis of Variance

[13.13] $F = \dfrac{MS_{\text{IV}}}{MS_{\text{ERROR}}}$ [13.14] $eta^2 = \dfrac{SS_{\text{IV}}}{SS_{\text{IV}} + SS_{\text{ERROR}}}$

[13.15] $eta^2 = \dfrac{(df_{\text{IV}})F}{(df_{\text{IV}})F + df_{\text{ERROR}}}$ [13.16] $CD = q\sqrt{\dfrac{MS_{\text{ERROR}}}{N}}$

Chi-Square Test

[15.1] $E_j = \left(\dfrac{CMF_j}{N}\right)(RMF_j)$ [15.2] $\chi^2 = \sum \dfrac{(O_j - E_j)^2}{E_j}$ [15.5] $V = \sqrt{\dfrac{\chi^2}{N(L - 1)}}$

Two-Way Between-Subjects Analysis of Variance

[17.19] $F_A = \dfrac{MS_A}{MS_{\text{WITHIN}}}$ [17.20] $F_B = \dfrac{MS_B}{MS_{\text{WITHIN}}}$ [17.21] $F_{A \times B} = \dfrac{MS_{A \times B}}{MS_{\text{WITHIN}}}$

[17.22] $eta_A^2 = \dfrac{SS_A}{SS_{\text{TOTAL}}}$ [17.23] $eta_B^2 = \dfrac{SS_B}{SS_{\text{TOTAL}}}$ [17.24] $eta_{A \times B}^2 = \dfrac{SS_{A \times B}}{SS_{\text{TOTAL}}}$

[17.25] $CD = q\sqrt{\dfrac{MS_{\text{WITHIN}}}{nb}}$ [17.26] $CD = q\sqrt{\dfrac{MS_{\text{WITHIN}}}{na}}$

[17.27] $SS_{A \times B(K)} = \dfrac{n(\overline{X}_a + \overline{X}_d - \overline{X}_b - \overline{X}_c)^2}{4}$ [17.28] $F_{A \times B(K)} = \dfrac{MS_{A \times B(K)}}{MS_{\text{WITHIN}}}$

THIRD EDITION

Statistics for the Behavioral Sciences

THIRD EDITION

Statistics for
the Behavioral
Sciences

James Jaccard
State University of
New York at Albany

Michael A. Becker
The Pennsylvania State
University at Harrisburg

Brooks/Cole Publishing Company

I(T)P® An International Thomson Publishing Company

Pacific Grove • Albany • Bonn • Boston • Cincinnati • Detroit • London • Madrid • Melbourne
Mexico City • New York • Paris • San Francisco • Singapore • Tokyo • Toronto • Washington

Sponsoring Editor: *Jim Brace-Thompson*
Marketing Team: *Gay Meixel and Romy Taormina*
Editorial Assistant: *Terry Thomas*
Production Coordinator: *Fiorella Ljunggren*
Production: *Greg Hubit Bookworks*
Manuscript Editor: *Carol Reitz*
Permissions: *Cathleen S. Collins*
Interior Design: *John Edeen*

Cover Design: *Cheryl Carrington*
Cover Photo: © 1996 *Pedro Lobo/Photonica Inc.*
Interior Illustrations: *Lotus Art*
Typesetting: *ETP/Harrison*
Cover Printing: *Color Dot Graphics, Inc.*
Printing and Binding: *R. R. Donnelley & Sons Company, Crawfordsville*
Credits continue on p. 646

For more information, contact:

BROOKS/COLE PUBLISHING COMPANY
511 Forest Lodge Road
Pacific Grove, CA 93950
USA

International Thomson Editores
Seneca 53
Col. Polanco
11560 México D. F. México

International Thomson Publishing Europe
Berkshire House 168-173
High Holborn
London WC1V 7AA
England

International Thomson Publishing GmbH
Königswinterer Strasse 418
53227 Bonn
Germany

Thomas Nelson Australia
102 Dodds Street
South Melbourne, 3205
Victoria, Australia

International Thomson Publishing Asia
221 Henderson Road
#05-10 Henderson Building
Singapore 0315

Nelson Canada
1120 Birchmount Road
Scarborough, Ontario
Canada M1K 5G4

International Thomson Publishing Japan
Hirakawacho Kyowa Building, 3F
2-2-1 Hirakawacho
Chiyoda-ku, Tokyo 102
Japan

Printed in the United States of America
10 9 8 7 6 5 4 3 2

Library of Congress Cataloging-in-Publication Data
Jaccard, James.
    Statistics for the behavioral sciences / James Jaccard, Michael A.
Becker. — 3rd ed.
        p. cm.
    Includes bibliographical references and index.
    ISBN 0-534-17406-X
    1. Psychology—Statistical methods. 2. Social sciences—
Statistical methods. 3. Psychometrics. I. Becker, Michael A.,
[date]- . II. Title.
BF39.J28 1997
519.5´0243—dc20   96-15744
                                                        CIP

Brief Contents

Part 1 Statistical Preliminaries 1

Chapter 1 Introduction and Mathematical Preliminaries 2
Chapter 2 Frequency and Probability Distributions 34
Chapter 3 Measures of Central Tendency and Variability 67
Chapter 4 Percentiles, Percentile Ranks, Standard Scores, and the Normal Distribution 100
Chapter 5 Pearson Correlation and Regression: Descriptive Aspects 125
Chapter 6 Probability 157
Chapter 7 Estimation and Sampling Distributions 181
Chapter 8 Hypothesis Testing: Inferences About a Single Mean 205

Part 2 The Analysis of Bivariate Relationships 239

Chapter 9 Research Design and Statistical Preliminaries for Analyzing Bivariate Relationships 240
Chapter 10 Independent Groups *t* Test 259
Chapter 11 Correlated Groups *t* Test 298
Chapter 12 One-Way Between-Subjects Analysis of Variance 323
Chapter 13 One-Way Repeated Measures Analysis of Variance 361
Chapter 14 Pearson Correlation and Regression: Inferential Aspects 392
Chapter 15 Chi-Square Test 421
Chapter 16 Nonparametric Statistics 451

Part 3 Additional Topics 483

Chapter 17 Two-Way Between-Subjects Analysis of Variance 484
Chapter 18 Overview and Extension: Selecting the Appropriate Statistical Test for Analyzing Bivariate Relationships and Procedures for More Complex Designs 531

Appendixes 547
Glossary of Major Symbols 607
References 611
Answers to Selected Exercises 619
Index 639

Contents

Preface   xi
To the Student   xviii

Part 1 Statistical Preliminaries   1

Chapter 1 Introduction and Mathematical Preliminaries   2

1.1   The Study of Statistics   2
1.2   Research in the Behavioral Sciences   4
1.3   Variables   5
1.4   Measurement   6
1.5   Discrete and Continuous Variables   14
1.6   Populations and Samples   16
          BOX 1.1 Biased Sampling   17
1.7   Descriptive and Inferential Statistics   18
1.8   The Concept of Probability   19
1.9   Mathematical Preliminaries: A Review   20
1.10   Statistics and Computers   26

Chapter 2 Frequency and Probability Distributions   34

2.1   Frequency Distributions for Quantitative Variables: Ungrouped Scores   34
2.2   Frequency Distributions for Quantitative Variables: Grouped Scores   38
2.3   Frequency Distributions for Qualitative Variables   41
2.4   Outliers   42
2.5   Frequency Graphs   43
2.6   Misleading Graphs   51
2.7   Graphs of Relative Frequencies, Cumulative Frequencies,
          and Cumulative Relative Frequencies   52
2.8   Probability Distributions   53
2.9   Empirical and Theoretical Distributions   56
2.10   Method of Presentation   57
2.11   Example from the Literature   59

Chapter 3 Measures of Central Tendency and Variability   67

3.1   Measures of Central Tendency for Quantitative Variables   68
3.2   Measures of Variability for Quantitative Variables   78
3.3   Computational Formula for the Sum of Squares   83
3.4   Relationship Between Central Tendency and Variability   84
3.5   Graphs of Central Tendency and Variability   86

3.6   Measures of Central Tendency and Variability for Qualitative Variables   89
3.7   Skewness and Kurtosis   89
3.8   Sample Versus Population Notation   90
3.9   Method of Presentation   91
3.10   Example from the Literature   92

Chapter 4   Percentiles, Percentile Ranks, Standard Scores, and the Normal Distribution   100

4.1   Percentiles and Percentile Ranks   101
4.2   Standard Scores   105
4.3   Standard Scores and the Normal Distribution   109
4.4   Standard Scores and the Shape of the Distribution   113
4.5   Method of Presentation   113
       APPENDIX 4.1   The Normal Distribution Formula   121

Chapter 5   Pearson Correlation and Regression: Descriptive Aspects   125

5.1   Use of Pearson Correlation   125
5.2   The Linear Model   126
5.3   The Pearson Correlation Coefficient   130
5.4   Correlation and Causation   138
5.5   Interpreting the Magnitude of a Correlation Coefficient   139
5.6   Regression   140
5.7   Additional Issues Associated with the Use of Correlation and Regression   145

Chapter 6   Probability   157

6.1   Probabilities of Simple Events   159
6.2   Conditional Probabilities   160
6.3   Joint Probabilities   161
6.4   Adding Probabilities   162
6.5   Relationships Among Probabilities   162
6.6   Sampling With Versus Without Replacement   164
       BOX 6.1   Beliefs and Probability Theory   165
6.7   Counting Rules   166
6.8   The Binomial Expression   169

Chapter 7   Estimation and Sampling Distributions   181

7.1   Finite Versus Infinite Populations   181
7.2   Estimation of the Population Mean   182
7.3   Estimation of the Population Variance and Standard Deviation   184
7.4   Degrees of Freedom   187
7.5   Sampling Distribution of the Mean and the Central Limit Theorem   188
       BOX 7.1   Polls and Random Samples   191
7.6   Types of Sampling Distributions   197

Chapter 8   Hypothesis Testing: Inferences About a Single Mean   205

8.1   A Simple Analogy for Principles of Hypothesis Testing   205
8.2   Statistical Inference and the Normal Distribution: The One-Sample z Test   206
8.3   Defining Expected and Unexpected Results   210

8.4   Failing to Reject Versus Accepting the Null Hypothesis   211
8.5   Type I and Type II Errors   212
8.6   Effects of Alpha and Sample Size on the Power of Statistical Tests   214
8.7   Statistical and Real-World Significance   215
8.8   Directional Versus Nondirectional Tests   216
8.9   Statistical Inference Using Estimated Standard Errors:
       The One-Sample t Test   218
8.10   Confidence Intervals   225
8.11   Method of Presentation   229
8.12   Examples from the Literature   230

Part 2   The Analysis of Bivariate Relationships   239

Chapter 9   Research Design and Statistical Preliminaries for Analyzing Bivariate Relationships   240

9.1   Principles of Research Design: Statistical Implications   240
       BOX 9.1   Confounding and Disturbance Variables   247
9.2   Selecting the Appropriate Statistical Test to Analyze a Relationship:
       A Preview   251

Chapter 10   Independent Groups t Test   259

10.1   Use of the Independent Groups t Test   259
10.2   Inference of a Relationship Using the Independent Groups t Test   261
10.3   Strength of the Relationship   271
10.4   Nature of the Relationship   277
10.5   Methodological Considerations   278
10.6   Numerical Example   279
10.7   Planning an Investigation Using the Independent Groups t Test   281
10.8   Method of Presentation   283
10.9   Examples from the Literature   284

Chapter 11   Correlated Groups t Test   298

11.1   Use of the Correlated Groups t Test   298
11.2   Inference of a Relationship Using the Correlated Groups t Test   299
11.3   Strength of the Relationship   304
11.4   Nature of the Relationship   306
11.5   Methodological Considerations   306
11.6   Power of Correlated Groups Versus Independent Groups t Tests   308
11.7   Numerical Example   309
11.8   Planning an Investigation Using the Correlated Groups t Test   311
11.9   Method of Presentation   312
11.10 Examples from the Literature   313
       APPENDIX 11.1   Computational Procedures for the
                       Nullified Score Approach   316

Chapter 12   One-Way Between-Subjects Analysis of Variance   323

12.1   Use of One-Way Between-Subjects Analysis of Variance   323
12.2   Inference of a Relationship Using One-Way
       Between-Subjects Analysis of Variance   324

12.3   Relationship of the *F* Test to the *t* Test   338
12.4   Strength of the Relationship   338
12.5   Nature of the Relationship   339
12.6   Methodological Considerations   343
12.7   Numerical Example   343
12.8   Planning an Investigation Using One-Way
        Between-Subjects Analysis of Variance   347
12.9   Method of Presentation   347
12.10   Examples from the Literature   348
        APPENDIX 12.1   Rationale for the Degrees of Freedom   354

Chapter 13   One-Way Repeated Measures Analysis of Variance   361

13.1   Use of One-Way Repeated Measures Analysis of Variance   361
13.2   Inference of a Relationship Using One-Way
        Repeated Measures Analysis of Variance   363
13.3   Strength of the Relationship   372
13.4   Nature of the Relationship   373
13.5   Methodological Considerations   374
13.6   Numerical Example   376
13.7   Planning an Investigation Using One-Way
        Repeated Measures Analysis of Variance   379
13.8   Method of Presentation   380
13.9   Examples from the Literature   380
        APPENDIX 13.1   Determining the Nature of the Relationship
                         Under Sphericity Violations   385

**Chapter 14   Pearson Correlation and Regression:
                Inferential Aspects   392**

14.1   Use of Pearson Correlation   392
14.2   Inference of a Relationship Using Pearson Correlation   393
14.3   Strength of the Relationship   397
14.4   Nature of the Relationship   397
14.5   Planning an Investigation Using Pearson Correlation   398
14.6   Method of Presentation for Pearson Correlation   398
14.7   Examples from the Literature   399
14.8   Regression   401
14.9   Numerical Example   403
14.10   Method of Presentation for Regression   407
        APPENDIX 14.1   Testing Null Hypotheses Other Than $\rho = 0$ 412

Chapter 15   Chi-Square Test   421

15.1   Use of the Chi-Square Test   421
15.2   Two-Way Contingency Tables   422
15.3   Chi-Square Tests of Independence and Homogeneity   423
15.4   Inference of a Relationship Using the Chi-Square Test   423
15.5 2×2 Tables   429
15.6   Strength of the Relationship   430
15.7   Nature of the Relationship   431
15.8   Methodological Considerations   432

15.9   Numerical Example   433
15.10   Use of Quantitative Variables in the Chi-Square Test   434
15.11   Planning an Investigation Using the Chi-Square Test   435
15.12   Method of Presentation   436
15.13   Examples from the Literature   437
15.14   Chi-Square Goodness-of-Fit Test   439
        APPENDIX 15.1   Determining the Nature of the Relationship
                        Using a Modified Bonferroni Procedure   444

Chapter 16   Nonparametric Statistics   451

16.1   Rank Scores   452
16.2   Nonparametric Statistics and Outliers   454
16.3   Analysis of Ranked Data Using Parametric Formulas   455
16.4   Rank Tests for Two Independent Groups   455
16.5   Rank Test for Two Correlated Groups   459
16.6   Rank Test for Three or More Independent Groups   462
16.7   Rank Test for Three or More Correlated Groups   465
16.8   Rank Test for Correlation   468
16.9   Examples from the Literature   471
        APPENDIX 16.1   Corrections for Ties for Nonparametric Rank Tests   474

Part 3   Additional Topics   483

Chapter 17   Two-Way Between-Subjects Analysis of Variance   484

17.1   Factorial Designs   485
17.2   Use of Two-Way Between-Subjects Analysis of Variance   486
17.3   The Concepts of Main Effects and Interactions   487
17.4   Inference of Relationships Using Two-Way
        Between-Subjects Analysis of Variance   492
17.5   Strength of the Relationships   502
17.6   Nature of the Relationships   502
17.7   Methodological Considerations   505
17.8   Numerical Example   506
17.9   Unequal Sample Sizes   513
17.10   Planning an Investigation Using Two-Way
        Between-Subjects Analysis of Variance   515
17.11   Method of Presentation   516
17.12   Examples from the Literature   518

**Chapter 18   Overview and Extension: Selecting the Appropriate
                Statistical Test for Analyzing Bivariate Relationships
                and Procedures for More Complex Designs**   531

18.1   Selecting the Appropriate Statistical Test
        for Analyzing Bivariate Relationships   531
18.2   Case I: The Relationship Between Two Qualitative Variables   532
18.3   Case II: The Relationship Between a Qualitative Independent
        Variable and a Quantitative Dependent Variable   532

18.4   Case III: The Relationship Between a Quantitative Independent
         Variable and a Qualitative Dependent Variable   536
18.5   Case IV: The Relationship Between Two Quantitative Variables   536
18.6   Procedures for More Complex Designs   537
18.7   Alternative Approaches to Null Hypothesis Testing   540

Appendix A   Table of Random Numbers   547
Appendix B   Proportions of Scores in a Normal Distribution   550
Appendix C   Factorials   560
Appendix D   Critical Values for the t Distribution   561
Appendix E   Power and Sample Size   563
Appendix F   Critical Values for the F Distribution   587
Appendix G   Studentized Range Values (q)   591
Appendix H   Critical Values for Pearson r 594
Appendix I   Fisher's Transformation of Pearson r (r')   596
Appendix J   Critical Values for the Chi-Square Distribution   598
Appendix K   Critical Values for the Mann–Whitney U Test   600
Appendix L   Critical Values for the Wilcoxon Signed-Rank Test   603
Appendix M   Critical Values for Spearman r 605

Glossary of Major Symbols   607
References   611
Answers to Selected Exercises   619
Index   639
Credits   646

Preface

While developing the outline for this text, we were haunted by our own reactions to new statistics books: "Oh no, not *another* introductory statistics text." Dozens of introductory statistics books are available, some of which are excellent. Several have been on the market for a decade and have had the benefit of two or more revisions. Introductory statistics, unlike content areas in the behavioral sciences, does not become dated quickly. Many of the concepts taught ten years ago are still relevant today. So why another text?

Despite the existence of some excellent books, we have been unable to locate any that meet our personal goals in teaching introductory statistics at the undergraduate level. What follows is an elaboration of some of these goals and of how this text differs from others currently available.

Application and Integration

In our opinion, most introductory statistics texts fail to integrate sufficiently the subject matter of statistics with what students will encounter in behavioral science journals. A statistics course should not only teach students basic skills for analyzing data but also make them intelligent consumers of scientific information. Students need to know how to make sense of research reports and journal articles, and a firm grasp of statistics is a major step in this direction. In our view, there is a large gap between what students learn in statistics classes and the way statistics are presented in research reports. Accordingly, one goal of this book is to teach the reader how to present the results of statistical analyses when writing research reports. This also conveys to the reader the form in which he or she will encounter statistical information when *reading* research reports. Most statistics texts emphasize the importance of stating null and alternative hypotheses, critical values, and formal decision rules for drawing inferences with respect to the null and alternative hypotheses. Yet these are rarely explicitly stated in research reports, and many students find this omission confusing. This book attempts to overcome this discrepancy. A special section, "Method of Presentation," found toward the end of most chapters, provides examples of how statistical analyses are typically presented in research reports (using the format of the American Psychological Association) and discusses the rationale underlying the format of such presentation.

Learning When to Use Statistical Tests

Because of the way chapters and exercises are organized in most texts, students are essentially told which statistical procedure to use on a given set of data. This state of affairs is simply unrealistic. It is just as important to teach students *when* to use a particular statistic and *why* it should be used as it is to teach them how to *compute* and *interpret* the statistic. Given a set of data, many students cannot determine where to begin in answering relevant research questions. We have attempted to address this problem. To be sure, it is impossible to state any hard-and-fast rules for data analysis; it is always possible to find an exception or alternative procedures that might give better insight into the data. But some rough guidelines can be given, and the issues involved in selecting a statistical test can be explicated. In this text, each statistical technique is introduced by listing the conditions under which the test is most typically applied, and an interesting research example is then presented. Chapter 18 develops in detail issues to consider when selecting a statistical test to analyze one's data.

Relevance

A common complaint among students is that statistics is irrelevant and boring. This view is fostered, in part, by the tendency of statistics texts to use examples and exercises that *are* irrelevant and boring. Yet, it is possible to provide interesting applications of statistics. We have accomplished this in three ways: First, in most chapters, we have included a section ("Examples from the Literature") that presents interesting published research findings that are based on the statistical method being developed. Second, where relevant, chapter exercises present problems for which students are asked to take representative data from research that has been reported in the behavioral science literature and perform statistical analyses using what they have learned. Third, we have included a feature in most chapters ("Applications to the Analysis of a Social Problem") in which we analyze data that we collected on the problem of unintended teen pregnancy in the United States.

Unifying Theme

During the years we have taught statistics, we have found that students do not readily understand the common focus of the various statistical tests. Most students simply cannot appreciate the conceptual relationships among the *t* test, analysis of variance, Pearson correlation, and the chi-square test. As a result, no unifying theme is developed and students lose sight of the purpose of statistical analysis. In the present book, a unifying structure is provided. The fact that each of the major statistical tests concerns the relationship between variables is made explicit: The *t* test and analysis of variance usually (but not always) are applied when analyzing the relationship between a qualitative independent variable and a quantitative dependent variable; Pearson correlation is applied when analyzing the relationship between two quantitative variables; and the chi-square test is applied when analyzing the relationship between two qualitative variables. Three questions serve as the organizing framework for each test: (1) Given sample data, can we infer that a relationship exists between two variables in the population? (2) If so, what is the strength of the relationship? and, (3) If so, what is the nature of the relationship? As an example, in analysis of variance, the first question can be addressed by the test of the null hypothesis in the form of the test of the F ratio, the second by eta-squared, and the third by the Tukey HSD test. By relating these three questions to each of the major tests, we create a unified framework.

Conceptual Versus Computational Emphasis

This book emphasizes a conceptual understanding of statistics. With rapid progress in the computer field and with the widespread use of hand calculators tailored for statistics, it seems unwise to spend considerable time on computational formulas and methods of calculation. Very few students who take introductory statistics ever find it necessary to calculate statistics by hand. Rather, they read about them in research reports or learn to program a computer to do the calculations. Because of this, we emphasize both conceptual *and* computational formulas; the latter formulas are presented after the former have been introduced and reinforced.

Research Design

Another unique characteristic of this text is a chapter on research methods. We have always believed that statistics and research design are intertwined and that statistics should be placed in the context of design concerns. For example, how can students really grasp the meaning of error variance without some elementary understanding of disturbance variables? Chapter 9 is intended to provide an appropriate research context. In addition, each research example used to develop a statistical technique is discussed in the context of its methodological constraints. We hope this will encourage the student to consider the results of statistical analysis in a broader sense than most statistics books convey.

Advanced Students

We have included a special feature for advanced students. Appendixes to several chapters explain in more detail certain advanced concepts referred to in the body of the text. These appendixes generally contain more complex information and can be easily excluded from class presentation.

Material Covered

At first glance, the table of contents might suggest that this text is more advanced than the typical introductory statistics book. This is not the case. We recognize that different instructors emphasize different material. An introductory class could not even begin to cover all 18 chapters of the present book. The chapters included are intended to provide the instructor with a useful set of topics from which to choose. The order of chapters is flexible, except for natural progressions (for example, no one would cover t tests before means and standard deviations). The material not covered in class will be available as reference for students who pursue graduate work or advanced undergraduate research projects.

In talking with statistics instructors, we have found that one of the main differences in teaching statistics is in the treatment of probability. Some instructors prefer to cover it in some detail (as we have done in Chapter 6), whereas others prefer to give it less emphasis. *All* instructors recognize that probability is a key concept in statistics. However, some feel that topics such as conditional probabilities, joint probabilities, and sampling with versus without replacement have little practical relevance for statistical applications (for example, for computing t tests or analysis of variance). Thus, we have written Chapter 6 so that it can be omitted without disrupting succeeding chapters. The concept of probability is discussed in Chapters 1–4 in sufficient detail to give students the necessary appreciation for later statistical tests.

A second major difference among instructors is in the introduction of correlation and regression. Some instructors introduce these concepts early in the course in the context of descriptive statistics, whereas others introduce the material in the context of inferential statistics. Our own preference is for the latter because it is rare that correlation and regression are used in a purely descriptive manner. We therefore omit Chapter 5 in the early part of the course and then assign it along with Chapter 14 after discussing analysis of variance. Instructors who prefer otherwise can present the chapters in sequence. We should note that Chapter 5 uses purely descriptive formulas (e.g., for the standard error of estimate), whereas some texts use inferential formulas when discussing correlation/regression in descriptive contexts. The inferential formulas are presented in Chapter 14.

When considering reactions to the second edition, we were faced with several difficult decisions regarding the current edition. Reviewers were split about our emphasis on the difference between sample statistics and statistics used for purposes of statistical inference. For example, we introduce the sample formulas for the variance and the standard deviation (using N in the denominator instead of $N - 1$) in the chapters on descriptive statistics. It is only when we consider statistical inference that we introduce the special adjustments (and rationale for them) that are necessary to apply the concepts to inferential statistics. Several reviewers indicated their preference for simply using $N - 1$ in the denominators from the outset, so as not to "confuse" students. Philosophically, we find it difficult to present inferential formulas within the context of traditional descriptive statistics, and we believe students have just as hard a time figuring out why the analyst is dividing by $N - 1$ as why he or she is dividing by N. After much trial and error using the two different approaches with our students, we decided to keep our original framework. It is conceptually correct, it lays a proper foundation for later courses, and students *can* follow the logic.

In our own one-semester courses (which are *very* introductory), we omit Chapters 6, 16, and 17. We try to give students a brief, one- or two-lecture overview of the remaining chapters and encourage them to read what we could not cover. Also, within certain chapters, we skip selected sections (for example, percentiles) so that we can emphasize material we think is more appropriate *for our particular students*. We have tried to structure sections within chapters so that instructors who want to skip a topic can easily do so.

We have also focused discussion on the most common techniques in the behavioral science literature. This is not to say that the omitted concepts are not important. The decision to exclude these reflects space demands and a cost-benefit analysis of what students need from the course more than anything else.

Chapter Structure

Chapters 10–17 develop the major statistical tests typically introduced in beginning statistics courses. We have imposed a common structure on these chapters in order to underscore the common focus of the tests.

Each chapter begins with a discussion of the conditions under which the test is typically applied. Attention then turns to inferring whether a relationship exists between the variables under study and, if so, to determining its strength and nature. These issues are developed in the context of a research example, and critical computational stages are highlighted with study exercises within the chapter. The test is then placed in context via a section on methodological considerations that underscores the importance of interpreting statistics relative to research design considerations. Then a numerical example takes the student through an application of the statistical test from start to finish. A discussion of planning an investigation using the test is presented, with explicit consideration of power and sample size selection. The "Method of Presentation" section then discusses how the statistical results are typically reported in journal articles, and several examples from the literature and a discussion of the application of the test to teen pregnancy issues follow. The exercises for each chapter are of two types: (1) exercises designed to review and reinforce concepts presented in the chapter and (2) exercises that require students to apply these concepts to real research situations.

Changes from the Second Edition

We have made several changes in the third edition. First, we have added the feature referred to above to most chapters. In this section we analyze data from a single, comprehensive database that we collected on the problem of unintended teenage pregnancies in the United States. This section has several purposes. First, we explicitly take the reader through data analytic issues that arise as one analyzes real data. We discuss what happens as we encounter missing data, nonnormal distributions, outliers, and other complications that are typical of real data. We try to place the reader in an apprenticeship mode, so that he or she can observe firsthand some of the issues that an analyst must confront when applying statistics. Second, we report the results of our statistical analyses using computer output (from SPSS). Our purpose is not to teach students how to program or use computers. There is precious enough time for teaching statistics in a first course, let alone computer programming. However, we want at least to expose the student to computer printouts and make them comfortable with their format. Third, by using the same database in every chapter, a thread of continuity is woven throughout the book. Given the importance of unintended teen pregnancy to our society at large and student interest in this topic, we believe that this new section will stimulate student interest in the power of statistics as a tool of the behavioral scientist. Our one regret is that space and practical limitations do not permit us to do justification to the complex, integrated theories of adolescent behavior that guided our research. Rather, we were forced to illustrate statistical methods using interesting but somewhat simplistic and isolated theoretical formulations. The relevant theoretical background would require a book in its own right. Finally, we purposely chose our examples within this section to illustrate one or more important points about the chapter contents or to highlight issues that the student might not otherwise consider. This helps somewhat abstract and mundane issues to take on relevance in the context of a meaningful database.

We have added new material to make the statistical content of the book more up to date. We expanded the sections on graphs and added material on exploratory data analysis. We discuss the use of computers in statistical analysis. We added new material on sphericity violations in the chapter on repeated measures analysis of variance. We added a section on interaction comparisons in the chapter on factorial analysis of variance, so that students can properly follow up a statistically significant interaction effect in order to correctly interpret it. We added material on inferential aspects of prediction when discussing correlation and regression in Chapter 14. We added formal procedures for breaking down a statistically significant chi-square given contingency tables that are larger than 2×2. And we added more material on diagnostics of statistical assumptions and the robustness of the different statistical tests to assumption violations.

On a pedagogical level, we have included many additional examples from the literature and we have expanded the problem sets for almost all of the chapters. We added more "Exercises to Apply Concepts" that use interesting, literature-based examples for the student to work through. We now include a minimum of ten multiple-choice questions at the end of each chapter. Finally, we have tried to tighten up the writing style and "fine-tune" the level of exposition to make it more readable and uniform.

This book is accompanied by a student study guide and an instructor's manual with test items, as well as test banks in electronic format.

Acknowledgments

Many individuals have helped in the development and publication of this book. First, we would like to acknowledge the excellent work of the staff at Brooks/Cole: Jim Brace-Thompson, our editor; Fiorella Ljunggren, who managed the production of the book; and Cheryl Carrington and Vernon T. Boes, who created the cover. We are also very grateful to Greg Hubit for his efforts in carrying out the production of the book; to Carol Reitz for her skillful editing of the manuscript; and to John Edeen for the elegant interior design.

We thank the reviewers of all three editions for their helpful comments and suggestions. For the first edition, they were Teresa Amabile, Brandeis University; Roger Baumgarte, Winthrop College; David Brinberg, State University of New York at Albany; Stephen Edgell, University of Louisville; Scott E. Graham, Allentown College; Alfred Hall, College of Wooster; Stephen W. Hinkle, Miami University, Ohio; John M. Knight, Central State University; Scott E. Maxwell, University of Notre Dame; Ervin M. Segal, State University of New York at Buffalo; Karyl Swartz, City University of New York Herbert H. Lehman College; Jerry W. Thornton, Angelo State University; Brian A. Wandell, Stanford University; and Arnold Well, University of Massachusetts, Amherst.

For the second edition, they were Raymond P. Carlson, Bemidji State University; John C. Jahnke, Miami University; William E. Jaynes, Oklahoma State

University; Jack Kirschenbaum, Fullerton College; Mary E. Kite, Ball State University; George O. Rogers, Oak Ridge National Laboratory; Lanna Ruddy, State University of New York at Geneseo; Kirk H. Smith, Bowling Green State University; and Robert B. Stewart, Oakland University.

For this edition, the reviewers were Stephen E. Edgell, University of Louisville; Nicholas Esposito, Professor of Psychology Emeritus, State University of New York College at Cortland; Karen Ford, Mesa State College; Dana K. Fuller, Middle Tennessee State University; Albrecht Inhoff, State University of New York at Binghamton; Mary E. Kite, Ball State University; Joseph Lappin, Vanderbilt University; David C. Moore, University of Nebraska-Omaha; Ralph Payne, Shippensburg University of Pennsylvania; Maureen Powers, Vanderbilt University; Mark Scerbo, Old Dominion University; John E. Sparrow, University of New Hampshire-Manchester; and Dirk Steiner, Louisiana State University.

If there is any merit in this book, it may well be the result of input from the above individuals. Of course, any shortcomings are our responsibility alone.

James Jaccard
Michael A. Becker

To the Student

This is an introductory text designed for a first course in statistics, and in writing it we have assumed that our readers possess a minimum of mathematical background (basic algebra). For most of you, much of the material in the book will be new. Our experience in teaching statistics has led us to conclude that one of the most difficult things for students is the amount of new material they must assimilate and use. Statistics courses are unique in several respects. Later material relies heavily on a clear understanding of previous material. There is a continual building process, and you *must* keep up with the pace your instructor sets. Statistics is not the kind of material you can put off until the night before an exam and then cram for a test on the next day.

We have developed a number of features in the text that should help you in your study of statistics. First, we have included examples (called "Study Exercises") that present problems based on the material just covered. Working through these examples will help you acquire many important statistical concepts. Second, key terms are **boldfaced** and should be reviewed after reading each chapter. Make sure you understand and can define each one. Third, we have provided extensive exercises. We strongly recommend that you work through *all* of these exercises because they reinforce much of what you read. Some of the exercises are worked out step by step in the Study Guide. Fourth, we have included examples of interesting research that use the concepts developed earlier in the chapter. If you read these carefully, not only will you learn a good deal about behavioral science research, but you will also be able to appreciate more fully the role of statistics. Fifth, we have included "Method of Presentation" sections that describe how statistical test results are reported in professional journals and reports. These sections should help you understand more fully the material in the book and also the material you encounter in your study of the behavioral sciences. Sixth, we have included a feature in most chapters called "Applications to the Analysis of a Social Problem." In these sections, we analyze data that we collected on the problem of unintended teen pregnancy. We show you firsthand how we approach data analysis using the tools presented in this text, illustrating the issues that must be considered as one tries to answer substantive questions. In this section, our attempt is to put you in an "apprentice" role as we take you through different data analytic scenarios. We also use this section to introduce you to computer printouts and how statistical analyses are usually presented on such printouts. Finally, we have endeavored to make the "Glossary of Major Symbols" and the index as complete as possible to allow for ready reference and have provided selected formulas and a summary of statistical tests on the endpapers.

THIRD EDITION

Statistics for
the Behavioral
Sciences

I always find that statistics are hard to swallow
and impossible to digest. The only one that
I can ever remember is that if all the people who
go to sleep in church were laid end to end,
they would be a lot more comfortable.

Mrs. William Howard Taft
Wife of American president

PART **1**

Statistical Preliminaries

CHAPTER 1

Introduction and Mathematical Preliminaries

1.1 The Study of Statistics

Behavioral science research has revealed many fascinating pieces of information about our society and everyday life. Consider the following examples:

A majority of children born in the United States today will not grow up in the traditional nuclear family environment. Over 50% of white children and over 80% of black children will spend some time in a single-parent family before they reach the age of 18.

Drunk driving is responsible for 15,000 to 25,000 deaths per year, a rate that over the period of two years equals the number of Americans killed throughout a decade of fighting in Vietnam.

In the context of dating and friendships, do opposites attract? Research suggests not. In general, the more similar your beliefs are to another individual's, the more you will tend to like that individual.

Some theorists have suggested that exposure to violence in the media may increase aggression in children and adults, whereas other theorists suggest that it may actually reduce aggression. Research tends to support the former position: The more media violence individuals watch as children, the more aggressive they tend to be as adults, everything else being equal.

Infertility in the United States is more widespread than people believe. Estimates are that as many as 20% of American couples who want to have (additional) children have difficulty doing so.

Many people believe that social problems such as crime, alcoholism, suicide, and divorce are more likely to occur in large urban cities. However, research suggests otherwise. For example, one analysis found that an individual is more likely to become a victim of crime in the parking lot of a suburban shopping mall than in the central section of a large city. Another analysis found that cities like New York, Chicago, Houston, Detroit, and Atlanta are *not* among the metropolitan areas with the highest rates of these phenomena, considered as a whole, but that Odessa (Texas), Reno (Nevada), and Lakeland (Florida) are among the 15 areas with the highest such rates in the nation.

These findings come from studies in different disciplines, but all have at least one thing in common: All the researchers used statistics to help reach their conclusions. In fact, the conclusions would have been impossible to make with any degree of scientific validity without the benefit of statistics. The behavioral scientists who conducted the research were not statisticians; rather, they were trained as psychologists, sociologists, health professionals, anthropologists, and economists. Nevertheless, they needed to use statistics as a tool to help them gain perspective on the particular social problems of interest to them.

It has become common for courses in statistics to be required of students who major in the behavioral sciences. Many students question why statistical training is necessary. There are several reasons. As noted above, statistics is an integral part of research activity. Important questions and issues are addressed in behavioral science research, and statistics can be a valuable tool in developing answers to these questions. For the student who makes a career of conducting research, statistical analysis will prove to be a useful aid in the acquisition of knowledge.

But in fact, many students who take statistics courses will not end up in careers that require them to play an active role in research. For example, many psychology students want to pursue a career in which they counsel others or conduct psychotherapy. They are uninterested in conducting formal psychological research. Although these students may not actually conduct research, they may still be required to read, interpret, and use research reports. These reports will usually rely on statistical analyses to draw conclusions and suggest courses of action. Knowledge of statistics is therefore important to help one understand and interpret these reports.

Research that uses statistical analysis is clearly having an impact on society, both in our everyday lives and in more abstract situations. On television we see commercials that report research "demonstrating" that brand A is three times as effective as brand X. In national magazines and newspapers, we read the results of surveys of public opinion and attitudes toward politicians. Many magazines include special sections designed to disseminate to the public at large the results of research in the physical and behavioral sciences. As our society becomes more technologically complex, greater demands are being placed on professionals to understand and use the results of research designed to solve applied problems. This generally requires a working understanding of statistical methods.

A knowledge of statistical analysis also helps to foster new and creative ways of thinking about problems. Several colleagues have remarked on the new insights they developed when they approached a problem from the perspective of statistical analysis. Statistical "thinking" can be a useful aid in suggesting alternative answers to questions and posing new ones. In addition, statistics helps to develop one's skills in critical thinking, with both inductive and deductive inference. These skills can be applied to any area of inquiry and hence are extremely useful.

1.2   Research in the Behavioral Sciences

The major concern with statistics in this book is how they are used in behavioral science research. As such, it will be useful to consider briefly the research process as it is commonly used in the behavioral sciences.

Most people do not view scientific research as a process but rather as a product. Reference is made to a "body of facts" that is known about some phenomenon. Scientific research is better characterized as an ongoing process that consists of five general activities. The first activity is the formulation of a question about some phenomenon. Why do people smoke marijuana? Why do some children do better in school than others? Why do some people fail to help another person who is in need of aid? The second activity is forming a **hypothesis** concerning the question. A hypothesis is a statement proposing that something is true about a given phenomenon. One might hypothesize that people smoke marijuana because of pressure from their peers to do so. Or one might hypothesize that children's school performance is influenced by the value placed on education in the home. The third activity involves designing an investigation to test the validity of the hypothesis. In such an investigation one makes systematic observations of individuals or groups of individuals in settings that are conducive to testing the hypothesis. The fourth activity is analyzing the data collected in the investigation in order to help the researcher draw the appropriate conclusions. This is generally done with the aid of statistics. The final activity is drawing a conclusion and thinking about the implications of the investigation for future research.

This characterization of the research process is somewhat oversimplified and does not do justice to the diverse ways in which scientific knowledge is gained. For example, sometimes theories and hypotheses are formed after empirical observations have been made, and then these theories are subjected to further evaluation in subsequent studies. Many investigations focus on a complex set of hypotheses that are organized in a multivariate fashion, rather than a simple, single hypothesis. Indeed, some of the most prominent behavioral theories ever proposed have evolved from a process that is different from that described above. For example, much of the theoretical work of Sigmund Freud was based on intensive case studies that were not subject to rigorous scientific testing and feedback. Science is best characterized as an interplay between theory and data, and the exact manner in which theory and data interact to advance our information base is complex. However, it is safe to say that statistics serves as a bridge between theory and data and, as such, it is an important tool for the behavioral scientist.

Although it may seem surprising, the most exciting aspect of research for many scientists is statistical analysis. This is because it is during the act of statistical analysis that the results of one's investigative efforts first become apparent, and the researcher gets his or her first insight into what the data are suggesting. On numerous occasions, we have had colleagues burst excitedly into our offices with a computer printout in hand, ready to discuss a fascinating result that has emerged from a statistical analysis.

The remainder of this chapter is designed to introduce you to some basic concepts that are central to statistical analysis. With this as a foundation, we will turn to formal aspects of data analysis in future chapters.

1.3 Variables

Most behavioral science research is concerned with **variables**. A variable is a phenomenon that takes on different values, or *levels*. For example, gender is a variable that takes on two values, male and female. The number of hours per week that someone watches violent television programs is a variable that takes on values of 1 hour, 2 hours, and so on. In contrast, a **constant** does not vary within given constraints. For instance, the value to four decimal places for the mathematical quantity π (pi) is always 3.1416. Because it takes on only one value that never changes, π is a constant. If an investigation is conducted with only females, then in the context of that investigation, gender is a constant because it takes on one and only one value (female).

Researchers distinguish between variables. One distinction in the behavioral sciences is between an **independent variable** and a **dependent variable.** Suppose an investigator is interested in the relationship between two variables: the effect of information about the gender of a job applicant on hiring decisions made by personnel managers. An experiment might be designed in which 50 personnel managers are provided with descriptions of a job applicant and asked whether they would hire that applicant. The applicant is described to all 50 managers in the same way on several pertinent dimensions. The only difference is that 25 of the managers are told that the applicant is a woman, and the other 25 managers are told that the applicant is a man. Each manager then indicates his or her hiring decision. In this experiment, the gender of the applicant is the independent variable and the hiring decision is the dependent variable. The hiring decision is termed the *dependent variable* because it is thought to "depend on" the information about the gender of the applicant. The gender of the applicant is termed the *independent variable* because it is assumed to influence the dependent variable and does not "depend" on the other variable (i.e., the hiring decision).

A useful tool for identifying independent and dependent variables is the phrase "The effect of _____ on _____." The variable name that fits into the first blank is the independent variable, and the variable name that fits into the second blank is the dependent variable. For example, in a study on the effect of psychological stress on blood pressure, the independent variable is the amount of psychological stress an individual is feeling and the dependent variable is the individual's blood pressure. Similarly, if the effect of child-rearing practices on intelligence is studied, the independent variable is the type of child-rearing practice and the dependent variable is the child's intelligence.

The term *independent variable* has assumed different meanings in various areas of the behavioral sciences. Some investigators restrict the definition of an independent variable to a variable that is explicitly manipulated in the context of an experiment (such as the information about the gender of the applicant in the hiring example). We adopt the more general definition of an independent variable as any variable that is presumed to influence a second variable (the dependent variable). According to this definition, it is not necessary for a variable to be experimentally manipulated in order to be an independent variable. Note that just because a researcher presumes that one variable influences another does not necessarily mean that it does. This is only a presumption made for purposes of the investigation.

The distinction between independent and dependent variables parallels cause-and-effect thinking, with the independent variable being the cause and the dependent variable being the effect. The distinction is central to theorizing in the behavioral sciences because so many of our theories are based on cause-and-effect reasoning. When reading studies or evaluating certain statistics, you will often find it useful to make distinctions between the presumed causes and the presumed effects. What is the cause and what is the effect in a theorist's analysis? Is it possible to logically reverse the order of the cause-and-effect relationship? For example, some theorists argue that the many hours people watch violent television programs lead to aggressive tendencies. Could it be that the reverse is true: that aggressive tendencies lead someone to watch more violent television programs? As it turns out, statistics and clever experimental design can help us tease out these competing explanations of an association between television viewing habits and aggression. We will say more about this in later chapters.

1.4   Measurement

A major feature of behavioral science research is **measurement.** Most empirical research involves measurement of some kind. For example, behavioral scientists who study intelligence have developed intelligence tests to measure intelligence. Organizational psychologists who study how happy people are at work have developed a variety of measures of job satisfaction. Some psychologists have even tried to measure such difficult concepts as love.

Developing valid measures of concepts can be difficult. Entire subdisciplines within psychology and sociology (called psychometrics and sociometrics) have evolved that specialize in identifying procedures for developing good measures. It is not surprising that there is some controversy about the best way to conceptualize the measurement process and the best way to develop measures. The issues are important not only because measurement is so central to scientific investigations, but also because measurement increasingly underlies major decisions that affect our lives. For example, a substantial number of elementary school districts in the

United States use scores on aptitude tests to place children into programs for advanced learning or study. But are these aptitude tests really sufficiently valid that we can justify using them as a basis for providing different educational experiences to children? Some behavioral scientists argue yes and others argue no.

Measurement involves translating empirical relationships between objects into numerical relationships. This frequently takes the form of assigning numbers to people (or objects) in such a way that the numbers have meaning and convey information about differences between people. For example, suppose we have measured the intelligence of 20 people using a standard IQ test. If two people are of equal intelligence, we would want them to have identical scores (i.e., numbers) on our measure of intelligence. If one person is smarter than another, then we would want the former individual to have a higher score (i.e., number) on our measure.

Obviously, there are many ways in which differences between people can be mapped onto a number system; hence, there are many different types (or levels) of measurement. We now discuss the four types of measurement typically used in the behavioral sciences: nominal, ordinal, interval, and ratio.

Nominal measurement involves using numbers merely as labels. An investigator might classify a group of people according to their religion—Catholic, Protestant, Jewish, and all others—and use the numbers 1, 2, 3, and 4 for these categories. In this case, the numbers have no special quality about them; they are used merely as labels. If we wished, we could have used any other set of numbers instead (for instance, 13, 48, 7, and 101). In behavioral science research, the basic statistics of interest for variables that involve nominal measurement are frequencies (for example, how many people are Democrats, how many are Republicans), proportions, and percentages.

A second level of measurement is **ordinal measurement.** A variable is said to be measured on an ordinal level when the categories can be *ordered* on some continuum or dimension. Suppose that a developmental psychologist is studying the effects of psychological stress during pregnancy on the growth of the fetus and uses height of the newborn as one index of physical growth. The psychologist takes four newborns who differ in height and assigns the number 1 to the shortest infant, the number 2 to the next shortest infant, the number 3 to the next shortest infant, and the number 4 to the tallest infant. In this case, height is measured on an ordinal level, which allows the infants to be ordered from shortest to tallest. Thus, with ordinal measurement, the researcher classifies individuals into different categories that, in turn, are ordered along a dimension of interest.

Note in the preceding example that height was *not* measured in terms of feet or inches. The shortest infant had a score of 1 on the measurement scale, the next shortest infant had a score of 2, and so on. This set of measures (that is, the rank order from shortest to tallest) exhibits ordinal characteristics. As we illustrate later, height can be measured in other ways whereby the measures have more than just ordinal characteristics.

A third level of measurement is **interval measurement.** Interval measures have all the properties of ordinal measures but allow us to do more than order objects on a dimension. They also provide information about the *magnitude* of the differences between the objects. For example, interval measures not only would tell us that one infant is taller than another, but also would convey a sense of how much taller one infant is than another. Stated somewhat more technically, interval measures have the property that numerically equal distances on the scale represent equal distances on the dimension being measured. For example, a psychologist might study the effect of temperature on aggression. When temperature is measured in degrees Fahrenheit, the difference in room temperature between 68° and 70° is the same as the difference in room temperature between 101° and 103°. In both instances, the difference of 2° corresponds to the same absolute amount of heat in the air. If we were to separately add each of these 2° to the air temperature, the level of heat would increase by the identical amount because any 2° represent the same amount of heat as any other 2°. Interval measures provide information about the magnitude of differences because of this useful property. Note that this was not true with the ordinal measure of height. It was not necessarily true that the difference in height between infants 3 and 4 was the same as the difference in height between infants 1 and 2. It was only true that infant 4 was taller than infant 3, who, in turn, was taller than infant 2, who, in turn, was taller than infant 1.

A fourth level of measurement is **ratio measurement.** Ratio measures have all the properties of interval measures (and, hence, ordinal measures as well) but provide even more information. Specifically, ratio measures map onto the underlying dimension in such a way that ratios between the numbers represent ratios of the dimension being measured. For example, if we use inches to measure the underlying dimension of height, it is the case that a child who is 50 inches tall is twice the height of a child who is 25 inches tall. Similarly, a child who is 60 inches tall is twice the height of a child who is 30 inches tall. Inch is a ratio level measure of height. A temperature of 80 °F is not twice as hot as a temperature of 40 °F in the sense that the amount of heat in the air at the former temperature is not twice that of the latter temperature.

We can provide an informal, intuitive appreciation of the differences between ordinal, interval, and ratio levels of measurement by examining three different ways of measuring the height of buildings. Suppose an environmental psychologist is interested in how people perceive the heights of buildings and as part of a study measures the exact heights of all buildings that are taller than 100 feet in a particular section of a city. Figure 1.1a shows graphically the heights of four such buildings and indicates how tall each one is. The first way of measuring the heights of these buildings is to assign the number 1 to the shortest building, the number 2 to the next shortest building, the number 3 to the next shortest building, and the number 4 to the tallest building (see Figure 1.1b). This assignment represents ordinal measurement. It allows us to order the buildings on the dimension of height, but it does not tell us anything about the magnitude of the heights. A second method is to measure by how many feet each building exceeds the 100-feet criterion. In this case, we find that building D is 2 feet taller than the criterion, building B is 4 feet taller than the criterion, building C is 80 feet taller than the criterion, and building A is 104 feet taller than the criterion (see Figure 1.1c). In contrast to the previous (ordinal) measurement, now not only can we order the buildings on a dimension of height, but also we have information about the relative magnitudes of the heights: Building B is 2 feet taller than building D (4 − 2 = 2),

FIGURE 1.1   **Three Different Ways of Measuring the Heights of Four Buildings**

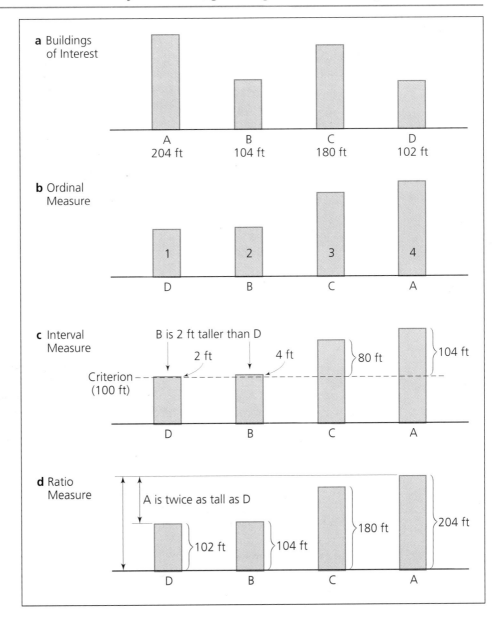

building C is 76 feet taller than building B (80 − 4 = 76), and so on. We have measured height on an interval scale. Note that on this scale, even though building B has a score of 4 (that is, it is 4 feet above the criterion) and building D has a score of 2 (it is 2 feet above the criterion), it is not the case that building B is twice as tall as building D. We cannot make a ratio statement because all measures were taken relative to an arbitrary criterion (100 feet). Finally, we can measure each building from the ground, which is a true zero point rather than an arbitrary criterion. Building D is 102 feet high, building B is 104 feet, building C is 180 feet, and building A is 204 feet (see Figure 1.1d). We can now state with confidence that building A is twice as tall as building D.

Care must be taken not to interpret this example too literally. For example, it is *not* the case that ordinal, interval, and ratio measures must occur on scales that are restricted to positive integers. On the contrary, all of the scale types can be used with integers that take on positive or negative values. Consider the ordinal measure of building height, where the four buildings have the scores (going from shortest to tallest) of 1, 2, 3, and 4. Suppose we perform a simple transformation on these scores; namely, we subtract the value 3 from each score. The new scores are −2, −1, 0, and 1. Notice that these four scores still preserve the ordinal characteristics of the measure: The higher the score, the taller the building. Also, although the new scores include a value of 0, this does not mean that there is a true zero point for the scores. A complete discussion of measurement properties is beyond the scope of this book, and interested readers are referred to Krantz, Luce, Suppes, and Tversky (1971).

Knowing whether a set of measures has nominal, ordinal, interval, or ratio properties is important because it affects the way scores can be interpreted. As an example, suppose we measure the math aptitudes of a group of children on a scale that ranges from 0 to 100. One child obtains a score of 50, and another child gets a score of 90. If the measures have only ordinal properties, then we know that the second child has more math aptitude than the first child, but we do not know how much more. Maybe the true difference in math aptitudes between the two children is trivial, or maybe it is quite substantial. Unless the measures possess or at least approximate interval or ratio level characteristics, their interpretation is restricted. Most measures in behavioral science research are at the nominal, ordinal, or interval level, with few being ratio level.

The Measurement Hierarchy   The four types of measurement can be thought of as a hierarchy. At the lowest level, nominal measurement allows us only to categorize phenomena into different groups. The second level, ordinal measurement, not only allows us to classify phenomena into different groups but also indicates the relative ordering of the groups on a dimension of interest. Interval measurement, the next level, possesses the same properties as ordinal measurement but, in addition, is sensitive to the magnitude of the differences in the groups on the dimension. However, ratio statements are not possible at this level. It is only at the final level, ratio measurement, that such statements are possible. Ratio measures have all of the properties of nominal, ordinal, and interval measures and also permit ratio judgments to be made.

Quantitative and Qualitative Measures   An important distinction can be made between nominal measurement, on the one hand, and ordinal, interval, and ratio measurements, on the other. A variable measured at one of the latter levels takes on an ordered set of values along some dimension. Scores can thus be ordered on the dimension in question depending on their values. In contrast, scores on a nominal level cannot be ordered; rather, they merely distinguish among categories.

Variables measured on the ordinal, interval, or ratio level are known as **quantitative variables**, whereas variables measured on a nominal level are called **qualitative variables.** Any variable can be classified as either quantitative or qualitative. As we will see later, the distinction between quantitative and qualitative variables is crucial in statistics.

Measures Versus Scales   When reading a research report, you may encounter references to interval scales, ordinal scales, and so on. Technically, the use of the word *scales* is somewhat misleading. Nominal, ordinal, interval, and ratio properties are characteristics of a set of measures, not just the scales used to generate those measures. A measure has as its referent not only a particular scale (for example, inches) but also an individual on whom the measure is taken, a time at which the measure is taken, and a setting in which the measure is taken. All of these must be considered when evaluating the properties of a set of measures. We can illustrate this idea using height as an example. Consider four individuals whose heights are 54, 53, 52, and 51 inches. We can rank order these individuals from shortest to tallest:

Individual	Height, in inches (X)	Rank order of height (Y)
1	54	4
2	53	3
3	52	2
4	51	1

We typically think of rank order measures, such as Y, as having ordinal but not interval properties. However, for this particular set of measures (that is, for these four individuals at this time and in this setting), the measures on Y have interval level properties. More specifically, a difference between scores of one unit between any two individuals (e.g., for individuals 1 and 2) corresponds to the same amount of underlying height difference as for any other two individuals who have a difference of one unit on Y. For this set of measures, Y has interval properties, even though it is measured on a scale that is traditionally thought of as yielding only ordinal level properties (i.e., a rank order scale).

Now suppose that we add to this set a 58-inch-tall individual who receives a rank of 5:

Individual	Height, in inches (X)	Rank order of height (Y)
1	54	4
2	53	3
3	52	2
4	51	1
5	58	5

Now Y no longer exhibits interval properties. It instead represents a set of measures with only ordinal properties. This is because it is no longer true that a one-unit difference in Y always corresponds to the same underlying height difference. When we compare individuals 1 and 2, a one-unit difference on Y ($4 - 3 = 1$) corresponds to a 1-inch difference on the underlying height dimension ($54 - 53 = 1$), whereas when we compare individuals 5 and 1, a one-unit difference on Y ($5 - 4 = 1$) corresponds to a 4-inch difference on the underlying height dimension ($58 - 54 = 4$). The point is that the concepts of nominal, ordinal, interval, and ratio properties are inherent in all the facets of measurement.

Suppose that instead of adding an individual who was 58 inches tall, we add an individual who is 55.1 inches tall to the initial four individuals. This individual receives a rank score of 5, just as the 58-inch-tall individual did:

Individual	Height, in inches (X)	Rank order of height (Y)
1	54	4
2	53	3
3	52	2
4	51	1
5	55.1	5

Strictly speaking, the rank order measure Y no longer has interval level properties. But it very closely approximates interval level characteristics because the difference in the underlying heights ($55.1 - 54 = 1.1$) between scores of 5 and 4 is almost equal to the underlying height difference between, say, scores of 2 and 1 ($52 - 51 = 1$). As we will see in later chapters, the validity of substantive conclusions made for some statistical tests assumes equal-interval measurement. Although statistics are sometimes applied to situations without strict interval properties, if the approximation to interval properties is reasonable (as in the above case), then the substantive conclusions are not affected. In fact, there is evidence to indicate that the approximation can, in some instances, be quite crude and still not affect conclusions.

Determining Levels of Measurement   The determination of whether a variable is measured on a nominal level is usually a straightforward matter in the behavioral sciences. This is not necessarily true for the other levels of measurement. For example, there is controversy as to whether intelligence test scores (such as the Wechsler Adult Intelligence Scale) reflect only an ordinal measure of intelligence or an interval measure of intelligence. The critical question is whether test score differences of a given magnitude *always* represent equivalent differences in intellectual ability; for instance, is the difference in *intelligence* between individuals with intelligence test scores of 110 and 120 the same as the difference in intelligence between individuals with test scores of 90 and 100? If so, scores on the Wechsler Adult Intelligence Scale represent an interval measure; if not, they represent an ordinal measure. What is clear is that intelligence test scores do not reflect a ratio measure; there is no evidence that an intelligence test has a true zero point.

Techniques for testing the assumptions of different measures have been developed in the area of psychophysical scaling (for example, Anderson, 1970). The majority of statistical techniques considered in this text assume that the dependent variables are measured on a level that at least approximates interval characteristics. This means that these variables must be measured on an ordinal level that approximates interval characteristics, an interval level, or a ratio level.

Study Exercise 1.1

For each of the following experiments, specify the independent variable and the dependent variable. Also identify any variables that are explicitly held constant by the experimenter. For both the independent variable and the dependent variable, indicate the level of measurement and whether a quantitative or a qualitative variable is represented.

Experiment I

Goldberg (1968) was interested in investigating gender bias among women. One hundred female college students were asked to rate an article in terms of its persuasiveness. The article was on the topic of education. Participants were assigned to one of two groups. One group of 50 women read the article and were told that it was authored by a woman named Joan McKay. The other group of 50 women read the same article but were told it was authored by a man named John McKay. After reading the article, each woman rated it on this seven-point scale:

| Not at all | | | | | | | Very |
| persuasive | 1 | 2 | 3 | 4 | 5 | 6 | 7   persuasive |

The average rating scores were compared for the two groups—that is, the group with a male author and the group with a female author. Results indicated that the average persuasive rating was higher when the article was attributed to a male author rather than to a female author.

Answer   The independent variable is the gender of the author of the article, male or female. It is a nominal measure and hence a qualitative variable. The dependent variable is the persuasiveness rating. This constitutes at least an ordinal measure because the higher the rating, the more persuasive the article was perceived to be. Because the dependent measure has at least ordinal characteristics, it is a quantitative variable. Numerous variables have been held constant. One of the most obvious ones is that the study was conducted with only women. Also, the content of the articles was held constant.

Experiment II

Research on extrasensory perception (ESP) has taken many different directions. Recently, attention has been given to the possibility that hypnosis may be helpful in fostering ESP in people. One standard ESP task involves Zener cards, special cards that have only five denominations and look like this:

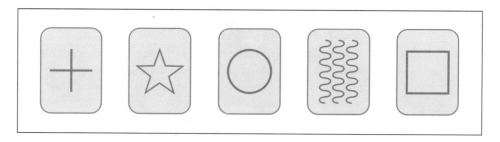

The standard task is to take a deck of 200 cards and have a "sender" shuffle them. The sender looks at the first card, thinks of the denomination of that card, and then the participant guesses what the card is. This process is repeated throughout the entire deck. ESP is measured by the number of correct guesses by the receiver.

Casler (1964) used this task with 100 female college students. Two conditions were used. In the first condition, 50 women were hypnotized and then given the task described above. In the second condition, an additional 50 women completed the task without being hypnotized. The average number of correct predictions was computed for the two groups. These averages turned out to be roughly equivalent, and it was concluded that hypnosis does not affect ESP.

Answer The independent variable is whether or not the participant was hypnotized. This could be conceptualized as either a qualitative variable (presence or absence of hypnosis) or a quantitative variable (degree of hypnosis: none versus some) and shows that the distinction between the two is sometimes arbitrary. If viewed as a quantitative variable, it would, strictly speaking, have interval level characteristics because there are only two points on the scale (and the comparison of the difference between the two points with any other two points is irrelevant). The dependent variable is the number of correct answers on the 200 trials. This represents at least an ordinal measure of ESP, and thus it is a quantitative variable. You might be inclined to view this measure as one with ratio characteristics because it appears that there is a true zero point (that is, none correct) and that ratio-type statements are possible (for instance, ten correct is twice as many as five correct). Actually, it is unclear whether this is the case. One can conceptualize the number of correct trials as an *index* of ESP ability. Given this, it may not be the case that ten correct trials relative to five correct trials reflects *twice* as much ESP ability. This example is important because it illustrates some of the complexities involved in identifying the properties of variables and the measures of them. It underscores the fact that conceptualizations and definitions are imposed by researchers, and sometimes these may differ from one person to the next. Numerous variables have been held constant. Again, the most obvious one is that all receivers were women. Try to specify some others.

1.5 Discrete and Continuous Variables

The Concept of Discrete and Continuous Variables

Another distinction made in statistics is between **discrete** and **continuous variables.** Often, the number of values that a variable can assume is relatively small and finite (such as the number of people in one's family). Or a variable may have a finite number of values that can occur between any two points. For example, consider the number of people who attend an anxiety clinic. For this variable, only one value can occur between the value of 1 person and 3 persons (namely, 2 persons). We do not think of there being 1.5 or 2.7 persons. Variables that can assume only a finite number of values or that have a finite number of values that can occur between any two points are called discrete variables. In contrast, a continuous variable can theoretically have an infinite number of values between any two points. Reaction time to a stimulus is an example of a continuous variable. Even between the values of 1 and 2 seconds, an infinite number of values could occur (1.001 seconds, 1.873 seconds, 1.874 seconds, and so on).

It should be emphasized that whether a variable is classified as discrete or continuous depends on the nature of the underlying theoretical dimension and not on the scale used to measure that dimension. Tests used to measure intelligence, for example, yield scores that are whole numbers (101, 102, and so on). Nevertheless, intelligence is still continuous in nature because it involves a dimension that permits an infinite number of values to occur, even though existing measuring devices are not sensitive enough to make such fine distinctions.

Real Limits of a Number

If a variable is continuous, then it follows that the measurements taken on that variable must be approximate in nature. When we say that a person reacted to a stimulus in 10 seconds, we do not usually mean exactly 10 seconds but only approximately 10 seconds because more refined measures are always possible, such as 10.02 seconds or 10.093 seconds. When we say the reaction was in 10 seconds, we actually mean it was somewhere between 9.5 seconds and 10.5 seconds, since any number less than 9.5 would be rounded to 9 and any number greater than 10.5 would be rounded to 11. Thus, the **real limits** of the number 10 are 9.5 and 10.5. The quantity 9.5 is called the *lower real limit* and the quantity 10.5 is called the *upper real limit. The real limits of a number are those points that fall one-half a measurement unit above that number and one-half a measurement unit below that number.* Figure 1.2 graphically presents the concept of real limits.

The real limits of a number can be stated not only with respect to whole numbers but also for numbers expressed as decimals. For example, consider the number 10.6. Since it is expressed in tenths, the unit of measurement is one-tenth, or .1. One-half a measurement unit is therefore .1/2 = .05. The lower real limit is

FIGURE 1.2   **Real Limits of 10**

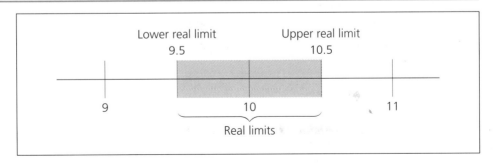

thus $10.6 - .05 = 10.55$, and the upper real limit is $10.6 + .05 = 10.65$. Similarly, the lower and upper real limits of 10.63 are $10.63 - .005 = 10.625$ and $10.63 + .005 = 10.635$, respectively.

Study Exercise 1.2

State the real limits of the following numbers, assuming they are measured in the units reported:

(a) 20
(b) 8.4
(c) 12.23
(d) 16.0478

Answers

(a) 19.5 and 20.5
(b) 8.35 and 8.45
(c) 12.225 and 12.235
(d) 16.04775 and 16.04785

1.6 Populations and Samples

In scientific research, we are often interested in making descriptive statements about a group of individuals or objects. For example, one might state that the average number of children desired by married men in the United States is 2.3 or that the average number of times that adult women in the United States go to a psychologist in a given year is 2.1. Such statements are made with reference to a **population**. A population is the aggregate of all cases to which one wishes to generalize statements. In the first example, the population consists of all married men in the United States. In the second example, the population consists of all adult women in the United States.

It may be the case that an investigator is unable to make observations on every member of the population about which he or she wishes to make a descriptive statement. Then the investigator must resort to using a **sample** of the population. A sample is simply a subset of the population. On the basis of observing the sample, the researcher makes generalizations to the population.

When we select a sample in order to make a statement about a population on a given dimension (for example, the average number of children), we want to ensure that we are using a **representative sample** of the population. If the population has 60% males and 40% females, we want our sample to reflect this. We do not want a *biased* sample that will lead us to make erroneous statements about the population. In selecting a sample, we want to use procedures that will yield a representative sample. Box 1.1 presents an interesting example of biased sampling as described by Darrell Huff in his excellent book *How to Lie with Statistics* (Huff, 1954).

One procedure for approximating representative samples is **random sampling**. The term *random* has a very precise meaning in scientific discourse. As applied to sampling problems, the essential characteristic of random sampling is that every member of the population has an equal chance of being selected for the sample.

BOX 1.1

Biased Sampling

"The average Yaleman, Class of '24," *Time* magazine noted once, commenting on something in the *New York Sun,* "makes $25,111 a year." Well, good for him! But wait a minute. What does this impressive [at the time of the report] figure mean? Is it, as it appears to be, evidence that if you send your boy to Yale, you won't have to work in your old age and neither will he? Two things about the figure stand out at first suspicious glance. It is surprisingly precise. It is quite improbably salubrious. . . .

Let us put our finger on a likely source of error, a source that can produce $25,111 as the "average income" of some men whose actual average may well be nearer half that amount. This is the sampling procedure, which is the heart of the greater part of the statistics you meet on all sorts of subjects. Its basis is simple enough, although its refinements in practice have led into all sorts of byways, some less than respectable. If you have a barrel of beans, some red and some white, there is only one way to find out exactly how many of each color you have: Count 'em. However, you can find out approximately how many are red in much easier fashion by pulling out a handful of beans and counting just those, figuring that the proportion will be the same all through the barrel. If your sample is large enough and selected properly, it will represent the whole well enough for most purposes. If it is not, it may be far less accurate than an intelligent guess and have nothing to recommend it but a spurious air of scientific precision. It is sad truth that conclusions from such samples, biased or too small or both, lie behind much of what we read or think we know.

The report on the Yale men comes from a sample. We can be pretty sure of that because reason tells us that no one can get hold of all the living members of that class of '24. There are bound to be many whose addresses are unknown twenty-five years later. And, of those whose addresses are known, many will not reply to a questionnaire, particularly a rather personal one. With some kinds of mail questionnaire, a five or ten per cent response is quite high. This one should have done better than that, but nothing like one hundred per cent. So we find that the income figure is based on a sample composed of all class members whose addresses are known and who replied to the questionnaire. Is this a representative sample? That is, can this group be assumed to be equal in income to the unrepresented group, those who cannot be reached or who do not reply?

Who are the little lost sheep down in the Yale rolls as "address unknown"? Are they the big-income earners—the Wall Street men, the corporation directors, the manufacturing and utility executives? No; the addresses of the rich will not be hard to come by. Many of the most prosperous members of the class can be found through *Who's Who in America* and other reference volumes even if they have neglected to keep in touch with the alumni office. It is a good guess that the lost names are those of the men who, twenty-five years or so after becoming Yale bachelors of arts, have not fulfilled any shining promise. They are clerks, mechanics, tramps, unemployed alcoholics, barely surviving writers and artists . . . people of whom it would take half a dozen or more to add up to an income of $25,111. These men do not so often register at class reunions, if only because they cannot afford the trip.

Who are those who chucked the questionnaire into the nearest wastebasket? We cannot be so sure about these, but it is at least a

(continued)

Box 1.1 *(continued)*

fair guess that many of them are just not making enough money to brag about.

It becomes pretty clear that the sample has omitted two groups most likely to depress the average. The $25,111 figure is beginning to explain itself. If it is a true figure for anything, it is merely for that special group of the class of '24 whose addresses are known and who are willing to stand up and tell how much they earn. Even that requires an assumption that the gentlemen are telling the truth.

Source: Huff & Geis, 1954.

Scientists use random sampling in a variety of ways. In order to obtain a random sample of a population, a survey researcher might make use of a very helpful resource known as a *random number table*. A random number table is a list of numbers generated by a computer that has been programmed to yield a set of truly random numbers. Computers are used to construct such tables because the typical person is not capable of generating random numbers. For example, a person might have a tendency to list mostly even numbers, or those ending in 5 or 0.

Appendix A presents a random number table. Suppose an investigator wanted to select a random sample of people from a population. This would involve obtaining a list of all members of the population and then arbitrarily assigning a number to each member. If the population consisted of 500 individuals, a list would be made of the names of these people numbered from 1 to 500. The investigator would then consult a random number table, such as the one in Appendix A. Using the directions provided, the investigator would draw a sample of, say, 50 individuals. The use of a random number table ensures that the selection will be random and not influenced by any unknown selection bias the investigator may have.

It should be emphasized that random sampling is an ideal that is seldom achieved in practice. This is not surprising given the difficulty of compiling a complete listing of a population and ensuring that all selected individuals will agree to participate. Furthermore, the use of random sampling procedures does *not* guarantee that a sample will be representative of the population. Random sampling will tend to yield representative samples, but sometimes nonrepresentative samples will result even when random sampling is used. Nevertheless, random sampling is an important concept in statistical theory.

Throughout this text, we will refer to various numerical indexes based on data from either populations or samples. When the indexes are based on data from an entire population, they will be referred to as **parameters.** When they are based on data from a sample, they will be referred to as **statistics.**

1.7   Descriptive and Inferential Statistics

The discipline of statistical analysis has traditionally been divided into two major subfields: descriptive statistics and inferential statistics. The two are highly related, and in some respects, the distinction is arbitrary. **Descriptive statistics** involves the use of numerical indexes to describe either a population (when measurements have been taken on all members of that population) or a sample.* In either case, the goal is to *describe* a group of scores in a clear and precise manner. **Inferential statistics,** in contrast, involves taking measurements on a sample and then, from the observations, inferring something about a population. In this instance, we are again attempting to describe a population. However, we do so not by taking measures on all cases in the population, but rather by selecting a sample, observing scores on the variable of interest for that sample, and then *inferring* something with respect to that variable for the entire population. As we will see, this involves the use of sample statistics to estimate population parameters.

1.8   The Concept of Probability

The concept of probability is an essential aspect of statistics, especially inferential statistics. For example, when making inferences about populations from samples, statisticians frequently report probability information to convey the likelihood that the sample results may be misleading about the populations.

All of us are somewhat familiar with the concept of probability in our everyday life. A weather forecaster tells us that the chances of rain tomorrow are 70%. A bettor at the racetrack knows that the odds on a given horse are 3 to 1. A student thinks it is "likely" he or she will get an A in a course.

In statistics, probability has a precise meaning. For a given task, there may be several different possible outcomes. If you roll a die, there are six possible outcomes: 1, 2, 3, 4, 5, and 6. If you draw a card from a standard deck of playing cards, there are 52 different possible outcomes. The **probability** of some outcome, A, can be defined as the ratio:

$$p(A) = \frac{\text{number of observations favoring outcome } A}{\text{total number of possible observations}}$$

On a die, what is the probability of rolling a 2? On a given roll, the total number of possible observations is six (that is, 1, 2, 3, 4, 5, and 6). There is only one 2 on a die and hence only one possible observation favoring the event "2." The probability is 1/6, or .17. If one draws a card at random from a standard deck, the probability of drawing an ace is 4/52, or .08. There are 52 possible observations, of which 4 favor an ace.

A given probability always ranges from 0 to 1.00. It can never be less than 0 or greater than 1.00. The probability of an impossible event is always 0. The probability of rolling an 8 on a single die is 0 because the number of observations favoring the event "8" is 0. It follows that 0/6 = 0. The probability of a completely

---

*As stated, numerical indexes based on data from entire populations are called *parameters*. Nevertheless, the term *descriptive statistics* has historically been used to encompass the description of both sample and population data.

certain event is always 1.00. For example, if an individual rolls a die, what is the probability that the individual will roll a 1, 2, 3, 4, 5, or 6? In this case there are six outcomes, of which all six satisfy the event 1, 2, 3, 4, 5, or 6. The probability is therefore 6/6 = 1.00.

Sometimes probabilities are derived on logical grounds, as in the examples above. Other times, they are estimated empirically based on data an investigator has collected. For example, when specifying the probability of a miscarriage during pregnancy, a researcher might calculate the number of miscarriages that occur over 100,000 pregnancies. If 20,000 miscarriages occur, the probability of a miscarriage is 20,000/100,000 = .20.

The probability of an event can also be interpreted in terms of a "long run" perspective. If we flip a coin ten times, it is unlikely that the coin will come up heads exactly five times and tails the other five times. As we continue to flip the coin, however, then across a large number of flips (that is, over the long run), the number of heads relative to the total number of flips will approach 1/2, or .50.

1.9 Mathematical Preliminaries: A Review

The purpose of this section is to review mathematical symbols and concepts that are used throughout this book.

Summation Notation

Suppose we have a measure of the number of months each of five individuals worked in the past year. The scores on this variable are as follows:

Individual	Number of months worked (X)
1	3
2	2
3	3
4	3
5	2

In statistical notation, the capital letter X is used as a general name for a variable. In this case, X stands for the variable "number of months worked." Sometimes the X has a number subscript to indicate that a particular individual's score is being represented. For instance, X_1 is the first individual's score on X, which in this case is 3; X_2 is the second individual's score on X, which is 2; and so on.

On some occasions, we want to consider simultaneously two variables, such as the number of months worked and monthly income:

Individual	Number of months worked (X)	Monthly income, in dollars (Y)
1	3	200
2	2	300
3	3	200
4	3	300
5	2	300

If we let X represent the number of months worked and Y represent monthly income, then we can refer to individual scores on X and Y using subscripts: $X_1 = 3$, $Y_1 = 200$, $X_2 = 2$, $Y_2 = 300$, and so on.

Suppose we want to sum the five scores on variable X to determine the total number of months worked by the five individuals. In statistics, we have a short-hand way of writing an instruction to sum a set of scores, called **summation notation.** The operation in this instance is written as follows:

$$\sum_{i=1}^{5} X_i$$

The summation operation is signaled by Σ (capital Greek S, called "sigma"). The notation below the sigma tells us to start with individual number 1, and the number above the sigma tells us to add through to individual number 5. The X_i to the right of the sigma is a general term that stands for the individual X scores. In this case,

$$\sum_{i=1}^{5} X_i = X_1 + X_2 + X_3 + X_4 + X_5 = 3 + 2 + 3 + 3 + 2 = 13$$

If the summation were written as

$$\sum_{i=2}^{4} X_i$$

it would mean to sum the scores of individuals 2 through 4 on variable X:

$$\sum_{i=2}^{4} X_i = X_2 + X_3 + X_4 = 2 + 3 + 3 = 8$$

Often we let the letter N represent the total number of cases. You might encounter this summation notation:

$$\sum_{i=1}^{N} X_i \qquad\qquad [1.1]$$

which is the same as having the number that represents the total number of cases above the sigma. In our example, $N = 5$ because there are five individuals. Thus,

$$\sum_{i=1}^{N} X_i = \sum_{i=1}^{5} X_i = 13$$

as calculated above. Similar terminology applies to the Y variable:

$$\sum_{i=1}^{N} Y_i = \sum_{i=1}^{5} Y_i = Y_1 + Y_2 + Y_3 + Y_4 + Y_5$$

$$= 200 + 300 + 200 + 300 + 300 = 1{,}300$$

In addition to expression 1.1, other summation terms are used in this book. We briefly review some of these. One such expression is

$$\sum_{i=1}^{N} X_i^2 \qquad\qquad [1.2]$$

This means that each X score should be first squared and then summed:

$$\sum_{i=1}^{N} X_i^2 = X_1^2 + X_2^2 + X_3^2 + X_4^2 + X_5^2$$

$$= 3^2 + 2^2 + 3^2 + 3^2 + 2^2 = 35$$

A third summation expression is

$$\left(\sum_{i=1}^{N} X_i \right)^2 \qquad\qquad [1.3]$$

This is *not* the same as expression 1.2. A general rule that we follow throughout this book is to *perform any mathematical operations within parentheses before performing the operations outside the parentheses.* In expression 1.3, the parentheses signal that the summation operation should be executed first (that is, the X scores should be summed) and then this sum should be squared:

$$\left(\sum_{i=1}^{N} X_i \right)^2 = (X_1 + X_2 + X_3 + X_4 + X_5)^2$$

$$= (3 + 2 + 3 + 3 + 2)^2 = 13^2 = 169$$

In short, then, *expression 1.2 means to sum the squared X scores, whereas expression 1.3 means to square the summed X scores.*

A fourth summation expression is

$$\sum_{i=1}^{N} X_i Y_i \qquad\qquad [1.4]$$

This means that for each pair of scores, each X score first should be multiplied by its corresponding Y score, and then these products should be summed:

$$\sum_{i=1}^{N} X_i Y_i = X_1 Y_1 + X_2 Y_2 + X_3 Y_3 + X_4 Y_4 + X_5 Y_5$$

$$= (3)(200) + (2)(300) + (3)(200) + (3)(300) + (2)(300)$$

$$= 600 + 600 + 600 + 900 + 600 = 3{,}300$$

Another summation term we will encounter is

$$\sum_{i=1}^{N} (X_i - c)^2 \qquad [1.5]$$

where c represents a constant. Suppose that $c = 2$. Then this expression indicates that we should subtract 2 from each X score, square each difference, and, last, sum these squared differences:

$$\sum_{i=1}^{N} (X_i - c)^2 = (X_1 - 2)^2 + (X_2 - 2)^2 + (X_3 - 2)^2 + (X_4 - 2)^2 + (X_5 - 2)^2$$

$$= (3 - 2)^2 + (2 - 2)^2 + (3 - 2)^2 + (3 - 2)^2 + (2 - 2)^2$$

$$= 1^2 + 0^2 + 1^2 + 1^2 + 0^2 = 3$$

The final summation expression we review is

$$\sum_{i=1}^{N} (X_i - c)(Y_i - k) \qquad [1.6]$$

where both c and k represent constants. This means for each individual, multiply the difference between X and c by the difference between Y and k, and then sum the resulting products. For instance, if $c = 2$ and $k = 100$, then

$$\sum_{i=1}^{N} (X_i - c)(Y_i - k) = (X_1 - c)(Y_1 - k) + (X_2 - c)(Y_2 - k) + (X_3 - c)(Y_3 - k)$$

$$+ (X_4 - c)(Y_4 - k) + (X_5 - c)(Y_5 - k)$$

$$= (3 - 2)(200 - 100) + (2 - 2)(300 - 100) + (3 - 2)(200 - 100)$$

$$+ (3 - 2)(300 - 100) + (2 - 2)(300 - 100)$$

$$= (1)(100) + (0)(200) + (1)(100) + (1)(200) + (0)(200) = 400$$

It is very important to understand these six summation expressions because we refer to them throughout this book. Frequently, a shorthand version of these terms is used. For instance, the first expression

$$\sum_{i=1}^{N} X_i$$

may also be written as

$$\sum X$$

Note that there is no subscript for X, no instruction below the sigma, and no symbol above the sigma. It is understood that X is subscripted with i, the instruction $i = 1$ applies below the sigma, and the letter N applies above it. Thus,

$$\sum_{i=1}^{N} X_i = \sum X$$

$$\sum_{i=1}^{N} X_i^2 = \sum X^2$$

$$\left(\sum_{i=1}^{N} X_i\right)^2 = \left(\sum X\right)^2$$

$$\sum_{i=1}^{N} X_i Y_i = \sum XY$$

$$\sum_{i=1}^{N} (X_i - c)^2 = \sum (X - c)^2$$

$$\sum_{i=1}^{N} (X_i - c)(Y_i - k) = \sum (X - c)(Y - k)$$

We use the shorthand versions of these terms wherever possible in this book.

Study Exercise 1.3

Given the following values for X and Y, complete the requested operations:

Individual	X	Y
1	3	3
2	4	3
3	2	3
4	7	5

(a) ΣX

(b) ΣX^2

(c) $\Sigma (X)^2$

(d) $\Sigma (X - 4)^2$

(e) ΣY

(f) ΣY^2

(g) $(\Sigma Y)^2$

(h) $\Sigma (Y - 3)^2$

(i) ΣXY

(j) $\Sigma (X - 1)(Y - 2)$

Answers

(a) $\Sigma X = 3 + 4 + 2 + 7 = 16$

(b) $\Sigma X^2 = 3^2 + 4^2 + 2^2 + 7^2 = 78$

(c) $(\Sigma X)^2 = 16^2 = 256$

(d) $\Sigma (X - 4)^2 = (3 - 4)^2 + (4 - 4)^2 + (2 - 4)^2 + (7 - 4)^2 = 14$

(e) $\Sigma Y = 3 + 3 + 3 + 5 = 14$

(f) $\Sigma Y^2 = 3^2 + 3^2 + 3^2 + 5^2 = 52$

(g) $(\Sigma Y)^2 = 14^2 = 196$

(h) $\Sigma\,(Y-3)^2 = (3-3)^2 + (3-3)^2 + (3-3)^2 + (5-3)^2 = 4$
(i) $\Sigma\,XY = (3)(3) + (4)(3) + (2)(3) + (7)(5) = 62$
(j) $\Sigma\,(X-1)(Y-2) = (3-1)(3-2) + (4-1)(3-2) + (2-1)(3-2) + (7-1)(5-2) = 24$

Rounding

It is often desirable to round numbers to a certain number of decimal places. For instance, the fraction 7/3 in decimal notation is equivalent to 2. followed by an infinite number of 3s (that is, 2.33333 . . .). In a case like this, we have to round off. The following commonly accepted rules for rounding are adopted in this book:

1. If the remainder to the right of the decimal place you wish to round to is greater than one-half a measurement unit, increase the last digit kept by one.

Suppose we wish to round to two decimal places. In this case, the unit of measurement is one-hundredth, or .01, and one-half a measurement unit is .01/2 = .005. Thus, according to this rule, 5.338 is rounded to 5.34 because the remainder after the second decimal place, .008, is greater than .005. The quantity 5.335001 is also rounded to 5.34 because .005001 is greater than .005.

2. If the remainder to the right of the decimal place you wish to round to is less than one-half a measurement unit, leave the last digit kept as it is.

Thus, 7/3 is represented in decimal notation as 2.33 because the remainder after the second decimal place, .00333 . . . , is less than .005. Similarly, 2.3348 is rounded to 2.33.

3. If the remainder to the right of the decimal place you wish to round to is exactly one-half a measurement unit, leave the last digit kept as it is if it is an even number, but increase it by one if it is an odd number. *Note that when this rule is used, the last digit of the answer will always be an even number.*

According to this rule, 10.345 is rounded to 10.34 because the 4 in the hundredths place is an even number, and 10.335 is also rounded to 10.34 because the 3 in the hundredths place is an odd number. The purpose of the rule is to avoid a bias in rounding up or down across a large set of numbers; with the rule, approximately half the time you will round up and half the time you will round down.

The number of decimal places that are used in reporting a statistic in a research report will differ depending on the nature of the variable being reported. The average annual income of a group of individuals might be rounded to the nearest whole number (for example, $10,030), whereas the average number of seconds it takes a group of rats to run a maze might be reported to two decimal places (for example, 5.32 seconds). The number of decimal places you should report depends on how precise you need to be in order to make your point. In practice, statistics in the behavioral sciences are most commonly reported to two decimal places.

When computing a statistic such as an average, you may need to do intermediate calculations before you arrive at a final answer. Again, the exact number of decimal places you should use in your calculations will depend on the variable you are studying and the nature of the calculations performed. No hard and fast rules can be given to reduce rounding error. As a rule of thumb, *intermediate calculations should be done using at least one decimal place beyond the number of decimal places you plan to report in your final answer.* If you are performing your computations on a calculator, rounding error can be substantially reduced by keeping all digits shown until you round the final result.

In this book, calculations are generally rounded to two decimal places. For clarity of presentation, we follow this strategy even when reporting intermediate values.

Study Exercise 1.4

Round the following numbers to two decimal places:

(a) 8.337
(b) 7.443
(c) 7.555001
(d) 10.54500
(e) 13.63500

Answers

(a) 8.34
(b) 7.44
(c) 7.56
(d) 10.54
(e) 13.64

1.10   Statistics and Computers

It is common for statistical analyses to be conducted on computers. Computers are useful because they can be programmed to do statistical computations, thereby freeing the analyst from the tedious task of calculating (as opposed to interpreting) statistical indexes. Indeed, some computer "packages" specialize in statistical computations. These packages usually are identified by an acronym. The more popular ones in the behavioral sciences are SPSS, BMDP, SAS, MINITAB, and STATISTICA.

If computers can be used for statistical analyses, then why must students be taught statistical formulas? In order to use statistics properly, one must understand the logic and theory underlying statistics. Statistical formulas are, in essence, summary statements of statistical theory, and a full understanding of the bases of a formula fosters an understanding of statistical theory. In this book, we discuss both conceptual formulas that underscore the inherent statistical theory being described and computational formulas that are used in performing calculations. In our experience, doing so helps students to appreciate the mathematical and logical bases by which statistical indexes are generated.

Applications
to the Analysis
of a Social Problem

Throughout this book, we illustrate the concepts that we develop using a database that we have collected for analyzing unintended teenage pregnancies. You will observe firsthand how we analyze real data designed to provide insight into this important social problem. We begin by providing some background on the problem of teenage pregnancy. We then discuss the data from the perspective of some of the issues raised in this chapter.

There are more than 1 million teenage pregnancies each year. Among teenagers in 1995, nearly 85% of such pregnancies were unintended. About 40% of all teenage pregnancies are terminated through abortion. Given current rates, it has been estimated that approximately four out of ten girls who are now 14 will get pregnant during their teenage years.

The social and economic implications of early childbirth for the teenager are well documented. Early childbirth is linked to lowered levels of educational attainment and career preparation. For example, teenage mothers and fathers are less likely to obtain high school diplomas than are teenagers who do not have children. With decreased educational opportunities, there is a higher chance of unemployment, greater reliance on public assistance, and, ultimately, a higher incidence of living in poverty. Teenage parents are more likely than those who delay childbearing to have low-paying jobs or to be unemployed. If teen mothers marry, they also are more likely to separate or divorce than couples who postpone childbearing until their 20s. It has been estimated that nearly 45% of women who give birth between the ages of 14 and 17 are separated or divorced within 15 years, a rate that is three times greater than for women who did not begin childbearing until age 20. Finally, early childbearing has consequences for the subsequent development of the children of teenage mothers. For example, data from the state of New York suggest that babies born to teenage mothers are more than twice as likely to die in the first year of life than those born to mothers over the age of 20. Similarly, teenage mothers are more

likely to have babies who are premature or of low birth weight. The forces that produce this state of affairs are complex and include biological, social, and economic mechanisms.

There have been many broad-based approaches to confronting the problem of unintended pregnancy in adolescents. One approach is sex education in the schools. A second approach is making family planning widely available to teens through the establishment and funding of community family planning clinics. A third approach is making abortion widely available to teens. All of these approaches are quite controversial. We have been developing a fourth approach—namely, influencing teen behavior through the parents of teens. Our work has focused on making parents better communicators with their teens with an eye toward preventing unintended pregnancy through more effective communication. In general, behavioral scientists are skeptical about such an approach. Adolescence is viewed as a period when teens are rejecting their parents and are heavily influenced by peers. In addition, parents tend to have limited knowledge about birth control and sexual behavior. Despite these perceptions, we undertook a study under the premise that parents can indeed influence the behavior of their adolescent children and that the issue needs to be explored in more depth.

The data we analyze throughout this book were collected in the inner city of Philadelphia with 14-, 15-, 16-, and 17-year-olds. A random sample of African American youths and their mothers was selected and interviewed by trained interviewers. Participants were asked a range of questions about general parent–teen relationships, sexual behavior, and birth control. We will provide more details about the study in later chapters. For now, let us consider selected issues relevant to the material covered in this chapter.

In terms of sampling, we restricted the sampling plan to adolescents between the ages of 14 and 17. This was primarily because of cost

(continued)

considerations. We could afford to conduct only approximately 700 interviews. Based on census data, we determined that a random sample would yield approximately equal numbers of males and females and about equal numbers of 14-, 15-, 16-, and 17-year-olds. The analysis of gender differences and age differences was central to many of our hypotheses, and if we covered a larger age range, it would be difficult to apply many of the statisical tests described in later chapters with any degree of confidence. The strategy used to obtain the random sample was complex and somewhat different from the idealized random sampling strategy described earlier. For example, it is not possible to obtain a list of all African-American youths between the ages of 14 and 17 living in the inner city of Philadelphia for purposes of selecting a formal random sample using a random number table like Appendix A. Instead, we used a strategy called area sampling, as described in the "Polls and Random Samples" box in Chapter 7.

Another set of issues concerns the validity of our measures. We asked a wide range of questions, some of them highly sensitive. For example, teens were asked to indicate whether or not they had engaged in sexual intercourse and, if so, how often during the past 6 months. One issue is whether such self-reports are valid. Perhaps teens do not tell the truth when asked such questions. Here are some of the strategies we used to make lying unlikely. First, all interviews were completely confidential, and teens were told that their names would never be associated with their responses. Second, teens recorded their answers to these questions on a separate sheet of paper, so that they would not have to reveal their answers face to face to the interviewer. Their answer sheets were placed in sealed envelopes that were then delivered directly to the project director. Third, the importance of honest responding for the scientific integrity of the project was stressed. Fourth, sensitive questions were asked only after good rapport had been established between the teen and the interviewer. Finally, we included in the broader interview a set of "social desirability" checks that permitted us to determine whether a teen was likely to distort answers to try to create a favorable impression. These were questions that generally are not true of anyone but that would be endorsed by someone who was trying to create a good impression. For example, in a larger battery of questions, teens were asked to indicate whether they agreed or disagreed with the following statements: "I never get sad," "I never criticize other people," "I never argue with others." If a teen consistently agreed with such statements, then the data were treated as being suspect. After obtaining self-reports, we performed several validation checks on the data using community-wide statistics from hospitals on the frequency of sexually transmitted diseases and teen pregnancies. In general, the data supported the validity of our measures.

Summary

Scientific research can be conceptualized as a five-step process involving the analysis of data collected to test the validity of hypotheses. Most hypotheses in the behavioral sciences concern relationships between variables. A variable is a phenomenon that takes on different values. Important distinctions can be made between independent and dependent variables, quantitative (ordinal, interval, and ratio levels of measurement) and qualitative (nominal level of measurement) variables, and discrete and continuous variables. A distinction can also be made between parameters, which are numerical indexes based on data from an entire population, and statistics, which are numerical indexes based on data from a sample.

The use of numerical indexes to describe either a sample or a population is referred to as descriptive statistics. The use of numerical indexes to infer something about a population from observation of a sample is referred to as inferential statistics. Samples should be representative of the populations from which they were selected. One procedure for approximating representative samples is to use random sampling.

Exercises

Answers to asterisked (*) exercises appear at the back of the book.

1. What are the five stages of the scientific research process?

*2. Identify each of the following as a variable or a constant. Explain the reasons for your choices.
   a. the number of hours in a day
   b. people's attitudes toward abortion
   c. the country of birth of presidents of the United States
   d. the value of a number divided by itself
   e. the total number of points scored in a football game
   f. the number of days in a month

*3. Identify each of the following as a qualitative or a quantitative variable:

   a. weight   d. age
   b. religion   e. gender
   c. income   f. eye color

For each of the studies described in Exercises 4–7, identify the independent variable and the dependent variable. Indicate whether each is a quantitative or a qualitative variable.

*4. Eron (1963) reported an investigation in which he examined the possible relationship between the exposure of young children to violent television shows and the amount of aggression they exhibited toward peers. Eron gathered information concerning the aggressive behavior and television viewing habits of 875 third-grade children. By questioning parents about their child's viewing habits, Eron developed a four-point scale to measure a child's preference for aggressive TV shows. The scale had the following categories: very low preference for aggressive TV shows, low preference for aggressive TV shows, moderately high preference for aggressive TV shows, and high preference for aggressive TV shows. Aggression was measured by peer ratings of each child by at least two other children. These ratings could range from 0 to 32, with higher scores indicating greater amounts of aggression.

*5. Touhey (1974) was interested in studying the relationship between various types of occupations and the prestige people associated with them. In this study, five different occupations were studied: architect, professor, lawyer, physician, and research scientist. A large number of individuals were asked to rate each of these occupations on a 60-point scale measuring perceptions of occupational prestige. Low scores indicated low levels of prestige and higher scores indicated increasingly higher levels of prestige.

6. Steiner (1972) discussed a series of experiments that studied the relationship between group size and how quickly a group could solve problems. In one experiment, six different group sizes were created: two members, three members, four members, five members, six members, and seven members. Each group was then given a series of problems to solve, and the time until solution was measured for each group and compared.

7. Rubovits and Maehr (1973) were interested in the effect of teachers' expectancies on their behavior toward students. Female

undergraduates who were enrolled in a teacher training class were asked to prepare a lesson for four seventh- and eighth-grade students. Just before meeting with the students, each teacher was told that two of the students were "gifted" and had high IQs, whereas the other two were "not gifted" and possessed average intelligence. In reality, all children were about equal in ability, and these labels were assigned in an arbitrary manner. The teachers were then observed during a 40-minute period while they interacted with the four students. Rubovits and Maehr measured the numbers of times the teacher interacted with each student. The average numbers of interactions were then compared for students who were labeled "gifted" versus those who were labeled "not gifted."

8. Indicate whether each measure is a nominal measure, ordinal measure, interval measure, or ratio measure. Explain the reasons for your choices.
   a. inches on a yardstick
   b. Social Security numbers
   c. dollars as a measure of income
   d. order of finish in a car race
   e. intelligence test scores

9. What is the difference between ordinal, interval, and ratio measures? Give an example of each.

*10. Indicate whether each of the following variables is discrete or continuous.
   a. grains of sand on a beach
   b. height
   c. the annual federal budget
   d. shyness

*11. State the real limits of the following numbers, assuming they are measured in the units reported:
   a. 21,384.11
   b. .689
   c. 13
   d. 13.0
   e. 13.00

12. What is the difference between a sample and a population? Give three examples of each.

*13. A newspaper conducted a survey in which readers were asked to indicate their preference for either of two mayoral candidates in an election, John Doe or Jane Smith. People were asked to cut out a ballot provided in the paper that day and send it to the newspaper with their preference indicated. One week later the newspaper reported it received 1,000 ballots, of which 800 favored Jane Smith. It stated that the "spirit of the community lies with Jane Smith" and predicted her victory in the upcoming election. A total of 100,000 people live in the community. Is the newspaper's sample a representative sample of the community in general? Why or why not? What implication does this have for the newspaper's conclusion about the election?

14. What is the essential characteristic of random sampling?

15. Suppose you have a list of all 1,000 members of a population of interest to you. Using the random number table in Appendix A, select a random sample of 25 individuals.

*16. Repeat the random sampling process described in Exercise 15. How many of the same individuals were selected to participate in *both* samples? What does this indicate about the use of random sampling to approximate representative samples?

17. How are descriptive and inferential statistics different?

18. Compute the probability of each of the following events:
   a. drawing an ace, king, or queen from a standard deck of 52 cards
   b. throwing a 2 or a 3 on a single die
   c. for a pair of dice, rolling a combination that totals 7
   d. drawing a face card from a standard deck of 52 cards

*19. Suppose you are considering whether to

have a particular type of operation. A total of 420 procedures of this nature have been performed in the past, of which 21 have been successful. Given only this information, what is the probability of success of the operation?

20. Consider the following data for eight individuals:

Individual	X	Y
1	4	5
2	1	7
3	2	3
4	8	2
5	8	1
6	2	1
7	5	4
8	7	7

Calculate the following sums:

a. ΣX
b. ΣY
c. $\sum_{i=1}^{4} X_i$
d. $\sum_{i=4}^{8} Y_i$
e. ΣXY
f. $(\Sigma X)/N$
g. $(\Sigma Y)/N$
h. $(\Sigma X)(\Sigma Y)$
i. ΣX^2
j. ΣY^2
k. $\Sigma (X-3)^2$
l. $\Sigma (Y-2)^2$
m. $(\Sigma X)^2$
n. $(\Sigma Y)^2$
o. $\Sigma (X-3)(Y-2)$
p. $\Sigma (X-2)(Y-3)$

*21. Express the following statements in summation notation for $N = 5$:

a. $X_1 + X_2 + X_3 + X_4 + X_5$
b. $X_3^2 + X_4^2 + X_5^2$
c. $(X_1 - 5) + (X_2 - 5) + (X_3 - 5) + (X_4 - 5) + (X_5 - 5)$
d. $(X_1^2 + X_2^2 + X_3^2 + X_4^2 + X_5^2)/N$
e. $(Y_1 + Y_2 + Y_3 + Y_4 + Y_5)^2$
f. $Y_1 + Y_2 + Y_3$
g. $(X_1 - 1)(Y_1 - 6) + (X_2 - 1)(Y_2 - 6) + (X_3 - 1)(Y_3 - 6) + (X_4 - 1)(Y_4 - 6) + (X_5 - 1)(Y_5 - 6)$
h. $X_1 Y_1 + X_2 Y_2 + X_3 Y_3 + X_4 Y_4 + X_5 Y_5$

*22. Consider the following data for five individuals:

Individual	X
1	4
2	6
3	2
4	2
5	4

Let k be a constant, with $k = 2$.

a. Compute $\Sigma X k$.
b. Compute $k \Sigma X$.
c. Compare your answer for **a** with your answer for **b**. What equation describes this relationship?
d. Compute $\Sigma (X/k)$.
e. Compute $(\Sigma X)/k$.
f. Compare your answer for **d** with your answer for **e**. What equation describes this relationship?

*23. Round each of the following numbers to three decimal places:

a. 4.8932
b. 8.9749
c. 1.4153
d. 4.1450
e. 6.245002
f. 2.615501
g. 6.3155
h. .39572
i. .9999
j. 3.6666
k. 12.2538
l. 9.724001
m. 1.9950
n. 2.0050

24. Round the numbers in Exercise 23 to two decimal places. Compare your answers with those you derived previously.

*25. Consider the following data for five individuals:

Individual	X
1	3.8753
2	4.2660
3	4.1156
4	3.4954
5	4.2061

Perform the following calculations on the given scores, keeping all digits shown until rounding the final answers to two decimal places. Then repeat the calculations, rounding the scores and all intermediate values to two decimal places. Compare the two sets of results. What accounts for the difference between them?

a. ΣX
b. $(\Sigma X)/N$
c. $(\Sigma X)^2$
d. ΣX^2

26. Consider the following data for five individuals:

Individual	X
1	5.4749
2	4.8348
3	4.2947
4	5.3650
5	4.7749

Perform the following calculations on the given scores, keeping all digits shown until rounding the final answers to two decimal places. Then repeat the calculations, rounding the scores and all intermediate values to two decimal places. Compare the two sets of results. What accounts for the difference between them?

a. ΣX
b. $(\Sigma X)/N$
c. $(\Sigma X)^2$
d. ΣX^2

Multiple-Choice Questions

*27. An investigator is studying the effect of religious upbringing on moral development. In this case, moral development is
a. the independent variable
b. the dependent variable
c. a constant
d. none of the above

28. Which of the following is *not* a quantitative variable?

a. time
b. number of children in a family
c. religion
d. height of a child

*29. What is the upper real limit of 20?
a. 19
b. 19.5
c. 20.5
d. 21

30. Calculate ΣX^2 for the following scores: 2, 2, 3, 2.
a. 2.25
b. 9
c. 21
d. 81

31. Round 2.34577 to two decimal places.
a. 2.33
b. 2.34
c. 2.35
d. none of the above

*32. Random sampling is a procedure for approximating _____ samples.
a. representative
b. nonrepresentative
c. biased
d. discrete

33. Which of the following does *not* constitute a quantitative level of measurement?
a. ordinal
b. ratio
c. nominal
d. interval

*34. When we want to summarize information (data) collected on a group of people, we use _____ statistics.
a. descriptive
b. inferential
c. summation
d. quantitative

35. $\Sigma X^2 = (\Sigma X)^2$
a. true
b. false

*36. Numerical indexes derived from population data are _____; numerical indexes derived from sample data are _____ .
a. parameters; parameters
b. statistics; statistics

c. statistics; parameters

d. parameters; statistics

37. The process of _____ sampling is one in which every member of the population has an equal chance of being selected for the sample.

a. biased

b. population

c. random

d. none of the above

CHAPTER 2

Frequency and Probability Distributions

In the behavioral sciences, we frequently want to convey to another person what a set of scores is like. For example, suppose we conduct an interview of 1,000 college students and ask them how many hours they spent studying during the past week. Or suppose we asked them whether they were satisfied or dissatisfied with their choice of a major, and to explain why. A researcher might interview 800 inner-city 13-year-olds and ask whether they have ever used drugs and, if so, how often during the past 6 months. In each of these instances, we want to convey a sense of how people responded and how frequently certain responses occurred. To do so, the statistical techniques discussed in this chapter can be of use.

2.1 Frequency Distributions for Quantitative Variables: Ungrouped Scores

A topic that has been studied extensively in child psychology is hyperactivity. Hyperactive children have inappropriately high activity levels and have difficulty controlling their energy in structured settings. Although hyperactivity is usually first diagnosed when children attend school (because of the structured demands that school places on children and the inability of the hyperactive child to deal with that structure), many hyperactive children show signs of heightened activity early in life, such as through irregular biological functions (eating and sleeping). Hyperactive children often do poorly in school and have difficulty forming friendships with other children. It is important to identify hyperactive children so that steps can be taken to ensure their optimal development before the hyperactivity creates a negative situation.

Suppose that an investigator administered to a class of 15 children a test designed to measure hyperactive tendencies. Scores on this test can range from 0 to 12, with higher values indicating greater activity levels. Based on past experience, the investigator was especially interested in children who scored 10 or more, because this is indicative of hyperactivity. The scores for the 15 children were as follows:

8	8	10	10	7
4	9	4	7	7
8	8	9	7	9

How can we best describe the scores on this test? One way is to list all 15 scores. By examination, we can then obtain an intuitive feel about what the scores tend to be like. But suppose that instead of 15 scores, there were 500 scores across a large number of classes. It then becomes impractical to list all the scores individually. A useful tool for summarizing a large set of data is a **frequency distribution.**

A frequency distribution is a table that lists the scores on a variable and the number of individuals who obtained each value. We begin by listing the obtained score values from highest to lowest. We then derive **absolute frequencies** (commonly referred to simply as *frequencies*) by counting the number of individuals who received each score, and we indicate these frequencies next to the corresponding score values. For the 15 scores listed above, we obtain the following frequency distribution:

Hyperactivity score	f
10	2
9	3
8	4
7	4
4	2

The symbol f is used to indicate absolute frequencies. We can see from the frequency distribution that two children obtained a score of 10, three children got a score of 9, four children got a score of 8, and so on. Adding together the frequencies for all the score values, we get a total of 15 cases, or $N = 15$.

If only a relatively small number of different scores are possible, researchers sometimes include all possible score values, even those that were not actually obtained, in the frequency table. When this is done, a frequency of 0 is indicated where appropriate. If we followed this approach, our frequency distribution would appear as follows:

Hyperactivity score	f
12	0
11	0
10	2
9	3
8	4
7	4
6	0
5	0
4	2
3	0
2	0
1	0
0	0

This illustrates an important point: Unlike most of the other statistical procedures we discuss, there are few hard and fast rules for presenting frequency information. Rather, several approaches are possible. The guidelines that we present are those that we find most useful. They are, however, only guidelines and, as such, might have to be modified depending on the specific characteristics of the data.

Considered alone, an index of frequency is not easily interpreted. Suppose you are told the results of a study showed that 200 people in a given town are prejudiced. This tells you little unless you also know the size of the town. With the information that the town population is 400, however, the frequency of 200 takes on more meaning: Half (200/400) of the town is prejudiced! This illustrates a more informative statistic used by researchers, called the **relative frequency**. A relative frequency is the number of scores of a given value (for example, scores of 10) divided by the total number of scores—that is, the proportion of times that a score occurred. The relative frequencies in a distribution will always sum to 1.00. Relative frequencies are indicated in a frequency distribution by the symbol rf. In the example on hyperactivity, the relative frequencies are computed as follows:

Hyperactivity score	f	rf
10	2	2/15 = .133
9	3	3/15 = .200
8	4	4/15 = .267
7	4	4/15 = .267
4	2	2/15 = .133

Relative frequency bears an important relationship with probability. Recall that the probability of an outcome, A, is the ratio of the number of observations that favor outcome A to the total number of possible observations. This is what a relative frequency reflects in a distribution: the number of individuals who obtain a particular score divided by the total number of scores. For example, the probability of randomly selecting a score of 7 in the preceding distribution is .267.

When a relative frequency is multiplied by 100, it reflects the **percentage** of times the score occurred. In our example, 13.3% of the children had a score of 10, 20.0% had a score of 9, and so on. As shown in the table below, percentages are indicated in a frequency distribution by the symbol %.

In addition to frequencies, relative frequencies, and percentages, we sometimes want to compute **cumulative frequencies** (symbolized by cf) and **cumulative relative frequencies** (symbolized by crf). For the hyperactivity example, these appear as follows:

Hyperactivity score	f	rf	%	cf	crf
10	2	.133	13.3	15	1.000
9	3	.200	20.0	13	.867
8	4	.267	26.7	10	.667
7	4	.267	26.7	6	.400
4	2	.133	13.3	2	.133

The entries in the cumulative frequency column are obtained by successive addition of the entries in the frequency column. For any given score (for example, 7), the cumulative frequency is the frequency associated with that score (for the score of 7, the frequency is 4) plus the sum of all frequencies below that score. For the score of 7, the cumulative frequency is $4 + 2 = 6$. For the score of 9, the cumulative frequency is $3 + 4 + 4 + 2 = 13$. Cumulative relative frequencies are computed in the same manner but use the column of relative frequencies instead of the column of frequencies. For the score of 7, the cumulative relative frequency is .267 $+ .133 = .400$. For the score of 9, the cumulative relative frequency is .200 $+ .267 + .267 + .133 = .867$.

The advantage of cumulative frequencies is that they allow us to tell at a glance the number of scores that are equal to or less than a given score value. We can readily see that 10 children had scores of 8 or lower and that 13 children had scores of 9 or lower. By looking at the cumulative relative frequency column, we can see that the proportion of children who had scores of 8 or lower was .667 and that the proportion of children who had scores of 9 or lower was .867. If we wished, we could also compute **cumulative percentages** to indicate the percentage of children who had scores equal to or lower than a given score value.

When we are concerned with a continuous variable, such as the degree of hyperactivity, frequencies and relative frequencies should be thought of in terms of the real limits of the scores. In the present example, although four children had scores of 8, this is more properly conceptualized as four individuals having scores between 7.5 and 8.5. Similarly, cumulative frequencies and cumulative relative frequencies are conceptualized with respect to the upper real limit of a score. The cumulative frequency in the hyperactivity data for a score of 8 is 10. Technically, this means that 10 individuals had scores of 8.5 or lower.

Study Exercise 2.1

Construct a frequency distribution containing frequencies, relative frequencies, percentages, cumulative frequencies, and cumulative relative frequencies for the following set of scores:

87	75	87	83	93
72	77	70	91	90
91	83	74	75	74
75	87	91	75	83

Answer

Score	*f*	*rf*	%	*cf*	*crf*
93	1	.050	5.0	20	1.000
91	3	.150	15.0	19	.950
90	1	.050	5.0	16	.800
87	3	.150	15.0	15	.750
83	3	.150	15.0	12	.600
77	1	.050	5.0	9	.450
75	4	.200	20.0	8	.400
74	2	.100	10.0	4	.200
72	1	.050	5.0	2	.100
70	1	.050	5.0	1	.050

2.2 Frequency Distributions for Quantitative Variables: Grouped Scores

The preceding analysis of frequency distributions examined the case in which a quantitative variable took on relatively few different values: 10, 9, 8, 7, and 4. Often, however, a quantitative variable takes on many different values. For example, an educational psychologist might be interested in describing the income of teachers in a large city. She obtains a random sample of 100 teachers, each having a different income. To construct a frequency table, it would be neither practical nor informative to list 100 different values, each with a frequency of 1. Rather, we would want to group the data before reporting it, perhaps as follows:

Income ($)	f	rf	%	cf	crf
30,000–34,999	14	.140	14.0	100	1.000
25,000–29,999	21	.210	21.0	86	.860
20,000–24,999	30	.300	30.0	65	.650
15,000–19,999	19	.190	19.0	35	.350
10,000–14,999	16	.160	16.0	16	.160

Since scores are grouped together in intervals, tables of this type are referred to as **grouped frequency distributions.** Note that as with the *ungrouped frequency distributions* discussed in Section 2.1, the scores in the preceding table are listed from high to low. Also note that the lower bound for each group of scores appears on the left and the upper bound on the right.

An important consideration in presenting grouped data is how to form the groups. Three questions are central: (1) How many groups should be reported? (2) What should the interval size be for each group? and (3) What should be the lowest value at which the first interval starts? No standard rules govern these issues. In large part, the nature of the grouping will depend on the particular characteristics of the data. Nevertheless, useful guidelines are available for each of the questions.

Number of Groups

In deciding how many groups to report, a researcher must strike a balance between having so many groups that the data are incomprehensible and having so few groups that the table is imprecise. The problem of too many groups would occur in its extreme in our previous example on income, should each of the 100 incomes be listed individually. The problem of too few groups is illustrated in the following table, which reports the scores of 65 individuals who completed a measure of how satisfied they were with their jobs. Scores could range from 0 to 100, with higher values indicating greater satisfaction.

Job satisfaction	f	rf	%	cf	crf
50–100	57	.877	87.7	65	1.000
0–49	8	.123	12.3	8	.123

This table is not very informative; it provides little insight into how the individuals differed in their job satisfaction. Too many of the individuals are grouped into a single category, and we simply do not have much appreciation for how they differ in their job satisfaction.

In general, if the number of possible score values is small, fewer groups can be used, whereas if the number of possible score values is large, more groups are required. As a rule of thumb, the use of 5 to 15 groups tends to strike the appropriate balance between imprecision and incomprehensibility in most instances.

Size of Interval

Once we have an idea of how many groups we wish to present, the question of the size of the interval arises (for example, should the interval for the income problem be $10,000, $5,000, or $1,000?). Typically, an interval size of two, three, or a multiple of five (5, 10, 15, and so on) is used. To determine the interval size for a particular set of data, first subtract the lowest score from the highest score. This difference should then be divided by the desired number of groups and the result rounded to the nearest of the commonly used interval-size values.

In the job satisfaction example, suppose that the lowest obtained score was 47 and the highest obtained score was 99. Further suppose that the decision was made to present the frequency analysis in five groups. In this case, $99 - 47 = 52$ and $52/5 = 10.40$. Since 10 is a multiple of five, we drop the .40 and use 10 as the interval size.

Beginning of Lowest Interval

We now know that we will present five groups with an interval of 10 units per group. The final question is where to begin the lowest interval (at 40? at 45? at 47?). The conventional starting point is the closest number evenly divisible by the interval size that is equal to or less than the lowest score. In our example, the lowest score is 47 and the interval size is 10. The closest number equal to or less than 47 that is evenly divisible by 10 is 40. This should be the starting point for the lowest interval.

The frequency distribution using the above guidelines appears as follows:

Job satisfaction	f	rf	%	cf	crf
90–99	3	.046	4.6	65	1.000
80–89	6	.092	9.2	62	.954
70–79	15	.231	23.1	56	.862
60–69	21	.323	32.3	41	.631
50–59	12	.185	18.5	20	.308
40–49	8	.123	12.3	8	.123

Note that the lower bound of each interval is a multiple of the interval size of 10. Also note that it was necessary to use six groups instead of five because of the determined interval size and lowest value. Again, the rules are only guidelines that might be useful in presenting grouped data.

Study Exercise 2.2

Construct a grouped frequency distribution containing frequencies, relative frequencies, percentages, cumulative frequencies, and cumulative relative frequencies for the following set of scores on a measure of self-esteem using ten groups:

9	29	39	5	34	43	39	26	22	4
29	14	28	22	15	26	25	44	24	16
36	8	39	16	21	7	24	17	23	21
15	24	49	23	35	44	19	25	45	34
37	13	27	6	11	49	17	31	27	14
12	6	31	46	32	28	5	42	28	35
38	48	9	38	2	20	33	43	3	13
19	19	26	4	19	41	32	12	18	36
37	2	18	49	29	1	27	3	42	21
47	25	1	39	11	41	23	33	22	46

Answer   The first step is to determine the interval size. The highest score is 49 and the lowest is 1. The difference between these values is $49 - 1 = 48$. Since 48 divided by 10, the number of groups, is 4.80, we define the interval size as 5. The next step is to determine the beginning of the lowest interval. The lowest score is 1. The closest number equal to or less than 1 that is evenly divisible by 5 is 0. This will be the starting point for the lowest interval. The frequency distribution appears as follows:

Self-esteem score	f	rf	%	cf	crf
45–49	8	.080	8.0	100	1.000
40–44	8	.080	8.0	92	.920
35–39	12	.120	12.0	84	.840
30–34	8	.080	8.0	72	.720
25–29	15	.150	15.0	64	.640
20–24	13	.130	13.0	49	.490
15–19	12	.120	12.0	36	.360
10–14	8	.080	8.0	24	.240
5–9	8	.080	8.0	16	.160
0–4	8	.080	8.0	8	.080

2.3   Frequency Distributions for Qualitative Variables

Frequency distributions for qualitative variables begin with a listing of the variable categories in the first column. This is followed by frequency, relative frequency, and/or percentage columns. The concepts of cumulative frequencies, cumulative relative frequencies, and cumulative percentages are not applicable because the "scores" for qualitative variables are not ordered on any dimension.

As an example of a frequency distribution for a qualitative variable, consider the results of a survey conducted by a researcher studying religious preference. He found the following distribution of religious affiliations in a random sample of individuals from the United States:

Religion	f	rf	%
Protestant	590	.590	59.0
Catholic	270	.270	27.0
Jewish	19	.019	1.9
Muslim	9	.009	.9
Eastern Orthodox	9	.009	.9
Other	31	.031	3.1
None	72	.072	7.2

The variable categories are "Protestant," "Catholic," "Jewish," "Muslim," "Eastern Orthodox," "Other," and "None." Adding up the individual frequencies, we find that the survey was based on a sample of 1,000 individuals. From this survey, we see that the majority of the U.S. population is either Catholic or Protestant, with a small minority being Jewish. Protestants outnumber Catholics by more than 2 to 1.

2.4 Outliers

Thus far, we have used frequency distributions to convey information about a set of scores. Another use of frequency analyses is to identify **outliers**. An outlier is a case that shows a very extreme score relative to the majority of cases in the data set—so extreme that the score is suspect. Consider the following frequency distribution of the numbers of times that college students engage in sexual intercourse during the first 3 months they are at college:

Incidence of intercourse	f
48	1
5	2
4	3
3	3
2	5
1	10
0	135

The score of 48 is an unusually extreme score relative to the others. When an outlier occurs, it is important to determine the reason for it. Sometimes it may just be a clerical error, as would be the case if you copied the score wrong from a questionnaire response. Alternatively, it may be that the person who gave the response is unique relative to the other people in the study. For example, the individual who indicated 48 instances of sexual intercourse may be an older, married student who is returning to college after many years away from school. In this case, the researcher might decide to exclude this "outlier" from further analyses and make explicit that any conclusions would apply only to a population that explicitly excludes such students. In later chapters on inferential statistics, we will discuss the importance of identifying outliers in the context of one's analysis.

2.5 Frequency Graphs

Frequency Graphs for Quantitative Variables

Instead of presenting tables of a frequency analysis, investigators sometimes report their data graphically. Consider the frequency analysis of the hyperactivity scores presented in Section 2.1:

Hyperactivity score	f
10	2
9	3
8	4
7	4
4	2

These data can also be presented in the form of a **frequency histogram,** such as Figure 2.1. The horizontal dimension in this graph is called the *X axis* or **abscissa,** and the vertical dimension is called the *Y axis* or **ordinate.** The abscissa lists the score values from low to high, extending from one unit below the lowest score to one unit above the highest score—in this case, from 3 to 11. A label that clearly names the variable under study should appear beneath the score values. Generally

FIGURE 2.1   **Frequency Histogram**

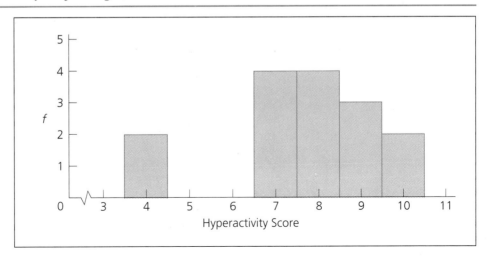

speaking, a graph should "stand alone" so that anyone who looks at it can readily interpret it. The ordinate represents the frequency with which each score occurred (corresponding to the second column of the frequency table) and is therefore labeled f, although sometimes the word "Frequency" is used instead. The bar for a test score of 4 goes up 2 units, indicating that two individuals had scores of 4; the bar for a test score of 7 goes up 4 units, indicating that four individuals had scores of 7; and so forth.

If a variable is continuous, the vertical boundaries of the bar for a given score represent the real limits of that score. Consider the bar for a score of 4. The leftmost boundary of the bar represents the lower real limit, 3.5, and the rightmost boundary of the bar represents the upper real limit, 4.5. The center of the bar corresponds to the midpoint of real limits 3.5 to 4.5, or 4. Notice also that there is a break in the abscissa in the form of a zigzag. This is traditionally done when the abscissa "jumps" from 0 to a larger number and is not drawn to scale. The same principle would hold for the ordinate.

A **frequency polygon** is similar to a frequency histogram and uses the same ordinate and abscissa. A frequency polygon of the hyperactivity data is shown in Figure 2.2. The major difference from the frequency histogram is that bars are not used, but rather solid dots corresponding to the appropriate frequencies are placed directly above the score values. The dots are then connected by solid lines. Frequency polygons are always "closed" with the abscissa in the sense that they always include a value that is a unit higher than the highest observed score and a value that is a unit lower than the lowest observed score, with a 0 frequency denoted for each. This is what forms the polygon. The similarity of the frequency histogram and frequency polygon can be seen in Figure 2.3, where one is superimposed on the other.

No specific rules govern when a frequency histogram as opposed to a frequency polygon should be used. Frequency polygons are typically used when the variables being reported are continuous in nature, whereas frequency

FIGURE 2.2   **Frequency Polygon**

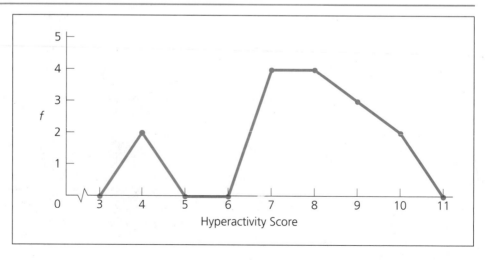

FIGURE 2.3   **Frequency Histogram and Frequency Polygon Superimposed**

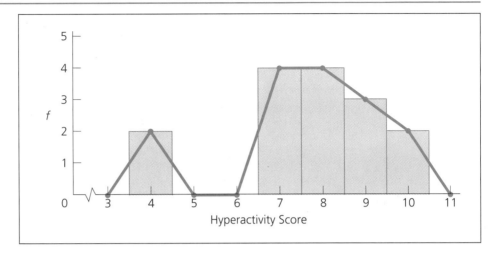

histograms are typically used when the variables being reported are discrete. The major reason is that, from a visual perspective, the frequency polygon tends to highlight the "shape" of the entire distribution more than the frequency histogram does. The frequency histogram, by comparison, tends to highlight the frequency of occurrence of specific scores rather than the entire distribution. The use of a frequency polygon for a continuous variable is thus more consistent with the notion of emphasizing a continuum.

A graph that is closely related to the frequency polygon is the **line plot**. The line plot is constructed exactly like the frequency polygon, except that it is not "closed." Rather, the left- and rightmost points of the line end on the lowest and highest scores, respectively. Figure 2.4 presents an example line plot for the hyperactivity data.[*] Line plots are particularly useful when one or more research participants receive the lowest or the highest score possible on a particular measure. For instance, one or more children could have obtained the maximum score of 12 on the hyperactivity test. If this were the case, it would not be meaningful to indicate that a score of 13 occurred with 0 frequency because 13 is not a valid score on the hyperactivity measure. In this case, one might use a line plot instead of a frequency polygon. This would also be true if one or more children obtained the minimum score of 0.

Frequency polygons and line plots are useful when one wants to compare distributions for two or more groups of individuals. Consider data for a group of 14-year-olds and a group of 17-year-olds who were asked to indicate how satisfied they are with their relationship with their mother, using a scale that ranged from 1 to 5. On this scale, a score of 1 indicates the teen is very dissatisfied with the relationship, a score of 2 indicates the teen is moderately dissatisfied with the

---

[*] Some investigators define line plots differently than this, but our use of the term is the more common.

FIGURE 2.4 Line Plot

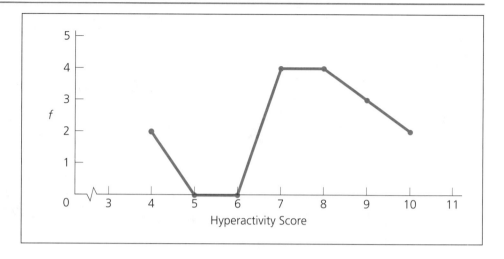

relationship, a score of 3 indicates the teen is neither satisfied nor dissatisfied with the relationship, a score of 4 indicates that the teen is moderately satisfied with the relationship, and a score of 5 indicates that the teen is very satisfied with the relationship. Some behavioral scientists have argued that the quality of the parent–teen relationship deteriorates with increasing age through the adolescent period. Figure 2.5 presents separate line plots for the samples of 14-year-olds ($N = 115$) and 17-year-olds ($N = 134$) on the same graph. A legend in the body of the graph identifies the line that describes each age group. The distributions are similar in that the majority of teens are very satisfied with their relationship with their mothers. However, relatively more 17-year-olds are moderately satisfied with their relationship with their mothers and relatively more 17-year-olds are very dissatisfied with their relationship, when compared with 14-year-olds. Later chapters in this book will discuss ways to examine these differences more formally in the context of inferential statistics.

It is also possible to construct frequency histograms with multiple groups, although this tends to become unwieldy for more than two groups. The data in Figure 2.5 are regraphed using a multiple-group frequency histogram in Figure 2.6. In this figure, the bars for the two groups are plotted side by side, but with a different color of "fills" to distinguish the two groups. Different patterns may also be used. A legend in the body of the graph identifies the groups.

Frequency histograms and frequency polygons can be constructed for grouped as well as ungrouped scores. When the scores are grouped, the abscissa might list the midpoints of the score intervals rather than the individual score values. Alternatively, the intervals for the groups may be presented. The procedures for indicating the frequency associated with each interval then parallel those outlined earlier.

FIGURE 2.5 **Line Plot with Two Groups**

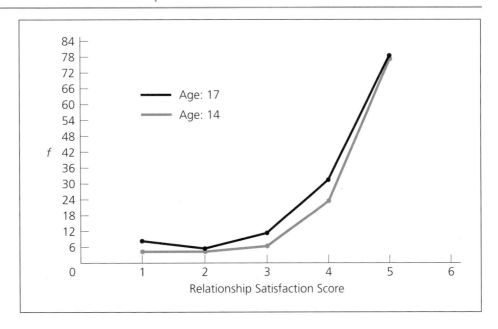

FIGURE 2.6 **Histogram with Two Groups**

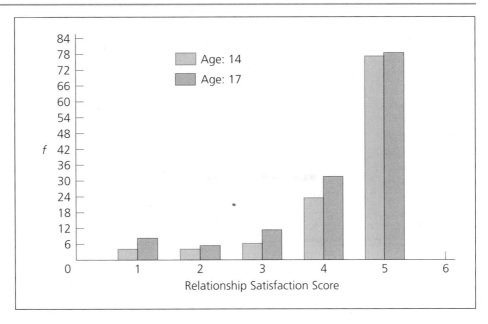

FIGURE 2.7   **Frequency Histogram of Grouped Data**

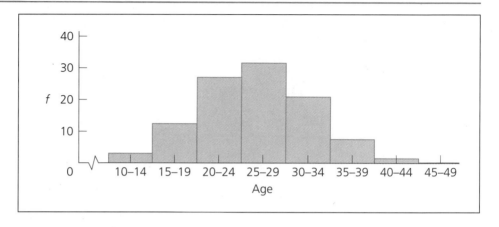

An example of a grouped frequency histogram is presented in Figure 2.7. This histogram presents data from a national survey of fertility behavior and focuses on women who gave birth to a child during 1989. Child psychologists have been keenly interested in the age of motherhood and how this affects child development. A substantial body of research documents potential negative effects of having children when the mother is too young (as in the case of adolescents) as well as risks associated with delayed motherhood (e.g., increased chances of a child with mental retardation). For these data, the histogram uses age intervals on the abscissa to show the ages of women who had a live birth. There were 4,040,958 women in the study, and the ordinate indicates percentages rather than frequencies. As we will discuss later, the use of percentages instead of frequencies does not change the relative heights of the bars in a frequency histogram and they often make it easier to interpret.

It can be seen that more births occurred in the 25–29 age group than in any other group. A substantial number of adolescents (ages 10–14 and 15–19) gave birth, representing just over 15% of all live births. This underestimates the "problem of teen pregnancy" because it excludes abortions, miscarriages, and stillborns, which are considerable in number. Relatively few women gave birth to a child when they were over 40, with this age group accounting for just over 1% of all live births.

Another useful technique for graphing the frequencies of quantitative variables is the **stem and leaf plot**. Consider the following example: A group of 50 first-year college students were asked to estimate what their scores would be on an intelligence test. They were told that the average score for first-year college students, in general, was 100 and that most students (about two-thirds) scored somewhere between 90 and 110. They were also shown a frequency distribution of the scores of 1,000 first-year college students from the previous year. The distribution showed that the most frequently occurring score was 100 and that very few students scored 115 or higher (only about 5%). With this as an orientation, they suggested what their scores would be. Here are the data:

95	95	100	100	100	100	100	105	105	105
110	110	110	110	110	110	110	110	110	115
115	115	115	115	115	115	115	115	115	115
120	120	120	120	120	120	120	120	125	125
125	125	130	130	130	130	135	135	140	140

A stem and leaf plot of these scores appears in Figure 2.8a. The digits that represent the numbers of "hundreds" and "tens" (e.g., the 10 from 100, the 11 from 110, the 12 from 120) are listed in the first column, and a vertical line separates them from the rest of the table. This constitutes the *base,* or the "stem," of the stem and leaf plot. The digits that represent the number of "ones" for each base value (e.g., the 0 from 100 representing zero "ones" and the 5 from 105 representing five "ones") are listed to the right of the line, one for each score

FIGURE 2.8   **Estimated Intelligence Data Displayed in (a) a Stem and Leaf Plot, (b) a Frequency Histogram, and (c) a Rotated Stem and Leaf Plot**

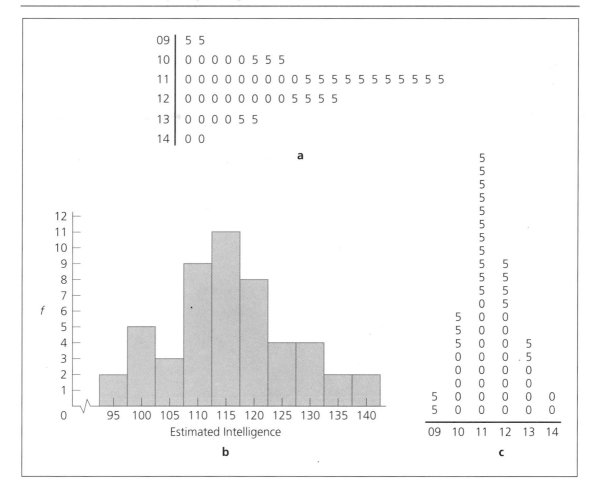

within that group. These are the "leaves." This is a compact way of conveying both the individual scores that occurred and the general "shape" of the frequency distribution. More specifically, the stem and leaf plot is somewhat like a frequency histogram turned on its side. To illustrate this, Figure 2.8b presents a traditional frequency histogram, and Figure 2.8c presents the stem and leaf plot rotated so that the tens and hundreds digits appear at the base rather than on the right. Note the similarity in the relative heights of the columns in the histogram and the rotated stem and leaf plot. For these data, even though the true average intelligence score is 100, most of the students felt they would score well above 100 on the intelligence test. The most frequent perceived intelligence score was 115, which is considerably above the true average (recall that students were told that only 5% of previous students had scores of 115 or higher). Apparently, these students have a high opinion of their own intellect!

Stem and leaf plots are useful as long as the number of scores is not too large and the number of different values of the base is reasonable (i.e., more than one or two but less than 20).

There are many other forms of frequency graphs that we do not describe here, including pie charts and three-dimensional histograms. Interested readers are referred to Fox and Long (1990).

Frequency Graphs for Qualitative Variables

Frequency histograms, frequency polygons, and line plots are used to graph frequency data for quantitative variables. Frequency graphs can also be constructed for qualitative variables. For instance, Figure 2.9 is a **bar graph** of the data from the survey of religious affiliation referred to in Section 2.3. The values of the variable are listed on the abscissa, and the frequencies are listed on the ordinate. The major difference from the frequency histogram is that the bars are drawn such that

FIGURE 2.9   Bar Graph

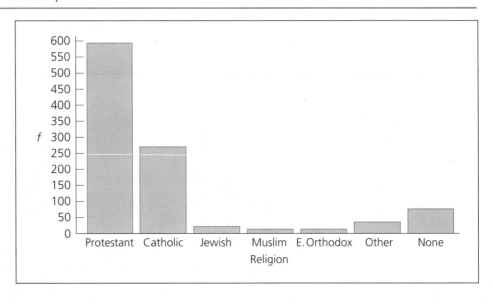

they do not touch one another. This is because each bar in a bar graph represents a distinct category. Aside from this feature, the basic principles in constructing a bar graph are the same as those for a frequency histogram.

2.6   Misleading Graphs

The presentation of data in graphic form can be highly informative, but it can also be misleading. Consider a consumer psychologist who has interviewed 100 people in order to determine how many prefer product A over product B, or vice versa. Suppose that 45 preferred A over B and 55 preferred B over A. This information could be depicted graphically as in either of the two frequency histograms in Figure 2.10.

The two graphs make the preferences for A and B appear different, even though the identical data were used in their construction. In Figure 2.10b, the difference in the number of people who prefer product B appears smaller than that in Figure 2.10a because the distance between the demarcations on the ordinate is smaller than in Figure 2.10a. As a result, even though the ordinates are the same physical height, a much smaller portion of the ordinate in Figure 2.10b is actually used to represent the observed frequencies.

Because frequency graphs can be misleading, depending on how the abscissa and ordinate are formatted, behavioral scientists have adopted a rule stating that the ordinate should be drawn such that its height at the demarcation for the highest frequency is approximately three-fourths to two-thirds the length of the abscissa (American Psychological Association, 1994, p. 157). In addition, the ordinate should start with a frequency of 0, with "jumps" to a larger number

FIGURE 2.10   **Examples of Potentially Misleading Graphs**

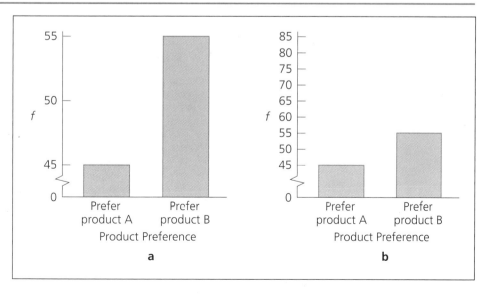

indicated by a break in the form of a zigzag if it is not drawn to scale. These rules ensure uniform, clearly interpretable presentation of graphed results. One should also be careful to examine the demarcation units on the ordinate. For instance, if you looked at Figure 2.10a without noting the values of the frequency labels, you might conclude that product B is more preferred over product A than it actually is.

2.7   Graphs of Relative Frequencies, Cumulative Frequencies, and Cumulative Relative Frequencies

It is interesting to note the nature of a graph of relative frequencies compared with a graph of frequencies. A polygon of the relative frequencies for the hyperactivity data from Section 2.1 appears in Figure 2.11. Notice that the ordinate is labeled *rf* and demarcated with relative frequency values. Also notice that the shape of the polygon is identical to the shape of the frequency polygon in Figure 2.2. Similarly, a relative frequency histogram for these data would take the identical shape as the frequency histogram contained in Figure 2.1. This should hardly be surprising because all we have done is divide each frequency by the constant *N*. It follows that graphs of percentages will also take the identical shape as the corresponding frequency graphs.

It is also possible to represent cumulative frequency information graphically. This has been done in Figure 2.12 for the hyperactivity example. The ordinate is labeled *cf* and demarcated with cumulative frequency values. Solid dots representing the cumulative frequencies are placed above the upper real limit of each score value. This contrasts with placing the dots directly above the score values in frequency and relative frequency graphs. The dots are placed above the upper real

FIGURE 2.11   **Graph of Relative Frequencies**

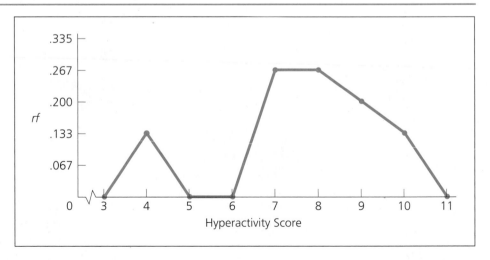

FIGURE 2.12   **Graph of Cumulative Frequencies**

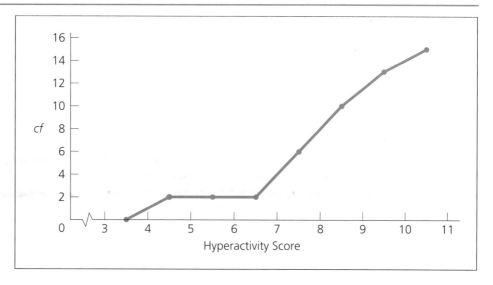

limits in Figure 2.12 because a cumulative frequency for a continuous variable en-compasses all scores up to the upper real limit of the specified score value. An ad-ditional aspect of Figure 2.12 should be noted: Because the cumulative frequency for a given score value is always equal to or greater than the cumulative frequency for the preceding score value, the cumulative frequency curve always remains level or increases as it moves from left to right.

Cumulative relative frequencies can be represented in a graph. Such a graph would take the identical shape as the corresponding cumulative frequency graph, but it would have its ordinate labeled *crf* and demarcated with cumulative relative frequency values.

2.8   Probability Distributions

In Section 2.1, we noted that the probability of a given outcome is equivalent to the relative frequency of that outcome. Thus, given a distribution of scores, the probability of randomly selecting a given score from that distribution equals the relative frequency of that score. However, there is an important conceptual difference between a probability and a relative frequency: *Whereas a relative frequency indicates the proportion of times that some score was previously observed, a probability represents the likelihood of observing that score in the future.* This distinction should be kept in mind as you read the following discussion.

Probability Distributions for Qualitative and Discrete Variables

Consider the case of a qualitative variable, gender, which has two possible values, male and female. Suppose we have a population of 200 individuals, 150 females and 50 males. The relative frequency for females is 150/200 = .75 and for males it is 50/200 = .25. The probability of randomly selecting a female from this population is .75, and the probability of randomly selecting a male is .25. When the potential values for a qualitative or discrete variable are such that a person can have one and only one score (for example, it is impossible for a person to be both male and female—he or she must be either a male *or* a female), the score values are said to be *mutually exclusive*. When the values considered are also *exhaustive* (that is, they include all possible values that could occur), the probabilities associated with the individual score values represent a **probability distribution** with respect to that variable. Figure 2.13 presents a graph of the probability distribution for gender for the population of 200 individuals just described. As with relative frequencies, a graph of probabilities takes the identical shape as the corresponding frequency graph. However, the ordinate in this instance is labeled *p*.

Probability Distributions for Continuous Variables

A probability distribution for a continuous variable is conceptualized somewhat differently than that for either a qualitative or a discrete variable. Recall from Chapter 1 that a number that represents a score on a continuous variable is properly considered as within the real limits of that number. When we say that a person solved a problem in 30 seconds, we do not mean exactly 30 seconds but rather somewhere between 29.5 and 30.5 seconds. Because it is always possible, in principle, to have a measuring device that is more accurate than the one we used, we cannot meaningfully talk about the probability of obtaining a score equal to an *exact* value for a continuous variable. Rather, probability is better conceptualized as being associated with a range of values, such as between 29.5 and 30.5.

FIGURE 2.13   **Graph of a Probability Distribution of a Qualitative Variable**

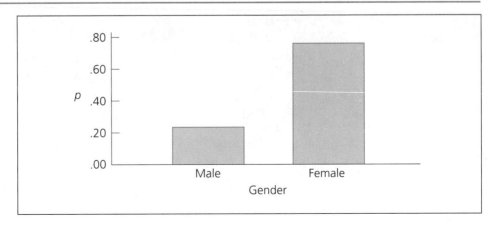

In contrast to qualitative and discrete variables, it is not possible to specify a probability distribution of a continuous variable by listing the possible values of the variable and their corresponding probabilities. This is because the number of possible values a continuous variable can have is, in principle, infinite. Statisticians therefore think of a probability distribution of a continuous variable in terms of a **probability density function.**

You can get an intuitive feel for this concept through graphs. Figure 2.14 presents a probability distribution for shyness (a continuous variable) in a population. A continuous probability distribution is always represented by a smooth curve over the abscissa. The abscissa represents the possible values of the continuous variable, with increasingly higher values going from left to right. The ordinate, though not formally demarcated, represents what statisticians call a density and is proportional to the probability of observing a given value (or, more technically, a range of values) on the abscissa. The higher the curve, the more "dense" are the scores and the more likely they are to occur.

The key to understanding a probability density function is to think of the probability as an *area* under the curve or, as it is more formally called, the **density curve.** The total area under the curve represents 1.00, or the probability that a given person will have *some* value on the dimension in question. In Figure 2.14, points *a* and *b* have been marked on the abscissa to represent the limits of an interval. The shaded portion between *a* and *b* is the area under the curve that corresponds to scores in that interval. The probability of obtaining a value between *a* and *b* is thus represented by this shaded area under the curve. Using appropriate mathematical procedures, we can compute the size of this area based on the case where the total area under the curve equals 1.00. In this way, it is possible to specify the probability of obtaining a set of values that fall within some interval for a continuous variable. Specifically, the probability of obtaining a score within a given interval equals the area corresponding to that interval under the density

FIGURE 2.14   **Graph of a Probability Distribution of a Continuous Variable**

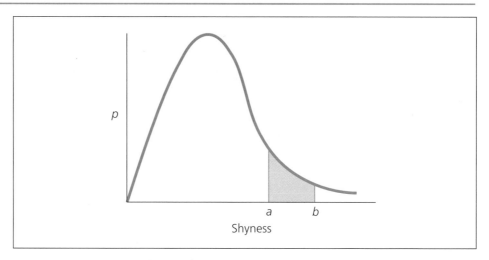

curve. In Figure 2.14, the probability of obtaining a score between points *a* and *b* is .11. We will discuss formal procedures for calculating such areas in Chapter 4.

2.9   Empirical and Theoretical Distributions

An important distinction in statistics is between **empirical distributions** and **theoretical distributions.** Empirical distributions are based on actual measurements collected in the real world. In contrast, theoretical distributions are not constructed by formally taking measurements but, rather, by making assumptions and representing these assumptions mathematically.

One very important type of theoretical distribution that has been studied extensively by statisticians is the **normal distribution.** The normal distribution was so named in the early 19th century by a French mathematician named Quetelet, who believed that it characterized the shape of a large number of phenomena, including height, weight, intelligence, and many psychological variables.

There is actually a family of normal distributions, each member of which is precisely defined by a mathematical formula given in Appendix 4.1. Figure 2.15 presents some examples of normal distributions. As can be seen, all distributions in this family are symmetrical and characterized by a "bell shape."

Figure 2.16a presents a relative frequency histogram for an empirical distribution of intelligence test scores for students in a particular high school, and Figure 2.16b presents a theoretical distribution of intelligence test scores for this same group. Notice that the shape of the latter distribution is normal and closely corresponds to the empirical distribution in Figure 2.16a. Theoretical distributions are sometimes used to represent reality when practical limitations make it impossible to construct empirical distributions. For instance, a distribution of the

FIGURE 2.15   **Examples of Normal Distributions**

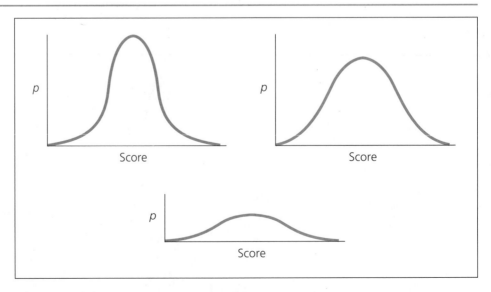

FIGURE 2.16   **(a) Empirical and (b) Theoretical Distributions of Intelligence Test Scores**

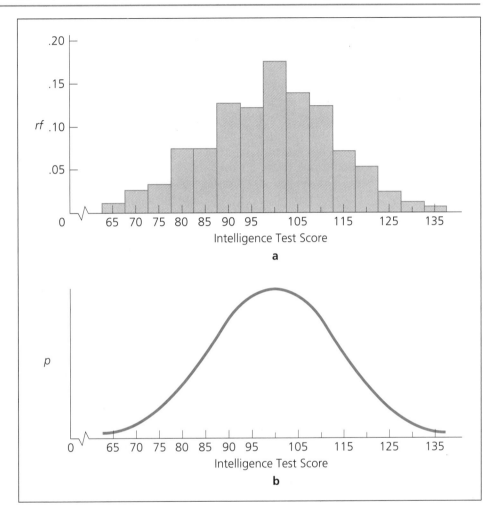

intelligence test scores of *all* adults in the United States could not be constructed empirically. However, we might be able to approximate it closely by reference to the normal distribution. If we are willing to make the assumption that the distribution of intelligence test scores is approximately normal, then our knowledge of this theoretical distribution can be used to help us gain insight into the nature of intelligence test scores in the real world. As we will see in later chapters, inferential statistics makes extensive use of theoretical distributions.

2.10 Method of Presentation

Many behavioral science journals follow the guidelines presented in the *Publication Manual of the American Psychological Association* (American

Psychological Association, 1994) for presenting the results of statistical analyses. As detailed in that manual, statistical information appears slightly different in published and unpublished versions of a manuscript. When describing statistical results in "Method of Presentation" sections, we follow the format used by authors when submitting manuscripts to journals for publication. Where relevant, we point out differences between this approach and that used in articles as they appear in published form.

It is rare for an investigator to report all the types of frequency information discussed in this chapter when presenting the results of a frequency analysis. The major reason for this is the cost of journal space. Consequently, the data must be presented as efficiently as possible. Most research reports focus on relative frequencies or percentages. When the total number of scores, N, is also provided, any of the other desired measures can be readily calculated.

An example of a table that might be presented is given here:

Percentage of Individuals Who Approve or
Disapprove of Nuclear Power Plants

Response category	%
Strongly approve	24.6
Moderately approve	19.5
Neither approve nor disapprove	8.4
Moderately disapprove	20.5
Strongly disapprove	27.0

Note: $N = 215$

This table reports responses to a question asked of 215 college students concerning their approval or disapproval of nuclear power plants. All the information is present to allow the reader to determine the frequencies, relative frequencies, cumulative frequencies, and cumulative relative frequencies. The relative frequencies are derived by dividing each percentage by 100. Thus, $24.6/100 = .246$, $19.5/100 = .195$, $8.4/100 = .084$, $20.5/100 = .205$, and $27.0/100 = .270$. The note to the table gives the total number of individuals who responded to the question; in this case, $N = 215$. The number of people in each category can then be obtained by multiplying each relative frequency by N and rounding to the nearest integer. Thus, $.246 \times 215 = 53$, $.195 \times 215 = 42$, $.084 \times 215 = 18$, $.205 \times 215 = 44$, and $.270 \times 215 = 58$. We now have the information to construct a more detailed table:

Response category	f	rf	%	cf	crf
Strongly approve	53	.246	24.6	215	1.000
Moderately approve	42	.195	19.5	162	.754
Neither	18	.084	8.4	120	.559
Moderately disapprove	44	.205	20.5	102	.475
Strongly disapprove	58	.270	27.0	58	.270

This example does not mean that you need to calculate all of this information each time you encounter an incomplete table. Usually what is presented will be sufficient for interpretation. However, if a reader wants to construct a more detailed table (for some reason the author had not anticipated), the information should be present that will allow him or her to do so. The format suggested here accomplishes this.

One difference between published and unpublished versions of a manuscript is the way statistical symbols are presented. In unpublished reports, these are typed in roman (nonitalic) letters and underlined, whereas in published form they are in italic type and not underlined. Thus, the symbol $\underline{N}$ from the original table would appear as N in a journal article.

Because of cost and space considerations, frequency graphs are typically used only when they are necessary to illustrate major trends in the data that might otherwise be difficult to note. It should be mentioned, however, that given an appropriately presented frequency table, the reader can, if he or she wishes, construct a graph from the table. When frequency graphs are presented, it is important that both the abscissa and the ordinate be clearly labeled and that the other instructions presented earlier in this chapter be followed.

2.11 Example from the Literature

The concept of intelligence is of major interest to behavioral scientists. Many attempts, some controversial, have been made to measure this variable. One of the most widely used tests for adults is the Wechsler Adult Intelligence Scale (WAIS). This test has been extensively studied and applied to several national samples in the United States. One such application is reported in Wechsler (1958). Based on a sample of 2,052 individuals, the following frequency analysis characterizes the scores on this test:

Intelligence	f	rf	%	cf	crf	Verbal description
Above 129	29	.014	1.4	2,052	1.000	Very superior
120–129	150	.073	7.3	2,023	.986	Superior
110–119	349	.170	17.0	1,873	.913	Bright normal
100–109	525	.256	25.6	1,024	.743	Average
90–99	504	.246	24.6	999	.487	Average
80–89	299	.146	14.6	495	.241	Dull normal
70–79	132	.064	6.4	196	.095	Borderline
Below 70	64	.031	3.1	64	.031	Mentally retarded

The verbal descriptions provided on the extreme right are labels often used by clinical psychologists for ranges of scores on the WAIS. Notice that scores labeled "average" (90–109) are the most common. Also, the proportion of

**Applications
to the Analysis
of a Social Problem**

The database described in Chapter 1 on parent–teen communication about teenage pregnancy and sexual behavior is amenable to interesting frequency analyses. We present only a few here. All of the analyses that we report were conducted on a personal computer (PC). We report the results of the analyses using the formal "printouts" or "output" from the SPSS for Windows statistical computer package that we used to analyze the data. We adopt this practice throughout those sections that analyze the parent–teen data in order to give you some exposure to statistical results on printouts. However, we do not discuss the more complex issue of computer programming used to produce the output, which could be the subject of a course in its own right.

One question of interest is just how sexually active teenagers are. Table 2.1 presents frequency distributions for three variables from the database. The first variable is a dichotomous (i.e., two-valued) variable that asked 14-year-olds whether or not they had ever had sexual intercourse. A score of 0 was assigned to each teen who reported never having engaged in sexual intercourse, whereas a score of 1 was assigned to each teen who reported having engaged in sexual intercourse. This variable is labeled "SEX" in the computer program, and the output is in the top portion of Table 2.1. The output is organized into six columns. The first column is called "Value Label" and is the label we told the computer to assign to each score. A score of 0 had the label "No sexual intercourse" assigned to it, and a score of 1 had the label "Had sexual intercourse" assigned to it. The second column (labeled "Value") lists the different values of the variable. The third column (labeled "Frequency") reports how frequently each score occurred. The fourth column converts the frequency to a percent. The fifth column is labeled "Valid Percent" and, in this case, is identical to the "Percent" column. We will explain this column later when discussing a second frequency analysis in which the two columns differ from each other. The final column (labeled "Cum Percent") is the cumulative percentage.

The labels "Valid cases" and "Missing cases" below the frequency information indicate the total number of cases included in the analysis and the number of cases for which no responses were provided, respectively. In this example, $N = 115$, with no missing data (i.e., a score was assigned to every individual).

Just fewer than 36% of the 14-year-olds in this sample had already engaged in sexual intercourse at least once. These were adolescents in, at most, eighth or ninth grade, and this represents a high degree of sexual activity for such young adolescents.

The second frequency analysis was conducted on a variable called "MSEX." We asked the mothers of the 14-year-olds whether they thought their teen had ever engaged in sexual intercourse. Of the 115 mothers, five did not answer this question; hence, there are "missing data" for these five mothers. Under the column "Value," missing data are indicated by a score of "." The "Percent" column indicates the percentage of mothers who had a given score relative to the total number of mothers, *including those who had missing data*. For example, $(94/115) \times 100$, or 81.7%, of the mothers did not think their teen had ever engaged in sexual intercourse. Also, $(5/115) \times 100$, or 4.3%, of the cases had missing data. The "Valid Percent" column calculates the percentage of mothers who had a given score, *eliminating those mothers with missing data*. For example, $(94/110) \times 100$, or 85.5%, of the mothers who responded thought their teen had never engaged in sexual intercourse.

Usually, researchers report the value in the "Valid Percent" column rather than the "Percent" column because it "adjusts" for missing data. However, care must be taken to ensure that any conclusions drawn from the "Valid Percent" value are qualified by the fact that some individuals chose not to answer the question. We will encounter situations in later chapters where both the "Percent" and the "Valid Percent" values are of

(continued)

TABLE 2.1   **Computer Output for Frequency Analysis**

SEX   Whether or not teen has ever had sexual intercourse

Value Label	Value	Frequency	Percent	Valid Percent	Cum Percent
No sexual intercourse	.00	74	64.3	64.3	64.3
Had sexual intercourse	1.00	41	35.7	35.7	100.0
	Total	115	100.0	100.0	

Valid cases 115   Missing cases 0

MSEX   Whether mother thinks teen has had sexual intercourse

Value Label	Value	Frequency	Percent	Valid Percent	Cum Percent
Thinks has not had sex	.00	94	81.7	85.5	85.5
Thinks has had sex	1.00	16	13.9	14.5	100.0
	.	5	4.3	Missing	
	Total	115	100.0	100.0	

Valid cases 110   Missing cases 5

AGEFIRST   Age at first intercourse

Value Label	Value	Frequency	Percent	Valid Percent	Cum Percent
	6	4	.5	1.0	1.0
	7	9	1.2	2.2	3.2
	8	6	.8	1.5	4.7
	9	13	1.7	3.2	7.9
	10	23	3.1	5.7	13.6
	11	27	3.6	6.7	20.3
	12	62	8.3	15.3	35.6
	13	101	13.4	25.0	60.6
	14	109	14.5	27.0	87.6
	15	43	5.7	10.6	98.3
	16	7	.9	1.7	100.0
	.	347	46.2	Missing	
	Total	751	100.0	100.0	

Valid cases 404   Missing cases 347

theoretical interest, with each providing a slightly different perspective on the data.

It is evident in this analysis that the mothers were underestimating the sexual activity of their teens. Whereas just over 35% of the teens reported having engaged in sexual intercourse, only about 14% of the mothers thought their teen had engaged in sexual intercourse. These results are consistent with numerous interviews we have conducted with parents. Parents are reluctant to acknowledge their child's sexual tendencies, especially during younger adolescence, when many parents insist on viewing the adolescent as a child rather than as an emerging adult. This tendency to deny an adolescent's sexual tendencies gets in the way of effective parent–teen communication about premarital pregnancy because many parents delay communicating about this important topic if they do not think their child is sexually oriented yet. Parents need to be educated that young teens are more sexually active than they think.

Actually, comparing percentages for different groups of individuals (in this case, mothers versus teens) is not as straightforward as this example portrays. Inferential statistics must be used to obtain more confidence that the differences we are observing in the samples also exist in the *populations* that the samples represent. We will address this issue in later chapters.

The third frequency analysis in Table 2.1 uses the entire sample of adolescents (i.e., 14-, 15-, 16-, and 17-year-olds) and focuses on the self-report of how old the teens were the first time they had sexual intercourse. The 347 cases of missing data represent those teens who have not yet engaged in sexual intercourse. Of those teens in the study who reported being sexually active, the most common ages for the first experience of intercourse were 13 and 14. Looking at the column labeled "Valid Percent," we see that 25.0% of the sexually active teens had their first intercourse at age 13 and 27.0% of the sample had their first intercourse at age 14. Of the *total* sample (i.e., including those who have not yet engaged in sexual intercourse), the percentage of teens who had sexual intercourse at ages 12 or younger was 19.2%. We obtained this figure by using the "Percent" column (which includes the missing data), and calculating the cumulative percentage (i.e., .5 + 1.2 + .8 + 1.7 + 3.1 + 3.6 + 8.3 = 19.2). The teens who reported first intercourse at very young ages (6–9) most certainly reflect instances of sexual abuse. If we eliminate these teens as being "atypical," we can see that 15.0% of the sample had intercourse by age 12 (i.e., 3.1% at age 10 + 3.6% at age 11 + 8.3% at age 12 = 15.0%). Overall, the data suggest that a substantial number of teens are sexually active at relatively young ages (at least for the population that the present samples represent), thereby increasing the risk of unintended pregnancy.

individuals with scores above 129 (very superior) is quite small. This is also true of individuals with scores below 70 (mentally retarded). Only 1.4% of the sample had scores above 129, and 3.1% had scores below 70.

In 1926, a behavioral scientist named Cox published an extensive study of eminent men in history. She attempted to estimate the IQ scores that these individuals would have achieved had they lived in a time when such a test could be administered. Some of these estimates follow:

Francis Galton, English scientist: 200
John Stuart Mill, English philosopher: 190
Johann Wolfgang von Goethe, German writer and philosopher: 185
Gottfried Wilhelm Leibniz, German philosopher and mathematician: 185
Samuel Taylor Coleridge, English writer and poet: 175
John Quincy Adams, American statesman and president: 165
David Hume, English philosopher: 155
Alfred Tennyson, English poet: 155

René Descartes, French philosopher and mathematician: 150
Wolfgang Amadeus Mozart, Austrian composer: 150
William Wordsworth, English poet and writer: 150
Francis Bacon, English philosopher and scientist: 145
Charles Dickens, English writer: 145
Benjamin Franklin, American inventor and statesman: 145
George Frideric Handel, German composer: 145
Thomas Jefferson, American statesman and president: 145
John Milton, English poet: 145
Daniel Webster, American statesman and senator: 145

The magnitude of these estimates is impressive, especially in light of the frequency analysis presented in the previous table.

Cox's research was based on an extensive analysis of case records and did not use intelligence testing as it is used today. In practice, it is difficult to extrapolate IQ scores from such archival data, and numerous scholars have questioned the accuracy of Cox's estimates (e.g., Fancher, 1985).

Summary

A frequency distribution is a table that conveys information about the frequencies, relative frequencies, percentages, cumulative frequencies, and cumulative relative frequencies of a set of scores. One use of frequency distributions is to succinctly summarize a set of scores. Another use is to identify very extreme scores, or outliers. When we present a frequency distribution for a quantitative variable, it is sometimes useful to group the scores. Then we must make three decisions: (1) the number of groups to report, (2) the size of the interval, and (3) the beginning of the lowest interval. Frequency information can also be presented graphically using a frequency histogram, frequency polygon, line plot, or stem and leaf plot for quantitative variables, and a bar graph for qualitative variables.

When the potential values on a qualitative or discrete variable are mutually exclusive and exhaustive, the probabilities associated with the individual scores represent a probability distribution with respect to that variable. Since the probability of an outcome is equivalent to the relative frequency of that outcome, the graph of a probability distribution in these cases takes the same shape as the corresponding relative frequency and frequency graphs. For a continuous variable, a probability distribution may be thought of as a probability density function. This is represented graphically as a smooth curve over the abscissa.

An important distinction can be made between empirical distributions and theoretical distributions. Empirical distributions are based on actual measurements collected in the real world, whereas theoretical distributions are derived by making assumptions and representing these assumptions mathematically. Normal distributions are a family of symmetrical and bell-shaped theoretical distributions that have been studied extensively by statisticians.

Exercises

Answers to asterisked (*) exercises appear at the back of the book.

Use the following information to complete Exercises 1–11:

An employer kept records of how many days her 20 employees reported in sick during the previous year. The scores on this variable were as follows:

8	7	6	4	3
6	3	7	6	6
4	6	6	6	7
6	6	8	7	6

*1. Construct a frequency distribution containing absolute frequencies.

*2. Compute the relative frequencies, percentages, cumulative frequencies, and cumulative relative frequencies.

*3. What proportion of employees were sick for 8 days? What proportion were sick for 8 or fewer days? What proportion were sick for fewer than 8 days?

*4. What proportion of employees were sick for 7 days? What proportion were sick for 7 or more days? What proportion were sick for more than 7 days?

 5. What proportion of employees were sick for 4 days? What proportion were sick for 4 or more days ? What proportion were sick for more than 4 days?

*6. Suppose you randomly selected a score from the 20 scores. What is the probability the selected score would be an 8? What is the probability the selected score would be a 6 or an 8? What is the probability the selected score would be 7 or less?

*7. Draw a frequency histogram for the set of scores.

 8. Draw a frequency polygon for the set of scores.

 9. Draw a line plot for the set of scores.

*10. Draw a polygon of the relative frequencies. Compare the shape of this graph

with that for the frequency polygon from Exercise 8.

11. Draw a graph of the cumulative frequencies.

*12. How does a grouped frequency distribution differ from an ungrouped frequency distribution?

13. What are the three questions that must be considered when deciding how to group data?

Use the following information to complete Exercises 14–25:

A principal in a small school measured the intelligence of fifth-grade students in her school. The intelligence test scores for these students were as follows:

129	99	98	113	103	128	102	110	80	105
93	98	109	109	100	111	106	96	108	90
104	94	92	119	127	89	95	92	105	108
83	100	107	106	101	118	84	119	105	111
118	106	122	120	102	117	103	117	103	88

14. Draw a frequency histogram for the set of scores.

15. Draw a stem and leaf plot for the set of scores. Rotate this so that the tens and hundreds digits appear at the base rather than on the right. Compare the relative heights of the columns in this graph with those in the histogram from Exercise 14.

*16. Construct a grouped frequency distribution containing absolute frequencies by grouping the scores into five groups.

17. Compute the relative frequencies, percentages, cumulative frequencies, and cumulative relative frequencies for the grouped scores.

*18. What proportion of students had scores of 109 or lower? What proportion had scores of 99 or lower? What proportion had scores of 110 or higher?

19. Draw a frequency histogram of the grouped data for the set of scores.

*20. Draw a frequency polygon of the grouped data for the set of scores.

21. Draw a histogram of the relative frequencies for the grouped data. Compare the shape of this graph with that of the frequency histogram from Exercise 19.

*22. Draw a graph of the cumulative frequencies for the grouped data. Compare the general shape of this graph with the cumulative frequency graph for the data on sick days from Exercise 11. What accounts for the similarity in their shapes?

*23. Suppose you wanted to compute a frequency distribution for the data by grouping the intelligence test scores into six groups. What size interval should you use? With what value should the lowest interval begin?

*24. Suppose you wanted to compute a frequency distribution for the data by grouping the scores into eight groups. What size interval should you use? With what value should the lowest interval begin?

25. Suppose you want to compute a frequency distribution for the data by grouping the scores into ten groups. What size interval should you use? With what value should the lowest interval begin?

Use the following information to complete Exercises 26–28.

Suppose you were commissioned to survey a small community to determine the marital status of all adults over 18 years of age. You select a random sample of 50 individuals and ask them if they are married (M), divorced (D), widowed (W), or single (S). The data for these individuals are as follows:

M M M M M S M W M S D S M M D W M
M S M W M M S M M M M M S M S D M
M S S D M M M D W M M M W M M S

26. Construct a frequency distribution containing absolute frequencies, relative frequencies, and percentages.

27. Draw a bar graph for the set of scores.

28. Draw a bar graph of the relative frequencies. Compare the shape of this graph with that of the bar graph from Exercise 27.

29. What does a break in the ordinate or abscissa of a frequency graph indicate?

*30. How tall should the ordinate of a frequency graph be relative to the abscissa? Why is this important?

31. Redraw the frequency histogram from Exercise 7 so that the height of the ordinate at the demarcation for the highest frequency is visually more than three-fourths the length of the abscissa. Now redraw it so that the height of the ordinate at the demarcation for the highest frequency is visually less than two-thirds the length of the abscissa. Compare the three graphs in terms of what they seem to suggest about employee sick days.

*32. What is a probability distribution? Why is the nature of probability distributions for qualitative and discrete variables different from that for continuous variables?

33. What is the difference between an empirical distribution and a theoretical distribution?

*34. A researcher surveyed 1,850 people on their attitudes toward capital punishment. One question asked respondents to indicate whether they thought the death penalty should be legal. Responses to this question could range from 1 ("definitely should not be legal") through 5 ("definitely should be legal"), where 3 represented a neutral stance. Compute the absolute frequencies, relative frequencies, cumulative frequencies, and cumulative relative frequencies for the set of responses based on the following percentage results:

Attitude	%
5	20.0
4	30.0
3	10.0
2	22.0
1	18.0

Multiple-Choice Questions

Refer to the following frequency distribution of the number of hours students studied for a test to do Exercises 35–38.

Study hours	f
8	20
7	30
6	30
5	10
4	10

*35. What proportion of students studied for 7 hours or longer?
   a. .300
   b. .500
   c. .800
   d. none of the above

36. The cumulative frequency for a score of 6 is
   a. 20
   b. 30
   c. 50
   d. 80

*37. What percentage of students studied for 5 hours?
   a. 10.0
   b. 20.0
   c. 30.0
   d. 90.0

38. A bar graph is a more appropriate way of graphing the data than is a frequency histogram.
   a. true
   b. false

39. Suppose you are constructing a grouped frequency distribution and you decide to use ten groups. If the highest score is 100 and the lowest score is 50, what interval size should you use?
   a. 2
   b. 3
   c. 5
   d. 10

40. A normal distribution is always bell shaped.
   a. true
   b. false

*41. The frequency of a score in comparison to the total number of scores in the group is called a
   a. cumulative frequency
   b. absolute frequency
   c. relative frequency
   d. cumulative relative frequency

*42. A probability density function
   a. can be graphically represented as a bar graph
   b. can be used only to represent variables that are measured on a ratio level
   c. is a smooth curve including all possible values of a continuous variable
   d. is always bell-shaped

*43. A stem and leaf plot is similar to a
   a. frequency histogram turned on its side
   b. frequency polygon
   c. bar graph
   d. line plot

44. A line plot is always closed with the abscissa.
   a. true
   b. false

*45. A case that shows a very extreme score relative to the majority of cases in the data set is known as a(n)
   a. descriptive statistic
   b. extremist
   c. outlier
   d. none of the above

CHAPTER 3

Measures of Central Tendency and Variability

Describing a set of scores with frequency distributions can be a highly informative way of presenting data about a variable. However, we often want to communicate information about data in a more succinct fashion. Statisticians have developed statistical indexes that are useful in characterizing scores for quantitative variables and that are more concise than frequency distributions. Many of these indexes are used in statistical tests described in later chapters of this book. The present chapter introduces you to these important indexes.

In the first section of this chapter, we focus on two categories of measures. The first category is measures of central tendency. Central tendency refers to the "average" score in a set of scores. When we specify a central tendency, we are, in essence, trying to specify a "representative" value of a set of scores. The second category is measures of variability. These measures indicate the extent to which scores within a set differ from one another. As an example, suppose that a researcher interviews a group of 100 14-year-olds who smoke cigarettes and asks them how many cigarettes they smoke on a typical day. The average number of cigarettes is 13.27, which conveys a sense of how much these youth smoke. However, do all of these teens smoke about 13 cigarettes a day, or is there variability about this average, with some of the smokers consuming only a few cigarettes a day and others consuming a large number of cigarettes? The heavy and light smokers may "average each other out," yielding a central tendency that is between the two extremes.

3.1 Measures of Central Tendency for Quantitative Variables

As noted, one piece of information that is useful when describing a set of scores is where the scores tend to fall on the numerical scale, or their **central tendency**. A central tendency refers to an *average,* or a score around which other scores tend to cluster. There are many indexes of central tendency; we consider three of them: the mode, the median, and the mean.

Mode

The **mode** of a distribution of scores is the most easily computed index of central tendency. It is simply the score that occurs most frequently. For the set of scores 6, 8, 8, 8, 10, 10, the mode is 8 because 8 occurs most frequently (three times). When the mode is used as the index of central tendency, we are using the most common score as the "representative" value for the set of scores. If we were to randomly select one score from a set of scores, the value of that score would most likely be equal to the value of the mode as opposed to any other value, since the mode occurs most frequently. In a graph of a distribution of scores, the modal score has the highest "peak" in the graph.

Although the mode is a relatively straightforward index of central tendency, it has some disadvantages. The major problem is that there can be more than one modal score. For example, consider the following scores: 6, 6, 6, 8, 8, 10, 10, 10. Both 6 and 10 are the most frequently occurring scores. In this case, there is ambiguity about which score is *the* mode because both occur with equal frequency. This set of scores is *bimodal*; it has two modes. When the mode is not equal to one unique value, it loses some of its effectiveness in characterizing the central tendency of a distribution of scores.

Median

Another measure of central tendency is the median. The **median** is the point in the distribution of scores that divides the distribution into two equal parts. In other words, 50% of the scores occur below the median and 50% of the scores occur above the median. In this sense, the median is a measure of central tendency. If the median is used as the index of central tendency, then we are using the "middle-most" score as the "representative" value for the set of scores.

There are three different approaches to computing the median of scores on a continuous variable. The first approach applies when there is an even number of scores; the second, when there is an odd number of scores; and the third, when there are duplications of the middle score(s), regardless of whether the total number of scores is odd or even. These are reviewed in turn.

1. *Computation of the median when there is an even number of scores.* Suppose we measured how long it took six students to complete a test. The times are 20, 22, 16, 18, 25, and 27 minutes. To compute the median, we first order the scores from lowest to highest:

16   18   20   22   25   27

The median is the arithmetic average of the two middle scores, or $(20 + 22)/2 = 21$. Figure 3.1a presents this graphically on a frequency histogram that also shows the real limits of each score value. Note that 50% of the scores occur below 21 and 50% of the scores occur above 21.

FIGURE 3.1   Illustration of the Median (a) When There Is an Even Number of Scores, (b) When There Is an Odd Number of Scores, and (c) When There Are Duplications of the Middle Score(s)

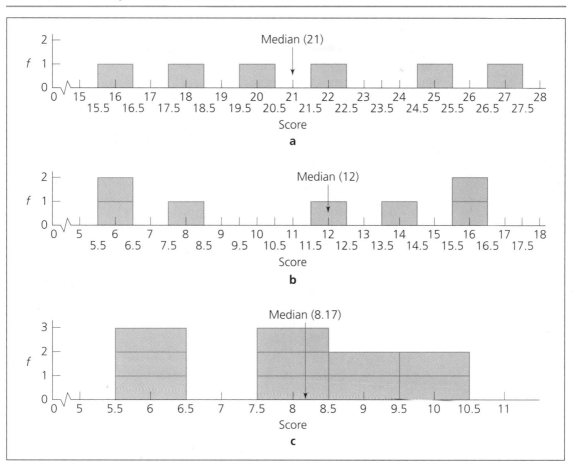

2. *Computation of the median when there is an odd number of scores.* Suppose we measured the test-taking times of seven students and their scores were 6, 8, 14, 12, 16, 6, and 16 minutes. To compute the median, we order the scores from lowest to highest:

    6   6   8   12   14   16   16

The median is simply the middle score, 12. Figure 3.1b presents this graphically. It also highlights the importance of considering the median in the context of the real limits of scores. An examination of the values 6, 6, 8, 12, 14, 16, and 16 shows that 50% of the scores are *not* less than 12. Only three of the seven scores are less than 12, and three are greater than 12. With an odd number of scores, the problem is what to do with the middle score. The answer lies in the concept of real limits. The individual who obtained the middle score of 12 minutes scored somewhere between 11.5 and 12.5 minutes. We do not know exactly where in the interval, so the most logical thing to do is to divide the interval in half to define the median. This yields the value of 12. Note in Figure 3.1b that the point 12 divides the distribution in half. There are 3½ frequency boxes below 12 and 3½ frequency boxes above 12.

3. *Computation of the median when there are duplications of the middle score(s).* This is the most common occurrence in the behavioral sciences. Suppose we measured the test-taking times of ten students and their scores were 6, 8, 6, 8, 6, 8, 9, 10, 9, and 10 minutes. We order the scores from lowest to highest:

    6   6   6   8   8   8   9   9   10   10

Applying the approach for when there is an even number of scores, we get 8 as the arithmetic average of the two middle scores. However, this is not the median; only three of the ten scores are less than 8. Statisticians have developed a formula for computing the median in cases of this type where there are duplications of the middle score(s). This formula is based on the assumption that the median occurs within the real limits of the middle score(s), and it is applicable regardless of whether the total number of scores is odd or even. To gain a conceptual understanding of this formula, consider the example in Figure 3.1c, in which there are ten scores and 8 is the middle score. The median occurs somewhere between 7.5 and 8.5 and represents the point where five scores are less than it and five scores are greater than it. There are already three scores less than 8 because three individuals had a score of 6. The median must therefore be defined so that two more scores are less than it. Individuals with a score of 8, technically, have scores between the real limits 7.5 and 8.5. Since three individuals had a score of 8 and we want to define the median so that two of these three individuals score less than the median, we specify a score two-thirds (or .67) greater than the lower real limit, or 7.5 + .67 = 8.17. Implicit in this approach is the assumption that the actual scores obtained by the three individuals were equally spaced from one another within the limits 7.5 and 8.5.

The formula for computing the median is

$$Mdn = L + \left[\frac{(N)(.50) - n_L}{n_W} \right] i \qquad\qquad [3.1]$$

where Mdn represents the median, L is the lower real limit of the category containing the median, N is the total number of scores in the distribution, n_L is the number of individuals with scores less than L, n_W is the number of individuals with scores within the category that contains the median, and i is the size of the interval of the category containing the median (that is, the difference between the upper real limit and the lower real limit of the category that contains the median). The .50 reflects the fact that we are seeking the point that divides the scores such that .50 (50%) of them fall below that point and .50 of them fall above that point. This formula can be applied to either grouped or ungrouped frequency information.

For the last example on test-taking times, the relevant portions of the frequency distribution are as follows:

Score	f	rf	crf
10	2	.20	1.00
9	2	.20	.80
8	3	.30	.60
6	3	.30	.30

To calculate the median time using Equation 3.1, we first refer to the cumulative relative frequency column. Starting from the bottom, we move up until we find the first number that is greater than or equal to .50. This number is .60 and occurs in the row for scores of 8. The lower real limit of this category is 7.5, so that is L. The number of scores *within* the category is 3, that is n_W. The number of scores *less than* 7.5 is 3, which is n_L. The interval size, i, is 8.5 − 7.5, or 1.0. Last, the total number of scores, N, is 10. Thus,

$$Mdn = 7.5 + \left[\frac{(10)(.50) - 3}{3} \right] (1.0)$$

$$= 7.5 + \left[\frac{2.00}{3} \right] (1.0)$$

$$= 7.5 + .67 = 8.17$$

This is the same value of the median we stated previously.

Sometimes when there are duplications of the middle score(s), a researcher might want to approximate the median quickly without using Equation 3.1. An approximation of the median can be obtained by following the first approach when N is even and the second approach when N is odd. In our example, $N = 10$, so we apply the approach for an even number of scores. This involves taking the arithmetic average of the two middle scores, which in this case is $(8 + 8)/2 = 8$. This value is quite similar to the exact value of the median (8.17) calculated using Equation 3.1.

In summary, when there are no duplications of the middle score(s), the median is equal to the arithmetic average of the two middle scores when N is even and equal to the middle score when N is odd. When there are duplications of the middle score(s), Equation 3.1 yields the value of the point that equally divides the scores in half, taking into consideration the real limits of the scores. The logic of Equation 3.1 is based on the assumption that scores in the middle interval (e.g., 7.5 to 8.5 in our example) are evenly distributed over that interval.

The median has an important statistical property that underscores its role as a measure of central tendency. Consider the following five scores, which have a median of 3: 1, 2, 3, 7, 8. If we subtract the median from each score and take the absolute values of the resulting differences, we get a set of unsigned **deviation scores.** Deviation scores are calculated by subtracting some constant (in this case, the median) from a set of scores. These deviations are said to be *unsigned* when their absolute values are taken. In our example, the absolute values of how far each score is from the median are

$$|1 - 3| = 2$$
$$|2 - 3| = 1$$
$$|3 - 3| = 0$$
$$|7 - 3| = 4$$
$$|8 - 3| = 5$$

If we add these deviation scores, we get 12. If any value other than the median were subtracted from the set of scores, the sum of the unsigned deviation scores would be greater than 12. In fact, across all individuals, scores will always be closer, in an absolute sense, to the median than to any other value. It is this property—the fact that the median minimizes the absolute difference between it and the scores in the distribution—that qualifies the median as a measure of central tendency.

Study Exercise 3.1

An organizational psychologist was interested in determining the median number of years that current employees had been working for a given company. Company records for the 200 employees were used to collect data. The obtained frequency distribution was as follows:

Number of years	f	rf	crf
6	20	.10	1.00
5	20	.10	.90
4	30	.15	.80
3	50	.25	.65
2	50	.25	.40
1	10	.05	.15
0	20	.10	.10

Compute the median number of years in the company.

Answer   Using Equation 3.1, we find that

$$Mdn = L + \left[\frac{(N)(.50) - n_L}{n_w}\right] i$$

$$= 2.5 + \left[\frac{(200)(.50) - 80}{50}\right] (1.0)$$

$$= 2.5 + \left[\frac{20.00}{50}\right] (1.0)$$

$$= 2.5 + .40 = 2.90$$

Thus, the median number of years in the company is 2.90.

Mean

The **mean** of a set of scores is familiar to all of us. It is simply the arithmetic average of the scores. The mean is computed by summing all the scores and then dividing the sum by the total number of scores. This can be represented by the equation

$$\overline{X} = \frac{\sum X}{N} \tag{3.2}$$

where $\overline{X}$ represents the mean of a set of scores on variable X. As an example, consider these ten scores: 2, 3, 3, 4, 5, 5, 5, 6, 8, 9. The sum of the ten scores is

$$\sum X = 2 + 3 + 3 + 4 + 5 + 5 + 5 + 6 + 8 + 9 = 50$$

Hence, the mean is

$$\overline{X} = \frac{\sum X}{N} = \frac{50}{10} = 5.00$$

The mean has an important statistical property that underscores its role as a measure of central tendency. Consider the following five scores, which have a mean of 3.00: 1, 1, 3, 4, 6. If we subtract the mean from each score and retain the signs of the resulting differences, we get a set of *signed* deviation scores:

$$1 - 3 = -2$$
$$1 - 3 = -2$$
$$3 - 3 = 0$$
$$4 - 3 = 1$$
$$6 - 3 = 3$$

These deviation scores sum to 0. If any value other than the mean were subtracted from the set of scores, the sum of the signed deviation scores would be greater than

FIGURE 3.2   **The Mean As a Balance Point**

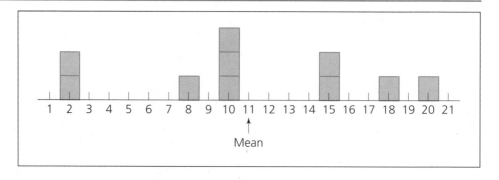

0 in absolute value. This is true for any set of scores: The sum of signed deviations about the mean will always equal 0. It is this property—the fact that the mean balances the deviations of scores above it with the deviations of scores below it—that qualifies the mean as an index of central tendency. Note that this is different from the median. The mean is the value that minimizes the sum of *signed* deviations, whereas the median is the value that minimizes the sum of *unsigned* deviations.

The mean parallels the idea of a center of gravity or a balance point in physics (Hays, 1994). Imagine a long board on which ten small blocks are stacked. Each block weighs the same amount. Twenty-one different positions are marked off on the board, as in Figure 3.2. Each block is placed in one of these positions, with some blocks being stacked on top of one another. Now suppose we wanted to place a fulcrum beneath the board that would perfectly balance it. At what position should we place the fulcrum? The point where the fulcrum would perfectly balance the board is the mean of the positions of the different blocks. For the example in Figure 3.2, this point would be $(2 + 2 + 8 + 10 + 10 + 10 + 15 + 15 + 18 + 20)/10 = 11.00$. In this analogy, the position of a block is analogous to the score of an individual within a set of scores, and each score is said to have equal "weight." The mean is like a center of gravity or a balance point and, thus, is a measure of central tendency.

Study Exercise 3.2

A researcher developed a test to measure reading ability. The test was intended to be used to help place students in remedial or advanced reading classes. A major concern of the researcher was how long it would take students to complete the test because it had to be administered during a short testing period. For 12 students in one class, the following completion times (in minutes) were recorded:

12	13	11	12	11	13
13	11	12	12	12	12

Compute the mean amount of time it took the class to complete the test.

Answer   We will compute the mean using Equation 3.2. The first step is to sum the 12 scores:

$$\Sigma X = 12 + 13 + 11 + 12 + 11 + 13 + 13 + 11 + 12 + 12 + 12 + 12 = 144$$

The mean is then:

$$\overline{X} = \frac{\Sigma X}{N} = \frac{144}{12} = 12.00$$

This means that the average amount of time it took the class to complete the test, as defined by the mean, was 12.00 minutes.

Comparison of the Mode, the Median, and the Mean

The mode, the median, and the mean are all indexes of central tendency. The mode is the most frequently occurring score in a distribution, the median is the point that divides the distribution into halves, and the mean is the arithmetic average of the scores. Sometimes the mode, the median, and the mean of a set of scores all have the same value. For instance, as illustrated in Figure 3.3, this is always true for normal distributions. But more often than not, the three measures of central tendency have different values. Which is the best index of central tendency? Ideally, when we are trying to characterize a set of scores, it is useful to report all three indexes. Each represents something slightly different, and the more information we can provide, the better. Most of the inferential statistics used by behavioral scientists make use of the mean, as will be explained in Chapter 7. Consequently, the mean is the most frequently encountered measure of central tendency in research reports.

FIGURE 3.3   **Illustration of the Equivalence of the Mode, Median, and Mean in a Normal Distribution**

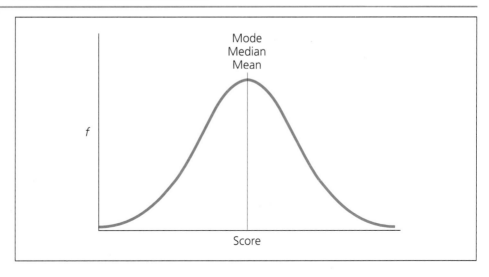

One way of contrasting the mode, the median, and the mean is in terms of a "best guess" interpretation of each. Suppose you know only the mode, median, and mean of a large set of scores. Each individual score is randomly selected in succession, and after each selection you are to guess the value of the score. What is your best guess? It depends. First, suppose you want the highest probability of being *exactly* correct. In this case, your best guess is the value of the mode because it is the most frequently occurring score and has the highest probability of predicting each score exactly. Second, suppose you are not interested in being exactly correct most often, but rather in making the smallest amount of *absolute* error, across all scores. In this case, the *sign* of the error is not important (that is, whether you overpredict or underpredict the value of the score is not critical), but the *size* of the error is important. Here, your best guess is the value of the median because it minimizes the absolute (unsigned) error across all scores. Finally, if your goal is to make the *signed* error be as close as possible to 0, then your best guess is the mean. As noted earlier, this is because across all scores, the amount of signed error from the mean will always equal 0.

There is no general rule as to which measure of central tendency is best; it depends on what one is trying to accomplish or communicate. For purely descriptive purposes, the median is a very useful measure because it minimizes the unsigned error. The mean is useful because it represents a "balance point" and has many desirable statistical properties that will be explained in Chapter 7. If one is primarily interested in representing the typical case, the mode is the measure of choice.

It might be instructive to consider an example of when the mode, the median, and the mean yield very different characterizations of a set of scores. Suppose you own a business that has seven employees, including yourself. The annual salaries are as follows:

Employee	Salary ($)
1	13,000
2	13,000
3	14,000
4	15,000
5	16,000
6	17,000
You	185,000

For the set of seven scores, the mode is 13,000, the median is 15,000, and the mean is 39,000. In characterizing the "average" score, one could portray the company differently depending on the measure of central tendency used. (For example, "Look how generous I am—the average salary is $39,000" as opposed to "Look how small my business is—the average salary is only $13,000.")

In general, when the distribution has one or more extreme scores (in the example, the salary of $185,000), the information conveyed by the mean can be distorted and it loses its value as a score that "represents" the set of scores. The mode and the median may provide better insight into the central tendency of the data in these cases. Note that in our example, these latter measures are not affected by the extremity of the seventh score; they would be the same if your salary were $185,000, $17,500, or $250,000. In contrast, the mean is based on the sum of all of the scores in a distribution and is therefore affected by the value of each score. The more extreme a score is relative to the other scores in the distribution, the more it will alter the mean. Thus, for instance, the mean salary of the seven employees is $39,000 when your own salary is $185,000, but the mean would be $15,071 if your salary were $17,500, and $48,286 if your salary were $250,000. In the behavioral science literature, income is a variable that frequently has extreme scores due to a small percentage of people who are extremely wealthy. For this reason, income is often reported in terms of medians rather than means.

Use of the Mode, the Median, and the Mean

We have presented three measures of central tendency in the context of analyzing quantitative variables. Some qualifications about the appropriateness of the indexes for describing central tendencies are now required. When a quantitative variable is measured on a level that at least approximates interval characteristics, all three measures of central tendency are meaningful. When a quantitative variable is measured on an ordinal level that departs markedly from interval characteristics, the mean is not an appropriate index of central tendency and the mode or median must be used instead. This is because the mean relies on calculating a sum, and a sum is meaningful only when the intervals between successive categories are approximately equal. Finally, the concepts of median and mean are meaningless for qualitative variables (that is, nominal measures) because these concepts require ordering objects along a dimension. In this case, the mode (that is, the most frequently occurring category) is the only applicable descriptor of central tendency.

Measures of Central Tendency for Discrete Variables

Our discussion of the mode, median, and mean has thus far assumed that the underlying variable being measured is continuous. However, these concepts also apply to discrete variables. For example, suppose we measure the number of children that families have. The number of children in a family is a discrete variable. Suppose we calculate the mean number of children and find it to be 2.07. It may seem awkward to think of families as having, on average, 2.07 children because a family cannot have ".07" of a child. However, fractions provide a better sense of how scores for a discrete variable tend to cluster and are ultimately more useful than a crude index that rounds the mean to the nearest integer. For example, a mean number of children of 2.07 and a mean number of children of 2.49 would both be rounded to 2. However, if a mean of 2.07 versus a mean of 2.49 were used to characterize the number of children in families in the United States, the differences between the two values would translate into hundreds of thousands of children. Rounding to the nearest integer loses this sensitivity. For this reason, you will frequently see fractions included in means and medians for discrete variables.

3.2 Measures of Variability for Quantitative Variables

Measures of central tendency indicate where scores tend to cluster in a distribution. A second important characteristic in analyzing quantitative variables is the extent to which scores are alike or different. Consider the two sets of scores in columns 1 and 4 of Table 3.1. In both cases, the mean is 5.00. However, the **variability** of the scores, or the extent to which they differ from one another, is quite different. Whereas the scores in Set II tend to be alike, the scores in Set I tend to differ from one another. Statisticians have developed indexes to measure such variability.

TABLE 3.1   **Two Sets of Scores with Different Variability**

	Set I			Set II	
X	$X - \overline{X}$	$(X - \overline{X})^2$	X	$X - \overline{X}$	$(X - \overline{X})^2$
2	−3	9	4	−1	1
3	−2	4	5	0	0
3	−2	4	5	0	0
5	0	0	5	0	0
7	2	4	5	0	0
7	2	4	5	0	0
8	3	9	6	1	1
$\Sigma X = 35$		SS = 34	$\Sigma X = 35$		SS = 2
$\overline{X} = 5.00$			$\overline{X} = 5.00$		

Range

One very simple index of variability is the **range,** which is the highest score minus the lowest score. In Set I the range is $8 - 2 = 6$, whereas in Set II it is $6 - 4 = 2$. The range is not a very good index of variability because it can be misleading. Consider the extreme case of 900 scores, of which 899 are equal to 100 and one is equal to 10. The range is quite large, $100 - 10 = 90$, even though almost all scores are identical.

Interquartile Range

A second index of variability is the **interquartile range,** which is frequently abbreviated IQR. The IQR is the difference between the highest and lowest scores (hence, it is a range) *after* the top 25% of the scores and the bottom 25% of the scores have been "trimmed," or eliminated, from the data. Stated another way, the IQR is the range of the middle 50% of the scores. As a simple example, consider the following 12 scores: 10, 13, 11, 6, 12, 9, 10, 8, 12, 10, 13, 14. If we order the scores from lowest to highest, we can divide them into four "quartiles" of three scores each:

6, 8, 9   10, 10, 10   11, 12, 12   13, 13, 14

If we eliminate the bottom and top groups, we have the middle 50% of the scores (10, 10, 10, 11, 12, 12). The range of these scores is $12 - 10 = 2$, which is the interquartile range for this problem.

Some researchers prefer the IQR to the range because it is not so sensitive to distortions from extreme cases. Recall our example involving the range of 900 scores, of which 899 are equal to 100 and one is equal to 10. In this case, the interquartile range is 0 and is not biased by the one extreme score.

The preceding calculation of the interquartile range was straightforward. However, computing the IQR can get complicated when there are duplicated scores (as was the case for the median) and when the number of scores is not evenly divisible by 4. Computational formulas are best developed in the context of percentiles and percentile ranks, which will be the topic of Chapter 4. We will defer until then a discussion of formal procedures for calculating the interquartile range. For now, you should concentrate on the conceptual definition of the interquartile range—namely, the range of the middle 50% of the observations.

Although the interquartile range can be useful, it has been criticized because it does not take into account all of the scores in a distribution. Consider the following sets of scores:

Set I:	1, 4, 5	9, 9, 9	11, 11, 13	24, 29, 33
Set II:	7, 8, 8	9, 9, 10	11, 11, 13	14, 14, 14

In both cases, the interquartile range is $13 - 9 = 4$, but the sets can hardly be thought of as having the same variability. The scores in Set I are much more diverse, taken as a whole, than the scores in Set II. Because of this, statisticians have developed measures of variability that take into account *all* of the scores in a distribution. These are now discussed.

Sum of Squares

One index of variability that takes into account all of the scores in a data set is called the **sum of squares.** We develop this measure using the two sets of scores in Table 3.1. We begin by asking how far scores tend to vary from the "typical" score. If we let the typical score be represented by the mean, then we are concerned with how much each score deviates from the mean. Columns 2 and 5 of Table 3.1 present deviation scores in which the mean of each data set (5.00 in both cases) has been subtracted from each original score in that set. If the scores in the distribution are similar, as in Set II, each will be near the mean and the deviation scores will therefore be close to 0. In contrast, if, as in Set I, the scores tend to differ from one another, the deviation scores will be relatively large.

We now want to combine the deviation scores in each data set in some way to derive a single numerical index of overall variability. We cannot use the sum of the deviation scores because, as noted in Section 3.1, the sum of the signed deviations about the mean will always equal 0. Statisticians have suggested a solution that is very desirable from a statistical perspective. The approach involves first squaring each deviation score and then summing the squares. Columns 3 and 6 of Table 3.1 do this. Note that the sum of squares for Set I is 34 and the sum of squares for Set II is 2. This reflects the greater variability in the Set I scores than in the Set II scores.

The sum of squares gets its name from the operations performed; it is a short-hand term for the *sum of the squared deviations from the mean*. This quantity can be represented symbolically as

$$SS = \sum (X - \overline{X})^2 \qquad\qquad [3.3]$$

where SS is the abbreviation for "sum of squares."

Students frequently ask why the mean and not the median is used as the measure of central tendency in defining the sum of squares. One reason is that when the mean and the median are different, the sum of the *squared* deviations from the mean will always be less than the sum of the *squared* deviations from the median. In fact, it turns out that the sum of squared deviations from the mean will always be less than the sum of squared deviations around any other value; that is, the mean minimizes the squared error. A second reason for preferring the mean concerns the important role that means and their sums of squares play when we make inferences about populations from sample data. This will be discussed in Chapter 7.

Study Exercise 3.3

A researcher was interested in the variability of hyperactivity test scores for five children. On a test with 15 points possible, the children's scores were 10, 8, 6, 4, and 2. Compute the sum of squares.

Answer   Using Equation 3.3, we find that SS = 40:

X	$X - \overline{X}$	$(X - \overline{X})^2$
10	4	16
8	2	4
6	0	0
4	−2	4
2	−4	16
$\sum X = 30$		SS = 40
$\overline{X} = 6.00$		

Variance

One problem with the sum of squares as an index of variability is that its size depends not only on the amount of variability among scores, but also on the *number* of scores (N). Consider two sets of data, where the scores in Set A are 2, 4, and 6 and the scores in Set B are 4, 4, 4, 4, 4, 6, 6, 6, 6, and 6. We can readily see that the scores in Set A tend to differ from one another more than do the scores in Set B. Nevertheless, the sum of squares for Set A is 8 and the sum of squares for

Set B is 10. Thus, although there appears to be greater variability in Set A than in Set B, the sum of squares for Set B is larger than the sum of squares for Set A. This is because the sum of squares for Set B is based on ten scores, whereas the sum of squares for Set A is based on only three scores.

To avoid inconsistencies of this type, an index that compares the variability of two or more sets of scores should take into account the number of cases in each set. One possibility is to divide the sum of squares by N—that is, to compute an average squared deviation from the mean. This is called the **variance** and is defined by the formula

$$s^2 = \frac{SS}{N} \tag{3.4}$$

The symbol used to represent variance is a lowercase s^2. Note that the variance for Set A is 8/3 = 2.67 and the variance for Set B is 10/10 = 1.00. Consistent with our eyeball interpretation, the mean squared deviation score (the variance) is greater in Set A than in Set B.

For the data in Table 3.1, the variance for Set I is

$$s^2 = \frac{34}{7} = 4.86$$

and the variance for Set II is

$$s^2 = \frac{2}{7} = .29$$

The finding of greater variability in Set I than in Set II is consistent with our earlier results for the sums of squares.

Standard Deviation

A fifth index of variability among scores is the **standard deviation.** The standard deviation is the positive square root of the variance and is denoted by the lowercase letter s:

$$s = \sqrt{s^2} \tag{3.5}$$

The standard deviation for Set I in Table 3.1 is

$$s = \sqrt{4.86} = 2.20$$

and the standard deviation for Set II is

$$s = \sqrt{.29} = .54$$

The standard deviation is the most easily interpreted measure of variability among a set of scores. Recall that the variance is the mean squared deviation score. Few of us feel comfortable interpreting squared deviation scores, and by taking the square root of the variance, we are, in essence, eliminating the square and returning to the original unit of measurement. *The standard deviation thus repre-sents an average deviation from the mean.** On the average, the Set I scores re-ported in Table 3.1 deviate 2.20 units from the mean and the Set II scores deviate .54 unit from the mean. Again, more variability is indicated in Set I than in Set II.

Students learning about the standard deviation frequently ask what value in-dicates a large standard deviation. The answer is that it depends on what is being measured. For example, suppose we determine the number of children that fami-lies have in a given country and find a mean of 4.20 and a standard deviation of 3.00. This represents considerable variability because, on the average, scores de-viate three "children" from the mean. In contrast, suppose we measure the annual income for people who live in a particular neighborhood and find a mean of $28,760.40 and a standard deviation of $3.00. In this case, there is very little vari-ability because scores deviate from the mean by an average of only $3.00. When the units are dollars and the concern is annual income, a standard deviation of 3.00 is trivial, but when the units are children and the concern is family size, a standard deviation of 3.00 is substantial.

Study Exercise 3.4

Compute the variance and the standard deviation for the scores from Study Exercise 3.3.

Answer Since the sum of squares was previously found to equal 40 and there are five scores, the vari-ance is equal to

$$s^2 = \frac{SS}{N}$$

$$= \frac{40}{5} = 8.00$$

The standard deviation is found by taking the positive square root of the variance:

$$s = \sqrt{8.00} = 2.83$$

Characteristics and Use of the Sum of Squares, the Variance, and the Standard Deviation

The sum of squares, the variance, and the standard deviation are all useful indexes of variability for quantitative variables. We will make extensive use of all three in this book. *It is always true that SS, s^2, and s are greater than or equal to 0.* These statistics can never be negative because they are all based on squared deviation scores and any number squared must be nonnegative. When the sum of squares

---

* As noted earlier, there are many types of averages. The standard deviation is an average that is the positive square root of the arithmetic average of the squared deviations from the mean.

equals 0, the variance and standard deviation will also equal 0. A value of 0 for these statistics means that there is no variability in the scores; they are all the same. As the values of the three statistics become increasingly greater than 0, more variability among the scores is indicated.

When variables are measured on a level that at least approximates interval characteristics, the sum of squares, the variance, and the standard deviation are appropriate indexes of variability. When variables are measured on an ordinal level that departs markedly from interval characteristics, these indexes are not applicable because they are based on summed deviation scores. For such measures, an appropriate index of variability is the range or the interquartile range. For qualitative variables (that is, nominal measures), no index of variability compares to those discussed above because the categories of scores cannot be ordered. (However, see Kirk, 1978, pp. 73–75 for one attempt to define a formal measure of variability for a qualitative variable.)

3.3 Computational Formula for the Sum of Squares

The determination of a sum of squares using Equation 3.3 involves computing a deviation score and a squared deviation score for each individual. Each time one of these scores is calculated, the opportunity exists for rounding error. There is an alternative formula for computing the sum of squares that does not require the computation of deviation scores. This formula is both more efficient and more precise because it requires fewer steps and presents fewer opportunities for rounding error. The formula is

$$SS = \sum X^2 - \frac{(\sum X)^2}{N} \qquad\qquad [3.6]$$

where $\sum X^2$ is the *sum of the squared X scores* and $(\sum X)^2$ is the *square of the summed X scores*. As demonstrated below, this formula is mathematically equivalent to Equation 3.3.

Consider the seven scores from Set I of Table 3.1. Application of Equation 3.6 requires that we first calculate the sum of the X scores and the sum of the squared X scores:

X	X^2
2	4
3	9
3	9
5	25
7	49
7	49
8	64
$\sum X = 35$	$\sum X^2 = 209$

Then

$$SS = 209 - \frac{(35)^2}{7}$$

$$= 209 - 175 = 34$$

which is identical to the value obtained in Table 3.1 using Equation 3.3.

Given its greater computational ease, we will use Equation 3.6 rather than Equation 3.3 when calculating sums of squares in the remainder of this book.

Study Exercise 3.5

Compute the sum of squares for the scores from Study Exercise 3.3 using Equation 3.6. Compare your result with the result you obtained in Study Exercise 3.3 using Equation 3.3.

Answer   We must first determine the sum of the X scores and the sum of the squared X scores:

X	X^2
10	100
8	64
6	36
4	16
2	4
$\Sigma X = 30$	$\Sigma X^2 = 220$

Then

$$SS = \Sigma X^2 - \frac{(\Sigma X)^2}{N}$$

$$= 220 - \frac{(30)^2}{5}$$

$$= 220 - 180 = 40$$

This result agrees with that obtained in Study Exercise 3.3.

3.4 Relationship Between Central Tendency and Variability

Central tendency and variability represent different characteristics of a distribution. For the purpose of illustration, we will focus on the mean and the standard deviation.

Suppose we assessed the depression levels of a large sample of "normal" (nonclinical) adults and found that their mean depression score on some measure was 16.00 and the standard deviation was 2.00. A frequency graph for these data might look like curve A of Figure 3.4a. Furthermore, suppose that we performed the same assessment on a large sample of individuals diagnosed as being clinically depressed and observed the frequency distribution depicted in curve B of Figure 3.4a. This distribution has the same shape as curve A and the same standard deviation, but the mean is equal to 32.00 rather than 16.00, indicating a higher overall level of depression. This illustrates an important point: Distributions of scores can have identical variabilities (in this case, standard deviations) but very different central tendencies (in this case, means).

Distributions can also have identical central tendencies but very different variabilities. For instance, Figure 3.4b presents frequency distributions for a large sample of "normal" adults and a large sample of adults diagnosed as having manic-depression. This is a psychological condition in which cycles of extreme

FIGURE 3.4   **Illustration of Distributions That Have (a) Identical Standard Deviations but Different Means, and (b) Identical Means but Different Standard Deviations**

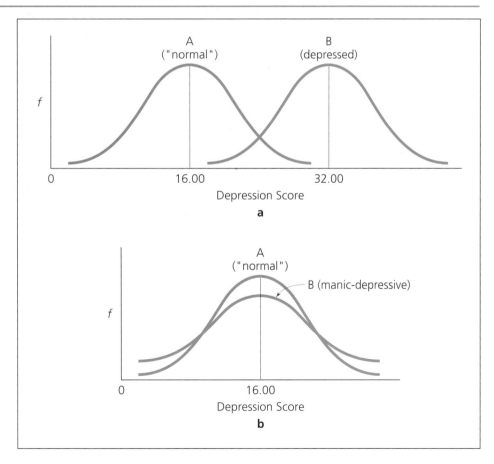

euphoria, sociability, and hyperactivity (mania) alternate with cycles of extreme despair and social withdrawal (depression). Although the mean depression rating is 16.00 for each group, the "normal" sample has a relatively small standard deviation of 2.00, indicating that people in this group tended to receive depression scores of right around 16 (see curve A), whereas the manic-depressive sample has a standard deviation of 5.00, indicating a relatively higher degree of variability in depression ratings (see curve B). This probably reflects the fact that some members of this group are currently in the manic stage and are thus not experiencing depression, whereas others are in the depression stage and still others are in remission. The greater variability of the manic-depressive group is illustrated graphically by the greater spread of curve B relative to curve A.

As illustrated by this second example, measures of variability can help us interpret measures of central tendency. To further demonstrate this point, think about where you would prefer to live if given a choice between a location that has a temperature with an annual mean of 70 °F and a standard deviation of 10 °F, and a location that has a temperature with an annual mean of 70 °F and a standard deviation of 25 °F.

3.5 Graphs of Central Tendency and Variability

Many behavioral scientists use graphs to describe central tendencies, especially when data are presented for more than one set of scores. As an example, consider the graph in Figure 3.5. This graph presents the mean Scholastic Aptitude Test (SAT) scores for students in the United States for 1960–1990, in five-year increments. The abscissa contains the different "groups," or levels, for the independent

FIGURE 3.5   Graph of Mean SAT Scores by Year

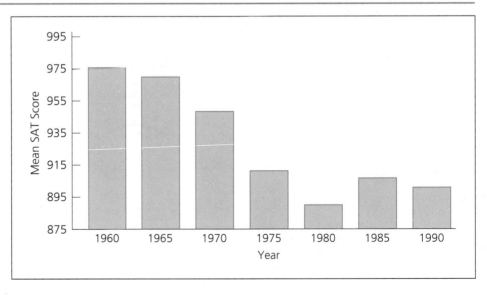

variable under study. In this case, the independent variable is the year of assessment, with the levels being 1960–1990. Each group is represented by a bar that extends to a height on the ordinate comparable to the mean score on the dependent variable (in this instance, the SAT scores). Unlike in frequency graphs, it is not necessary for the ordinate for this type of graph to start at 0. In 1960, the mean SAT score was 975, so the bar reaches to 975 on the ordinate. In 1965, the mean SAT score was 969; in 1970, the mean SAT score was 948; and so on. Usually the bars for the different groups do not touch one another, to indicate that the means are from different sets of scores. Some investigators permit the bars to touch if the independent variable (in this case, year) is quantitative, and they keep the bars from touching if the independent variable is qualitative. However, in the case of mean plots, there is no set rule in this regard. It can be seen that SAT scores have declined rather dramatically from 1960 to 1990—a decline of about 75 points, on average. Statistical procedures discussed in Chapter 12 can be used to study mean differences in the context of inferential statistics.

Another way of graphically presenting the same information is with a line plot. Figure 3.6 presents such a plot, with the addition of a vertical line for each group. The vertical line conveys information about the standard deviation associated with each mean. Specifically, it extends one-half standard deviation above the mean and one-half standard deviation below the mean. Sometimes plots are structured so that the vertical line is one full standard deviation above the mean and one full standard deviation below the mean. The choice is arbitrary. The ordinate of Figure 3.6 is slightly different from that of Figure 3.5 because of the rescaling necessary to accommodate the standard deviations.

Another type of graph is the **boxplot**, also called a **box and whisker** plot.

FIGURE 3.6   **Line Plot of Mean SAT Scores by Year**

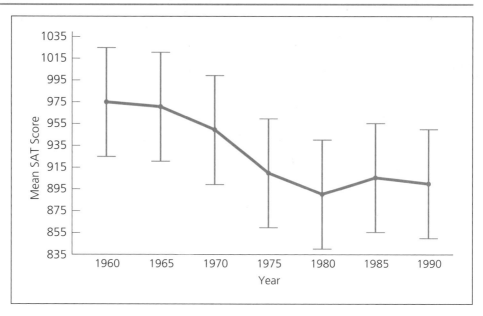

Boxplots typically display information about central tendency and variability. Figure 3.7 presents a boxplot for the number of hours that a random sample of 100 adults from a small rural community spent watching television per day over a 3-month period. The small square in the middle of the box represents the mean, which in this case is 7.04 hours. The top of the box represents one standard deviation above the mean, and the bottom of the box represents one standard deviation below the mean. In this case, the standard deviation is 2.01; hence, the top of the box is at 7.04 + 2.01 = 9.05 and the bottom of the box is at 7.04 − 2.01 = 5.03. The "Ts" extending away from the box, called "whiskers," are the criteria that the investigator uses to define outliers. Suppose, for example, that an investigator decided that any value above 13 or below 1 is an outlier, based on theoretical and practical criteria. This is reflected in Figure 3.7. The dots above and below the "Ts" reflect outlier cases. Two individuals had scores above 13 (hence, two circles above the upper "T"); one watched television 15 hours a day and the other 17 hours a day. One individual had a score below 1. This individual watched television a half hour per day.

Boxplots can also be used to present the median and the interquartile range, rather than the mean and the standard deviation. In fact, boxplots were originally proposed using the median and the interquartile range and only recently have been adapted to present information about means and standard deviations. Figure 3.8 presents a boxplot for the median time served in prison for three different offenses (rape, robbery, and assault) for a sample of 600 convicts (200 per type of offense) in 1992. When medians are plotted, the height of the box reflects the interquartile range (i.e., the top of the box is the upper bound of the interquartile range and the bottom of the box is the lower bound of the interquartile range). For these data, there were no outliers. The median number of years served in prison for rape was 3, for robbery was 2, and for assault was 1.5. These medians are similar to those observed for convicts in the United States during 1992, as reported by the FBI.

FIGURE 3.7   **Boxplot with Mean and Standard Deviation**

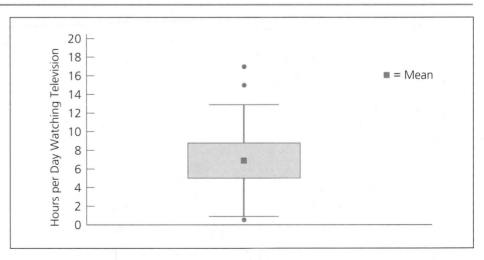

FIGURE 3.8   **Boxplot with Medians and Interquartile Ranges**

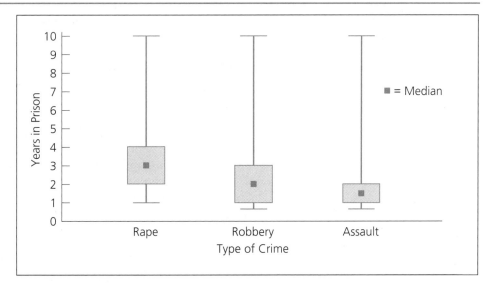

3.6 Measures of Central Tendency and Variability for Qualitative Variables

To this point, all the measures of central tendency and variability we have described apply to quantitative variables. The indexes are not appropriate for qualitative variables. For example, for a variable like political party affiliation (does one classify oneself as a Democrat, Republican, or Independent?), we cannot meaningfully define a range because the categories cannot be ordered to define a low and a high score. Similarly, it makes no sense to talk about a "mean" political party preference. The closest concept to central tendency is the modal category, or that category that occurs most frequently. For example, in the United States, the political party that the most people identify with is the Democratic party, which is the modal value for this qualitative variable. In terms of variability for a qualitative variable, there is none if all individuals fall into a single category. For example, if everyone was a Democrat, there would be no variability on political party identification. Maximum variability occurs when the number of individuals is evenly divided among the possible categories (e.g., there are equal numbers of Democrats, Republicans, and Independents). Probably the best way to characterize the distribution of a qualitative variable is to simply use the entire frequency distribution or to report frequencies (or percentages) for the categories that are of theoretical interest for the questions being addressed.

3.7 Skewness and Kurtosis

Thus far, we have considered two ways in which distributions of scores can differ: central tendency and variability. Two other differences are *skewness* and *kurtosis*. **Skewness** refers to the tendency for scores to cluster on one side of the mean.

Figure 3.9 presents two graphs of scores that illustrate this concept. The graph in Figure 3.9a is said to be **positively skewed** because the "tail" is toward the right, or positive, end of the abscissa. In positively skewed distributions, most scores are below the mean and only a relatively few extreme scores are above it. Thus, the mean will always be greater than the median in a positively skewed distribution. The graph in Figure 3.9b represents a distribution that is **negatively skewed** because the "tail" is toward the left, or negative, end of the abscissa. Most scores in negatively skewed distributions are above the mean and only a relatively few extreme scores are below it. This is reflected in the fact that the mean of a negatively skewed distribution will always be less than the median.

Normal distributions, such as that depicted in Figure 3.3, are not skewed because equal numbers of scores are above and below the mean, which is also the median (and the mode). As shown in Figure 3.9, these three measures of central tendency all take on different values in skewed distributions.

Kurtosis refers to the flatness or peakedness of one distribution relative to another. If a distribution is less peaked than another, it is said to be more *platykurtic,* and if it is more peaked than another, it is said to be more *leptokurtic.* It is conventional to label a distribution as being either platykurtic or leptokurtic depending on whether it is more or less peaked than the normal distribution. Figure 3.10 presents graphs of platykurtic, leptokurtic, and normal distributions.

Statisticians have derived numerical indexes of skewness and kurtosis similar to the indexes of central tendency and variability already discussed. These are rarely reported in the behavioral science literature and, hence, are not described here. Interested readers are referred to Ferguson (1976).

3.8 Sample Versus Population Notation

The descriptive statistics we use most frequently are the mean, the variance, and the standard deviation. Reference will be made to these indexes for both samples

FIGURE 3.9   **Frequency Graphs of (a) Positively Skewed and (b) Negatively Skewed Distributions**

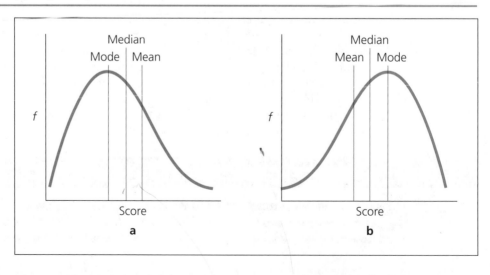

FIGURE 3.10   **Frequency Graphs of (a) Platykurtic, (b) Leptokurtic, and (c) Normal Distributions**

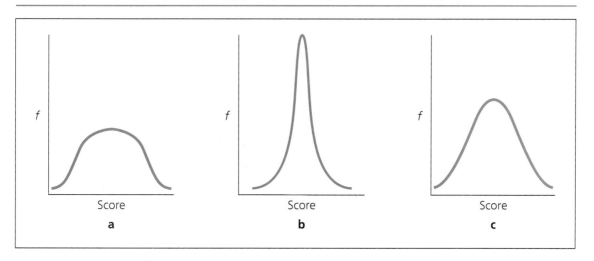

and populations. It is traditional in statistics to denote population-derived indexes with Greek letters. A sample mean is typically denoted as $\overline{X}$, whereas a population mean is symbolized as μ (lowercase Greek m, called "mu"). Of course, the formula for computing the two indexes is identical: Sum the scores and divide by N. The Greek notation, however, makes it explicit that we are describing a population, whereas the $\overline{X}$ notation makes it explicit that we are describing a sample. The Greek notations for variance and standard deviation are σ^2 and σ (lowercase Greek s, called "sigma"), respectively, paralleling the symbols s^2 and s for a sample variance and standard deviation.

3.9 Method of Presentation

When presenting measures of central tendency and variability, most researchers report means and standard deviations. These are undoubtedly the most frequently encountered descriptive statistics. Occasionally, means are supplemented with reports of the median, especially when the mean can be a misleading index of central tendency (for example, when there are extreme scores). A common format used in behavioral science journals presents means and standard deviations in a table like the following:

Means and Standard Deviations for Intelligence Scores of Males and Females

Gender	M	SD	n
Males	101.31	10.62	120
Females	102.48	10.31	115

In this table, *M* represents the mean and *SD* the standard deviation according to the American Psychological Association (APA) format. It is also common to report the sample sizes of the groups. This is indicated by *n* because *N* is reserved to designate the size of the overall data set when there is more than one group. In this example, $N = 120 + 115 = 235$.

Note that it is possible to compute the variances and the sums of squares from the information provided. For instance, the variance for males is the square of the standard deviation of 10.62, or $10.62^2 = 112.78$. The sum of squares can then be computed by multiplying the variance by n: $(112.78)(120) = 13,533.60$.

Often measures of central tendency and variability are presented in the text of a Results section rather than in a table. Here is an example of how the information about male and female intelligence scores might appear:*

Results

The mean intelligence score for males (M = 101.31, SD = 10.62) was comparable to the mean intelligence score for females (M = 102.48, SD = 10.31).

This example uses the APA conventions M and SD to denote the means and standard deviations. Even though the statement refers only to mean scores, the text follows the APA practice of reporting standard deviations as well. Note that sample size information is not given. Usually the sample sizes are identified in the section of the report that provides details about how the study was conducted, known as the Method section. However, if the sample sizes are not presented in the Method section, they would be added to the information in parentheses (e.g., M = 101.31, SD = 10.62, n = 120).

3.10 Example from the Literature

Psychologists have studied how we form impressions of others and how we perceive various traits and characteristics. For example, Anderson (1968) asked 100 individuals to rate a number of traits in terms of how much they would like a person who had each characteristic. The ratings were made on a seven-point scale, where 0 indicated that the trait was "least favorable or desirable" and 6 indicated that the trait was "most favorable or desirable." The means and standard deviations for six of the traits are as follows:

Trait	$\overline{X}$	s
Sincere	5.73	.55
Honest	5.55	.69
Narrow-minded	.80	.76
Selfish	.82	.81
Cunning	2.62	1.48
Inexperienced	2.62	.81

* Although we use single-spacing in the text of all "Results" sections throughout the book, APA format dictates that double-spacing be used between *all* lines of the manuscript (including tables, footnotes, references, etc.).

Applications

to the Analysis

of a Social Problem

Many variables within the parent–teen data set on teenage pregnancy described in Chapter 1 can be explored using the descriptive statistics from this chapter. We highlight two here.

A major purpose of the parent–teen study was to examine communication between mother and teen. One variable that is important to consider is the age of the mother. Past research has suggested that the style of communication between mother and teen differs as a function of this dimension. Because of this, the way in which one structures educational materials to foster better parent–teen communication should differ, depending on the mother's age. Thus, it is important to document the age of the mother in the present sample.

One issue that arose early in the planning of the research was how to define who the teen's mother is. The traditional approach is to use the teen's biological mother. However, the teen's pri-

mary mother figure is not always the biological mother. Sometimes the primary female caretaker is a stepmother. For inner-city, African American teens, it is not uncommon for the primary female caretaker to be a grandmother, an aunt, or even an older sister. Ultimately, the "mother" that we decided to interview was that person who the teen identified (based on a series of questions) as the primary female caretaker in the household. This "mother" was not always the biological mother of the teen.

Table 3.2 presents a computer printout that summarizes basic descriptive statistics for maternal age. The mean age was 40.76, the median age was 39.00, and the modal age was 38. The youngest mother figure was 19 years old (indicated by "Minimum"), and the oldest was 81 (indicated by "Maximum"), which yields a range of 62. The standard deviation was 8.23, indicating that, on

(continued)

TABLE 3.2   **Computer Output for Descriptive Statistics for Mother's Age and Use of Birth Control**

MOTHAGE   Mother's age

Mean	40.759	Median	39.000	Mode	38.000
Std dev	8.234	Variance	67.801	Range	62.000
Minimum	19.000	Maximum	81.000		

Percentile	Value	Percentile	Value	Percentile	Value
25.00	36.000	50.00	39.000	75.00	44.000

Valid cases	634	Missing cases	15		

BCFIRST   Used Birth Control at First Intercourse

Mean	.416	Median	.000	Mode	.000
Std dev	.494	Variance	.244	Range	1.000
Minimum	.000	Maximum	1.000		

Percentile	Value	Percentile	Value	Percentile	Value
25.00	.000	50.00	.000	75.00	1.000

Valid cases	425	Missing cases	326		

average, mother figures tended to deviate about 8 years from the mean age of 40.76. The interquartile range is the difference between the 75th percentile and the 25th percentile, or 44 − 36 = 8. Note that the median equals the 50th percentile. Figure 3.11 presents a grouped frequency histogram of maternal age with a normal distribution superimposed on it. Recall that normal distributions have no skewness and that the distribution peaks at the mean. It is evident from the histogram that the distribution of maternal age is positively skewed. This is also reflected by the fact that the mean is greater than the median.

We also calculated descriptive statistics for a self-report of whether or not the teen used birth control at the first instance of sexual intercourse. This is an important variable because teens who engage in unprotected intercourse are at risk of unintended pregnancy. In fact, statistics indicate that an unusually large proportion of unintended pregnancies in teenagers occurs within the first 6 months of sexual initiation. Self-reports of birth control use must be treated with caution because the teen may not accurately recall this information. In addition, social desirability tendencies may cause

teens to overreport use of birth control (although we did not find any relationship between this self-report and our measure of social desirability tendencies). This birth control variable was dichotomous (i.e., it had only two values) and was scored as 1 if the teen reported using birth control at first intercourse and as 0 if the teen reported not using birth control at first intercourse. The analysis was performed for only those teens who indicated they had engaged in sexual intercourse, with the other teens being treated as "missing data."

It turns out that the mean of a dichotomous variable that is scored 1 and 0 will simply equal the proportion of people who have a score of 1. Multiplying the mean by 100 yields a percentage. In this sample, the mean was .416, which yields a percentage of 41.6%. Thus, just over 40% of the teens reported using some form of birth control at first intercourse, and 58.4% (100.0 − 41.6) reported that they engaged in unprotected intercourse (and were thus at risk for an unintended pregnancy). For dichotomous variables, the other descriptive statistics either are not meaningful or are not informative for our purposes; hence, we do not consider them here.

FIGURE 3.11   Grouped Frequency Histogram of Maternal Age

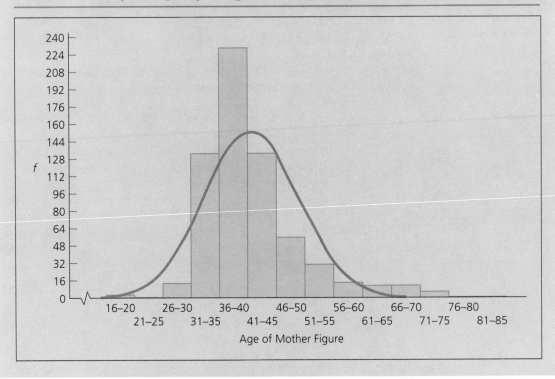

The first two traits were rated very positively by the individuals in the study. The means were quite high (5.73 and 5.55) and the standard deviations were relatively small, indicating that most individuals rated the traits using the upper points of the scale. The traits "narrow-minded" and "selfish" were similarly rated, but in a negative fashion. The traits "cunning" and "inexperienced" both got relatively neutral mean scores, 2.62. However, notice the large difference between the standard deviations for the two traits. The small standard deviation for "inexperienced" indicates that people consistently tended to rate this trait near the neutral point, whereas the large standard deviation for "cunning" indicates that there was considerable variability in these ratings, with many individuals perceiving "cunning" as being a positive trait and many perceiving "cunning" as a negative trait. When averaged, the mean was near the neutral point. By showing that the perceptions of "cunning" were not really all that neutral but rather exhibited considerable variability across individuals, the standard deviation helps us interpret the mean.

Summary

Two important characteristics of a distribution of scores are where the scores tend to fall on the numerical scale (central tendency) and the extent to which the scores are alike or different (variability). Central tendency and variability are different, independent aspects of a distribution. Distributions can also be described in terms of skewness and kurtosis.

Three measures of central tendency are the mode, the median, and the mean. The mode is the most frequently occurring score, the median is the point that divides the distribution into halves, and the mean is the arithmetic average.

Five measures of variability are the range, the interquartile range, the sum of squares, the variance, and the standard deviation. The range is the difference between the highest score and the lowest score. The interquartile range is the range after trimming off the top 25% and the bottom 25% of the scores. The sum of squares is the sum of the squared deviations from the mean. By dividing the sum of squares by N, we obtain the variance. The standard deviation is the positive square root of the variance and represents an average deviation from the mean.

Several methods are available for graphing central tendencies. These include the use of vertical bars or line plots to indicate means for different levels of the independent variable, and boxplots. When boxplots are used to display means, they also provide information about the corresponding standard deviations and outliers. When they are used to present medians, boxplots also provide information about the interquartile range and outliers.

Exercises

Answers to asterisked (*) exercises appear at the back of the book. Answers to exercises with two asterisks are also worked out step by step in the Study Guide.

1. Identify and define the three measures of central tendency.

2. What are the three approaches to computing the median? When is each appropriate?

**3. Compute the median for the following scores: 4, 4, 4, 6, 9.

4. Compute the median for the following scores: 2, 3, 5, 5, 7, 12.

5. Given the impact of television on children's attitudes and behavior, an important concern of behavioral scientists is the amount of time children of various ages spend watching television. The following data are the weekly viewing times (in hours) of 12-year-olds. Compute the mode, the median, and the mean for these scores.

```
18  17  22  20  25  20  16
19  18  22  26  23  23  23
24  24  22  21  19  20  20
```

*6. Compute the mode, the median, and the mean for the following scores: −3, −3, −2, −2, −2, −1, −1, −1, 0, 0, 0, 0, 1, 1, 1, 2, 2, 2, 3, 3.

7. What are unsigned deviation scores? What are signed deviation scores?

*8. Compute the mean for the following five scores: 10, 11, 12, 13, 14. Now, generate a new set of five scores by adding a constant of 3 to each original score. Compute the mean for the new scores. Do the same for another set of five scores generated by subtracting a constant of 10 from each original score. What are the effects on the mean of adding a constant to or subtracting a constant from each score in a set of scores?

9. Suppose you measured the mean amount of time it took ten children to solve a problem and found it to be 35.31 seconds. You later discovered, however, that your timing device (a watch) was 2 seconds too slow for each child. What was the real mean score?

*10. Compute the mean for the following five scores: 10, 20, 30, 40, 50. Now, generate a new set of five scores by multiplying each original score by a constant of 3. Compute the mean for the new scores. Do the same for another set of five scores generated by dividing each original score by a constant of 10. What are the effects on the mean of multiplying or dividing each score in a set of scores by a constant?

*11. Repeat the procedures outlined in Exercises 8 and 10, but compute the median rather than the mean. What are the effects on the median of adding a constant to or subtracting a constant from each score in a set of scores? What are the effects on the median of multiplying or dividing each score in a set of scores by a constant? What effects do you think these operations would have on the mode?

12. What are the advantages and disadvantages of using the mean as a measure of central tendency? The median? The mode?

*13. Consider these two sets of data:

Set I		Set II	
20	21	53	55
20	10	54	54
19	21	54	53
21	300	55	54

For which set is the mean a poorer descriptor of central tendency? Why?

14. Identify and define three measures of variability.

*15. Under what condition is the range a misleading index of variability?

16. Define the interquartile range.

17. What is the problem with the sum of squares as a measure of variability?

*18. If the variance of a set of scores is 100.00, what is the standard deviation?

19. If the standard deviation of a set of scores is 7.00, what is the variance?

*20. Without actually calculating it, what must the standard deviation of the following scores equal: 6, 6, 6, 6, 6? What must the variance equal? What must the sum of squares equal? Why?

*21. Why is the standard deviation more "interpretable" than the variance? That is, what is the advantage of reporting statistics in terms of the standard deviation as opposed to the variance?

22. Compute ΣX^2 and $(\Sigma X)^2$ for the following scores: 3, 6, 12, 5, 9, 2, 3, 6, 11.

**23. For the scores in Exercise 22, compute the sum of squares using the defining formula (Equation 3.3). Recalculate the sum of squares using the computational formula (Equation 3.6). Compare the two results. Which approach did you find more efficient?

*24. Compute the range, the sum of squares, the variance, and the standard deviation for the data in Exercise 5.

25. Compute the range, the sum of squares, the variance, and the standard deviation for each of the two sets of data in Exercise 13.

*26. Compute the variance and the standard deviation for the following five scores: 1, 2, 3, 4, 5. Now, generate a new set of five scores by adding a constant of 3 to each original score. Compute the variance and standard deviation for the new scores. Do the same for another set of five scores generated by subtracting a constant of 2 from each original score. What are the effects on the variance of adding a constant to or subtracting a constant from each score in a set of scores? What are the effects on the standard deviation?

27. Suppose you measured the weights of 100 people and found that $\overline{X} = 180.29$ and $s = 10.36$. Then you learned that your scale was 1 pound too heavy. What were the correct mean and standard deviation?

*28. Compute the variance and the standard deviation for the following eight scores: 6, 6, 8, 8, 8, 8, 10, 10. Now, generate a new set of eight scores by multiplying each original score by a constant of 3. Compute the variance and the standard deviation for the new scores. Do the same for another set of eight scores generated by dividing each original score by a constant of 2. What are the effects on the variance of multiplying or dividing each score in a set of scores by a constant? What are the effects on the standard deviation?

*29. Generate two sets of scores with equal means but unequal standard deviations.

30. Generate two sets of scores with unequal means but equal standard deviations.

*31. How accurate are eyewitness reports of accidents? Behavioral scientists have studied this question in some detail. In one experiment, people viewed a film of an accident in which a car ran through a stop sign and hit a parked car. The speed of the car was 30 miles per hour. After viewing the film, the people were asked to estimate the speed of the car. Fifteen participants gave the following estimates:

15	18	37	40	25
40	35	35	20	30
30	20	28	32	25

Compute the mean and the standard deviation for these data. How accurate are the estimates considering the mean score across all participants? How does the standard deviation help to interpret the mean?

32. An organizational psychologist studied how satisfied employees were in two different companies. All employees were given a job satisfaction test on which scores could range from 1 to 7, with higher values indicating greater satisfaction. The scores were as follows:

Company A	Company B
4	4
6	4
5	4
4	4
3	4
2	4

Compute the mean and the standard deviation for each company. Based on the results, compare employee satisfaction in the two companies. How do the standard deviations help to interpret the means?

*33. A consultant you hired to assess the public's attitude toward your company told you that the mean evaluation on a seven-point scale (where 1 = "extremely negative" and 7 = "extremely positive") was 5.16 and the standard deviation was −1.43. What should you conclude?

34. What are the three methods for graphing central tendencies?

35. Draw a boxplot for the data in Exercise 5.

*36. Given a set of scores for which the mode is 12, the median is 15, and the mean is 20.68, are these scores skewed? If so, how? If the mode were 33, the median were 25, and the mean were 20.17, would the scores be skewed? If so, how? If the mode were 20, the median were 20, and the mean were 20.00, would the scores be skewed? If so, how?

37. What does it mean to say that a distribution of scores is more leptokurtic than another distribution? What does it mean to say that a distribution of scores is more platykurtic than another distribution?

38. Identify each of the following symbols in terms of the index it represents and whether it is a sample or a population value:
   a. σ^2 d. σ
   b. $\overline{X}$ e. μ
   c. s f. s^2

Multiple-Choice Questions

*39. Which of the following measures is appropriate for use with qualitative variables?
   a. range
   b. modal category
   c. median
   d. sum of squares

40. μ is the symbol for a
   a. population mean
   b. sample mean
   c. population standard deviation
   d. sample standard deviation

*41. Which measure of central tendency is most influenced by extreme scores?
   a. mean
   b. median
   c. mode
   d. The mean, the median, and the mode are equally influenced by extreme scores.

42. Which of the following is *not* an index of variability?
   a. standard deviation
   b. range
   c. sum of squares
   d. skewness

*43. The vertical lines on a line plot convey information about the _____ associated with each mean.
   a. sample size
   b. sum of squares
   c. variance
   d. standard deviation

44. Which measure of variability is the most easily (intuitively) interpreted?
    a. interquartile range
    b. sum of squares
    c. standard deviation
    d. variance

45. If a set of scores is negatively skewed, this means that it is less peaked than the normal distribution.
    a. true   b. false

*46. The sum of a set of signed deviations around the mean will always equal 1.00.
    a. true   b. false

47. What is the interquartile range for the following scores: 1, 1, 3, 4, 6, 7, 8, 9?
    a. 3   c. 7
    b. 4   d. 8

*48. When boxplots are used to display means, they also provide information about the interquartile range and outliers.
    a. true   b. false

CHAPTER **4**

Percentiles, Percentile Ranks, Standard Scores, and the Normal Distribution

Suppose we tell you that we have developed a measure of dominance and administered it to a group of 75 college students. One of these students, Mary, obtained a score of 50. This information is relatively useless when it stands alone. Suppose we tell you further that the highest possible score is 100. Now you might infer that Mary is not very dominant. But suppose you are also told that the mean score on the test for the 75 college students was 30. Mary is beginning to look a little more dominant. Now we tell you that the standard deviation for this group was 5. This is even more enlightening. The mean score was 30 and, on the average, scores deviated from the mean by 5 units. Mary's score is $50 - 30 = 20$ units *above* the mean, or four standard deviations above the mean, indicating that, relative to the rest of the students, her score is very high.

The point to be made is that, in general, observations are meaningful only in relation to other observations. When you are told that a person is 7 feet tall, this is meaningful because you have an intuitive grasp of the distribution of heights in humans. You know that most people are between 5 and 6 feet tall and that someone who is 7 feet tall is unusually tall. In essence, you are intuitively using information about the mean and the standard deviation of height. In this chapter we consider several measures used to identify the location of a specified score within a set of scores.

4.1 Percentiles and Percentile Ranks

One approach for expressing the relative standing of a score in a distribution uses percentage as a basis. The percentage of scores in the distribution that are at or below a given value, X, is the **percentile rank** of that value.* Consider a national survey of 2,000 adults in the United States who responded to the question: Approximately how many hours per week do you watch television? If 60% of the responses are at or below a score of 30 hours, then the percentile rank of 30 hours is 60. Thus, one way to convey the relative position of a score is to compute its percentile rank. On the other hand, we might want to know the reverse—namely, at what value is it that a certain percentage of scores are at or below? For instance, we might want to know the number of hours of television viewing that defines the point at which 60% of the reported viewing times are at or below. The value corresponding to a given percentile rank is called a **percentile.** For instance, the value corresponding to a percentile rank of 60 is referred to as the 60th percentile. In our example, the 60th percentile is 30. We now consider how to compute percentiles and percentile ranks.

Computation of Percentiles

Suppose we administered a test designed to measure intelligence to a group of 200 children. A frequency analysis of the children's scores is presented in Table 4.1. Our task is to specify the score that corresponds to a given percentile, P. If P

TABLE 4.1   Frequency Analysis of Intelligence Test Scores for 200 Children

Score	f	rf	cf	crf
105	9	.045	200	1.000
104	16	.080	191	.955
103	20	.100	175	.875
102	26	.130	155	.775
101	29	.145	129	.645
100	30	.150	100	.500
99	25	.125	70	.350
98	21	.105	45	.225
97	14	.070	24	.120
96	10	.050	10	.050

* A percentile rank is different from a *cumulative percentage*, which, following the logic outlined in Chapter 2, is found by adding the percentage of all scores that are equal to X to the percentage of scores that are less than X. As will be seen shortly, when percentile ranks are determined, half of the scores that have a value of X are considered to be less than X and half are considered to be greater than X, yielding a different computational approach.

is 70, then we want to specify the score that defines the 70th percentile. Actually, we have already considered a special case of the formula for solving this problem in Equation 3.1 for calculating the median. Recall that the median is that score in the distribution for which 50% of the scores are greater and 50% of the scores are less. The median corresponds to the 50th percentile. Thus, the procedures for calculating the 50th percentile are the same as those for calculating the median. The formula for calculating the median of a measure of a continuous variable, as presented in Equation 3.1, is

$$Mdn = L + \left[\frac{(N)(.50) - n_L}{n_W} \right] i$$

where Mdn denotes the median, L is the lower real limit of the category containing the median, N is the total number of scores in the distribution, n_L is the number of individuals with scores less than L, n_W is the number of individuals with scores within the category containing the median, and i is the size of the interval of the category that contains the median. The .50 in the numerator refers to the percentile of interest (the 50th) expressed in the form of a proportion.

If we were interested in the 70th percentile, the formula for the median would be adapted as follows:

$$X_{70} = L + \left[\frac{(N)(.70) - n_L}{n_W} \right] i$$

where X_{70} denotes the score value defining the 70th percentile, L is the lower real limit of the category containing the 70th percentile, N is the total number of scores in the distribution, n_L is the number of individuals with scores less than L, n_W is the number of individuals with scores within the category containing the 70th percentile, and i is the size of the interval of the category that contains the 70th percentile. The method of computation is then analogous to that used in computing the median. We start at the bottom of the column of cumulative relative frequencies (see Table 4.1) and move up until we find the first number that is greater than or equal to .70 (the percentile of interest, expressed in proportion form). In this case, it is .775 and occurs at the score value 102. The lower real limit (L) of 102 is 101.5. The number of individuals (n_L) whose scores are less than L is 129. The number of individuals (n_W) with scores of 102 (or, more precisely, with scores between 101.5 and 102.5) is 26. Thus,

$$X_{70} = 101.5 + \left[\frac{(200)(.70) - 129}{26} \right] (1.0)$$

$$= 101.5 + \left[\frac{11.00}{26} \right] (1.0)$$

$$= 101.5 + .42 = 101.92$$

In this instance, a value of 101.92 reflects the 70th percentile. Note that the value 101.92 did not actually occur in the set of scores listed in Table 4.1. This is because the real limits of the numbers were considered, as discussed in Chapter 3. Technically, this value should *not* be rounded up to 102 because doing so undermines the idea that the individuals who scored 102 are evenly dispersed between the limits of 101.5 and 102.5. The percentile rank for the "rounded up" value of 102 is actually slightly above 70. Despite this fact, some research reports (and some computer programs) round percentile calculations to units of the original scale, trading off technical precision for ease of presentation. Indeed, some analysts are uncomfortable with the assumption of individuals being evenly dispersed across the real limits of a score, noting that this may not be true (which would undermine the logic of the approach described here). These critics also contend that carrying calculations to two decimal places beyond the original units implies a level of precision that is not justified. If one took this latter position, the 70th percentile would be reported as 102 because 101.92 is rounded to 102. We believe that there are merits to both positions (i.e, rounding to the nearest integer versus carrying calculations to two additional decimal places), and we do not take a formal position on which one you should use. We merely point out that you will encounter both in the literature. In either case, it is important to recognize that both approaches yield only an approximation to the true 70th percentile. For consistency, we follow the strategy of carrying calculations to two additional decimal places. In cases where computer output rounds percentiles to the nearest whole number, we use the value provided in the output.

The general formula for computing a score that defines a given percentile is

$$X_P = L + \left[\frac{(N)(P) - n_L}{n_W} \right] i \qquad [4.1]$$

where X_P represents the score value defining the percentile of interest, referred to as the *P*th percentile; *L* is the lower real limit of the category containing the *P*th percentile; *N* is the total number of scores in the distribution; *P* is the *P*th percentile expressed in the form of a proportion; n_L is the number of individuals with scores less than *L*; n_W is the number of individuals with scores within the category containing the *P*th percentile; and *i* is the size of the interval of the category that contains the *P*th percentile.

Formula 4.1 can be used to calculate the interquartile range discussed in Chapter 3. The interquartile range is the 75th percentile minus the 25th percentile. For the data in Table 4.1, the 75th percentile is 102.31 and the 25th percentile is 98.70. The interquartile range is therefore 102.31 − 98.70 = 3.61.

Study Exercise 4.1

A researcher was interested in determining the size of babies at birth. A group of 1,000 newborn infants was studied, and their lengths and weights at birth were measured. The distribution of lengths (measured to the nearest inch) was as follows:

Length	f	rf	cf	crf
24	27	.027	1,000	1.000
23	48	.048	973	.973
22	77	.077	925	.925
21	125	.125	848	.848
20	226	.226	723	.723
19	225	.225	497	.497
18	124	.124	272	.272
17	73	.073	148	.148
16	52	.052	75	.075
15	23	.023	23	.023

Compute the score that defines the 90th percentile.

Answer Using Equation 4.1, we find that

$$X_{90} = 21.5 + \left[\frac{(1,000)(.90) - 8.48}{77} \right] (1.0)$$

$$= 21.5 + \left[\frac{52.00}{77} \right] (1.0)$$

$$= 21.5 + .68 = 22.18$$

We conclude that 90% of the newborn infants studied were 22.18 inches long or less. Stated another way, the 90th percentile is 22.18.

Computation of Percentile Ranks

Just as Equation 4.1 can be used to compute the score that defines a given percentile, it is also possible to specify a formula to compute the percentile rank for any given score. This formula is

$$PR_X = \left[\frac{(.5)(n_W) + n_L}{N} \right] (100) \tag{4.2}$$

where PR_X represents the percentile rank of the score X, n_W is the number of individuals with scores equal to X, n_L is the number of individuals with scores less than X, and N is the total number of scores in the distribution. For example, the percentile rank for a score of 101 from Table 4.1 is

$$PR_{101} = \left[\frac{(.5)(29) + 100}{200} \right] (100)$$

$$= \left[\frac{114.50}{200} \right] (100)$$

$$= (.572)(100) = 57.20$$

For this distribution, a score of 101 reflects a percentile rank of 57.20; that is, 57.20% of the 200 children studied obtained an intelligence test score of 101 or less.

Study Exercise 4.2

For the data presented in Study Exercise 4.1, compute the percentile rank for 20 inches.

Answer   Using Equation 4.2, we get:

$$PR_{20} = \left[\frac{(.5)(226) + 497}{1,000} \right](100)$$

$$= \left[\frac{610.00}{1,000} \right](100)$$

$$= (.610)(100) = 61.00$$

The percentile rank for a length of 20 inches is 61.00; that is, 61.00% of the newborn infants in the study were 20 inches long or less at birth.

4.2 Standard Scores

A percentile rank is one index of the relative position of a score in a set of scores. However, a percentile rank reflects only an *ordinal* measure of relative standing. To say that a score has a percentile rank of 80 simply states that 80% of the individuals scored at or below that score. But *how much* lower did these other individuals score? Consider the following scores on two tests:

Test 1	Test 2
100	100
99	44
98	43
96	40
96	40
95	39

In both cases, a score of 100 reflects the same percentile rank. However, the score of 100 on Test 2 is certainly more distinctive than the score of 100 on Test 1. An index of relative standing that reflects this difference is **standard scores**.

The Concept of Standard Scores

Recall the example at the beginning of this chapter concerning the dominance measure. Mary got a score of 50 out of 100. When we were told that the average score was 30, a score of 50 took on more meaning: Mary scored 20 units above the mean. With the additional information that the standard deviation was 5, the significance of Mary's score was even clearer. By comparing Mary's score to the mean and standard deviation, we gained considerable insight into its relative position. A standard score does just this: It converts a score from its original, or *raw,* form to a form that takes into consideration its standing relative to the mean and standard deviation of the entire distribution of scores.

Columns 1 and 5 of Table 4.2 present two sets of scores that will be used for illustration. If the data in Table 4.2 represent sample information, then the raw scores in each set can be converted to standard scores by the following formula:

$$\text{Standard score} = \frac{X - \overline{X}}{s} \tag{4.3}$$

If the data in Table 4.2 represent population information, the standard score formula is

$$\text{Standard score} = \frac{X - \mu}{\sigma} \tag{4.4}$$

Note that the only difference between these two formulas is the notation used for the mean and standard deviation of the distribution. For the purpose of our demonstration, we assume that we are dealing with sample data.

A standard score is the difference between the original score and the mean, divided by the standard deviation. The numerator of the standard score formula reflects the number of units the score is above or below the mean. When the numerator is divided by the standard deviation, the result expresses the number of *standard deviations* the score is above or below the mean. For instance, Mary's score of 50 is $50 - 30 = 20$ units above the mean. The standard deviation was 5, and hence Mary's score is $20/5 = 4$ standard deviations above the mean. In other words, her raw score corresponds to a standard score of 4. Suppose John obtained a dominance score of 25. This score is $25 - 30 = -5$, or 5 units below the mean; or, in terms of standard deviations, $-5/5 = -1$, or 1 standard deviation below the mean. John's standard score is thus -1. *A standard score represents the number of standard deviation units that a score falls above or below the mean.* It summarizes the individual's relative standing, taking into consideration the mean and standard deviation of the distribution.

The standard score equivalents of the raw scores in columns 1 and 5 of Table 4.2 can be found, respectively, in columns 4 and 8. To obtain these values, we first computed the mean for each data set. Next, we used the computational formula for the sum of squares to separately calculate the sum of squares for Set I and the sum of squares for Set II. The sums of squares were then used to calculate the standard deviations. Finally, we subtracted the mean for each data set from the constituent scores (see columns 3 and 6) and divided the resulting values by the appropriate standard deviation (see columns 4 and 8). If we compare the standard score for the first raw score in Set I with the standard score for the first raw score in Set II, we find that although the raw scores are the same (both have a value of 1), the standard scores are different, -1.49 as compared to -2.25. The negative signs indicate that in both cases the raw score is below the mean of its distribu-

TABLE 4.2   **Raw Scores and Standard Scores for Two Sets of Scores**

	Set I				Set II		
X	X^2	$X - \overline{X}$	$\dfrac{X - \overline{X}}{s}$	X	X^2	$X - \overline{X}$	$\dfrac{X - \overline{X}}{s}$
1	1	−2	−1.49	1	1	−2	−2.25
3	9	0	.00	3	9	0	.00
5	25	2	1.49	3	9	0	.00
2	4	−1	−.75	3	9	0	.00
3	9	0	.00	5	25	2	2.25
4	16	1	.75	3	9	0	.00
1	1	−2	−1.49	3	9	0	.00
3	9	0	.00	3	9	0	.00
5	25	2	1.49	3	9	0	.00
3	9	0	.00	3	9	0	.00
$\Sigma X = 30$	$\Sigma X^2 = 108$		Sum = 0	$\Sigma X = 30$	$\Sigma X^2 = 98$		Sum = 0
$\overline{X} = 3.00$			Mean = 0	$\overline{X} = 3.00$			Mean = 0

$$SS = \Sigma X^2 - \frac{(\Sigma X)^2}{N} \qquad\qquad SS = \Sigma X^2 - \frac{(\Sigma X)^2}{N}$$

$$= 108 - \frac{30^2}{10} = 18 \qquad\qquad = 98 - \frac{30^2}{10} = 8$$

$$s^2 = \frac{SS}{N} = \frac{18}{10} = 1.80 \qquad\qquad s^2 = \frac{SS}{N} = \frac{8}{10} = .80$$

$$s = \sqrt{s^2} = \sqrt{1.80} = 1.34 \qquad\qquad s = \sqrt{s^2} = \sqrt{.80} = .89$$

tion. For Set I, a score of 1 is only 1.49 standard deviation units below the mean of its distribution, whereas for Set II, it is 2.25 standard deviation units below the mean of its distribution. Thus, as reflected by the standard scores, a score of 1 is more distinctive in the context of Set II scores than in the context of Set I scores.

Study Exercise 4.3

For a set of scores with $\overline{X} = 20.00$ and $s = 2.00$, what standard score corresponds to a raw score of 17?

Answer   Using Equation 4.3, we have

$$\text{Standard score} = \frac{X - \overline{X}}{s}$$

$$= \frac{17 - 20.00}{2.00} = -1.50$$

This means that a score of 17 is 1.50 standard deviations below the mean of its distribution.

Properties of Standard Scores

Standard scores have several important properties. A positive standard score indicates that the original score is greater than the mean, and a negative standard score indicates that the original score is less than the mean. A standard score of 0 indicates that the original score is equal to the mean.

If we sum the standard scores for Set I in Table 4.2, the result is 0 (see column 4). The same result occurs for Set II (see column 8). In fact, the sum of a set of standard scores will always be 0. This is because standard scores reflect signed deviation scores, and as discussed in Section 3.1, the sum of signed deviations about the mean always equals 0. It follows that if the sum of a set of standard scores is always equal to 0, *the mean of a set of standard scores is also always equal to 0.*

Also, if we were to compute the standard deviation for either set of standard scores, the result would be 1.00. *The standard deviation (and the variance) of a set of standard scores is always equal to 1.00.*

Use of Standard Scores

As alluded to earlier, one important use of standard scores is to compare scores on distributions that have different means and standard deviations. For instance, suppose you are contemplating employment in one of two different fields (management or advertising). Unable to make a decision between the two, you decide to enter the field for which you have the greater aptitude. As assessed by the most widely accepted aptitude tests in the respective fields, your management aptitude score turns out to be 73 (out of 100) and your advertising aptitude score turns out to be 82 (out of 100). At first glance, it appears that your advertising aptitude is substantially greater than your management aptitude. However, before rushing off in search of an advertising position, you should examine the two scores in the context of their respective distributions.

Suppose, for instance, that the mean aptitude score for the population of individuals who have previously taken the management test is 62.00 and the standard deviation is 4.00. Furthermore, suppose that the mean aptitude score for the population of individuals who have previously taken the advertising test is 78.00 and the standard deviation is 6.00. Now it is not quite so clear that your advertising aptitude is superior to your management aptitude.

Because the mean (0) and standard deviation (1.00) of a set of standard scores are always the same, it is possible to compare raw scores from different distributions by converting them to standard scores. In our example, your standard score for management aptitude is

$$\text{Standard score} = \frac{73 - 62.00}{4.00} = 2.75$$

and your standard score for advertising aptitude is

$$\text{Standard score} = \frac{82 - 78.00}{6.00} = .67$$

Thus, compared with other individuals who have taken the two aptitude tests, a management aptitude score of 73 represents relatively greater aptitude than an advertising aptitude score of 82. In the first instance, the obtained aptitude score is 2.75 standard deviation units above the mean of its distribution, whereas in the second instance, the obtained aptitude score is only .67 standard deviation unit above the mean of its distribution.

4.3   Standard Scores and the Normal Distribution

A standard score yields considerable information about the relative position of a score in a distribution. Such scores are even more meaningful when they occur in a normal distribution. In Section 2.9, it was noted that there is a family of normal distributions, each member of which is precisely defined by a mathematical formula.* For our purposes, the important points are that (1) there is a different normal distribution for each unique combination of values of the mean and standard deviation for a distribution, and (2) all normal distributions share similar characteristics. For instance, as discussed in Chapters 2 and 3, all normal distributions are symmetrical about the mean, all are characterized by a "bell shape," and in all cases, the mean, the median, and the mode are equal. Another important feature of normal distributions is that they are theoretical in nature. If we are able to assume that a set of scores approximates a normal distribution, then we can invoke certain statistical properties of normal distributions to aid in interpreting the data.

One useful property of normal distributions is that the proportion of scores that occur above or below a given standard score is the same in all such distributions, as is the proportion of scores that occur between two specified standard scores. It is always the case, for example, that .50 of the scores in a normal distribution occur above the mean and .50 of the scores in a normal distribution occur below the mean. Thus, knowing that a set of scores approximates a normal distribution allows us to make probability statements with respect to those scores.

Figure 4.1 presents a normal distribution and the proportions of scores that occur between selected standard scores. The proportion of scores between standard scores of 0 and 1 is .3413, the proportion of scores between standard scores of 1 and 2 is .1359, and the proportion of scores between standard scores of 2 and 3 is .0215. The proportion of scores greater than or equal to a standard score of 3 is .0013. Note that these proportions are symmetrical about the mean, which, as stated previously, always equals 0 for a set of standard scores. For instance, the proportion of scores that occur between standard scores of 0 and −1 is the same as the proportion of scores that occur between standard scores of 0 and +1. Thus, in a normal distribution, .6826, or approximately 68%, of all scores fall between standard scores of −1 and +1; .9544, or approximately 95%, of all scores fall between standard scores of −2 and +2; and .9974, or over 99%, of all scores fall

---

* The normal distribution formula is given in Appendix 4.1.

FIGURE 4.1   **Proportion of Scores Between Selected Standard Scores in a Normal Distribution**

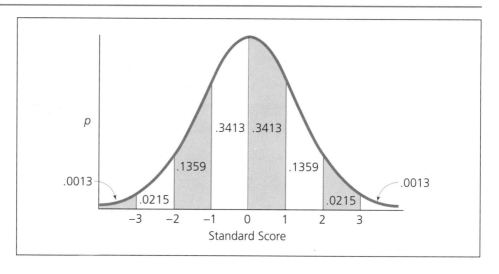

between standard scores of −3 and +3. Before proceeding further, take a moment to verify these proportions from the information presented in Figure 4.1.

A standard score in a normal distribution is referred to as a *z* **score.** Some texts refer to any standard score as a *z* score. However, in traditional statistics, a *z* score is used only to represent a standard score *in a normal distribution.* This distinction will be maintained in the present book.

Appendix B presents a table that lists the proportions of scores in a normal distribution that are greater than or equal to selected *z* scores (for example, greater than or equal to 1.00). This table also lists the proportions of scores in a normal distribution that occur between selected points. Instructions for using the table are presented at the start of Appendix B and should be read at this time.

Let us now explore how the table in Appendix B, coupled with our knowledge of the normal distribution, can give us insight into scores that approximate normality (that is, approximate a normal distribution). For the sake of illustration, try to estimate how many hours per week you spend watching television. This might be easiest to do if you think of each day of the week (Sunday through Saturday) individually and estimate how many hours you typically watch television on those days. Then sum across the seven days to obtain an estimate of how much you tend to watch per week. For us, the estimate is 14 hours per week. How does this compare with others'? Let us compare our score with results suggested by national surveys of adults in the United States.

Suppose the mean number of hours of television watched per week by adults in the United States is 25.40 and the standard deviation is 6.10. Suppose also that the scores in this distribution closely approximate a normal distribution. How does a score of 14 hours per week compare? When we substitute the more specific *z* notation for "Standard score" in Equation 4.3, the formula for converting a score in a sample to a *z* score is

$$z = \frac{X - \overline{X}}{s} \qquad\qquad [4.5]$$

Similarly, the formula for converting a score in a population to a z score is

$$z = \frac{X - \mu}{\sigma} \qquad\qquad [4.6]$$

In our example,

$$z = \frac{14 - 25.40}{6.10} = -1.87$$

A weekly television viewing time of 14 hours is 1.87 standard deviations *below* the national average. From column 3 of Appendix B, we find that the proportion of scores less than or equal to a z score of −1.87 is .0307. Thus, 14 hours defines the 3.07th percentile. Stated another way, 3.07% of adults in the United States watch 14 hours or less of television per week. Approximately 1.00 (the total proportion of all scores) − .0307 = .9693, or 96.93%, of adults in the United States must therefore watch 14 hours or more of television per week. Relative to most American adults, the amount of time we spend watching television is quite low. Do a similar analysis for your own estimated viewing behavior.

Suppose we want to know what percentage of American adults watch between 30 and 40 hours of television per week. We can use our knowledge of the normal distribution and Appendix B to estimate this. First, we convert the scores of 30 and 40 into z scores:

$$z = \frac{30 - 25.40}{6.10} = .75$$

FIGURE 4.2   **Proportion of Scores Between *z* Scores of .75 and 2.39**

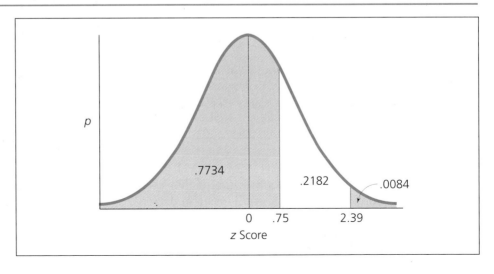

and

$$z = \frac{40 - 25.40}{6.10} = 2.39$$

Examining Appendix B, we find that .7734 of the scores in a normal distribution are less than or equal to a z score of .75. This is illustrated in Figure 4.2. There are several ways that this value can be obtained from Appendix B. For instance, we could refer to column 3, where we find that .2266 of the scores are greater than or equal to a z score of .75. Thus, 1.00 − .2266 = .7734 of the scores must be .75 or less. Alternatively, realizing that .50 of the scores occur below the mean (z score of 0), we could refer to column 5, where we find that an additional .2734 of the scores fall between a z score of 0 and a z score of .75. The sum of these figures (.50 + .2734) is .7734. Appendix B also indicates that .0084 of the scores are greater than or equal to a z score of 2.39. The most direct way to obtain this value is to refer to column 3. The proportion of scores that occur *between* .75 and 2.39 is 1.00 minus the .0084 of the scores equal to or greater than a z score of 2.39 and minus the .7734 of the scores equal to or less than a z score of .75, or 1.00 − .0084 − .7734 = .2182. Thus, approximately 21.82% of American adults watch between 30 and 40 hours of television per week.

Study Exercise 4.4

Given a set of scores that are normally distributed with a mean of 100.00 and a standard deviation of 10.00, what proportion of scores is greater than or equal to 120? Less than or equal to 90?

Answer   We begin by converting the raw score of interest into a z score. For the proportion of scores greater than or equal to 120, we have

$$z = \frac{120 - 100.00}{10.00} = 2.00$$

Using column 3 of Appendix B, we find that the proportion of scores greater than or equal to a z score of 2.00 is .0228. Thus, the proportion of scores greater than or equal to 120 is .0228.

For the proportion of scores less than or equal to 90,

$$z = \frac{90 - 100.00}{10.00} = -1.00$$

Again using column 3 of Appendix B, we find that the proportion of scores less than or equal to a z score of −1.00 is .1587. Thus, the proportion of scores less than or equal to 90 is .1587.

In some applications, we might want to reverse the previous procedures and determine the raw scores that define specified proportions of an approximately normal distribution. If the mean and standard deviation of this distribution are known, this can be accomplished by substituting these values and an appropriate value of z into one of the following equations (depending on whether we are dealing with sample data or population data) and solving for X:

$$X = \overline{X} + (z)(s) \tag{4.7}$$

$$X = \mu + (z)(\sigma) \tag{4.8}$$

These equations are merely rearrangements of Equations 4.5 and 4.6, respectively. The value of z of interest is the value that cuts off the specified proportion of the distribution. For instance, in the television-viewing example, we might want to determine the number of hours of television viewing at or above which 5% of all weekly viewing times fall. From column 3 of Appendix B, we find that .0505 of all scores in a normal distribution are greater than or equal to a z score of 1.64 and .0495 of all scores in a normal distribution are greater than or equal to a z score of 1.65. Thus, the value of z that defines the upper .05, or 5%, of scores in a normal distribution is approximately halfway between these two values, or 1.645. Since the mean in our example is 25.40 and the standard deviation is 6.10, the number of hours of television viewing at or above which 5% of all weekly viewing times fall is

$$X = 25.40 + (1.645)(6.10) = 35.43$$

4.4 Standard Scores and the Shape of the Distribution

As noted in Section 4.2, when a set of scores is transformed to standardized scores, the standard scores will always have a mean of 0 and a standard deviation of 1.00. However, the process of standardizing scores does not change the fundamental shape of the distribution. A set of scores that are positively skewed will remain positively skewed after having been converted to standard scores. A set of scores that are platykurtic will still be platykurtic after having been converted to standard scores. Standardization affects the central tendency and the variability of the scores but not the skewness or kurtosis of the scores. Other transformations are possible that affect the shape of the distribution; these are discussed in Gulliksen (1960).

4.5 Method of Presentation

Percentile information is most commonly encountered in manuals available to educators and professionals for interpreting scores on educational and psychological tests. Such manuals generally list selected percentile ranks and corresponding raw score values. For instance, the Miller Analogies Test is designed to measure the general aptitude of applicants for graduate and professional schools. The test consists of a series of items in the form of analogies. An example item might appear as follows (where you are to fill in the blank with the appropriate word from those given):

A *book* is to *trees* as a *skirt* is to _____ .

(a) shoes   (b) sheep   (c) dresses   (d) women

Scores on the Miller Analogies Test can range from 0 to 100. A list of the raw score equivalents defining selected percentiles for different groups of graduate students might appear in the test manual as follows:

Percentile Rank	Physical Sciences	Medical Science	Social Sciences	English	Law	Social Work
99	93	92	90	87	84	81
90	88	78	82	80	73	67
80	82	74	76	74	63	61
70	78	67	69	68	58	58
60	74	60	64	65	53	54
50	68	57	61	59	49	50
40	63	53	56	53	45	46
30	58	47	51	46	40	41
20	51	43	46	41	35	37
10	43	34	39	35	30	27
1	28	24	18	7	18	9

Although not all possible values are included in this table, the relative standing of selected test scores can be determined. Focusing on law students, for instance, we find that a score of 49 has a percentile rank of 50, a score of 58 has a percentile rank of 70, and so forth.

Percentile ranks must always be interpreted relative to the group upon which the scores are based. Someone who scores at the 95th percentile on the Miller Analogies Test, for instance, probably has greater general aptitude than someone who scores at the 95th percentile on the Scholastic Aptitude Test (an aptitude test given to applicants for undergraduate schools). This is because only the brighter students who have done well in college tend to take the Miller Analogies Test, whereas the Scholastic Aptitude Test is taken by a more general population.

It might be interesting to contrast your own major with those of others in terms of which scores define various percentiles in the above table. For example, the 50th percentile is 68 for graduate students in the physical sciences, 61 for graduate students in the social sciences, 59 for graduate students in English (literature and language), 57 for medical students, 50 for graduate students in social work, and 49 for law students.

To aid in interpreting specified scores, most educational and psychological test manuals report standard score equivalents of raw scores in addition to percentile tables. To avoid the confusion associated with decimals and negative values, standard scores are frequently further transformed into *T scores*. *T* scores are directly analogous to standard scores, but instead of having a mean of 0 and a standard deviation of 1.00, they have a mean of 50.00 and a standard deviation of 10.00. The transformation is accomplished through this formula:

$$T = 50 + 10(\text{standard score}) \qquad [4.9]$$

The resultant value is usually rounded to the nearest whole number. For instance,

**Applications
to the Analysis
of a Social Problem**

The parent–teen communication data described in Chapter 1 can be explored in the context of percentiles, percentile ranks, and standard scores. A critical variable in these data is the amount of sexual activity of the adolescents. This was measured by asking adolescents a series of questions about their sexual behavior. One question asked the teens to indicate the number of times that they had engaged in sexual intercourse during the past 6 months. Table 4.3 presents a computer printout of the frequency distribution (as discussed in Chapter 2) for this variable as well as basic descriptive statistics (as discussed in Chapter 3). In addition, the computer program calculated percentiles, in ten-unit increments. Figure 4.3 presents a grouped frequency histogram of the data.

There were 49 teens who chose not to respond to the question and, hence, were treated as "missing cases." The mean number of times the teens engaged in sexual intercourse during the past 6 months was 4.38, the median was 0, and the mode was 0. In terms of variability, the lowest score was 0 and the highest was 90, yielding a range of 90 and a standard deviation of 9.82. The frequency histogram reveals that the distribution is highly positively skewed, and this is reflected in the

fact that the mean is greater than the median. This particular computer program computes percentiles "rounding to the original metric" and, hence, the percentiles are always whole numbers. For these data, the 10th percentile is 0, as are the 20th, 30th, 40th, and 50th. The 60th percentile is 2, the 70th percentile is 3, the 80th percentile is 6, and the 90th percentile is 12. Very few of the adolescents had sexual intercourse more than 12 times in the past 6 months.

If desired, we can convert the raw scores to standard scores. For example, a single instance of sexual intercourse translates into a standard score of $(1 - 4.38) / 9.82 = -.34$, which is about a third of a standard deviation below the mean. Ten instances of sexual intercourse represent a standard score of $(10 - 4.38) / 9.82 = .57$, which is just over half a standard deviation above the mean. Because the original scores are highly positively skewed, the distribution is decidedly nonnormal. Therefore, we cannot use the normal distribution to help us interpret what these standard scores mean. Rather, we must rely on the percentages and cumulative percentages in the frequency distribution. In fact, some analysts might question whether

(continued)

a standard score of $-.53$ is equivalent to a T score of $50 + (10)(-.53) = 44.7$. This is typically reported as 45.

A portion of a test manual report of raw scores and their standard score and T score equivalents might appear as follows:

Raw score	Standard score	T score
100	2.90	79
99	2.70	77
98	2.50	75
97	2.30	73
96	2.10	71

In practice, all possible raw score values and corresponding standard and T scores would be included in this listing.

TABLE 4.3 Computer Output for Frequency Distribution and Descriptive Statistics for Instances of Sexual Intercourse for Overall Data Set

Value	Freq	Percent	Valid Percent	Cum Percent
0	371	49.7	53.2	53.2
1	34	4.6	4.9	58.0
2	51	6.8	7.3	65.3
3	40	5.4	5.7	71.1
4	34	4.6	4.9	75.9
5	19	2.5	2.7	78.7
6	17	2.3	2.4	81.1
7	15	2.0	2.1	83.2
8	10	1.3	1.4	84.7
9	3	.4	.4	85.1
10	21	2.8	3.0	88.1
11	2	.3	.3	88.4
12	14	1.9	2.0	90.4
13	3	.4	.4	90.8
14	4	.5	.6	91.4
15	6	.8	.9	92.3
16	3	.4	.4	92.7
18	1	.1	.1	92.8
19	1	.1	.1	93.0
20	18	2.4	2.6	95.6
22	2	.3	.3	95.8
24	1	.1	.1	96.0
25	3	.4	.4	96.4
26	1	.1	.1	96.6
28	1	.1	.1	96.7
29	1	.1	.1	96.8
30	9	1.2	1.3	98.1
32	1	.1	.1	98.3
35	2	.3	.3	98.6
40	3	.4	.4	99.0
48	1	.1	.1	99.1
60	1	.1	.1	99.3
63	1	.1	.1	99.4
70	1	.1	.1	99.6
90	3	.4	.4	100.0
.	49	6.6	Missing	
Total	747	100.0	100.0	

Mean	4.384	Median	.000	Mode	.000
Std dev	9.816	Variance	96.349	Range	90.000
Minimum	.000	Maximum	90.000		

Percentile	Value	Percentile	Value	Percentile	Value
10.00	.000	20.00	.000	30.00	.000
40.00	.000	50.00	.000	60.00	2.000
70.00	3.000	80.00	6.000	90.00	12.000

Valid cases   698   Missing cases   49

standard scores are meaningful at all in the present analysis, given the substantial positive skew in the data.

Recall from Chapter 3 that the mean is influenced by extreme scores and may not be the most "representative" value for a set of scores when the data are skewed with extreme scores at one end of the distribution. In the present case, the mean is near 4, but the frequency distribution and the median and mode make it clear that most adolescents have not been sexually active during the past 6 months. The median and modal values of 0 probably best characterize a "representative" value for the data. Standard scores, however, use

the mean as the basis for describing how extreme a given score is. If the mean is a poor "representative" value, then standard scores must be interpreted cautiously. Standard scores (and the mean) *are* informative in any distribution if the analyst keeps in mind their fundamental statistical properties and how they are affected by skewed data and extreme scores.

When data are highly skewed, some analysts prefer to calculate a variant of standard scores that uses the median and an index of variability that does not rely on the mean. Let us briefly develop this concept. A general formula for describing the uniqueness of a given individual's raw score is

$$u = \frac{\text{score for an individual} - \text{measure of central tendency}}{\text{measure of variability}} \qquad [4.10]$$

(continued)

FIGURE 4.3 **Grouped Frequency Histogram of Instances of Sexual Intercourse for Overall Data Set**

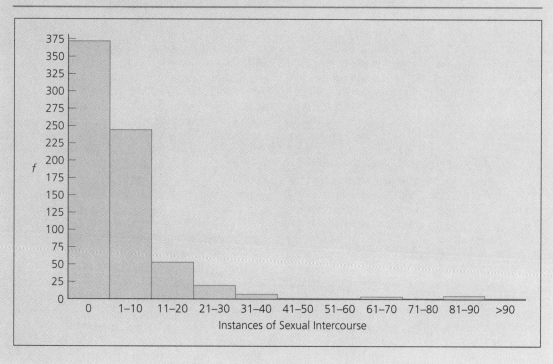

This is, in fact, our formula for a standard score, where the measure of central tendency is the mean and the measure of variability is the standard deviation (see Equations 4.3 and 4.4). With skewed data, an alternative index of uniqueness uses the median as the measure of central tendency and either the interquartile range or a measure called the *median of absolute deviations*, commonly referred to as MAD, as the measure of variability (see Hoaglin, Mosteller & Tukey, 1983). (MAD is calculated by first computing for each individual the absolute value of the individual's score minus the median. MAD is the median of these absolute values. The MAD index reflects how far, on average, scores deviate from the median.)

When Equation 4.10 is applied using the median and the interquartile range (IQR), the resulting *u* score is the number of IQR units that an individual's score is above or below the median. For the data on sexual intercourse, the median value that we use is 0 and the IQR is 4. A single instance of sexual intercourse translates into a *u* score of $(1 - 0) / 4 = .25$. One act of sexual intercourse is one-fourth of an IQR above the median. Note how this contrasts with the standard score for one instance of sexual intercourse, $-.34$. When the median and IQR are used in Equation 4.10, a single instance of sexual intercourse is somewhat higher than the "central tendency" of the data, whereas when the mean and standard deviation are used, a single instance of sexual intercourse is somewhat lower than the central tendency of the data. Which index do you think is a better descriptor of what a score of "1" on this variable reflects?

Many problem behaviors in adolescence have frequency distributions that are similar in shape to Figure 4.3—namely, highly skewed. For example, most teens do not use hard drugs, drink excessively, attempt suicide, run away from home, or drive drunk. However, usually a significant minority of teens do these things, sometimes excessively so, which yields distributions that have a shape similar to the one we observe here.

The responses to the question about the frequency of sexual intercourse should not be interpreted too literally. For example, the teen who reported 28 instances of sexual intercourse probably did not engage in exactly 28 instances of sexual intercourse. Rather, this response probably represents an approximate number of instances of sexual intercourse based on some cognitive heuristic that the teen used for answering the question.

It would be a serious error to think of these data as having ratio level properties. As indicators of sexual activity, the measures probably reasonably approximate interval level properties, although some researchers might contend even this is questionable.

Our analysis has focused on the total sample. Now we examine the data for the age groups 14–17 separately. Table 4.4 presents a computer printout with basic descriptive statistics for each group. In terms of central tendency, the mode was 0 for all age groups. The median was 0 for 14- and 15-year-olds but shifted to 2.00 for 16- and 17-year-olds. The mean tended to increase with each age group: 1.91 for 14-year-olds, 2.80 for 15-year-olds, 5.84 for 16-year-olds, and 7.66 for 17-year-olds. Chapter 12 will discuss formal procedures for comparing multiple means in the context of inferential statistics. The standard deviations tended to be lower for the 14- and 15-year-olds (4.94 and 6.58, respectively) than for the 16- and 17-year-olds (12.28 and 12.71, respectively), which suggests that there is more variability in sexual activity for older as opposed to younger adolescents. Chapter 12 will also discuss procedures for testing the equivalence of variability across multiple groups in an inferential context. Some of the percentiles differ as a function of age. For example, the 80th percentile is 3 for 14-year-olds, 4 for 15-year-olds, 9 for 16-year-olds, and 13 for 17-year-olds.

We will explore these age trends in greater depth in later chapters. We use the data in the present chapter to underscore an important point about percentile ranks and standard scores: Percentile ranks and standard scores represent an individual's performance on some variable *relative to some group of individuals*. They do not convey performance in an absolute sense. For example, if a teen engaged in sexual intercourse three times, this would translate into a standard score of .22 for 14-year-olds, .03 for 15-year-olds, $-.23$ for 16-year-olds, and $-.37$ for 17-year-olds. Three instances of sexual intercourse represent a percentile rank of approximately 80 for 14-year-olds, 75 for 15-year-olds, 60 for 16-year-olds, and 53 for 17-year-olds. The absolute level of sexual activity (three instances of sexual intercourse) takes on different meaning in the context of different reference groups. What is "above average" sexual activity in one group may be "below average" sexual activity in another group. This relative nature of standard scores and percentile ranks is important to keep in mind.

TABLE 4.4 Computer Output for Descriptive Statistics for Instances of Sexual Intercourse for Four Age Groups

AGE: 14.00

NUMBER6   Number of Times Had Intercourse in Past Six Months

Mean	1.911	Median	.000	Mode	.000
Std dev	4.935	Variance	24.350	Range	30.000
Minimum	.000	Maximum	30.000		

Percentile	Value	Percentile	Value	Percentile	Value
10.00	.000	20.00	.000	30.00	.000
40.00	.000	50.00	.000	60.00	.000
70.00	.000	80.00	3.000	90.00	7.000

Valid cases   112 Missing cases   3

AGE: 15.00

NUMBER6   Number of Times Had Intercourse in Past Six Months

Mean	2.797	Median	.000	Mode	.000
Std dev	6.581	Variance	43.315	Range	70.000
Minimum	.000	Maximum	70.000		

Percentile	Value	Percentile	Value	Percentile	Value
10.00	.000	20.00	.000	30.00	.000
40.00	.000	50.00	.000	60.00	1.000
70.00	2.000	80.00	4.000	90.00	9.000

Valid cases   256 Missing cases   17

AGE: 16.00

NUMBER6   Number of Times Had Intercourse in Past Six Months

Mean	5.838	Median	2.000	Mode	.000
Std dev	12.279	Variance	150.768	Range	90.000
Minimum	.000	Maximum	90.000		

Percentile	Value	Percentile	Value	Percentile	Value
10.00	.000	20.00	.000	30.00	.000
40.00	.000	50.00	2.000	60.00	3.000
70.00	5.000	80.00	9.000	90.00	15.000

Valid cases   210 Missing cases   13   (continued)

```
        Table 4.4 (continued)

  AGE:       17.00

      NUMBER6    Number of Times Had Intercourse in Past Six Months


      Mean          7.661  Median       2.000  Mode           .000
      Std dev      12.706  Variance   161.440  Range        90.000
      Minimum        .000  Maximum     90.000


      Percentile  Value  Percentile   Value  Percentile   Value

      10.00        .000  20.00         .000  30.00          .000
      40.00       1.000  50.00        2.000  60.00         5.000
      70.00       7.000  80.00       13.000  90.00        25.000


      Valid cases    118  Missing cases     16
```

Summary

Percentile ranks, percentiles, and standard scores are measures used to identify the location of a specified score within a set of scores. A percentile rank is the percentage of scores in a distribution that occur at or below a given value. A percentile is a score value that corresponds to a given percentile rank. A standard score is the difference between a score in a distribution and the mean of the distribution, divided by the distribution's standard deviation. A standard score thus represents the number of standard deviation units that a score occurs above or below the mean. When data are highly skewed, an alternative to standard scores is a measure of uniqueness that uses the median as the measure of central tendency and either the interquartile range or the median of absolute deviations as the measure of variability.

The mean of a set of standard scores is always equal to 0, and the standard deviation (and the variance) of a set of standard scores is always equal to 1.00. The process of standardizing scores does not change the fundamental shape of the distribution.

A standard score in a normal distribution is called a z score. If we are able to assume that a set of scores approximates a normal distribution, we can use Appendix B to determine the proportion of scores that occur above or below a given z score and the proportion of scores that occur between two different z scores. We can also reverse these procedures and determine the raw scores that define specified proportions of a distribution.

Appendix 4.1   The Normal Distribution Formula

The formula for a normal distribution is

$$d(X) = \frac{e^k}{\sqrt{2\pi\sigma^2}} \qquad [4.11]$$

where $k = -(X - \mu)^2/2\sigma^2$, $d(X)$ is the probability density function associated with a particular value of X, π (pi) is a constant approximately equal to 3.1416, e is a constant approximately equal to 2.7183, μ is the mean of the distribution, and σ^2 is the variance of the distribution. When μ and σ^2 are specified, different values of X can be substituted into the equation and the corresponding values of $d(X)$ obtained. If paired values of X and $d(X)$ are plotted graphically, they will form a normal curve. Thus, as noted in the text, there is a different normal curve for every unique combination of μ and σ^2.

Exercises

Answers to asterisked (*) exercises appear at the back of the book. Answers to exercises with two asterisks are also worked out step by step in the Study Guide.

1. What is the relationship between percentile ranks and percentiles?

Use the following information to complete Exercises 2–4:

A researcher administered a questionnaire to 500 people that was designed to measure knowledge of the positions that two presidential candidates held on major issues. Scores on the test could range from 0 to 10, with higher values indicating more knowledge. The following frequency analysis resulted:

Score	f	rf	cf	crf
8	6	.012	500	1.000
7	44	.088	494	.988
6	51	.102	450	.900
5	49	.098	399	.798
4	200	.400	350	.700
3	98	.196	150	.300
2	52	.104	52	.104

**2. Compute the scores that define the following percentiles:

    a. 20th   d. 80th
    b. 40th   e. 99th
    c. 60th   f. 50th

3. Compute the interquartile range.
**4. Compute the percentile ranks that correspond to the following scores:
    a. 7   b. 5   c. 3   d. 2
5. For the following set of data, compute the score that defines the 50th percentile: 100, 100, 99, 99, 99, 98, 98, 97, 97, 96, 95, 94.
6. Compute the interquartile range for the data in Exercise 5.
7. What are the two approaches to reporting percentiles? What rationale underlies each position?
8. What is the advantage of standard scores compared with percentile ranks as an index of relative standing?
*9. Given a distribution with a mean of 20.00 and a standard deviation of 3.00, compute the standard score equivalents of the following scores:
    a. 21.87 c. 18.91 e. 15.63
    b. 23.00 d. 20.08 f. 24.30
10. Given a distribution with a mean of −4.00 and a standard deviation of 2.50, compute the standard score equivalents of the following scores:
    a. −6.83 c. 2.84 e. 1.00
    b. 0   d. −6.50 f. .87

*11. For the following distribution, convert a score of 50 to a standard score: 60, 55, 50, 50, 55, 60.

12. For the following distribution, convert a score of 6 to a standard score: 14, 6, 13, 17, 11, 13, 14, 9.

*13. What does a positive standard score indicate about the original score's position relative to the mean? What does a negative standard score indicate about the original score's position relative to the mean?

*14. What are the values of the mean and the standard deviation for any set of standard scores?

15. How does standardizing a set of scores affect the shape of the distribution?

*16. John received 90 out of 100 points on an English exam and 60 out of 100 points on a math exam. The overall class performance was $\overline{X} = 70.00$ and $s = 20.00$ on the English exam, and $\overline{X} = 40.00$ and $s = 3.00$ on the math exam. On which exam was John's performance better (relative to his classmates')?

17. If a person got a score of 80 on a test, which one of the following distributions allows for the most favorable interpretation of that score (assuming that higher values are more favorable)?
   a. $\overline{X} = 60.00, s = 5.00$
   b. $\overline{X} = 60.00, s = 10.00$
   c. $\overline{X} = 60.00, s = 1.00$
   d. $\overline{X} = 60.00, s = 20.00$

*18. Describe a situation in which it would be useful to convert scores from two different distributions to standard scores before comparing them.

19. What are the major characteristics of a normal distribution?

20. What is a z score?

21. Given a normal distribution with a mean of 24.87 and a standard deviation of 6.00, compute the z score equivalents of the following scores:
   a. 13.78   c. 26.81   e. 37.90
   b. 29.42   d. 12.87   f. 33.35

*22. What proportion of z scores in a normal distribution are
   a. 2.38 or less
   b. 1.17 or greater
   c. −1.17 or less
   d. between 0 and 2.05
   e. between −2.05 and 0
   f. between .37 and 3.19
   g. between −3.19 and −.37
   h. between −1.24 and +1.24

*23. Given a set of normally distributed scores with a mean of 20.00 and a standard deviation of 5.00, what proportion of scores are
   a. 25 or higher
   b. 15 or less
   c. between 15 and 28
   d. between 8 and 32
   e. 20 or higher
   f. 23 or less

24. Suppose IQ scores in a population are approximately normally distributed with $\mu = 100.00$ and $\sigma = 10.00$. What proportion of individuals have IQ scores of
   a. 100 or higher
   b. 100 or less
   c. between 105 and 112
   d. 103 or less
   e. 95 or higher
   f. 95 or less

*25. Lie detectors, or polygraphs, are used to help determine whether a person has knowledge of a crime. These devices are based on autonomic changes in a person's nervous system. The assumption is that lying will be reflected in physiological changes that are not under the voluntary control of the individual. Measurements of the individual's physiological response while answering selected questions can be used to make inferences about the veridicality of his or her answers. Lie detection tests typically involve answering a series of neutral questions (for example, What is your name? or Where do you work?) among which the critical questions are embedded. Consider the case where an individual has been asked a series of questions, including a critical item. Physiological measurements in the form of galvanic skin responses are taken for each question. The galvanic skin response scores approximate a normal distribution with a mean of 49.40 and a standard deviation of 3.00. For the critical question, the score was 61.40. Convert this into a standard score and draw a conclusion.

26. A major form of identification in criminal investigations is fingerprints. Fingerprints vary on many different dimensions, one of which is called the ridgecount. Suppose you know that the ridgecounts of human beings follow a normal distribution with a mean of 165.00 and a standard deviation of 10.00. Suppose furthermore that a set of fingerprints was found at the scene of a crime and it was determined that the ridgecount was at least 200 (the exact value being in question because of smudging). Finally, suppose that a particular suspect has a ridgecount of 225. What should you conclude and why?

*27. Given a set of normally distributed scores with $\overline{X} = 100.00$ and $s = 10.00$, what score corresponds to a z score of
   a. 2.86   c. 0   e. 1.59
   b. −2.44   d. −1.50   f. .75

*28. Given a normal distribution with a mean of 100.00 and a standard deviation of 5.00, what score would
   a. 33% of the cases be greater than or equal to
   b. 5% of the cases be greater than or equal to
   c. 50% of the cases be less than or equal to
   d. 2.50% of the cases be greater than or equal to
   e. 2.50% of the cases be less than or equal to

29. Suppose income in a sample is approximately normally distributed with $\overline{X} =$ $20,000.00 and $s =$ $2,000.00. What income level defines the
   a. top 2.5% of salaries
   b. bottom 2.5% of salaries
   c. top 5% of salaries
   d. bottom 5% of salaries
   e. top 33% of salaries
   f. bottom 50% of salaries

*30. Convert each of the following standard scores to a T score:
   a. .87   c. 1.56   e. 0
   b. 2.00   d. −1.56   f. 4.04

Multiple-Choice Questions

*31. John received a standard score of −2.37 on an exam. This means that John
   a. did better than the class average
   b. did worse than the class average
   c. received a score equal to the class average
   d. received a raw score of −2.37

32. If a set of scores is normally distributed with a mean of 50.00 and a standard deviation of 10.00, what standard score corresponds to a raw score of 60?
   a. −1.00   c. 1.00
   b. −2.00   d. 2.00

33. If a set of scores has a mean of 100.00 and a standard deviation of 5.00, what is the variance of the standard scores?
   a. 1.00
   b. 5.00
   c. 25.00
   d. This cannot be determined

34. A normal distribution is always bell-shaped.
   a. true
   b. false

*35. On average, it takes a person 5.00 hours to clean a house, with a standard deviation of 1.00. If Bill has a standard score of −2.00, how many hours does it take him to clean his house?
   a. 2.00   c. 5.00
   b. 3.00   d. 7.00

36. In a normal distribution
    a. the mean is usually greater than the mode
    b. the mode is usually greater than the median
    c. the mode, median, and mean are always equal to one another
    d. the mode and median are always equal to each other and usually greater than the mean

*37. The percentile rank that corresponds to Chester's score on a class test is 30. This means that
    a. Chester answered 30% of the questions on the test correctly
    b. 30% of the obtained scores were greater than or equal to Chester's
    c. 30 students obtained scores that were less than or equal to Chester's
    d. 30% of the obtained scores were less than or equal to Chester's

*38. Which of the following is a problem with percentile ranks as an index of relative standing?
    a. They cannot be precisely calculated.
    b. They can be used only with samples of at least size 30.
    c. They don't tell us anything about the *magnitude* of the score of interest relative to the other scores in the distribution.
    d. They can be used only with ratio data.

39. A negative standard score indicates that its corresponding raw score is
    a. below the mean of the distribution
    b. above the mean of the distribution
    c. negative in value

    d. more than one standard deviation from the mean (in either direction)

*40. Standard scores provide information about a score's location relative to the mean in _____ units of the distribution.
    a. cumulative frequency
    b. variance
    c. raw score
    d. standard deviation

*41. What proportion of scores in a normal distribution lie between z scores of -3.12 and $+3.12$?
    a. .0009   c. .9982
    b. .5018   d. .9991

42. What proportion of scores in a normal distribution lie between the mean and a z score of $-.44$?
    a. .1700   c. .3400
    b. .3300   d. .6700

43. What proportion of scores in a normal distribution are less than or equal to a z score of 1.86?
    a. .0314   c. .9372
    b. .4686   d. .9686

*44. Given a set of normally distributed scores with a mean of 12.00 and a standard deviation of 4.00, what proportion of scores are greater than or equal to 14?
    a. .3085   c. .6170
    b. .3830   d. .6915

*45. The median of the absolute values of the difference between each score in a data set and the median is referred to as the
    a. IQR
    b. u score
    c. MAD
    d. none of the above

CHAPTER 5

Pearson Correlation and Regression: Descriptive Aspects

5.1 Use of Pearson Correlation

To this point, we have emphasized ways of summarizing and describing scores on a single variable. Research in the behavioral sciences, however, often involves the measurement of two variables for the same individuals. A common question in this situation concerns the way in which scores on the first variable are related to scores on the second variable. For instance, an investigator interested in the relationship between women's traditionalism and their ideal family size might conduct a study in which each participant is asked to indicate her ideal number of children and to respond to a traditionalism questionnaire. The question of interest is whether there is a relationship between women's traditionalism scores and their ideal family sizes. Actually, there are many different ways in which two variables might be related; however, research in the behavioral sciences is often concerned with *linear* relationships. When both variables under study are quantitative, have many values, and are measured on a level that at least approximates interval characteristics, the statistical technique of *Pearson product-moment correlation,* known more simply as **Pearson correlation,** can be used to determine the extent to which they are linearly related.* To lay the groundwork for an in-depth discussion of this technique, we now review the basic characteristics of linear relationships.†

5.2 The Linear Model

Consider two variables: number of hours worked (X) and amount of money paid (Y). Each of four individuals works at a rate of $1 per hour. Their scores on X and Y are as follows:

Individual	X (hours worked)	Y (amount paid, $)
1	1	1
2	4	4
3	3	3
4	2	2

The relationship between X and Y is illustrated in Figure 5.1 on a graph called a **scatterplot.** The abscissa represents the values of X, and the ordinate represents the values of Y. For each individual, we find the value on the X axis corresponding to his or her score on X and the value on the Y axis corresponding to the score on Y, and we place a dot where the two values intersect. For instance, individual 1 had a score of 1 on X and a score of 1 on Y. We therefore place a dot at the intersection of $X = 1$ and $Y = 1$. The same procedure is repeated for each individual and the resulting dots connected, as in Figure 5.1. As indicated by the straight line on the scatterplot, there is a linear relationship between X and Y. This relationship can be stated as

$$Y = X$$

In other words, the number of dollars paid equals the number of hours worked.

Suppose the individuals were not paid $1 per hour, but instead were paid $2 per hour. The scores on X and Y would be as follows:

---

* The condition that both variables be measured on a level that at least approximates interval characteristics means that they must be measured on an ordinal level that approximates interval characteristics, on an interval level, or on a ratio level.

† Because this chapter is concerned with descriptive statistics, we present formulas that are appropriate for descriptive but not inferential use. Inferential aspects of Pearson correlation and regression will be discussed in Chapter 14.

FIGURE 5.1   **Example of a Scatterplot**

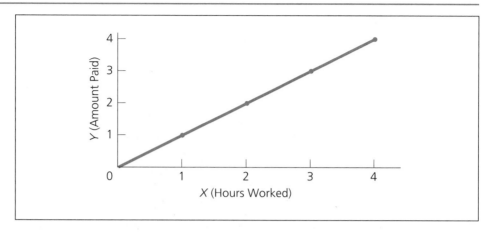

Individual	X (hours worked)	Y (amount paid, $)
1	1	2
2	4	8
3	3	6
4	2	4

In this case, the relationship between X and Y can be stated as

$$Y = 2.00X$$

In other words, the number of dollars paid equals 2 times the number of hours worked. Figure 5.2 presents a scatterplot of these data (line B) as well as the data from Figure 5.1 (line A). Notice that we still have a straight line (and, hence, a linear relationship) but, in the case of $2 per hour, the line rises much faster than with $1 per hour; that is, the **slope** of the line is greater. Technically, the slope of a line indicates the number of units variable Y changes as variable X changes by one unit. When people are paid $2 per hour, an individual who works 1 hour is paid $2, one who works 2 hours is paid $4, and so forth. When X goes up by one unit (for instance, from 1 to 2 hours), Y goes up by two units (from $2 to $4). The slope that describes this linear relationship is therefore 2. In contrast, the slope that describes the linear relationship $Y = X$ is 1, reflecting that as X changes by one unit, so does Y. Thus, linear relationships can differ in terms of the slopes that describe them.

The slope that describes a linear relationship can be determined from a simple algebraic formula. This formula involves first selecting the X and Y scores of any two individuals. The slope is computed by dividing the difference between the

FIGURE 5.2   **Three Examples of a Linear Relationship**

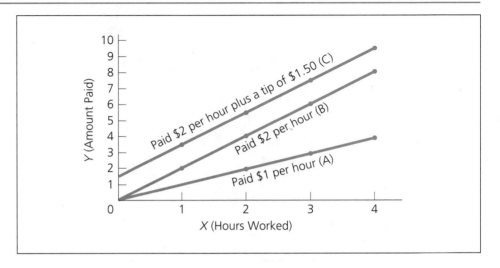

two Y scores by the difference between the two X scores; in other words, the change in Y scores is divided by the change in X scores. Symbolically,

$$b = \frac{Y_1 - Y_2}{X_1 - X_2}$$ [5.1]

where b represents the slope, X_1 and Y_1 are the X and Y scores for any one individual, and X_2 and Y_2 are the X and Y scores for any other individual. Inserting the scores for individuals 1 ($X = 1$, $Y = 2$) and 2 ($X = 4$, $Y = 8$), we find that the slope for line B is

$$b = \frac{2 - 8}{1 - 4} = 2.00$$

This is consistent with what was stated previously.

The value of a slope can be positive, negative, or 0. Consider the following scores:

Individual	X	Y
1	2	3
2	1	4
3	4	1
4	3	2

Inserting the scores for individuals 2 and 4 into Equation 5.1, we find that the slope is

$$b = \frac{4 - 2}{1 - 3} = -1.00$$

Figure 5.3 presents a scatterplot of this relationship. The relationship is still linear, but now the line moves downward as we move from left to right on the X axis. This downward direction characterizes a negative slope, whereas an upward direction characterizes a positive slope. A slope of 0 is represented by a horizontal line because the value of Y is constant for all values of X.

A positive slope indicates a **positive** or **direct relationship** between X and Y, whereas a negative slope indicates a **negative** or **inverse relationship** between X and Y. In the case of a positive relationship, as scores on X *increase*, scores on Y also *increase*. In the case of a negative relationship, as scores on X *increase*, scores on Y *decrease*. For instance, the slope in the present example is -1.00, meaning that for every unit X increases, Y decreases by one unit.

Let us return to the example where individuals are paid \$2 per hour worked. Suppose that in addition to this wage, each individual is given a tip of \$1.50. Now the relationship between X and Y is

$$Y = 1.50 + 2.00X$$

Line C of Figure 5.2 plots this case for the four individuals. If we compute the slope of this line, we find it to be 2.00, as before. Notice that lines C and B are parallel but that line C is higher than line B. The amount of separation between these two lines can be measured at the Y axis, where $X = 0$. When $X = 0$, the Y value is 1.50 for line C and 0 for line B. Thus, line C is raised 1.50 units above line B.

The point at which a line intersects the Y axis when $X = 0$ is called the **intercept**, and its value is denoted by the letter a. Linear relationships can differ in the values of their intercepts as well as the values of their slopes, as indicated in the previous problem in which the intercept of line C is 1.50 and the intercept of line B is 0. The general form of the **linear model** is thus

FIGURE 5.3   **Example of a Linear Relationship with a Negative Slope**

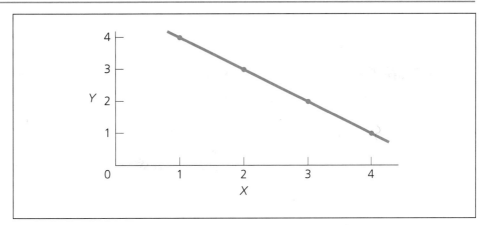

$$Y = a + bX \qquad\qquad [5.2]$$

Any line can be represented by Equation 5.2. A slope and an intercept will always describe the linear relationship between two variables. Given these values, we can substitute scores on X into the **linear equation** to determine the corresponding scores on Y. For example, the linear equation $Y = 1.50 + 2.00X$ tells us that an individual who worked for 2 hours was paid $5.50 because the Y score associated with an X score of 2 is

$$Y = 1.50 + 2.00X$$

$$= 1.50 + (2.00)(2) = 5.50$$

Similarly, an individual who worked for 3 hours was paid $Y = 1.50 + (2.00)(3) = \$7.50$, and an individual who worked for 4 hours was paid $Y = 1.50 + (2.00)(4) = \$9.50$.

Study Exercise 5.1

Suppose you are told that for a group of 20 students, there is a perfect linear relationship between grade point average (Y) and scores on an intelligence test (X). Suppose you are also told that the equation describing the relationship is

$$Y = 1.00 + .025X$$

If a student obtained a score of 100 on the intelligence test, what must his or her grade point average be? What must a student's grade point average be if he or she obtained an intelligence test score of 97? Of 108?

Answer   The grade point average associated with an intelligence test score of 100 is

$$Y = 1.00 + (.025)(100) = 3.50$$

The grade point average associated with an intelligence test score of 97 is

$$Y = 1.00 + (.025)(97) = 3.42$$

Last, the grade point average associated with an intelligence test score of 108 is

$$Y = 1.00 + (.025)(108) = 3.70$$

5.3   The Pearson Correlation Coefficient

It is rare in the behavioral sciences to observe a perfect linear relationship between two variables. Far more common is for the relationship between two variables to *approximate* a linear one. The extent of linear approximation between two variables is indexed by a statistic known as the **Pearson correlation coefficient**. The *correlation coefficient*, represented by the letter *r*, can range from −1.00 through 0 to +1.00. The *magnitude* of the correlation coefficient, as indexed by its absolute value, indicates the *degree to which a linear relationship is approximated*: The further *r* is in either a positive or a negative direction from 0, the better is the approximation. The *sign* of the correlation coefficient indicates the *direction* of the linear approximation. A correlation coefficient of +1.00 means the two variables form a perfect linear relationship that is direct in nature (that is, the higher the score an individual obtains on X, the higher the score that individual obtains on Y). A correlation coefficient of −1.00 also means the two variables form a perfect linear relationship, but it is inverse in nature (that is, the higher the score an individual obtains on X, the lower the score that individual obtains on Y). A correlation coefficient of 0 means there is no linear relationship between the two variables.

Pearson Correlation and Regression

Figure 5.4 presents some scatterplots for correlation coefficients of different magnitudes. As can be seen, the more strongly the two variables are correlated, the more closely the data points form a line. In fact, when the relationship between X and Y is perfectly linear (that is, when $r = +1.00$ or -1.00), all the data points fall exactly on a straight line. As discussed in Section 5.2, this line can be represented by an equation of the form $Y = a + bX$. When the correlation is not perfect, the statistical technique of *regression* can be used to identify a line that, though imperfect, will fit the data points better than any other line that we could try to fit to them, as determined by a statistical criterion known as *least squares*. Regression is closely related to Pearson correlation. Together, the two procedures allow us to determine the extent to which two variables approximate a linear relationship and the line that describes this relationship.

We discuss regression and the least squares criterion in detail in Section 5.5. First, however, we consider issues related to the calculation and interpretation of the Pearson correlation coefficient.

Calculation of the Pearson Correlation Coefficient

Consider Table 5.1, which presents data illustrating (a) a perfect positive linear relationship, (b) a perfect negative linear relationship, and (c) a complete lack of linear relationship between two variables, X and Y. For each example, we have converted the raw scores on X and Y to standard scores, z_X and z_Y. When a linear relationship is positive, the z scores on X will tend to be similar to the z scores on Y and they will also tend to be alike in sign. In fact, when a positive linear relationship is perfect, the standard scores on X and Y will be identical (see columns 4 and 5 of Table 5.1a). This means, for example, that an individual who has a score 1 standard deviation above the mean on X ($z_X = 1.00$) will also have a score 1 standard deviation above the mean on Y ($z_Y = 1.00$). When the z scores for a given individual are multiplied by one another, the product will be positive (unless $z_X = z_Y = 0$, in which case $z_X z_Y$ will also equal 0). When these products are summed across individuals, a relatively large positive value of $\Sigma\, z_X z_Y$ results (see column 6 of Table 5.1a). Positive, though less extreme, values of $\Sigma\, z_X z_Y$ will similarly result for a nonperfect positive linear relationship between two variables.

When a linear relationship is negative, the z scores on X will also tend to be

FIGURE 5.4 **Scatterplots for (a) a Perfect Negative Relationship, (b) a Perfect Positive Relationship, (c) No Relationship, (d) a Strong Negative Relationship, and (e) a Strong Positive Relationship (Adapted from Johnson & Liebert, 1977)**

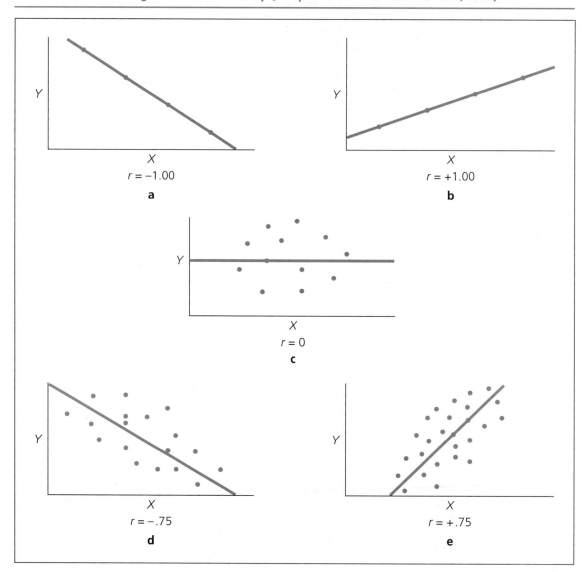

similar to the z scores on Y, but they will generally be opposite in sign. For instance, large positive z scores on X will tend to be associated with large negative z scores on Y. For a perfect negative linear relationship, the standard scores on X and Y will have different signs but will be identical in size (see columns 4 and 5 of Table 5.1b). For example, an individual who has a score 1 standard deviation *above* the mean on X ($z_X = 1.00$) will have a score 1 standard deviation *below* the mean on Y ($z_Y = -1.00$). When the z scores for a given individual are multiplied,

TABLE 5.1   **Examples of a Perfect Positive Linear Relationship, a Perfect Negative Linear Relationship, and No Linear Relationship**

(a) Perfect Positive Linear Relationship

Individual	X	Y	z_X	z_Y	$z_X z_Y$	
1	8	10	1.41	1.41	2.00	
2	7	9	.71	.71	.50	
3	6	8	.00	.00	.00	$r = \dfrac{5.00}{5} = +1.00$
4	5	7	−.71	−.71	.50	
5	4	6	−1.41	−1.41	2.00	
					$\Sigma z_X z_Y = 5.00$	

(b) Perfect Negative Linear Relationship

Individual	X	Y	z_X	z_Y	$z_X z_Y$	
1	8	6	1.41	−1.41	−2.00	
2	7	7	.71	−.71	−.50	
3	6	8	.00	.00	.00	$r = \dfrac{-5.00}{5} = -1.00$
4	5	9	−.71	.71	−.50	
5	4	10	−1.41	1.41	−2.00	
					$\Sigma z_X z_Y = -5.00$	

(c) No Linear Relationship

Individual	X	Y	z_X	z_Y	$z_X z_Y$	
1	3	7	1.12	1.20	1.34	
2	1	3	−1.12	−.98	1.10	
3	2	4	.00	−.44	.00	$r = \dfrac{.00}{5} = .00$
4	1	7	−1.12	1.20	−1.34	
5	3	3	1.12	−.98	−1.10	
					$\Sigma z_X z_Y = .00$	

the product will be negative (unless $z_X = z_Y = 0$), and when these products are summed across individuals, a relatively large negative value of $\Sigma z_X z_Y$ results (see column 6 of Table 5.1b). Negative, though less extreme, values of $\Sigma z_X z_Y$ will similarly result for a nonperfect negative linear relationship between two variables.

Finally, when there is *no* linear relationship, the z scores on X will bear no consistent relationship to the z scores on Y, in either size or sign. The product of the z scores will be positive for some individuals and negative for others (and for still others, equal to 0), and when summed, the positive and the negative $z_X z_Y$ values will cancel each other out, yielding a $\Sigma z_X z_Y$ value of 0 (see column 6 of Table 5.1c).

To summarize, when a linear relationship is positive, the sum of the products of z scores will also be positive; when a linear relationship is negative, the sum of the products of z scores will also be negative; and when there is a complete lack of a linear relationship, the sum of the products of z scores will be 0. These observations are consistent with the nature of the correlation coefficient as described earlier. We can therefore use the sum of the products of z scores as an index of the correlation between two variables. There is, however, one complication. When the correlation between two variables is nonzero, the value of the sum of z score products is influenced not only by the size of the correlation but also by the sample size (N). For a positive correlation, for example, the larger the number of observations, the greater is the sum of the products, everything else being equal. Since we want an index of correlation that is independent of N, we can divide $\Sigma z_X z_Y$ by N. Dividing by N is also advantageous because of a certain property of standard scores. Recall that when the relationship between two variables is positive and perfect, the z scores are equal. In this case, $\Sigma z_X z_Y = \Sigma z_X^2 = \Sigma z_Y^2$ because $z_X = z_Y$. It turns out that the sum of a set of squared standard scores will always equal N. Thus, when the correlation is positive and perfect, dividing $\Sigma z_X z_Y = N$ by N will always yield a value of $N/N = +1.00$. For a perfect negative relationship, $\Sigma z_X z_Y$ will always equal $-N$ such that a perfect negative correlation will always yield a $\Sigma z_X z_Y/N$ value of $-N/N = -1.00$. Nonperfect linear relationships will yield $\Sigma z_X z_Y/N$ values somewhere between the two extremes of -1.00 and $+1.00$. These properties suggest the following equation for calculating a Pearson correlation coefficient:

$$r = \frac{\Sigma z_X z_Y}{N} \qquad [5.3]$$

Although Equation 5.3 gives you an intuitive understanding of the correlation coefficient, it is not computationally efficient. We now present a formula that is easier from a computational standpoint:

$$r = \frac{SCP}{\sqrt{SS_X SS_Y}} \qquad [5.4]$$

where SS_X is the sum of squares for variable X, SS_Y is the sum of squares for variable Y, and SCP is the **sum of cross-products.** Recall that the defining formula for the sum of squares for variable X is $\Sigma (X - \overline{X})^2$. This can also be written as

$$\Sigma (X - \overline{X})(X - \overline{X})$$

In other words, we simply multiply each individual's deviation score by itself and then sum these products across individuals. Similarly, the sum of squares for variable Y can be represented as

$$\Sigma (Y - \overline{Y})(Y - \overline{Y})$$

A sum of cross-products is similar to a sum of squares but, rather than indicating the extent to which a set of scores varies from its mean, a sum of cross-products indicates the extent to which two sets of scores vary from each other, or *covary*. The sum of cross-products is defined as

$$SCP = \sum (X - \overline{X})(Y - \overline{Y}) \qquad\qquad [5.5]$$

In other words, we multiply an individual's deviation score for X by that individual's deviation score for Y and then sum these *cross-products* across individuals.* Unlike a sum of squares, a sum of cross-products can be negative because scores are not being squared in its calculation.

Phrased in terms of defining formulas, Equation 5.4 is thus equivalent to

$$r = \frac{\sum (X - \overline{X})(Y - \overline{Y})}{\sqrt{\sum (X - \overline{X})^2 \sum (Y - \overline{Y})^2}} \qquad\qquad [5.6]$$

Consider a group of ten individuals. Two variables have been measured for each: traditionalism as assessed by responses to a ten-item questionnaire (X) and ideal family size (Y). Scores on the traditionalism questionnaire can range from 0 to 10, with higher values indicating a more traditional orientation. The scores on X and Y are presented in columns 2 and 5 of Table 5.2. The first step in computing the correlation coefficient is to calculate the sum of squares for the X scores and the sum of squares for the Y scores. This has been done in columns 4 and 7 of Table 5.2; $SS_X = 80$ and $SS_Y = 60$. The sum of cross-products is computed in column 8 of Table 5.2. For instance, the first individual's deviation score on X is 4 (column 3) and her deviation score on Y is 5 (column 6). Her cross-product score

TABLE 5.2   Data and Calculation of SS_X, SS_Y, and SCP for Traditionalism and Ideal Family Size Study

Individual	X	$X - \overline{X}$	$(X - \overline{X})^2$	Y	$Y - \overline{Y}$	$(Y - \overline{Y})^2$	$(X - \overline{X})(Y - \overline{Y})$
1	9	4	16	10	5	25	20
2	7	2	4	6	1	1	2
3	5	0	0	3	−2	4	0
4	3	−2	4	6	1	1	−2
5	1	−4	16	3	−2	4	8
6	1	−4	16	3	−2	4	8
7	3	−2	4	5	0	0	0
8	7	2	4	6	1	1	2
9	5	0	0	1	−4	16	0
10	9	4	16	7	2	4	8
	$\sum X = 50$		$SS_X = 80$	$\sum Y = 50$		$SS_Y = 60$	SCP = 46
	$\overline{X} = 5.00$			$\overline{Y} = 5.00$			

---

* If we wanted to compute the variance of X or the variance of Y, we would divide the respective sum of squares by N. If we divide the sum of cross-products by N, we obtain a statistic called the *covariance*.

is thus $(4)(5) = 20$. Summing the cross-product scores for the entire data set, we find that SCP $= 46$. The correlation coefficient can now be calculated using Equation 5.6:

$$r = \frac{46}{\sqrt{(80)(60)}} = .66$$

A correlation coefficient of .66 indicates that there is some degree of a positive linear relationship between the two variables. The linear trend can be seen in Figure 5.5 (on page 140), the scatterplot of the two variables.

Computational Formula

The expression for the correlation coefficient presented in Equation 5.6 includes the defining formulas for the sum of squares for variable X, the sum of squares for variable Y, and the sum of cross-products. However, the sum of squares for variable X can also be derived using the computational formula

$$SS_X = \sum X^2 - \frac{(\sum X)^2}{N}$$

and the sum of squares for variable Y can be derived using the computational formula

$$SS_Y = \sum Y^2 - \frac{(\sum Y)^2}{N}$$

Analogous to the computational formulas for the sums of squares, the computational formula for the sum of cross-products is:

$$SCP = \sum XY - \frac{(\sum X)(\sum Y)}{N} \qquad [5.7]$$

where $\sum X$ is the sum of individuals' X scores, $\sum Y$ is the sum of individuals' Y scores, and $\sum XY$ is the sum of the products of individuals' X scores and Y scores.

Substituting the computational formulas for the sum of squares of variable X, the sum of squares of variable Y, and the sum of cross products into the general expression for a correlation coefficient, $r = SCP/\sqrt{SS_X SS_Y}$, yields the following computational formula for the correlation coefficient:

$$r = \frac{\sum XY - \frac{(\sum X)(\sum Y)}{N}}{\sqrt{\left(\sum X^2 - \frac{(\sum X)^2}{N}\right)\left(\sum Y^2 - \frac{(\sum Y)^2}{N}\right)}} \qquad [5.8]$$

This formula is both more efficient and more precise than Equation 5.6 because it requires fewer steps and presents fewer opportunities for rounding error.

The calculation of intermediate statistics for Equation 5.8 is demonstrated in Table 5.3 for the traditionalism and ideal family size data from Table 5.2. As can be seen, $\sum X^2 = 330$, $\sum Y^2 = 310$, and $\sum XY = 296$. Thus,

$$r = \frac{296 - \frac{(50)(50)}{10}}{\sqrt{\left(330 - \frac{50^2}{10}\right)\left(310 - \frac{50^2}{10}\right)}} = \frac{46}{\sqrt{(80)(60)}} = .66$$

This is the same value of r that was calculated using Equation 5.6.

TABLE 5.3   **Computational Formula Statistics for Correlation Coefficient for Traditionalism and Ideal Family Size Study**

Individual	X	Y	X^2	Y^2	XY
1	9	10	81	100	90
2	7	6	49	36	42
3	5	3	25	9	15
4	3	6	9	36	18
5	1	3	1	9	3
6	1	3	1	9	3
7	3	5	9	25	15
8	7	6	49	36	42
9	5	1	25	1	5
10	9	7	81	49	63
	$\Sigma X = 50$	$\Sigma Y = 50$	$\Sigma X^2 = 330$	$\Sigma Y^2 = 310$	$\Sigma XY = 296$

Study Exercise 5.2

A political psychologist was interested in the relationship between voters' perceptions that a candidate supported labor unions (X) and their willingness to vote for that candidate (Y). Individuals were asked to indicate on a ten-point scale the extent to which they thought the candidate supported labor unions, with higher scores indicating greater perceived support. They also indicated on a similar ten-point scale their willingness to vote for the candidate. The scores on these variables are presented for nine individuals. Compute the Pearson correlation coefficient for these data.

Individual	X	Y
1	6	5
2	7	4
3	8	4
4	8	5
5	8	3
6	9	6
7	9	7
8	7	5
9	10	6

Answer The correlation coefficient is most readily calculated using Equation 5.8. The intermediate statistics necessary for the application of this equation are computed as follows:

Individual	X	Y	X²	Y²	XY
1	6	5	36	25	30
2	7	4	49	16	28
3	8	4	64	16	32
4	8	5	64	25	40
5	8	3	64	9	24
6	9	6	81	36	54
7	9	7	81	49	63
8	7	5	49	25	35
9	10	6	100	36	60
	$\Sigma X = 72$	$\Sigma Y = 45$	$\Sigma X^2 = 588$	$\Sigma Y^2 = 237$	$\Sigma XY = 366$

Thus,

$$
\begin{aligned}
r &= \frac{\Sigma XY - \dfrac{(\Sigma X)(\Sigma Y)}{N}}{\sqrt{\left(\Sigma X^2 - \dfrac{(\Sigma X)^2}{N}\right)\left(\Sigma Y^2 - \dfrac{(\Sigma Y)^2}{N}\right)}} \\[2ex]
&= \frac{366 - \dfrac{(72)(45)}{9}}{\sqrt{\left(588 - \dfrac{72^2}{9}\right)\left(237 - \dfrac{45^2}{9}\right)}} \\[2ex]
&= \frac{6}{\sqrt{(12)(12)}} = .50
\end{aligned}
$$

A correlation coefficient of .50 indicates that some degree of a positive linear relationship exists between the two variables.

5.4 Correlation and Causation

The fact that two variables are correlated does not necessarily imply that one variable *causes* the other to vary as it does. It is entirely possible for two variables to be related to each other but have no causal relationship. In fact, there are many reasons two variables, X and Y, might be correlated. Three possibilities are that (1) X might cause Y, (2) Y might cause X, or (3) some additional variable(s) might cause both X and Y. For instance, although we might find a positive correlation between the number of hours college students spend working for pay and the number of campus organizations they belong to, it is unlikely that working *causes* students to join organizations or that membership in organizations *causes* students to work. Rather, the correlation between hours of work and group membership is probably attributable to the desire to achieve and related personality characteristics; as the desire to achieve increases, individuals might work more as they pursue their financial and occupational goals and join more organizations as a means of achieving in the social realm. As this example illustrates, one must be cautious when drawing causal inferences from correlational analyses.

Numerous examples occur in the literature where the causal relationship underlying a correlation is ambiguous. For example, there is a correlation between the amount of violence watched on television and aggressive behavior in children. Does watching violent programs on television make children more aggressive? Or do children who are more aggressive (for reasons other than television viewing habits) prefer to watch violent television shows? Could it be that older children tend to both be more aggressive and watch more television, thus producing a spurious correlation between television viewing and aggression?

Spurious relationships sometimes result in surprising correlations between variables. For example, in the U.S. population, there is a moderate positive correlation between shoe size and verbal ability: People with larger feet tend to have more verbal ability. Surely there is not a causal relationship between these variables. As it turns out, the population of Americans includes a sizeable number of children. Very young children tend to have small feet and poor verbal ability. As children grow, they acquire more verbal skills and their feet get larger. Age is a common "cause" of both increased verbal ability and increasingly larger feet, and this common cause produces a spurious correlation between shoe size and verbal ability.

5.5   Interpreting the Magnitude of a Correlation Coefficient

When students first learn about the correlation coefficient, they frequently ask what represents a "large" correlation, a "moderate" correlation, and a "small" correlation and what magnitude of correlations is typical in the behavioral science literature. These questions are difficult to answer. A correlation of .50 might be considered a "large" correlation in one context but a "small" correlation in another. For example, suppose we were studying the reliability and validity of an intelligence test and we administered the test to the same people twice, with a 3-week interval between the test administrations. Because intelligence should be stable over a 3-week period, we would expect a valid test to yield highly similar results at the two testing times. In this case, we would expect a correlation in the .80 to .90 range. If we found a correlation of .50, we would not trust the test.

By contrast, suppose we were trying to determine whether there is a relationship between parental income and children's intelligence test scores. We know that intelligence is a complicated construct that is influenced by many variables. It would be rather remarkable if a single variable, considered in isolation, showed much of a correlation with intelligence, given all the factors that impinge on intelligence. A correlation of .50 would be considered substantial when viewed in these terms.

As an interesting example of interpreting correlation coefficients, Rosenthal (1995) asked physicians to identify what they considered to be the most important medical breakthroughs of recent times, and most physicians mentioned the introduction of a drug called cyclosporine. This drug is given to patients who are to receive organ transplants to reduce the likelihood of organ rejection. A calculation of the relationship between patient survival and use of the drug revealed a correlation of .15. Although the correlation is near 0, it nevertheless means that thousands of lives are being saved.

In behavioral science research, where complex behaviors are studied, correlations of .20 to .30 (and −.20 to −.30) are often considered important. The interpretation of such magnitudes, however, is complicated by inferential considerations that we will not introduce until Chapter 14.

5.6   Regression

We noted earlier that when two variables are perfectly correlated, all the data points will fall exactly on a straight line defined by an equation of the form $Y = a + bX$. However, we will almost certainly never encounter such a situation in the behavioral sciences. In fact, correlations for the types of variables typically studied seldom exceed −.40 or +.40 and are often considerably smaller. When two variables are not perfectly correlated, the statistical technique of **regression** can be used to identify a line that, though imperfect, will fit the data points better than any other line that we could try to fit to them, as determined by a criterion known as *least squares*, to be discussed shortly. This line describes the nature of the linear relationship between the two variables.

We begin our discussion of regression by returning to the example on women's traditionalism and ideal family size. The correlation between these two variables was found to be .66, indicating some degree of a direct linear relationship. This linear trend is illustrated in Figure 5.5. The linear relationship between X and Y can be formally represented by a **regression line** that takes the general form

FIGURE 5.5   Scatterplot for Traditionalism and Ideal Family Size Study

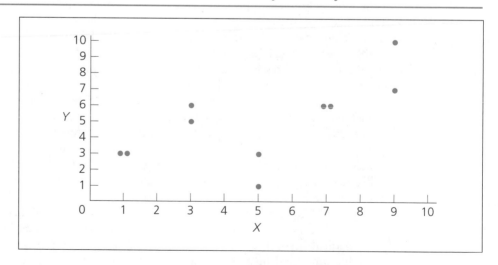

$$\hat{Y} = a + bX \qquad\qquad [5.9]$$

This equation, known as the **regression equation,** is similar to the linear model, $Y = a + bX$, but because the data points in our example do not form a straight line, different Y scores might be associated with the same X score. For instance, one of the research participants had a traditionalism (variable X) score of 3 and indicated a preference for five children (variable Y). A second individual also had a traditionalism score of 3 but perceived six children, rather than five, as the ideal family size. Because of this and related issues, the symbol $\hat{Y}$ (read "predicted Y") is used in the framework of regression to indicate the value of Y that is *predicted* to be paired with a specified value of X. This value of Y is the point on the line $a + bX$ corresponding to a person's score on X.

Calculation of the Slope and Intercept

The slope and intercept of a regression line are readily calculated from formulas. We present the formulas and then discuss the logic underlying them. For the slope, the formula is*

$$b = \frac{\text{SCP}}{\text{SS}_X} \qquad\qquad [5.10]$$

Turning again to the example on traditionalism and ideal family size, we find that

$$b = \frac{46}{80} = .58$$

since SCP was previously calculated to be 46 and SS_X was calculated to be 80. The formula for computing the intercept is

$$a = \overline{Y} - b\overline{X} \qquad\qquad [5.11]$$

Inserting the value of b from above and the values of $\overline{Y}$ and $\overline{X}$ from Table 5.2, we find that

$$a = 5.00 - (.58)(5.00) = 2.10$$

The regression equation that describes the relationship between traditionalism (X) and ideal family size (Y) is thus

$$\hat{Y} = 2.10 + .58X$$

The regression line described by this equation has been drawn in Figure 5.6. Note that this line intersects the Y axis at the value of the intercept (2.10). Furthermore, the slope of this line is such that when X increases by 1 unit, Y increases by .58 unit. Lines with arrowheads have been drawn from the data points and extending to the regression line. These illustrate visually the rationale in defining the values of the slope and intercept: The slope and intercept are defined so as to *minimize the squared vertical distances that the data points, considered collectively,*

---

* An alternative formula for the slope is $b = r(s_Y / s_X)$.

FIGURE 5.6   **Scatterplot and Regression Line for Traditionalism and Ideal Family Size Study**

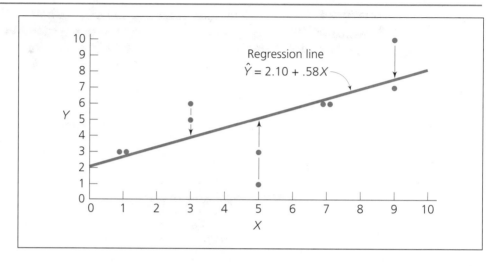

are from the regression line. This is what we mean when we say that the regression line fits the data points better than any other line that we could try to fit to them. This is accomplished in our example by the equation $\hat{Y} = 2.10 + .58X$. Any other linear equation would result in a larger sum of squared distances from the generated line.

The criterion for deriving the values of the slope and intercept is formally known as the **least squares criterion.** It can be illustrated algebraically as well as visually. Earlier we noted that a regression equation can be used to identify the value of Y that is predicted to be paired with an individual's score on X. This is done by substituting a person's score on X into the regression equation. For the traditionalism and ideal family size example, this equation is

$$\hat{Y} = 2.10 + .58X$$

Let us substitute the X scores of each of the ten research participants into this equation. These scores are listed in column 2 of Table 5.4. The score on X for the first individual is 9. The predicted Y score ($\hat{Y}$) for this individual is $2.10 + (.58)(9)$ = 7.32. The score on X for the second individual is 7. The predicted Y score for this individual is $2.10 + (.58)(7) = 6.16$. The remaining predicted scores are obtained similarly.

Columns 3 and 4 of Table 5.4 present individuals' actual Y scores and their predicted Y scores based on the regression equation. Inspection of these scores indicates that there are discrepancies between Y and $\hat{Y}$. These differences are listed in the last column of Table 5.4. Note that the discrepancies are generally rather small, reflecting the approximation of a linear relationship ($r = .66$) between X and Y. If the correlation had been smaller, there would be larger discrepancies between actual and predicted Y scores. The least squares criterion concerns itself with the *squares* of the discrepancy scores and formally defines the values of the slope and intercept so as to minimize the sum of these squares; that is, the least squares criterion defines the regression line such that $\Sigma (Y - \hat{Y})^2$ is minimized.

TABLE 5.4   Scores on *X* and *Y*, Predicted Scores, and Discrepancy Scores

Individual	X	Y	$\hat{Y}$	$Y - \hat{Y}$
1	9	10	7.32	2.68
2	7	6	6.16	−.16
3	5	3	5.00	−2.00
4	3	6	3.84	2.16
5	1	3	2.68	.32
6	1	3	2.68	.32
7	3	5	3.84	1.16
8	7	6	6.16	−.16
9	5	1	5.00	−4.00
10	9	7	7.32	−.32

The Standard Error of Estimate

Unless two variables are perfectly correlated, some degree of error will result when scores on *Y* are predicted from scores on *X* using a regression equation. The amount of error for a given individual can be represented by the discrepancy between that person's actual and predicted *Y* scores. However, the usefulness of discrepancy scores as a summary measure of predictive error is limited by the fact that the sum of the discrepancies between actual *Y* scores and *Y* scores predicted from the regression equation will always equal 0. This is demonstrated in column 5 of Table 5.4 for the traditionalism and ideal family size example.

A more useful index of predictive error is provided by the **standard error of estimate,** defined as:

$$s_{YX} = \sqrt{\frac{\sum (Y - \hat{Y})^2}{N}}$$

[5.12]

where s_{YX} represents the standard error of estimate and all other symbols are as previously presented. Equation 5.12 is conceptually similar to the defining formula for a standard deviation, which, phrased in terms of *Y*, is

$$s_Y = \sqrt{\frac{\sum (Y - \overline{Y})^2}{N}}$$

The only difference between the two equations is in the numerator. For the standard deviation, the numerator reflects the deviation of the *Y* scores from their mean. For the standard error of estimate (Equation 5.12), the numerator reflects the deviation of the *Y* scores from the predicted *Y* scores. The standard error of estimate thus represents an average error across individuals in predicting scores on *Y* from the regression equation.

Although Equation 5.12 has the advantage of being conceptually clear, a more efficient computational formula is available for calculating the standard error of estimate:

$$s_{YX} = s_Y \sqrt{1 - r^2}$$

[5.13]

Phrased in terms of the sum of squares of Y rather than the standard deviation, this is equivalent to

$$s_{YX} = \sqrt{\frac{SS_Y(1 - r^2)}{N}}$$ [5.14]

Since SS_Y was previously calculated to be 60 and r is .66, the value of the standard error of estimate for the traditionalism and ideal family size example is, according to Equation 5.14,

$$s_{YX} = \sqrt{\frac{60(1 - .66^2)}{10}} = 1.84$$

which indicates that, on average, predicted Y (ideal family size) scores deviate from actual Y scores by 1.84 units.

There are two perspectives in interpreting the standard error of estimate. First, its absolute magnitude is meaningful. In the present example, the average error in predicting scores on Y from scores on X was calculated to be 1.84 units. Given the range of possible ideal family sizes, this degree of error is not unreasonable. Second, the standard error of estimate can be compared with the standard deviation of Y. The standard deviation of Y indicates what the average error in prediction would be if one were to predict for each individual a Y score equal to the mean of Y. If X helps to predict Y, then the standard error of estimate will be smaller than the standard deviation of Y. The better the predictor X is, the smaller the standard error of estimate will be. In the present example, $SS_Y = 60$ and $N = 10$, so $s_Y = \sqrt{SS_Y/N} = \sqrt{60/10} = 2.45$. The reduction in error from 2.45 (the standard deviation of Y) to 1.84 (the standard error of estimate) in predicting scores on Y when scores on X are considered reflects the approximation of a linear relationship ($r = .66$) that exists between traditionalism and ideal family size.

Study Exercise 5.3

Compute the regression equation and the standard error of estimate for the data in Study Exercise 5.2.

Answer The sum of cross-products and the sum of squares for X were calculated in Study Exercise 5.2 to equal 6 and 12, respectively. From Equation 5.10, the slope of the regression line is

$$b = \frac{SCP}{SS_X} = \frac{6}{12} = .50$$

Since $\Sigma X = 72$, $\Sigma Y = 45$, and $N = 9$, the mean of X is $72/9 = 8.00$ and the mean of Y is $45/9 = 5.00$. From Equation 5.11, the intercept of the regression line is

$$a = \overline{Y} - b\overline{X} = 5.00 - (.50)(8.00) = 1.00$$

This yields a regression equation of $\hat{Y} = 1.00 + .50X$.

The standard error of estimate can be calculated using Equation 5.14. For $SS_Y = 12$, $r = .50$, and $N = 9$, this is found to equal

$$s_{YX} = \sqrt{\frac{SS_Y(1-r^2)}{N}}$$

$$= \sqrt{\frac{12(1-.50^2)}{9}} = 1.00$$

This indicates that, on average, predicted Y (willingness to vote for the candidate of interest) scores deviate from actual Y scores by 1.00 unit.

5.7   Additional Issues Associated with the Use of Correlation and Regression

Nonlinear Relationships

There are many ways in which two variables might be related. For instance, Figure 5.7 illustrates a *curvilinear relationship* between two variables—for example, the association between anxiety and test performance. At low levels, increasing amounts of anxiety are associated with better performance on tests (perhaps due to heightened arousal). However, as anxiety becomes more elevated, test performance starts to decrease (perhaps due to the resultant difficulty in concentrating).

Two variables might be related, but if they are related in a fashion that is *nonlinear*, Pearson correlation will not be sensitive to this. A Pearson correlation coefficient computed on the data in Figure 5.7 would be near zero even though there is clearly a strong curvilinear relationship. This is because Pearson correlation assesses only linear relationships. Nonlinearity can be effectively modeled using a technique known as *curvilinear* or *polynomial regression*, which is described in Pedhazur (1982).

FIGURE 5.7   **Scatterplot of a Curvilinear Relationship Between Two Variables**

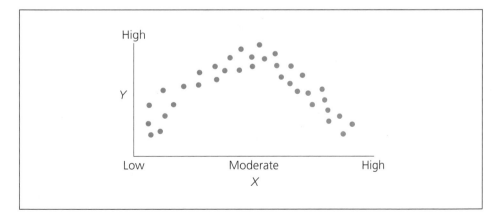

Predicting X from Y

A second consideration to bear in mind when contemplating the application of regression procedures is that *the regression equation for predicting X from Y is not the same as the regression equation for predicting Y from X.* An intuitive appreciation for this can be gained by considering an analogy to foreign-currency exchange rates. Currency in Spain is the peseta. At present, one U.S. dollar is worth 140 pesetas. One can convert dollars to pesetas using a linear model. The equation appears as follows:

Pesetas = 0 + 140 Dollars

Using the above equation, 10 dollars is equivalent to 1,400 pesetas. For every additional dollar that something costs, the number of pesetas it costs increases by 140. If we reverse the X and Y variables, we want to convert pesetas to dollars, rather than vice versa. The linear equation is:

Dollars = 0 + .00714 Pesetas

One peseta is worth about seven-tenths of a penny. Ten pesetas is worth .0714 of a dollar, or just over seven cents. For every additional peseta that something costs, the number of dollars it costs increases by .00714. The slope of the linear equation differs, depending on which is the X variable and which is the Y variable.

We earlier calculated the following regression equation to describe the relationship between traditionalism (variable X) and ideal family size (variable Y):

Ideal family size = 2.10 + .58 traditionalism

If we reverse the variables such that ideal family size is the X variable and traditionalism is the Y variable, the regression equation would be

Traditionalism = 1.17 + .77 ideal family size

Using traditionalism to predict how many children an individual thinks is ideal, for every 1 unit that traditionalism increases, we predict the ideal family size will increase by .58 child. In terms of predicting how traditional someone is from the number of children she desires, for every additional child that someone desires, traditionalism is predicted to increase .77 unit on the traditionalism scale. In contrast to the slope and intercept, the correlation coefficient is the same in both analyses because it reflects the extent to which the two variables, considered together, approximate a linear relationship.

From a *statistical* perspective, the designation of one variable as X and one variable as Y is arbitrary. It is as easy to derive the line for predicting the first variable from the second as it is to derive the line for predicting the second variable from the first. We merely reverse which variable is labeled X and which variable is labeled Y and apply the usual formulas for the slope and intercept. From a *conceptual* perspective, however, the tradition is to designate the independent variable as X and the dependent variable as Y. The choice of the labels therefore has important implications. A personnel director in a large company would obviously be more interested in predicting job success from an individual's score on an aptitude test than the reverse. The use of regression presupposes an underlying rationale for making predictions about variable Y from variable X. If interest is merely in whether a given variable is (linearly) *related* to another variable, Pearson correlation is applied.

Restricted Range

Another issue of note is the effect on the observed correlation of examining only a portion of the range of a variable. Depending on the particular circumstances, the magnitude of the correlation when a limited portion of this range is considered might be either less or greater than if the range had not been so restricted. Suppose, for instance, that we were interested in the correlation between anxiety and test performance. Research has shown that these variables generally tend to be curvilinearly related. However, suppose the sample were selected such that the research participants were all moderately to highly anxious. Figure 5.7 suggests that the correlation between anxiety and test performance for this range of anxiety would be a relatively strong negative one, even though the linear approximation across the whole range of motivation scores is extremely poor. In this case, the restricted range on anxiety yields an increased correlation between anxiety and test performance.

On the other hand, if two variables are linearly related, then restricting the range of one of them will often reduce the magnitude of the observed correlation coefficient. For instance, Figure 5.8 illustrates a strong positive correlation between two variables. If only individuals who have moderate to high scores on X had been included in the study, the correlation between X and Y would be substantially reduced. In general, the effect of restricting the range of a variable that is linearly related with another variable is to reduce the magnitude of the observed correlation. This might explain the relatively weak correlations that have been

FIGURE 5.8   **The Effect of a Restricted Range on the Correlation Coefficient When Two Variables Are Linearly Related (Based on Howell, 1985)**

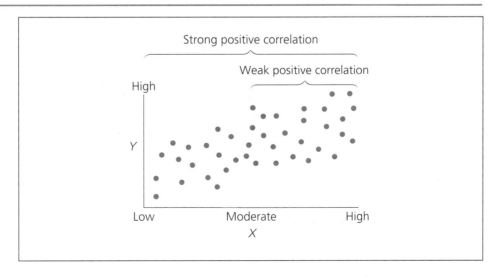

reported between Scholastic Aptitude Test (SAT) scores and college grade point averages. Typically, only students who are interested in pursuing a college education take the SAT. To the extent that this group of students is more academically inclined than the general population of high school students (and, thus, more likely to obtain high SAT scores), the correlations that are observed between SAT scores and grade point averages are based on a restricted range of SAT scores. If *all* graduating high school students took the SAT and went on to college, then the correlation between SAT scores and grade point averages might reflect a strong positive association.

Because the effect of a restricted range on the Pearson correlation coefficient can never be known with certainty, we must be careful to select our sample such that the entire range of values of interest is studied. We must also be careful not to extend our interpretation of correlational results *outside* the range of the original data set. The conclusions drawn from a correlational analysis apply only to the range of the variables on which the correlation was based. This is because the Pearson correlation coefficient represents the extent to which two variables approximate a linear relationship *for the range of the variables included in its calculation*. For instance, it would be meaningless to try to generalize about the relationship between anxiety and task performance at low to moderate levels of anxiety from the relationship between these variables at moderate to high levels of anxiety.

As you might expect, a similar caution holds for regression analysis: Prediction of Y from X is meaningful for the range of X values that formed the basis for the calculation of the regression equation. Suppose, for instance, that a researcher is interested in predicting the vividness with which people are able to recall their dreams from the length of time they sleep. One hundred volunteers spend the night in a sleep laboratory. In the morning, the vividness with which they are able to recall their night's dreams is measured on a scale of 1 to 9, where higher scores indicate more vivid dream recall. The range of sleep duration for the participants in this study is from 5.3 to 8.6 hours, and the regression equation for predicting the vividness of dream recall from the amount of sleep is found to be $\hat{Y} = 8.71 - .68X$. This equation indicates, for example, that the vividness of dream recall for an individual who sleeps 6 hours is predicted to be $8.71 - (.68)(6) = 4.63$, and the vividness of dream recall for an individual who sleeps 8 hours is predicted to be $8.71 - (.68)(8) = 3.27$. This equation also indicates that the vividness of dream recall for an individual who fails to fall asleep is predicted to be a very vivid $8.71 - (0)(.68) = 8.71$ even though a person cannot dream if he or she is not sleeping! This inconsistency reflects that the regression equation was established on an X value range of 5.3 to 8.6 and does not apply to the relationship between sleep and vividness of dream recall for an X score of 0.

The Regression Equation for Standardized Scores

All of the examples we have considered with regression have used raw scores. Sometimes researchers are interested in predicting standard scores rather than raw scores. For example, a teacher may not be interested in predicting the absolute performance of a student but rather the relative performance of the student compared with other students in the school. In this case, the analyst would first standardize the X and Y scores and then apply the usual formulas (Equations 5.10 and 5.11) to calculate the slope and the intercept. When standard scores are analyzed in this manner, the intercept will always equal 0, and the slope will always equal the correlation coefficient. Both of these results can be proved with simple algebra, although we do not do so here. The behavior of the slope for standardized scores provides us with yet another perspective for interpreting a correlation coefficient: A correlation coefficient tells us the number of standard scores that one variable is predicted to change given a change of one standard score in the other variable, everything else being equal. Stated another way, a correlation coefficient conveys the number of standard deviations that one score is predicted to change given a 1-standard deviation change in the other variable. The correlation between traditionalism and ideal family size is .66. For every standard deviation (of traditionalism) that traditionalism changes, ideal family size is predicted to change .66 standard deviation (of ideal family size).

Outliers

The magnitude and sometimes even the sign of a correlation coefficient can be influenced by outliers, and it is important for the analyst to consider this possibility. Consider the nine scores on the scatterplot in Figure 5.9a. It is evident that there is no linear trend in these scores and the correlation coefficient is, in fact, 0. Now suppose that a single outlier case is added, as in Figure 5.9b. The correlation between the two variables with this one case added is .84. The addition of a single case changed a 0 correlation to a .84 correlation! This example is extreme but illustrates that even a single outlier can yield a correlation that masks the basic trend in the data.

Just as outliers can turn a weak correlation into a strong correlation, so outliers can turn a strong correlation into a weak one. It is important for the analyst

FIGURE 5.9   **Scatterplots Illustrating an Outlier Effect**

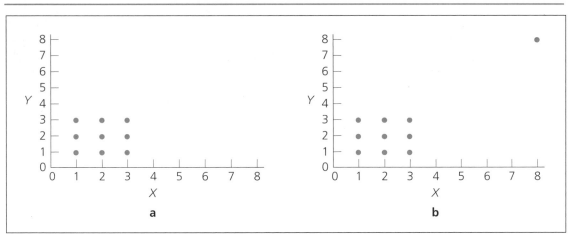

**Applications
to the Analysis
of a Social Problem**

Correlation and regression analysis, like the statistics we discussed in previous chapters, are usually pursued in an inferential context. However, all of the statistics we have considered thus far are descriptive, not inferential. We will begin our discussion of inferential statistics in Chapter 7, and we will cover correlation and regression using inferential methods in Chapter 14. It is important to recognize that the statements given below are purely descriptive of the sample and do not permit inferences to the broader population of individuals who our sample represents.

Based on the analyses in Chapter 4, we know that many of the teens in the parent–teen study described in Chapter 1 are sexually active. It is also evident that there is variability in the amount of sexual activity: Some teens are more sexually active than others. One question of interest is whether the amount of sexual activity is associated with how conscientious the teen is about using birth control. Are teens who engage in relatively more sex more careful about not getting (someone) pregnant and therefore using birth control more consistently? Or are teens who engage in relatively little sex just as (un)likely to use birth control as teens who engage in more sexual activity?

Recall that teens were asked the number of times they had engaged in sexual intercourse dur-

ing the past 6 months. Those teens who indicated sexual activity were also asked the following:

> When you had intercourse during the past 6 months, how much of the time, if ever, did you or your partner use any birth control (for example, the pill, a condom, etc.) so that you wouldn't get pregnant? Circle the number to the right of the one statement that best describes the use of birth control by you and your partner(s) during the past 6 months.
>
> | | |
> |---|---|
> | Always (100% of the time) | 6 |
> | Most of the time (71% to 99% of the time) | 5 |
> | Often (51% to 70% of the time) | 4 |
> | Sometimes (31% to 50% of the time) | 3 |
> | Occasionally (1% to 30% of the time) | 2 |
> | Never (0% of the time) | 1 |

Considerable research suggests that this measure is a reasonably valid indicator of the consistency with which an individual uses birth control. For example, the measure has been found to predict unintended pregnancy in both adult and teen populations. However, psychometric studies have also indicated that the percentages associated with the verbal labels should not be interpreted literally,

to explore the potential effect of outliers on the correlation coefficient (as well as on regression statistics, if they are the focus of analysis). Outliers are most likely to complicate interpretation when sample sizes are small (which is the case in Figure 5.9).

Statisticians have developed numerous procedures for identifying outliers. One way to determine whether a given case is an outlier is to compute the correlation for the total sample and then to compare this with the correlation when the case in question is omitted from the analysis (i.e., using $N - 1$ cases). If the correlation changes relatively little, then the case is not an outlier. However, if the deletion of the case dramatically changes the correlation, then the case *is* deemed an outlier. Statisticians have developed more sophisticated approaches to outlier identification than the "delete a case" approach described here. These are discussed in Norusis (1992).

in a ratio scale sense. The mental processes used by teens in generating responses to the question do not strictly follow the percentages listed (e.g., a teen who has, in fact, used a birth control method 50% of the time may not indicate a response of "3" on the scale, in part because the teen's *perception* of use consistency may not map perfectly onto actual use consistency, although the approximation is reasonably close). Furthermore, even though the percentages that accompany the verbal descriptors suggest a scale that is not equal interval, psychometric studies suggest that the heuristics used by teens for generating responses produce measures that reasonably approximate interval level characteristics.

The frequency histograms for the use consistency measure and the number of instances of sexual intercourse (using only those teens who were sexually active) are presented in Figures 5.10 and 5.11, with a normal curve superimposed on each. The use consistency measure shows negative skewness, and the number of instances of sexual intercourse shows positive skewness.

Although one might use several methods to analyze these data, we focus on how the data are analyzed using Pearson correlation to gain per-

spective on our central question: Are higher levels of sexual activity associated with greater consistency in birth control use? Table 5.5 presents computer output for the correlation analysis. The results are organized on the printout in the form of a **correlation matrix**. A *matrix* is a set of numbers organized into rows and columns. In this case, the two variables (use consistency, labeled CONSIST6, and instances of sexual intercourse, labeled NUMBER6) are listed in the two rows and the same two variables are listed in the two columns. The number in the matrix where a row intersects a column is a correlation coefficient. For example, the correlation between the variable in the first row and the variable in the first column is 1.00. This is not surprising because it is the correlation of use consistency with itself, which must be 1.00. Table 5.5 reveals that the correlation between the number of instances of sexual intercourse and use consistency is near 0—namely −.09. Figure 5.12 presents a scatterplot for the two variables, with the least squares regression line indicated. The scatterplot shows minimal linear trend in the data. We also conducted formal outlier analyses to determine whether the correlation was unduly influenced by
(continued)

FIGURE 5.10   Frequency Histogram of Birth Control Use Consistency

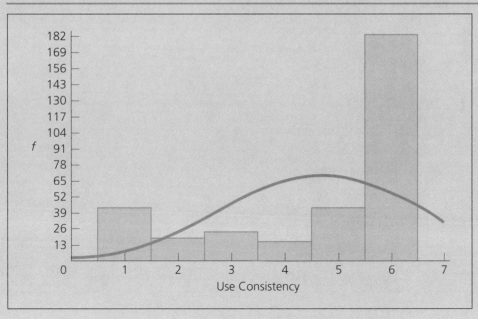

a few aberrant cases. No outliers were identified that altered the observed correlation coefficient in any meaningful way. From the perspective of Pearson correlation, there appears to be little linear relationship between how sexually active the teens in the sample were, considered as a whole, and how diligent they were about protecting themselves from an unintended pregnancy.

FIGURE 5.11 Grouped Frequency Histogram of Sexual Intercourse

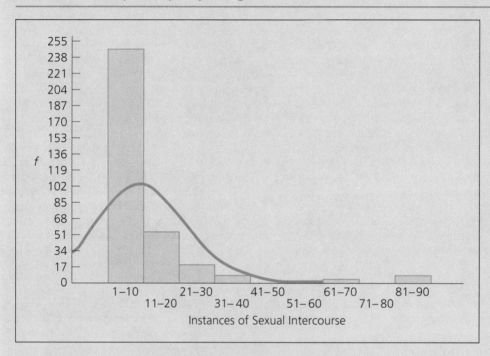

TABLE 5.5   **Computer Output for Pearson Correlation Between Use Consistency and Sexual Intercourse**

CONSIST6	Birth control use consistency in past 6 months
NUMBER6	Number of instances of intercourse in past 6 months

```
                        - -  Correlation Coefficients   - -

                 CONSIST6    NUMBER6

CONSIST6          1.0000     -.0857
NUMBER6           -.0857      1.0000
```

FIGURE 5.12   Scatterplot of Use Consistency and Sexual Intercourse

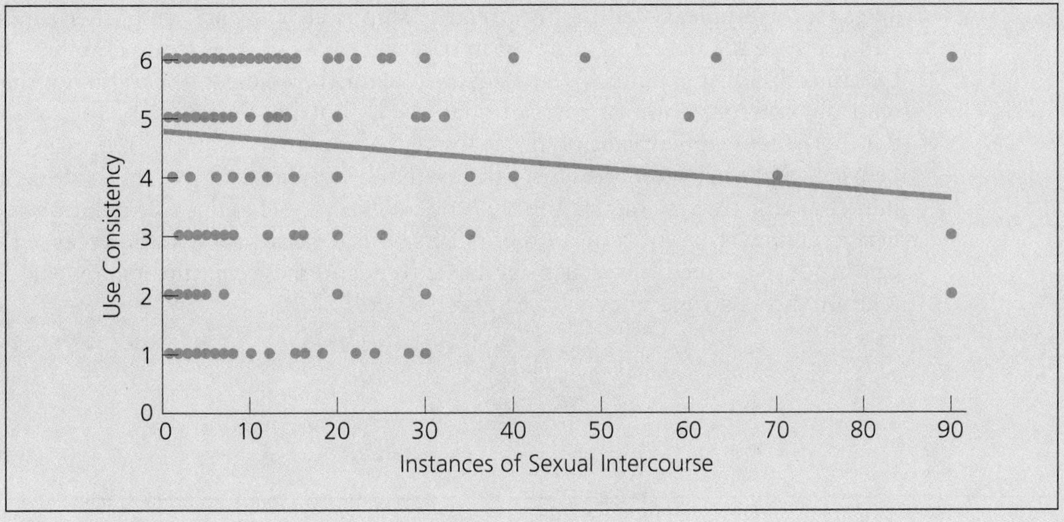

Summary

Pearson correlation is based on the linear model and indexes the extent of linear relationship between two quantitative variables that are measured on a level that at least approximates interval characteristics. The correlation coefficient can range from −1.00 through 0 to +1.00. The magnitude of the correlation coefficient indicates the degree to which the variables approximate a linear relationship, and the sign of the correlation coefficient indicates whether this relationship is direct or inverse. The fact that two variables are correlated does not necessarily imply that one variable causes the other to vary as it does, because both variables might be caused by some additional variable(s).

The least squares criterion can be used to identify the slope and intercept of the line that fits the data points better than any other line that we could try to fit to them. This line, known as the regression line, describes the nature of the linear relationship between the two variables and can be used to identify the value of Y that is predicted to be paired with an individual's score on X. This is accomplished by substituting a person's score on X into the regression equation. A measure of the average error across individuals in predicting scores on Y in this manner is provided by the standard error of estimate.

Several issues are associated with the use of correlation and regression. First, Pearson correlation assesses only linear relationships. If two variables are related in a nonlinear fashion, this relationship will not be reflected in the correlation coefficient. A second consideration is that the regression equation for predicting X from Y is not the same as the regression equation for predicting Y from X. A third issue concerns the effect of a restricted range on the observed correlation. Sometimes a restricted range will increase the observed correlation between two variables (for example, when the true relationship is curvilinear), and other times it will decrease the observed correlation between two variables (for example, when the true relationship tends toward linearity). Fourth, we must be careful not to extend our interpretation of correlational results outside the range of the original data set, and a similar caution holds for regression analysis. Fifth, regression can be used to predict standard scores rather than raw scores. When this is done, the intercept will always equal 0 and the slope will always equal the correlation coefficient. A final issue concerns outlier effects. Outliers are problematic because they can produce correlations that mask basic trends in the data, turning a weak correlation into a strong one, or vice versa.

Exercises

Answers to asterisked (*) exercises appear at the back of the book. Answers to exercises marked with two asterisks are also worked out step by step in the Study Guide.

1. Draw a scatterplot for the following data:

Individual	X	Y
1	10	9
2	8	7
3	6	5
4	9	8
5	10	8
6	8	8
7	5	6

*2. What information is conveyed by the slope of a line?

3. What information is conveyed by the intercept of a line?

4. What is the general form of the linear model?

*5. Given a perfect linear relationship between two variables, X and Y, and a slope of 3.00, by how many units will Y change if X changes by 1 unit? If X changes by 2 units? If X changes by 7 units?

*6. What information is conveyed by the magnitude of a correlation coefficient?

7. What information is conveyed by the sign of a correlation coefficient?

*8. For each pair of correlation coefficients, indicate which coefficient represents a better approximation to a linear relationship:
   a. +.37 or +.18
   b. −.37 or −.18
   c. +.52 or −.76
   d. 0 or +.26
   e. 0 or −.44
   f. +.61 or +1.07

*9. Draw a scatterplot for two variables that are negatively correlated.

10. Draw a scatterplot for two variables that are positively correlated.

*11. Give an example of two variables that are probably positively correlated.

12. Give an example of two variables that are probably negatively correlated.

**13. Compute the Pearson correlation coefficient for the following data:

Individual	X	Y
1	3	7
2	8	9
3	3	3
4	2	8
5	6	8
6	6	9
7	8	6
8	5	4
9	7	2
10	2	4

14. Compute the Pearson correlation coefficient for the following data:

Individual	X	Y
1	6	10
2	7	9
3	7	7
4	7	8
5	8	8
6	8	8
7	9	8
8	9	9
9	9	7
10	10	6

15. Why must one be cautious when drawing causal inferences from correlational analyses?

*16. Give an example of two variables that are probably correlated but not causally related. What additional factor(s) might account for the correlation between these two variables?

*17. What is the general form of the regression equation? How does this differ from the linear model?

*18. How does the least squares criterion define the values of the slope and intercept of the regression line?

**19. Compute the regression equation that describes the relationship between X and Y for the data in Exercise 13.

*20. What is the predicted Y score for individual 1 in Exercise 13? What is the predicted Y score for individual 5? What is the predicted Y score for individual 10?

21. Compute the regression equation that describes the relationship between X and Y for the data in Exercise 14.

22. What is the predicted Y score for individual 2 in Exercise 14? What is the predicted Y score for individual 4? What is the predicted Y score for individual 6?

23. What information is conveyed by the standard error of estimate?

*24. Compute the standard error of estimate for the data in Exercise 13.

25. Compute the standard error of estimate for the data in Exercise 14.

*26. What are the effects on the correlation coefficient of restricting the range of one of the variables?

27. When regression is used with standard scores, what will the intercept equal? What will the slope equal? What implication does this have for the interpretation of a correlation coefficient?

28. Why are outliers a problem for correlation analysis?

Multiple-Choice Questions

29. Which of the following correlation coefficients represents the strongest approximation to a linear relationship?
    a. −.60 c. .10
    b. 0   d. .50

*30. The regression equation for predicting X from Y is the same as the regression equation for predicting Y from X.
    a. true b. false

31. According to the linear equation $Y = 2.16 + 4.37X$, for every 1 unit that X changes, Y will change _____ units.
    a. 2.16 c. 6.53
    b. 4.37 d. cannot be determined

*32. Outliers are most likely to raise interpretational complexities for the correlation coefficient when sample sizes are small.
a. true   b. false

33. A correlation of .60 between height and weight is an example of a(n) _____ relationship.
a. direct   c. negative
b. inverse   d. curvilinear

*34. Given that $r = -.83$, as scores on X increase, scores on Y
a. increase   c. remain the same
b. decrease   d. cannot be determined

*35. If SCP $= 0$, the correlation coefficient will be
a. negative   c. 0
b. positive   d. cannot be determined

36. A graph illustrating the linear relationship between two variables, X and Y, is called a
a. bar graph
b. boxplot
c. scatterplot
d. three-dimensional histogram

*37. Correlations of .20 to .30 (and −.20 to −.30) are often considered important in behavioral science research.
a. true   b. false

Refer to the following data set to do Exercises 38–41:

X	Y
5	4
4	2
3	3

38. What is the sum of cross-products?
a. 0   c. 2.00
b. 1.00   d. 3.00

*39. What is the correlation between X and Y?
a. 0   c. .50
b. .25   d. .75

40. What is the slope of the regression line?
a. 0   c. .50
b. .25   d. .75

*41. What is the intercept of the regression line?
a. 1.00   c. 5.00
b. 2.50   d. 5.50

42. Pearson correlation is sensitive to both linear and nonlinear relationships.
a. true   b. false

CHAPTER 6

Probability

The concept of **probability** forms the foundation of inferential statistics as well as several descriptive statistical methods. The purpose of this chapter is to provide a background in elementary probability theory that will serve as the basis for your understanding a wide array of statistical problems.

When we flip a coin, we get one of two possible outcomes: We can get a head, or we can get a tail. In probability theory, the act of flipping the coin is called a **trial,** and each unique outcome is called an **event.** There are two different conceptualizations of probability. The first is the classical viewpoint, which is based on logical analysis. The probability of an event, A, is formally defined as the number of observations favoring event A divided by the total number of possible observations:

$$p(A) = \frac{\text{number of observations favoring event } A}{\text{total number of possible observations}} \qquad [6.1]$$

In the coin flipping example, there are two possible observations. On any given trial, one observation favors a head and one observation favors a tail. The probability of obtaining a head is ½, or .50. Suppose you randomly select an individual from a population that consists of 60 men and 40 women. What is the probability you will select a man? There are 60 observations favoring the event "man," and a total of 100 observations. According to Equation 6.1, the probability of selecting a man is $60/100 = .60$.

The second conceptualization of probability focuses on the long run. According to this interpretation, if we flip a coin a large number of times, then over the long run, the proportion of heads relative to the total number of observations should approach .50. Over an infinite number of trials, the proportion of heads will be .50.

We develop additional concepts in probability theory using the following example. Suppose an investigator classifies all married men in a community in terms of two variables: (1) satisfaction with one's marriage and (2) satisfaction with one's job. The first variable has two values: men who are satisfied with their marriages versus men who are dissatisfied with their marriages. The second variable also has two values: men who are satisfied with their jobs versus men who are dissatisfied with their jobs. These two variables can be combined to classify each man into one of four groups: (1) those who are satisfied with both their marriages and their jobs, (2) those who are satisfied with their jobs but not their marriages, (3) those who are satisfied with their marriages but not their jobs, and (4) those who are dissatisfied with both their marriages and their jobs.

The results of this study can be displayed in a **contingency table** (also called a *frequency* or *crosstabulation table*), as in Table 6.1. Each unique combination of variables in a contingency table is referred to as a **cell**. The entries within the cells represent the number of individuals who are characterized by the corresponding values of the variables. In our example, 156 men are satisfied with both their marriages and their jobs, 54 men are satisfied with their jobs but not their marriages, 52 men are satisfied with their marriages but not their jobs, and 148 men are dissatisfied with both their marriages and their jobs. The numbers in the last column and the bottom row of Table 6.1 indicate how many individuals have each separate characteristic. For example, collapsing across marital satisfaction, we find that 210 men are satisfied and 200 men are dissatisfied with their jobs. Collapsing across job satisfaction, we find that 208 men are satisfied and 202 men are dissatisfied with their marriages. These frequencies are called **marginal frequencies** and are the sum of the frequencies in the corresponding row (for example, $156 + 54 = 210$) or column (for example, $156 + 52 = 208$). Finally, the num-

TABLE 6.1   Contingency Table for Marital and Job Satisfaction Study

Job Satisfaction	Marital Satisfaction		Totals
	Satisfied with marriage	Dissatisfied with marriage	
Satisfied with job	156	54	210
Dissatisfied with job	52	148	200
Totals	208	202	410

ber in the lower right-hand corner is the total number of observations in the study (in this case, 410).

6.1 Probabilities of Simple Events

In the language of probability theory, the variable of job satisfaction has two possible outcomes: (1) being satisfied with one's job or (2) being dissatisfied with one's job. These outcomes are said to be **mutually exclusive** because it is impossible for both outcomes to occur for a given individual: If a person is classified as being satisfied with his job, he cannot also be classified as being dissatisfied with his job. The variable of marital satisfaction also has two possible outcomes: (1) being satisfied with one's marriage or (2) being dissatisfied with one's marriage. These outcomes are also mutually exclusive.

If we were to select a married man at random from the community in the preceding example, what would the probability be that he is satisfied with his job? According to Equation 6.1, it is the number of men who are satisfied with their jobs (210) divided by the total number of men (410):

$$p(\text{satisfied with job}) = \frac{210}{410} = .512$$

The probability that a man is dissatisfied with his job is the number of men who are dissatisfied with their jobs (200) divided by the total number of men (410):

$$p(\text{dissatisfied with job}) = \frac{200}{410} = .488$$

Note that the sum of these two probabilities is 1.00 (.512 + .488 = 1.00). Given a set of outcomes that are mutually exclusive and **exhaustive** (that is, includes all outcomes that could occur), the sum of the probabilities of the outcomes will always equal 1.00. As noted in Chapter 2, such a set of outcomes and their associated probabilities are known as a **probability distribution.**

Study Exercise 6.1

For the data in Table 6.1, what is the probability of randomly selecting a man who is satisfied with his marriage? A man who is dissatisfied with his marriage?

Answer   From Equation 6.1, the probability that a man is satisfied with his marriage is

$$p(\text{satisfied with marriage}) = \frac{208}{410} = .507$$

and the probability that a man is dissatisfied with his marriage is

$$p(\text{dissatisfied with marriage}) = \frac{202}{410} = .493$$

Thus, for these data, the likelihood that a man is satisfied with his marriage is roughly equal to the likelihood that he is dissatisfied with his marriage.

6.2 Conditional Probabilities

A useful concept for analyzing the data in Table 6.1 is **conditional probability**. A conditional probability indicates the likelihood that an event will occur given that some other event occurs. In the context of Table 6.1, this issue might be expressed as follows: Given that a person is satisfied with his job, what is the probability that he is satisfied with his marriage? A total of 210 men are satisfied with their jobs. Of these 210 men, 156 are satisfied with their marriages. The probability that a man is satisfied with his marriage given that he is satisfied with his job is thus

$$p\left(\text{satisfied with marriage | satisfied with job}\right) = \frac{156}{210} = .743$$

The general symbolic form for a conditional probability is

$$p(A \mid B) = \frac{\text{number of observations favoring } both \text{ event } A \text{ } and \text{ event } B}{\text{number of observations favoring event } B} \qquad [6.2]$$

where $p(A \mid B)$ is read "the probability of event A, given event B." The conditional probability in this example was computed by considering only those individuals who are satisfied with their jobs (event B) and then deriving the proportion of these individuals who are also satisfied with their marriages (event A).

Conditional probabilities are useful for understanding relationships between events. This can be illustrated by considering the concept of **independence**. An event A is said to be *independent* of an event B if $p(A) = p(A \mid B)$—that is, if the occurrence of event A is unrelated to the occurrence of event B.* Consider the two events "being satisfied with one's marriage" and "being satisfied with one's job." The probability of the first event was computed in Study Exercise 6.1 as .507, and the conditional probability of being satisfied with one's marriage given that one is satisfied with his job was computed above as .743. It is clear that being satisfied with one's marriage is not independent of being satisfied with one's job for the married men in this town. The occurrence of event A in this case *is* related to the occurrence of event B because its probability is substantially raised (from .507 to .743) given event B.

It is important to realize that even though two events are related (nonindependent), this does not necessarily mean that one causes the other. We will return to this issue in Chapter 9.

---

* If two events are independent, not only does $p(A) = p(A \mid B)$ but the reverse is also true; that is, $p(B) = p(B \mid A)$.

Study Exercise 6.2

For the data in Table 6.1, compute the conditional probability of being dissatisfied with one's marriage given being dissatisfied with one's job.

Answer   A total of 200 men are dissatisfied with their jobs, of whom 148 are also dissatisfied with their marriages. Thus, from Equation 6.2, the conditional probability of being dissatisfied with one's marriage given being dissatisfied with one's job is

$$p\left(\text{dissatisfied with marriage} \mid \text{dissatisfied with job}\right) = \frac{148}{200} = .740$$

6.3   Joint Probabilities

Another type of probability an investigator might be interested in is **joint probability.** This refers to the likelihood of observing each of two events. A joint probability is represented as

$$p(A , B) = \frac{\text{number of observations favoring } \textit{both} \text{ event } A \textit{ and } \text{event } B}{\text{total number of observations}} \qquad [6.3]$$

where $p(A, B)$ stands for the probability of *both* event A *and* event B occurring. For the data in Table 6.1, we might want to determine the probability that an individual is satisfied with his marriage (event A) *and* satisfied with his job (event B). Since there are 156 such men out of the total of 410 men in the study, the joint probability of being satisfied with both one's marriage and one's job is (using Equation 6.3)

$$p\left(\text{satisfied with marriage, satisfied with job}\right) = \frac{156}{410} = .380$$

Study Exercise 6.3

For the data in Table 6.1, compute the joint probability of being dissatisfied with both one's marriage and one's job.

Answer   There are 148 men who are dissatisfied with both their marriages and their jobs out of a total of 410 men. According to Equation 6.3, this yields a joint probability of being dissatisfied with both one's marriage and one's job of

$$p\left(\text{dissatisfied with marriage, dissatisfied with job}\right) = \frac{148}{410} = .361$$

6.4 Adding Probabilities

A fourth probability of interest focuses on the likelihood of observing *at least one* of two events. For the data in Table 6.1, we might be interested in specifying the probability that a man is satisfied with his marriage or satisfied with his job, or both. One way to determine this is as follows: First, determine the probability that a man is satisfied with his marriage (previously calculated to be .507) and the probability that a man is satisfied with his job (calculated to be .512). We cannot merely add these two values because their sum does *not* represent the probability that a person is satisfied with at least his marriage or satisfied with at least his job; we would have counted men who are satisfied with both their marriages and their jobs twice. So we subtract out this duplicate probability (previously calculated to be .380). This yields a probability that a man is satisfied with his marriage or satisfied with his job, or both, of

$$p(\text{satisfied with marriage or satisfied with job}) = .507 + .512 - .380$$
$$= .639$$

In general notation, the probability of observing *at least one* of event A and event B is

$$p(A \text{ or } B) = p(A) + p(B) - p(A, B) \qquad [6.4]$$

Study Exercise 6.4

For the data in Table 6.1, compute the probability of being dissatisfied with one's marriage or being dissatisfied with one's job, or both.

Answer The proportion of men who are dissatisfied with their marriages is .493, the proportion of men who are dissatisfied with their jobs is .488, and the proportion of men who are dissatisfied with both their marriages and their jobs is .361. Applying Equation 6.4, the probability of being dissatisfied with one's marriage or being dissatisfied with one's job, or both, is

$$p(\text{dissatisfied with marriage or dissatisfied with job}) = .493 + .488 - .361$$
$$= .620$$

6.5 Relationships Among Probabilities

We have discussed the probability of a simple event, conditional probability, and joint probability. Each of these provides useful information. The probability of a simple event refers to the likelihood that a given event will occur. A conditional probability tells us the likelihood that a given event will occur given some other event. Finally, a joint probability indicates the likelihood of two events occurring together. Let us explore the insights such concepts can give when considered together.

Suppose you are a counselor and are discussing with a couple their decision whether to have a child. You are exploring the advantages and disadvantages. You know that the incidence of a certain type of birth defect is generally relatively low. In fact, the probability of the defect occurring is estimated to be .001. However, the offspring of couples who are in racial group X are particularly susceptible to the defect. The conditional probability that a baby will have the birth defect given that its parents are in racial group X is estimated to be .020. In this instance, the occurrence of the birth defect is not independent of the race of the parents and your advice to the couple might differ depending on their race.

Let us consider another example. As you might have suspected, there are certain mathematical relationships among probabilities of simple events, conditional probabilities, and joint probabilities. One important relationship is characterized by the following equation:

$$p(A, B) = p(B)p(A \mid B) \qquad [6.5]$$

In our earlier example, we were concerned with men's satisfaction with their marriages and jobs. Let event A be "satisfied with one's marriage" and event B be "satisfied with one's job." Suppose we were particularly interested in men who are satisfied with both their marriages and their jobs. Equation 6.5 states that the probability that someone is satisfied with *both* his marriage *and* his job [$p(A, B)$] is equal to the probability that he is satisfied with his job [$p(B)$] weighted by the conditional probability that being satisfied with his job implies that he is also satisfied with his marriage [$p(A \mid B)$]. This can be demonstrated by substituting the relevant probabilities from our earlier calculations into Equation 6.5. In Section 6.3 we found that $p(A, B) = .380$, in Section 6.1 we found that $p(B) = .512$, and in Section 6.2 we found that $p(A \mid B) = .743$. Indeed, consistent with Equation 6.5,

$$.380 = (.512)(.743)$$

The fact that $p(A, B)$ is low implies that men in our study are generally not satisfied with both their marriages and their jobs. The probability $p(B)$ is moderate, indicating that only about half of the men are satisfied with their jobs. The conditional probability, $p(A \mid B)$, is high, which suggests that if the incidence of job satisfaction were to increase, the incidence of marital satisfaction might also increase. Equation 6.5 can thus provide insight into the nature of joint probabilities.

It is possible to derive other formulas that relate different probabilities to one another. An extended discussion of these is beyond the scope of this book. However, for the sake of completeness, we list the formulas that are most frequently used in the behavioral sciences:

1. In Equation 6.5, it was noted that $p(A, B) = p(B)p(A \mid B)$. If A and B are independent, then $p(A \mid B) = p(A)$. It follows by simple substitution and rearrangement of terms that *when A and B are independent,*

$$p(A, B) = p(A)p(B) \qquad [6.6]$$

2. In Equation 6.4, it was noted that $p(A \text{ or } B) = p(A) + p(B) - p(A, B)$. If A and B are mutually exclusive, then $p(A, B)$ will equal 0. It follows that *when A and B are mutually exclusive,*

$$p(A \text{ or } B) = p(A) + p(B) \qquad [6.7]$$

3. Since $p(A, B) = p(B)p(A \mid B)$ (Equation 6.5), it is possible to express a conditional probability in terms of the following formula, often referred to as *Bayes' theorem*:

$$p(A \mid B) = \frac{p(A, B)}{p(B)} \qquad [6.8]$$

Equations 6.1–6.8 have wide applicability to a number of problems. As an example, suppose a student applied to two medical schools, A and B, and suppose further that she assessed the probability that she would get into school A as .200 and the probability that she would get into school B as .300. What is the probability that the student will be admitted to *at least one* of the two schools? Let A be "being admitted to school A" and B be "being admitted to school B." According to Equation 6.4, the probability of observing at least one of event A and event B is

$$p(A \text{ or } B) = p(A) + p(B) - p(A, B)$$

Assuming A and B are independent, we can use Equation 6.6 to compute $p(A, B)$:

$$p(A, B) = p(A)p(B)$$
$$= (.200)(.300) = .060$$

Then, using Equation 6.4, we find the probability that the student will be admitted to at least one of the two schools to be

$$p(A \text{ or } B) = p(A) + p(B) - p(A, B)$$
$$= .200 + .300 - .060 = .440$$

For an additional application of probability principles, see Box 6.1.

6.6 Sampling with Versus Without Replacement

The preceding discussion of probability theory has important implications for researchers who draw random samples from populations. When cases are sampled from a population, two different situations are possible. First, a given case can be randomly selected, the measurement of interest taken, and the case then returned to the population. Then, with the population fully intact, the random selection procedure can be performed again. This process is called **sampling with replacement** because each case is "replaced" back into the population before another case is randomly selected. In contrast, **sampling without replacement** involves selecting a case at random and then, without replacing this case, selecting another case at

BOX 6.1

Beliefs and Probability Theory

A major concern of social psychologists has been the relationships among the beliefs that an individual holds. How are a person's beliefs structured? If we change a belief, will this produce changes in other beliefs? If so, can we predict these changes?

One approach to the study of relationships among an individual's beliefs has been the application of probability theory. In this approach, beliefs are characterized as subjective probabilities (that is, probability judgments made by an individual). The belief that "nuclear energy is dangerous" is characterized by the individual's subjective probability that this statement is true. Some individuals might believe that nuclear energy is *definitely* dangerous, whereas others might believe that nuclear energy is *probably* dangerous. Still others might believe that it is not dangerous at all. In accord with probability theory, a subjective probability can range from .00 to 1.00. A subjective probability of .00 means that the individual is completely certain that a given statement is not true, whereas a subjective probability of 1.00 means that the individual is completely certain that a given statement is true. Increasing numbers between these values represent increasing degrees of certainty that the belief statement is true.

Some psychologists have suggested that, if beliefs are characterized as subjective probabilities, it may be possible to predict and understand these beliefs using probability theory. For example, consider the following two statements:

B: Nuclear power plants are dangerous.
$A \mid B$: Nuclear power plants should be shut down if they are dangerous.

It is possible to measure an individual's subjective probability concerning each of these statements. If subjective probabilities are organized in accord with probability theory, then knowledge of these two subjective probabilities should allow us to predict the individual's belief in the statement

A, B: Nuclear power plants are dangerous and they should be shut down.

Recall from Equation 6.5 that

$$p(A, B) = p(B)p(A \mid B)$$

Thus, the belief in statement A, B should be predictable from the product of the beliefs in statements B and $A \mid B$. If the individual believes the probability that nuclear power plants are dangerous is .90, and if he also believes with a probability of .95 that nuclear power plants should be shut down given that they are dangerous, then his belief that nuclear power plants are dangerous and should be shut down should be $(.95)(.90) = .855$.

Research on the relationship between subjective and actual probabilities has provided some interesting insights into belief organization and change. Although there is not always a strong correspondence between them, the degree of agreement has been striking in many respects. Interested readers are referred to Wyer and Goldberg (1970) for a discussion of this research.

random, and so on. The method of sampling—with versus without replacement—can affect the probability of observing some event.

Consider the data in Table 6.1 and suppose we are interested in determining the probability of randomly selecting first someone who is satisfied with his job and then someone who is dissatisfied with his job. In this case, event B is "selecting someone who is satisfied with his job as the first case" and event A is "selecting someone who is dissatisfied with his job as the second case."

Let us first consider sampling with replacement. Before any case is selected, the probability of selecting someone who is satisfied with his job as the first case [$p(B)$] is 210/410 = .512. After this has been done, the case is returned to the population. This means that the outcome of the first random selection in no way affects the outcome of the second random selection; that is, the two events are independent. The probability of selecting someone who is dissatisfied with his job as the second case [$p(A)$] is thus 200/410 = .488. According to Equation 6.6, the joint probability of two independent events is

$$p(A, B) = p(A)p(B)$$

In our example,

$$p(A, B) = (.488)(.512) = .2499$$

Now consider sampling without replacement. The outcome of the second selection is no longer independent of the first selection. The probability that we will select someone who is satisfied with his job as the first case and someone who is dissatisfied with his job as the second case is now equal to

$$p(A, B) = p(B)p(A \mid B)$$

per Equation 6.5. The probability of selecting someone who is satisfied with his job as the first case [$p(B)$] is, as before, 210/410 = .512. Since 200 of the 409 remaining men are dissatisfied with their jobs, the probability of selecting someone who is dissatisfied with his job as the second case given that we selected and did not replace someone who is satisfied with his job as the first case [$p(A \mid B)$] is 200/409 = .489. Hence,

$$p(A, B) = (.512)(.489) = .2504$$

a value that is slightly higher than the probability computed on the basis of sampling with replacement.

If the size of the sample is small relative to the size of the population (for example, a ratio of 1 case in the sample for every 20 cases in the population), sampling with versus without replacement will not affect probabilities appreciably. Such was the case in our example because the sample size was 1 and the population size was 410. However, when the sample is large relative to the size of the population, the different sampling procedures can produce very different results. One should keep in mind the sampling procedure that was used when interpreting the results of probability studies.

6.7 Counting Rules

We have defined a probability as the number of observations favoring some event divided by the total number of observations. For some complex events, it would be very tedious to enumerate all the possible outcomes in order to compute a probability. For example, if we wanted to state the number of ways we could seat five different people in each of five positions at a table in a study of social interaction, the task of enumerating each sequence would be quite laborious. Fortunately, there are *counting rules* that permit such computations to be easily performed.

Suppose we have a set of three objects, A, B, and C, and we want to specify for this set all possible subsets of two objects. There are two different conditions under which we can attempt this task. One condition is where the ordering of the two objects matters, such that the subset AB is considered different from the subset BA because the two objects occur in a different order. The other condition is where ordering does not matter, and AB and BA are considered instances of the same event. All possible subsets of two objects under the two conditions are listed here:

Permutations (ordering matters)	Combinations (ordering does not matter)
AB BA	AB
AC CA	AC
BC CB	BC

As can be seen, a **permutation** of a set of objects or events is an *ordered* sequence, whereas a **combination** of a set of objects or events is a sequence in which the internal ordering of elements is irrelevant. The general notation for the number of permutations (P) of n things taken r at a time is

$$_nP_r$$

In our problem, n = 3 because there are three objects (A, B, and C), and r = 2 because we are concerned with how we can order these three objects taking two at a time. The general notation for the number of combinations (C) of n things taken r at a time is

$$_nC_r$$

Before specifying the formulas for permutations and combinations, we must introduce one other concept—a **factorial** of a number. The factorial of a number, n, is formally defined as

$$n! = (n)(n-1)(n-2)(n-3)\cdots(1) \qquad [6.9]$$

where n! stands for "n factorial."* Consider the following examples:

$$6! = (6)(5)(4)(3)(2)(1) = 720$$

$$5! = (5)(4)(3)(2)(1) = 120$$

$$3! = (3)(2)(1) = 6$$

---

* The notation "···" means that although we have not listed all relevant values, the inclusion in the multiplication of all values between n − 3 and 1 is implicit.

In factorial notation, the expression 0! equals 1.

The formula for calculating the number of *permutations* of n things taken r at a time is

$$_nP_r = \frac{n!}{(n-r)!}$$ [6.10]

This is our first counting rule. In our problem, we have three objects (A, B, and C) and we want to specify the number of ordered sequences of two objects that can be derived from these. Using Equation 6.10, we find that

$$_3P_2 = \frac{3!}{(3-2)!} = \frac{(3)(2)(1)}{1} = 6$$

This is consistent with the list presented in the table; there are six permutations of three objects taken two at a time (AB, BA, AC, CA, BC, CB).

The formula for calculating the number of *combinations* of n things taken r at a time is

$$_nC_r = \frac{n!}{(n-r)!\,r!}$$ [6.11]

This is our second counting rule. For the problem with three objects taken two at a time, we find that

$$_3C_2 = \frac{3!}{(3-2)!\,2!} = \frac{(3)(2)(1)}{(1)(2)(1)} = 3$$

Thus, consistent with the list presented in the table, there are three combinations of three objects taken two at a time (AB, AC, BC).

Appendix C presents factorials for the numbers 0 through 20. Use of this appendix can greatly simplify the calculation of the number of permutations or combinations of a set of objects or events.

A third useful counting rule can be specified as follows: If any one of r_i mutually exclusive and exhaustive events can occur on trial i, then the number of different sequences of events that can occur across n trials is $(r_1) \cdots (r_n)$. Suppose you toss a coin on the first trial (two possible outcomes) and roll a die on the second trial (six possible outcomes). Then the total number of different sequences that can result is (2)(6) = 12 (for example, a head and a 1, a head and a 2, and so on).

Although these three counting rules might seem abstract, they can be used to good effect in many instances. Consider the following example on crime and statistics by Zeisel and Kalven (1978), which makes use of the third counting rule:

This simple example comes from a Swedish trial on a charge of overtime parking. A policeman had noted the position of the valves of the front and rear tires on one side of the parked car, in the manner pilots note directions: One valve pointed, say, to one o'clock, the other to six o'clock, in both cases to the closest "hour." After the allowed time had run out, the car was still there, with the two valves still pointing toward one and six o'clock. In court, however, the accused denied any violation. He had left the parking place in time, he claimed, but had returned to it later, and the valves just happened to come to rest in the same positions as before. The court had an expert compute the probability of such a coincidence by chance. The answer was that the probability is 1 in 144 (12 × 12) because there are 12 positions for each of two wheels. In acquitting the defendant, the judge remarked that if all four wheels had been checked and found to point in the same directions as before, then the coincidence claim would have been rejected as too improbable and the defendant convicted: Four wheels with 12 positions each can combine in 20,736 (12 × 12 × 12 × 12) different ways, so the probability of a chance repetition of the original position would be only 1 in 20,736. Actually, these formulas probably understate the probability of a chance coincidence because they are based on the assumption that all four wheels rotate independently of each other, which, of course, they do not. On an idealized straight road all rotate together, in principle. It is only in the curves that the outside wheels turn more rapidly than the inside wheels, but even then the front and rear wheels on each side will presumably rotate about the same amount. (p. 140)

For those readers who attend horse races, the calculation of permutations could prove illuminating. A common bet at horse races is the "trifecta," or predicting which horses will come in first, second, and third in a given race. In a ten-horse race, in how many ways could the finishes of three horses be ordered, considering the entire field? From Equation 6.10,

$$_{10}P_3 = \frac{10!}{(10-3)!} = \frac{(10)(9)(8)(7)(6)(5)(4)(3)(2)(1)}{(7)(6)(5)(4)(3)(2)(1)}$$

$$= (10)(9)(8) = 720$$

Only 1 of these 720 orders will actually occur. Thus, in the absence of any information about the horses, the probability of accurately predicting a trifecta in a ten-horse race is 1/720, or .0014. Even for individuals with racetrack savvy, betting a trifecta is likely to be a losing proposition.

6.8   The Binomial Expression

Consider an experiment on extrasensory perception (ESP) in which an individual's claim that she possesses psychic powers is tested by asking her to predict the outcome of each of ten tosses of a coin. The individual could, in principle, correctly predict none out of ten, one out of ten, two out of ten, and so on, through all ten out of ten tosses. If we assume that the person has no psychic powers, then the probability that she will correctly guess the outcome of a given toss is .50. Another way of saying this is that the probability of a "success," p, is .50 and the probability of a "failure," q, is $1 - p = .50$. Under the assumption that $p = .50$ and $q = .50$, what is the probability that the person would get each of the above scores (none out of ten guesses correct, one out of ten guesses correct, and so on)? The **binomial expression** is a formula that allows us to answer this question. It can formally be stated as follows: In a sequence of n independent trials, each of which has only two possible outcomes (arbitrarily called a "success" and a "failure"), with the probability p of success and the probability q of failure, the probability of r successes in n trials is

$$p(r \text{ successes}) = \frac{n!}{r!\,(n-r)!}\, q^r q^{n-r} \qquad\qquad [6.12]$$

As an example, let us compute the probability of correctly predicting ten out of ten coin tosses. In this instance, the number of coin tosses, or trials (n), is 10 and the number of successes (r) is also 10. For $p = .50$ and $q = 1 - p = .50$, we find that

$$p(\text{ten correct}) = \frac{10!}{10!\,(10-10)!}\,(.50^{10})(.50^{10-10})$$

$$= (1)(.001)(1) = .001$$

The probability of correctly predicting nine out of ten coin tosses is

$$p(\text{nine correct}) = \frac{10!}{9!\,(10-9)!}\,(.50^9)(.50^{10-9})$$

$$= (10)(.002)(.50) = .010$$

Column 2 of Table 6.2 presents the probability of each possible score in the ESP experiment under the assumption that $p = q = .50$.

Use of the Binomial Expression in Hypothesis Testing

Before undertaking the ESP experiment, we want to specify what scores are consistent with the conclusion that the individual possesses psychic powers. We begin by assuming that she does *not* possess such ability. If this is true, we would expect

TABLE 6.2   Probabilities of Scores in ESP Experiment

Number of correct predictions	Probability	Probability of making the indicated number or more correct predictions
10	.001	.001
9	.010	.011
8	.044	.055
7	.117	.172
6	.205	.377
5	.246	.623
4	.205	.828
3	.117	.945
2	.044	.989
1	.010	.999
0	.001	1.000

her to accurately predict the outcome of the coin toss .50 of the time, or five out of ten trials. We realize, of course, that due to chance, she might be slightly more accurate than this figure even if she does not actually possess the claimed powers. Thus, we would be unlikely to conclude that she is paranormal even if she were to accurately predict six, or even seven, of the ten coin tosses.

As the number of correct predictions increases, however, the probability that these outcomes are caused by chance alone decreases. We note from column 3 of Table 6.2 that the probability of correctly guessing eight or more tosses is .044 + .010 + .001 = .055, that the probability of correctly guessing nine or more tosses is .010 + .001 = .011, and that the probability of correctly guessing all ten tosses is .001. At some point, the probability that a set of outcomes is due to chance alone will be so small that, lacking evidence of chicanery, we will accept the claim of psychic powers.

In the context of **hypothesis testing,** the proposal, or *hypothesis,* that the individual does not possess psychic ability is called a **null hypothesis.** Assuming the null hypothesis is true, we can specify an expected result of an investigation. In our example, this takes the form that the individual will accurately predict five of the ten coin tosses. If our observations are so discrepant from the expected result that the difference cannot be attributed to chance, then we will reject the null hypothesis in lieu of a competing proposal, referred to as an **alternative hypothesis.** In our example, the alternative hypothesis is that the individual possesses the claimed psychic ability. On the other hand, if the observed result is similar enough to the outcome stated in the null hypothesis such that it can reasonably be attributed to chance, we will *fail to reject* the null hypothesis.

The problem thus becomes determining what constitutes chance versus nonchance results under the assumption that the null hypothesis is true. This distinction is made with reference to a probability value known as an **alpha level.** For example, when the alpha level is .05, a result is defined as *nonchance* (that is, reflective of factors other than chance) if the probability of obtaining that result, assuming the null hypothesis is true, is less than .05. Applying the .05 criterion to our example, we decide that correctly predicting nine or more coin tosses will cause us to conclude that the results are consistent with an individual having psychic powers.[*] Fewer correct predictions will be attributed to chance guessing. Try flipping a coin ten times and see if you can meet this criterion.

The Binomial and Normal Distributions

In experiments such as the ESP experiment, more than ten trials are typical. Suppose 500 trials had been used and 280 correct predictions were made. Assuming $p = .50$, is this sufficient evidence for us to conclude that an individual possesses psychic powers, or can this outcome be attributed to chance? The application of

---

[*] There may, however, be alternative explanations for such an outcome. All we really know is that something other than chance is probably operating and this "something" might be psychic ability. However, the observed outcome could also be the result of something else. This issue will be considered in Chapter 9.

the binomial expression would be an arithmetic nightmare in this instance. Fortunately, when certain conditions are met, the normal distribution can be used to obtain a very close approximation to relevant binomial probabilities. For instance, Figure 6.1 presents a histogram of the probability distribution in Table 6.2 for the ten-trial ESP experiment. A normal distribution has been superimposed over the exact binomial probabilities of the score values. Note the similar shapes of the distributions. It turns out that the binomial and normal distributions are closely related, with the correspondence between them depending on the values of n and p. The correspondence improves as n increases and as p becomes closer to .50. For small n, statisticians have developed a correction factor (called the **correction for continuity**) that, when applied to the data, yields even better correspondence. Table 6.3 illustrates the relationship between the two distributions for our ESP example.

With the advent of powerful computers, the reliance on the normal approximation for binomial problems is less frequent, if it is necessary at all. Computers can do the calculations necessary to apply the binomial rule even in computationally demanding situations. Nevertheless, the binomial–normal approximation is an important principle about binomial distributions in general.

We now demonstrate how the similarity between normal and binomial probabilities can be applied to hypothesis testing by addressing the question, "Is the fact that an individual correctly predicts 280 of 500 coin tosses sufficient evidence for us to conclude that she possesses psychic powers, or can this outcome be attributed to chance?" To answer this question, we must first determine the mean

FIGURE 6.1   **Relationship Between the Binomial and Normal Distributions for $n = 10$ and $p = .50$**

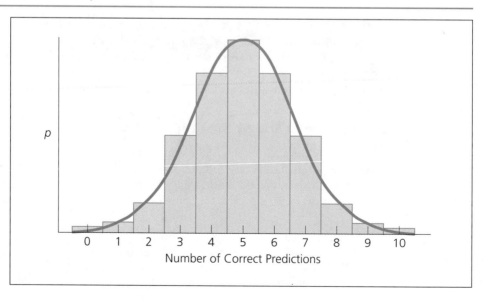

TABLE 6.3   **Comparison of Binomial Probabilities and Corresponding Normal Approximations for ESP Experiment**

Number of correct predictions	Binomial probability	Normal approximation
10	.001	.002
9	.010	.011
8	.044	.044
7	.117	.114
6	.205	.205
5	.246	.248
4	.205	.205
3	.117	.114
2	.044	.044
1	.010	.011
0	.001	.002

and standard deviation of the relevant binomial distribution. The mean of a binomial distribution is equal to

$$\mu = np \tag{6.13}$$

where n is the number of trials and p is the probability of success. In the present instance,

$$\mu = (500)(.50) = 250.00$$

The standard deviation of a binomial distribution is equal to

$$\sigma = \sqrt{npq} \tag{6.14}$$

where $q = 1 - p$. In the present example,

$$\sigma = \sqrt{(500)(.50)(.50)} = 11.18$$

Since the distribution of the number of correct predictions for $n = 500$ and $p = .50$ approximates a normal distribution, we can convert a score of 280 to a z score using Equation 4.6:

$$z = \frac{X - \mu}{\sigma}$$

$$= \frac{280 - 250.00}{11.18} = 2.68$$

Referring to Appendix B, we find that the probability of obtaining a z score of 2.68 or greater is .0037. This is less than the criterion value (alpha level) of .05. Thus, 280 correct predictions is consistent with the hypothesis that the person possesses psychic powers.

Applications
to the Analysis
of a Social Problem

As in previous chapters, the material on probability in this chapter involves primarily descriptive statistics rather than inferential statistics. Our discussion therefore focuses on sample descriptions of the parent–teen data presented in Chapter 1. Chapter 15 will extend these concepts to an inferential context.

Using the parent–teen data described in Chapter 1, we found in Chapter 2 that parents tend to underestimate the sexual activity of their young (14-year-old) teenage sons and daughters. Specifically, we found that whereas 35.7% of the 14-year-olds had engaged in sexual intercourse, only 14.5% of the mothers thought their teens had engaged in sexual intercourse. To explore some of the probability concepts developed in this chapter, we decided to form a 2 × 2 contingency table. The

rows are whether or not the teen reported having engaged in sexual intercourse, and the columns are whether or not the mother thought the teen had engaged in sexual intercourse. We focused on only 14-year-olds and eliminated teens from the analysis who were sexually abused (i.e., teens who reported that their age at first intercourse was younger than 10 or who indicated abuse in other questions that we asked). These latter teens were omitted from the analysis because our primary interest is in how aware mothers are of their teens' "normal" sexual activity as opposed to maternal awareness of past sexual abuse. Table 6.4 presents the computer printout for the contingency table.

The columns of the 2 × 2 table represent whether the mother thinks the teen has ever had sexual intercourse. A score of 0 means no and a

---

Study Exercise 6.5

A person who claims to have psychic powers tries to predict the outcome of a roll of a die on each of 100 trials. He correctly predicts 21 rolls. Using an alpha level of .05 as a criterion, what should we conclude about the person's claim?

Answer   Since there are six numbers on a die, the probability of success on any one trial is $p = 1/6 = .167$, and the probability of failure is $q = 1 - p = 1 - .167 = .833$. The mean of the relevant binomial distribution is thus

$$\mu = np = (100)(.167) = 16.70$$

and the standard deviation is

$$\sigma = \sqrt{npq} = \sqrt{(100)(.167)(.833)} = 3.73$$

The binomial probability associated with a score of 21 can be approximated by converting this score to a z score and deriving the corresponding normal probability from Appendix B. A score of 21 translates into a z score of

$$z = \frac{X - \mu}{\sigma} = \frac{21 - 16.70}{3.73} = 1.15$$

From Appendix B, the probability of obtaining a z score of 1.15 or greater is .1251. Since .1251 is not less than the criterion value of .05, we conclude that the person's performance does not allow us to say that he has the claimed powers. In other words, there is insufficient evidence to conclude that he performed at an above-chance level.

TABLE 6.4   Computer Output for Contingency Table for Sexual Intercourse

```
SEX    Whether or not teen has ever had sexual intercourse
MSEX   Whether mother thinks teen has had sexual intercourse
```

		MSEX		
Count				
Row Pct				
Col Pct		No	Yes	Row
Tot Pct		.00	1.00	Total
No	.00	67	5	72
		93.1	6.9	65.5
		71.3	31.3	
		60.9	4.5	
Yes	1.00	27	11	38
		71.1	28.9	34.5
		28.7	68.8	
		24.5	10.0	
Column		94	16	110
Total		85.5	14.5	100.0

(SEX label appears at left of the table rows)

score of 1 means yes. The rows of the table represent whether the teen reported having had sexual intercourse. A score of 0 means no and a score of 1 means yes. The first entry under "Row Total" indicates the number of teens who said no (72), and directly beneath this is the percentage who said no (65.5%). Corresponding entries below are for the number of teens who said yes (38) and the percentage who said yes (34.5). The "Column Total" indicates comparable information but for the mothers. The number of mothers who said no (i.e., who did not think their 14-year-old had engaged in sexual intercourse) was 94, or 85.5%. The number of mothers who said yes was 16, or 14.5%.

There are four entries in each cell of the table. Consider the cell where both the mother and the teen said no. The four numbers in this cell are 67, 93.1, 71.3, and 60.9. The first number (67) is the absolute frequency, or the number of cases where both the mother and the teen said no. The second number (93.1) is the row percent and is the absolute frequency (67) divided by the row total (72) multiplied by 100. The third number (71.3) is the column percent and is the absolute frequency (67) divided by the column total (94) multiplied by

100. The final number (60.9) is the total percent and is the absolute frequency (67) divided by the total sample size (110) multiplied by 100. The order of information is keyed in the upper left corner of the contingency table.

If we divide a given percentage by 100, we obtain a probability. This permits us to specify some interesting probabilities based on the discussion in this chapter. For example, the probability that a teen has engaged in sexual intercourse is 34.5/100 = .345. According to mothers, however, this probability is only 14.5/100 = .145.

Let us consider two conditional probabilities: (1) the probability that the mother thinks the teen has engaged in sex given that the teen has, in fact, engaged in sex, and (2) the probability that the mother thinks the teen has engaged in sex given that the teen has not, in fact, engaged in sex. These are:

p(mother says yes/teen says yes) = 11/38
= .289
p(mother says yes/teen says no) = 5/72
= .069

(continued)

It can be seen that mothers are more likely to say their teens are sexually active if the teens are, in fact, sexually active than if the teens are not. Indeed, mothers are about 4 times more likely (.289/.069 = 4.19) to say their teen is sexually active if the teen, in fact, is sexually active than if not. This suggests that, although there is a general tendency for mothers to underestimate sexual activity, there is at least some sensitivity on the part of mothers to the sexual activity of their teens.

Let us consider two additional conditional probabilities: (1) the probability that the mother thinks the teen has not engaged in sex given that the teen has, in fact, engaged in sex, and (2) the probability that the mother thinks the teen has not engaged in sex given that the teen has not, in fact, engaged in sex. These are:

$$p(\text{mother says no}/\text{teen says yes}) = 27/38$$
$$= .711$$
$$p(\text{mother says no}/\text{teen says no}) = 67/72$$
$$= .931$$

Analogous to the results for the previous conditional probabilities, mothers are more likely to say their teens are not sexually active if the teens are not, in fact, sexually active than if the teens are. Apparently, the mothers in this sample are not completely in the dark about the sexual activity of their 14-year-old sons and daughters.

Summary

Among the probabilities of interest to behavioral scientists are the probability of a simple event, conditional probability, and joint probability. The probability of a simple event refers to the likelihood that a given event will occur. A conditional probability tells us the likelihood that a given event will occur given that some other event occurs. A joint probability indicates the likelihood of observing each of two events. A fourth probability of interest focuses on the likelihood of observing at least one of two events. These concepts can be used to gain considerable insight into relationships between events.

When cases are sampled from a population, two different situations are possible. In sampling with replacement, each case is "replaced" back into the population before another case is randomly selected. In contrast, sampling without replacement involves selecting a case at random and then, without replacing this case, selecting another case at random, and so on. Whether sampling is done with or without replacement can affect the probability of observing some event.

The number of subsets of a set of objects or events can be specified using either of two counting rules. First, we can determine the number of permutations, or ordered sequences. Second, we can determine the number of combinations, or sequences in which the internal ordering of elements is irrelevant. A third counting rule specifies the number of different sequences of mutually exclusive and exhaustive events that can occur across trials.

The binomial expression can be used to determine the probability of various numbers of "successes" and "failures" across trials in situations where trials are independent and each has only two outcomes. One important use of the binomial expression is in hypothesis testing. This involves specifying an expected result of an investigation under the assumption that some proposal, known as a null hypothesis, is true. If the observations are so discrepant from the expected result that the difference cannot be attributed to chance, then the null hypothesis will be rejected in lieu of a competing proposal, known as an alternative hypothesis. Otherwise, we will fail to reject the null hypothesis. The decision whether to reject the null hypothesis is made with reference to a probability value known as an alpha level.

When the application of the binomial expression is not practical, under certain conditions the normal distribution can be used to obtain a very close approximation to relevant binomial probabilities. Given the mean and standard deviation of the binomial distribution of interest, our knowledge of the normal distribution can thus be used to test whether an observed outcome is due to chance or nonchance factors.

Exercises

Answers to asterisked (*) exercises appear at the back of the book. Answers to exercises with two asterisks are also worked out step by step in the Study Guide.

1. What are the two conceptualizations of probability?

Use the following information to complete Exercises 2–10:

When the Equal Rights Amendment (ERA) was proposed as an amendment to the constitution, it was the subject of considerable controversy in the United States. The proposed amendment was concerned with equal rights for both genders and consisted of three statements: "Equality of rights under the law shall not be denied or abridged by the United States or by any State on account of sex. The Congress shall have power to enforce, by appropriate legislation, the provisions of this article. This amendment shall take effect two years after the date of ratification."

A behavioral scientist was interested in the extent to which people in a particular city supported the ERA and, accordingly, interviewed 175 men and 175 women from this city. The numbers of men and women favoring or opposing the ERA are summarized in the following contingency table:

Position	Gender	
	Men	Women
Favor ERA	60	120
Oppose ERA	115	55

*2. What is the marginal frequency for people who favor the ERA?

3. What is the marginal frequency for women?

*4. How many observations are represented?

*5. What is the probability that an individual favors the ERA? What is the probability that an individual opposes the ERA? What is the probability that an individual is a man? A woman?

*6. What is the probability that an individual favors the ERA given that the individual is a man? What is the probability that an individual favors the ERA given that the individual is a woman? What is the probability that an individual opposes the ERA given that the individual is a man? What is the probability that an individual opposes the ERA given that the individual is a woman?

*7. Are being a man and favoring the ERA independent? Explain. Be specific.

*8. What is the probability that an individual is a man who favors the ERA? What is the probability that an individual is a woman who opposes the ERA?

*9. How many joint probabilities can be computed? Name them.

*10. What is the probability that an individual is a man or favors the ERA, or both? What is the probability that an individual is a woman or opposes the ERA, or both?

Use the following information to complete Exercises 11–16:

A behavioral scientist interviewed 500 people to study the relationship between political party identification and attitudes toward legal abortions. The following contingency table was observed:

Party identification	Attitude	
	Favors legal abortions	Opposes legal abortions
Democrat	160	40
Republican	40	160
Independent	75	25

11. What is the probability that an individual favors legal abortions? What is the probability that an individual opposes legal abortions? What is the probability that an individual is a Democrat? A Republican? An Independent?

12. What is the probability that an individual favors legal abortions given that the individual is a Democrat? What is the probability that an individual is a Democrat given that the individual favors legal abortions? Why are these probabilities different?

13. Are being a Democrat and opposing legal abortions independent? Why or why not? Be specific.

14. What is the probability that an individ-

ual is a Republican who favors legal abortions? What is the probability that an individual is an Independent who opposes legal abortions?

15. How many joint probabilities can be computed? Name them.

16. What is the probability that an individual is a Republican or favors legal abortions, or both? What is the probability that an individual is an Independent or opposes legal abortions, or both?

*17. If the probability of some event, B, is .542 and the probability of some second event, A, given event B, is .896, what is the probability of observing both event A and event B?

18. If two events, A and B, are independent and the probability of A is .40 and the probability of B is .30, what is the probability of observing both event A and event B?

19. If two events, A and B, are mutually exclusive and the probability of A is .278 and the probability of B is .581, what is the probability of observing one of event A and event B?

*20. If the probability of some event, B, is .349 and the probability of observing both B and some second event, A, is .180, what is the probability of event A, given event B?

21. Explain the difference between sampling with and without replacement.

22. Explain the difference between a permutation and a combination.

*23. Compute the following permutations:
   a. $_5P_3$ c. $_4P_2$ e. $_4P_4$
   b. $_6P_5$ d. $_5P_5$

*24. Compute 5!. Compare your answer with the answer for **d** of Exercise 23. Compute 4!. Compare your answer with the answer for **e** of Exercise 23. What generalization can you draw from this?

25. Suppose you are conducting an experiment that involves showing people five slides. You are concerned that the order in which you present the slides could af-

fect the outcome of the experiment, so you decide to present them to each person in a different order. In how many ordered sequences can the slides be presented?

*26. Compute the following combinations:
   a. $_5C_3$ c. $_4C_2$ e. $_4C_4$
   b. $_6C_5$ d. $_5C_5$

27. A researcher was interested in the effects of four different independent variables on a dependent variable. For practical reasons, she could study only two of the variables. How many different combinations of two variables could she potentially study? Suppose she could study three variables. How many different combinations of three variables could she potentially study?

*28. How many different sequences of events can occur on three rolls of a die?

29. How many different sequences of events can occur if one card is drawn from each of three standard bridge decks?

*30. If a student simply makes random responses on a ten-item true–false test, what is the probability he or she:
   a. gets nine correct
   b. gets eight correct
   c. gets seven or more correct
   d. gets one correct
   e. gets two correct

31. Sixty-five percent of people eligible to vote in the 1976 presidential election did not do so. Compute the probability for each possible number of people who did not vote in a random sample of ten eligible voters.

32. Define each of the following:
   a. null hypothesis
   b. alternative hypothesis
   c. alpha level

*33. What factors influence the correspondence between the binomial and normal distributions?

*34. Calculate the mean and the standard deviation of the binomial distribution for $n = 150$ and $p = .60$.

35. Calculate the mean and the standard deviation of the binomial distribution for $n = 156$ and $p = .80$.

**36. Twenty percent of individuals who seek psychotherapy will recover from their symptoms irrespective of whether they receive treatment (a phenomenon called *spontaneous recovery*). A researcher finds that a particular type of psychotherapy is successful with 30 out of 100 clients. Using an alpha level of .05 as a criterion, what should she conclude about the effectiveness of this psychotherapeutic approach?

37. A student takes a multiple-choice quiz of ten items that has four response options for each question and gets five items correct. Using an alpha level of .05 as a criterion, can we conclude that the student performed at an above-chance level?

Multiple-Choice Questions

Refer to the following contingency table to do Exercises 38–43:

Place of residence	Marital Status	
	Divorced	Married
Large city	50	50
Small town	20	80

*38. What is the probability that an individual is divorced?
   a. .10   c. .35
   b. .25   d. .50

39. What is the probability that an individual lives in a large city?
   a. .10   c. .50
   b. .20   d. .65

*40. What is the probability that an individual is divorced given that the individual lives in a large city?
   a. .20   c. .50
   b. .35   d. .65

41. What is the probability that an individual is divorced given that the individual lives in a small town?
    a. .20   c. .35
    b. .25   d. .50

*42. What is the probability that an individual lives in a small town and is divorced?
    a. .10   c. .25
    b. .20   d. .65

43. What is the probability that an individual lives in a large city and is married?
    a. .10   c. .35
    b. .25   d. .65

*44. When p is small, the correspondence between the binomial and normal distributions increases when a correction factor is applied to the data.
    a. true
    b. false

45. An interviewer wants to vary the order in which she presents the three major sections of a questionnaire. How many different orders can she use?
    a. 4   c. 8
    b. 6   d. 10

46. Each unique combination of variables in a contingency table is called a *cell*.
    a. true
    b. false

*47. Given 100 trials and a probability of success of .40, the mean of the binomial distribution is _____ and the standard deviation is _____.
    a. 40; 4.90   c. 60; 4.90
    b. 40; 24   d. 60; 24

*48. When the size of the sample relative to the size of the population is large, sampling with or without replacement can produce very different results.
    a. true
    b. false

CHAPTER 7

Estimation and
Sampling Distributions

Suppose you wanted to describe the annual income of all married women in the United States. It would be impossible to contact each of these women to determine her income level. Instead, you might decide to select a random sample and then try to make generalizations about the population based on the sample. The problem of using sample data to make inferences about populations is fundamental to statistical techniques used in the behavioral sciences. This chapter is concerned with estimating population parameters from sample statistics.

7.1 Finite Versus Infinite Populations

Population parameters can be estimated with reference either to small, finite populations or to populations that are so large that for all practical purposes they can be considered infinite. Behavioral science research is typically conducted with the goal of explaining the behavior of large numbers of individuals, often including people who have lived previously or who have yet to be born, as well as those residing in the present. For instance, if we were studying the course of a particular brain disease, we might think of the relevant population as all people, whether currently living or not, who have ever had or will have this illness. As such, the statistics used in most behavioral science disciplines are applicable to extremely large, if not infinite, populations.

Given that behavioral science research is concerned with very large populations, it is impossible for investigators to select truly random samples from such populations. Recall from Chapter 1 that a random sample requires listing all members of the population and then using a random number table to select a sample. Because this is not possible for very large populations, behavioral scientists typically select a set of individuals to study, and then *assume* that these individuals are randomly drawn from *some* population. Exactly what population the individuals are assumed to represent depends, in part, on the characteristics of the participants in the study. If a learning experiment was conducted on 200 college students at a midwestern university, the investigator might want to generalize his or her results to people in general. In this case, the population is conceptualized as consisting of "all people," and the college students are said to represent a random sample from this population. Obviously this is a very questionable assumption. Perhaps the population should be conceptualized as "all college students." Or maybe it should be conceptualized as "all college students at midwestern universities." Maybe it should be considered in even more specific terms. Behavioral scientists sometimes disagree about what the appropriate population is for purpose of generalization based on an investigation of a small set of individuals. When interpreting research, you should always keep in mind who the results should generalize to (in other words, exactly what the relevant population is).

For the sake of illustration, the examples in this chapter use small, finite populations. Sampling from a small, finite population *with replacement* (that is, where each sample member is returned to the population before the next sample member is selected) is analogous to sampling from a very large or infinite population without replacement, as behavioral science research is conceptualized as doing. Accordingly, when we develop principles for estimating population parameters using finite populations, sampling is done with replacement to mimic the case of infinite populations.*

7.2 Estimation of the Population Mean

Consider the case of 100 families in a small town. An investigator wants to describe the average number of children in each family. Because of practical limitations, the investigator is unable to interview all 100 families, so she instead resorts to a sample. Let us assume that the sample size is ten. In this case, the *population* is the 100 families in the town, and the *sample* is the ten families who are selected to be interviewed. Table 7.1 lists the number of children in each family of the population. Although the 100 family sizes are unknown to the investigator, the fact remains that there *are* 100 families in the population and that each one has the indicated number of children. In other words, if the investigator *were* able to inter-

---

* For a discussion of statistical applications to finite populations, see Hays (1981).

TABLE 7.1 **Number of Children in Families for a Hypothetical Population**

3	6	5	4	4	1	3	4	4	5
4	4	6	4	3	3	5	3	4	4
4	3	4	3	5	3	4	3	5	2
2	3	2	5	4	2	3	5	2	3
0	2	4	3	2	4	2	2	1	3
4	3	3	2	4	5	4	4	2	4
3	5	4	5	2	3	5	3	4	3
2	4	3	7	5	1	4	4	5	4
5	2	2	3	7	4	7	3	3	1
2	0	7	4	3	6	5	2	3	1

$$\mu = 3.50$$
$$\sigma^2 = 2.09$$

view all 100 families, she would obtain the scores (family sizes) listed in Table 7.1. The true population mean, μ, is 3.50, and the variance, σ^2, is 2.09. Again, the investigator is unaware of the values of μ and σ^2.

Let us randomly select ten families from this set of 100, and let these represent the ten families interviewed by the investigator. Suppose the scores for these families are found to be 3, 4, 4, 5, 2, 4, 1, 1, 4, 3, with a mean of 3.10. Note that this value is *not* equal to the true population mean. A sample statistic may not equal the value of its corresponding population parameter because of **sampling error.** More specifically, sample values are likely to differ from population values because they are based on only a portion of the overall population. It does not imply that mistakes have been made in the collection and analysis of the data. The amount of sampling error can be represented as the difference between the value of a sample statistic (in the present context, $\overline{X}$) and the value of the corresponding population parameter (in the present context, μ). In our example, the amount of sampling error is $3.10 - 3.50 = -.40$. In practice, an investigator does not know the value of the population parameter, so it is impossible to compute the exact amount of sampling error that occurs.

In the absence of any other information, the sample mean that one observes is the "best estimate" of the value of the population mean. One justification for this is based on an important statistical property of the mean. This property can be illustrated with reference to the family size example. Suppose, in addition to the first sample, we select another random sample of size ten from the population of 100 families. The scores for this sample are 1, 3, 3, 4, 3, 7, 5, 1, 3, 4, with a mean of 3.40. Suppose this process is repeated over and over until we have computed the mean score for *every possible random sample of size ten* that could be selected from this population. Table 7.2 lists 15 of the many sample means that could be observed and the amount of sampling error in each instance (that is, $\overline{X} - \mu$). If we were to compute the average (mean) amount of sampling error that resulted across all possible samples of size ten, it would equal 0. In other words, some of the sample means overestimate the true population mean, whereas others

TABLE 7.2   **Means for Samples of Size Ten Randomly Selected**
                   from the Population in Table 7.1

Sample scores	$\overline{X}$	$\overline{X} - \mu$
3, 4, 4, 5, 2, 4, 1, 1, 4, 3	3.10	−.40
1, 3, 3, 4, 3, 7, 5, 1, 3, 4	3.40	−.10
3, 5, 4, 3, 2, 4, 5, 4, 3, 2	3.50	.00
0, 2, 7, 2, 5, 2, 5, 5, 4, 3	3.50	.00
6, 3, 4, 4, 4, 2, 3, 3, 4, 3	3.60	.10
3, 3, 4, 4, 5, 3, 2, 2, 3, 3	3.20	−.30
7, 4, 7, 3, 3, 3, 5, 4, 5, 2	4.30	.80
5, 4, 5, 2, 2, 3, 4, 3, 5, 3	3.60	.10
6, 7, 4, 4, 4, 5, 4, 4, 2, 2	4.20	.70
2, 2, 3, 3, 1, 2, 3, 2, 3, 4	2.50	−1.00
3, 4, 3, 4, 3, 4, 3, 4, 3, 4	3.50	.00
1, 3, 3, 2, 4, 3, 4, 3, 2, 0	2.50	−1.00
4, 6, 4, 3, 3, 4, 4, 5, 4, 1	3.80	.30
2, 2, 2, 2, 5, 5, 5, 3, 5, 4	3.50	.00
6, 7, 7, 4, 3, 0, 5, 5, 4, 2	4.30	.80

underestimate it. Across all samples of size ten, however, the overestimations cancel the underestimations and the *average* of the many sample means equals the true population mean. This property of the sample mean makes it a useful estimate of the population mean. In statistical terms, the sample mean is said to be an *unbiased estimator* of the population mean. An **unbiased estimator** of a population parameter is a statistic whose average (mean) across all possible random samples of a given size equals the value of the parameter. It is in this sense that the observed sample mean is the "best estimate" of the value of the population mean.

7.3 Estimation of the Population Variance and Standard Deviation

In the previous section, we selected a sample that had the following family sizes from the population of 100 families: 3, 4, 4, 5, 2, 4, 1, 1, 4, 3. If we were to calculate the variance for this set of scores, we would find 1.69. This value is different from the true population variance ($\sigma^2 = 2.09$) and, again, reflects sampling error. Unlike the sample mean, however, the sample variance is *not* our best estimate of the true population variance. Statisticians have determined that the sample variance is a *biased estimator* of the population variance because it underestimates (is smaller than) the population variance across all possible samples of a given size.

Equation 3.4 defined the variance of a set of scores as the sum of squares divided by N:

$$s^2 = \frac{SS}{N}$$

As noted above, this is not our best estimate of the population variance because it is biased. However, an unbiased estimator can be obtained from sample data by modifying the denominator of Equation 3.4:

$$\hat{s}^2 = \frac{SS}{N-1} \qquad\qquad [7.1]$$

where $\hat{s}^2$ (read "s-hat squared") is the symbol used to represent a sample estimate of the population variance or, as it is more commonly called, a **variance estimate.** Equation 7.1 is identical to Equation 3.4 except that $N-1$ is the denominator instead of N. This "correction" involving the subtraction of 1 from N makes the variance estimate larger than the sample variance and, hence, corrects for the tendency of the sample variance to underestimate the population variance.

Consider the ten family size scores for the sample described at the beginning of this section. The sample variance of these scores is 1.69. This reflects that the sum of squares for the data set is 16.90. Applying Equation 7.1, we find the variance estimate is

$$\hat{s}^2 = \frac{SS}{N-1}$$

$$= \frac{16.90}{10-1} = 1.88$$

Our best estimate of the value of the population variance, based on these data, is thus 1.88. Again, we recognize that there is sampling error and that the variance estimate will usually not exactly equal the true population variance. Note that the variance estimate ($\hat{s}^2 = 1.88$) in this example is closer to the population variance ($\sigma^2 = 2.09$) than is the sample variance ($s^2 = 1.69$).

The sample standard deviation is defined as the positive square root of the variance, or $\sqrt{s^2}$. By the same token, the estimate of the population standard deviation is the positive square root of $\hat{s}^2$, or

$$\hat{s} = \sqrt{\hat{s}^2} \qquad\qquad [7.2]$$

In our example, $s = \sqrt{1.69} = 1.30$ and $\hat{s} = \sqrt{1.88} = 1.37$. Our best estimate of the standard deviation of the population is 1.37. The symbol $\hat{s}$ is referred to as the **standard deviation estimate.***

---

* Technically, $\hat{s}$ is not an unbiased estimator of σ, and a correction factor is necessary to make $\hat{s}$ an unbiased estimator (Hays, 1981, p. 189). However, if $N > 10$, the amount of bias in $\hat{s}$ tends to be small. Given this and the fact that the statistics to be used in later chapters circumvent potential problems introduced by this bias, we use $\hat{s}$ as the sample estimate of the population standard deviation.

Study Exercise 7.1

A sample of ten individuals took an intelligence test on which scores could range from 0 to 150. This sample yielded a mean score of 100.00 and a sum of squares of 50. Compute the variance and the standard deviation for the sample, and estimate the variance and the standard deviation in the population.

Answer   The sample variance is the sum of squares divided by N, whereas the variance estimate is the sum of squares divided by $N - 1$. Thus,

$$s^2 = \frac{50}{10} = 5.00$$

$$\hat{s}^2 = \frac{50}{9} = 5.56$$

The respective standard deviations are the square roots of the variances:

$$s = \sqrt{5.00} = 2.24 \qquad \hat{s} = \sqrt{5.56} = 2.36$$

The preceding relationships are summarized in the following table, where calculations for sample values and estimated population values are based on N *sample* scores and calculations for population values are based on N *population* scores:

Statistical term	Sample value	Population value	Sample estimate of population value
Mean	$\overline{X} = (\Sigma\, X)/N$	$\mu = (\Sigma\, X)/N$	$\overline{X} = (\Sigma\, X)/N$
Variance	$s^2 = SS/N$	$\sigma^2 = SS/N$	$\hat{s}^2 = SS/(N - 1)$
Standard deviation	$s = \sqrt{s^2}$	$\sigma = \sqrt{\sigma^2}$	$\hat{s} = \sqrt{\hat{s}^2}$

If we had scores for *all* members of a population, we would use the formulas for population values. This situation is extremely rare in the behavioral sciences. If we had scores for a subset of a population and were interested in describing only that subset *without making inferences to the population,* we would use the formulas for sample values. This situation is also extremely rare in the behavioral sciences. By far the most common occurrence in the behavioral sciences is estimating population parameters from sample data. To do this, we use the formulas for sample estimates of population values.

Some texts do not distinguish between the sample variance and sample standard deviation, on the one hand, and the variance estimate and standard deviation estimate, on the other. Rather, the sample estimates, $\hat{s}^2$ and $\hat{s}$, are introduced as *the* measures of variability for a sample (and usually labeled s and s^2). For some applications, this might lead to confusion and, hence, the more traditional distinctions are maintained in this book.

7.4 Degrees of Freedom

An important concept in statistical estimation is *degrees of freedom.* Suppose you are told that a friend got a score of 95 on a statistics exam. Suppose you are also told that there were 100 points possible and that your friend missed 5 points. You have been given three pieces of information:

1. Your friend received a score of 95 on a test.

2. The test had 100 points possible.

3. Your friend missed 5 points.

If you had been told any two of the above pieces of information, you could have deduced the third (a score of 95 out of 100 means 5 points were missed). Thus, you have actually been given two independent pieces of information and one piece of information that is dependent on (follows from) the other two. Which two of the three are called "independent" and which of the three is called "dependent" are arbitrary. The fact of the matter is that only two pieces of information are independent. In statistics, the phrase **degrees of freedom** is used to indicate the number of pieces of information that are "free of each other" in the sense that they cannot be deduced from one another. In the above example, there are two degrees of freedom because there are two pieces of independent information.

Statistical indexes such as the sum of squares and the variance have a certain number of degrees of freedom because they are based on a certain number of pieces of information (scores on the variable of interest) that are independent of one another. Let us consider the sum of squares. When we compute a sum of squares, it is necessary to derive deviation scores about the mean. Consider the following four scores: 8, 10, 10, 12. The mean of these scores is 10.00, and if we were to compute the deviation scores of the first three scores, we would get –2, 0, and 0. Recall from Chapter 3 that the sum of signed deviation scores about the mean will always equal 0. If you are told that there are four scores in a distribution and that the signed deviation scores for three of these are –2, 0, and 0, you know that the last deviation score must be 2. The last deviation score to be computed is not *free to vary* but is determined by the other deviation scores. Thus, there are four scores and $4 - 1 = 3$ degrees of freedom in this example. This illustrates that a sum of squares around a sample mean will always have $N - 1$ degrees of freedom associated with it.

The above leads us to a basic principle in estimation techniques: *As the degrees of freedom associated with an estimate increase, the accuracy of the estimate also tends to increase.* Of concern here is the distinction between the number of degrees of freedom and the total number of pieces of information. When an estimate is based on a large number of independent pieces of information, it will tend to be relatively accurate. However, if the pieces of information tend to be dependent on one another, there is greater room for error. Technically, the accuracy of a variance estimate is not a function of the sample size (N) but rather a function of the degrees of freedom $(N - 1)$—that is, the number of independent pieces of information—used in calculating such an estimate.

In later chapters, we will encounter types of sums of squares in which deviations are taken around entities other than the sample mean. The degrees of freedom for these sums of squares will typically not be $N - 1$.

Irrespective of its specific computation, any sum of squares divided by its associated degrees of freedom is referred to as a mean sum of squares, or **mean square.** Symbolically,

$$MS = \frac{SS}{df} \qquad\qquad [7.3]$$

where MS represents a mean square, SS represents the sum of squared deviations around some entity, and df represents the degrees of freedom. As we have seen, the mean square derived by dividing the sum of squares around a sample mean by its corresponding degrees of freedom, $N - 1$, is formally known as a variance estimate.

To summarize thus far, a sample statistic may differ from the value of its corresponding population parameter because of sampling error. The mean and variance estimate are unbiased estimators of the population mean and population variance, respectively. An unbiased estimator is a statistic whose average (mean) across all possible random samples of a given size is equal to the value of the population parameter. In general, as the degrees of freedom associated with a sample statistic increase, the more accurate that statistic will be in estimating the corresponding population parameter.

7.5 Sampling Distribution of the Mean and the Central Limit Theorem

As previously noted, a sample mean will often differ from the corresponding population mean because of sampling error. For a given population of scores, a useful index of the degree to which sampling error affects the accuracy of the sample mean as an estimator of the population mean can be obtained by selecting all possible samples of a specified size, calculating the mean score for each sample, and then computing the standard deviation of the sample means across all samples.

For the sake of illustration, suppose we are trying to estimate the mean number of children for the population of 100 families referred to in Section 7.2 from a sample of ten cases. As shown in Table 7.1, the overall population mean is 3.50. If we randomly select a sample of ten families, we might observe a mean score of, say, 3.10. Although we would not actually do so in practice, suppose we were to select another random sample of size ten from the 100 families and find the mean to be 3.70. Furthermore, suppose that we repeated this process over and over until we computed the mean score for all possible samples of size ten. We would then have a distribution of "scores" composed of the mean scores for all possible samples of size ten. A distribution of scores consisting of the means for all possible samples of a given size is called a **sampling distribution of the mean.** If we wanted, we could compute a standard deviation for the scores constituting the sampling distribution of the mean just as we could for any other set of scores. We could also compute a mean for these scores (that is, the mean of the sample means), if we so desired.

In practice, we would never actually construct a sampling distribution of the mean. It is not necessary to do so because statisticians have determined that sampling distributions of this type possess certain common characteristics that allow us to gain perspective on the extent of sampling error, as well as other important information. Thus, a sampling distribution of the mean can be formally defined as a *theoretical distribution consisting of the mean scores for all possible random samples of a given size that can be drawn from a population.*

Perspectives on the mean and standard deviation of a sampling distribution of the mean, as well as its shape, are derived from an important formulation called the **central limit theorem.** This theorem has been stated in different ways within the statistical literature (e.g., see Freund, 1962), but the essence of it is captured by the following:

> *Given a population with a mean of μ and a standard deviation of σ, the sampling distribution of the mean has a mean of μ and a standard deviation of $\sigma/\sqrt{N}$, and approaches a normal distribution as the sample size on which it is based, N, approaches infinity.*

The theorem assumes that the sampling distribution is based on random samples of independent observations. Although the theorem may seem abstract, it has important implications for characterizing the sampling distribution of $\overline{X}$. We now develop these implications.

The Mean of the Sampling Distribution of the Mean

As stated in the the central limit theorem, *the mean of a sampling distribution of the mean is always equal to the population mean* (of the raw scores). The reason for this is intuitive: When we select samples of a given size, some of the sample means will overestimate the true population mean and others will underestimate it. However, when we average all of these, the underestimations will cancel the overestimations, with the result being the true population mean. If we were, for instance, to compute the mean of all possible sample means in the family size example, it would equal the population mean of 3.50. This characteristic of a sampling distribution of the mean is very important for the statistical concepts developed later.

The Standard Deviation of the Sampling Distribution of the Mean

The standard deviation of a sampling distribution of the mean is called the **standard error of the mean.** *The standard error of the mean reflects the accuracy with which sample means estimate a population mean.* Recall from Chapter 3 that a standard deviation represents an average deviation from the mean of a distribution. Because the mean in this case is the true population mean, μ, the standard error of the mean represents an average deviation of the sample means from the population mean. If the standard error of the mean is small, then the sample means based on a given sample size (N) will tend to be similar and all will tend to be close to the population mean. If the standard error of the mean is large, then the sample means based on samples of a given size will tend to differ from one another, and only some will be close to the population mean.

If the standard error of the mean for samples of size ten is 1.30 for the family size example, this would mean that, on the average, the sample means differ 1.30 units (in this case, "children") from the true population mean. This indicates a sizeable amount of error, and one would have to be concerned with how similar the particular observed sample mean is to the actual population mean. If, on the other hand, the standard error of the mean for samples of size ten is only .10, this would mean that, on the average, sample means deviate only .10 unit (or "child") from the true population mean. You would now have reason to believe that the sample mean is a fairly accurate estimator of the population mean. The utility of the concept of the standard error of the mean is that statisticians have developed a method for estimating it based on sample data. The estimated value can then be used to help interpret the sample mean relative to the population mean.

As stated in the central limit theorem, the standard deviation of a sampling distribution of the mean is equal to

$$\sigma_{\overline{X}} = \frac{\sigma}{\sqrt{N}} \qquad\qquad [7.4]$$

where $\sigma_{\overline{X}}$ represents the standard error of the mean, σ is the standard deviation of scores in the population, and N is the sample size. Equation 7.4 indicates that two factors influence the size of the standard error of the mean. The first is the sample size. As the sample size increases, the standard error becomes smaller, other things being equal. If, in the example on family sizes, we had used a sample size of 99 instead of 10, it is clear that the sampling distribution would consist of sample means that are very similar to one another because each would be based on 99 of the same 100 scores. This would result in a relatively small value of $\sigma_{\overline{X}}$. However, with smaller sample sizes, there is likely to be greater variability among the sample means. For a further discussion of the effect of sample size on statistical outcomes, see Box 7.1.

The second factor that influences the size of the standard error of the mean is the variability of scores in the population. As σ becomes smaller, so does the standard error, other things being equal. To take an extreme example, if all of the scores in the population were identical (that is, if $\sigma = 0$), then every sample would yield a mean exactly equal to the population mean. For instance, if all the scores were 4, then the mean of any sample would also be 4, as would the population mean. This would result in a value of $\sigma_{\overline{X}}$ of 0. However, when there is considerable variability in the population, the mean of a given sample might be influenced by one or more extreme scores. This would result in greater variability among the sample means and, consequently, a larger standard error of the mean.

In practice, the population standard deviation, σ, is rarely known. Rather, we typically have data based on one sample. We must therefore estimate the stan-

BOX 7.1

Polls and Random Samples

Most of us have encountered the concept of sampling error in the context of political polls published in newspapers. A consideration of sampling procedures used by major polling agencies and how sampling error affects their results may be useful. Recall from Chapter 1 that a random sample requires a list of every member of the population. Such a list is often not feasible, or even possible, to construct. For example, no master list of the names of all people in the United States currently exists, and even if it did, such a list would become dated very quickly. Because of this, many polling agencies use a variant of random sampling, called *area sampling,* when doing national polls. With this method, the United States is first divided into a large number of homogeneous geographic regions. A subset of these regions is then randomly selected using a random number table. Each of these regions is then subdivided into a large number of geographic areas (based on census tracts or districts). A subset of these areas is then randomly selected. Within each area, a list of the city or town streets is compiled. A subset of streets is then randomly selected. Finally, on each of the chosen streets, the house addresses are listed, and a subset of these is, in turn, randomly selected. The interviewer then approaches each house and randomly selects an individual within it. This completes the process. The overall region is selected at random. The tract is selected at random. The streets are selected at random. The houses are selected at random. And the individuals are selected at random. The end result is that every individual ideally has an equal chance of being included in the sample.

Although this is true in theory, it does not really hold in practice. The chosen dwelling might be vacant. Or the individual within the dwelling may not agree to participate in the poll. No nationwide survey has ever reached every person designated in a random sample and, accordingly, special correction factors must be adopted. A detailed consideration of these is presented in Gallup (1976). The process of area sampling is an extremely expensive one and is usually carried out only by major polling firms. Many smaller companies, newspapers, and magazines obtain samples in ways that call their representativeness into question. The results of public opinion polls must always be interpreted in light of their sampling procedures.

One of the most famous polling errors in presidential elections was the prediction made by the *Literary Digest* in 1936. The polling procedure used by this magazine had accurately predicted previous election results. It consisted of mailing out millions of postcard ballots to individuals who were listed in the phone book or who were on lists of automobile owners. This system was effective as long as the more affluent (people with phones or cars) were as equally likely to vote Democratic or Republican as the less affluent (people without phones and cars). With the advent of the New Deal, however, a shift in the American electorate occurred, with the more affluent tending to vote Republican and the less affluent tending to vote Democratic. The result was that the *Literary Digest* sample was overrepresented with Republican voters, leading to a prediction that the Republican candidate (Landon) would defeat the Democratic candidate (Roosevelt) by a margin of 57% to 43%. Of course, Landon did not win and was, in fact, defeated by Roosevelt by a margin of 62.5% to 37.5%.

Even when random procedures are used, a sample may not be representative of its population because unrepresentative samples can still occur by chance. What factors influence whether a random sample will truly reflect *(continued)*

Box 7.1 *(continued)*

the population? One factor noted in this chapter is the size of the sample. In general, the larger the sample, the closer a sample estimate will be to the corresponding population parameter. Most reputable polling agencies report margins of error (ranges that have a high probability of containing the actual population values) associated with their polls, and these are directly related to the size of the sample. In most cases, sample sizes of approximately 1,500 have been found to produce acceptable margins of error for political polls in the United States.

dard error of the mean by substituting the sample estimate of the population standard deviation, $\hat{s}$, into Equation 7.4 for σ:

$$\hat{s}_{\overline{X}} = \frac{\hat{s}}{\sqrt{N}} \qquad [7.5]$$

where $\hat{s}_{\overline{X}}$ is the estimate of the standard error of the mean and all other terms are as previously defined. Let us consider a numerical example.

Suppose we have two populations, A and B. A sample of ten individuals is interviewed from population A, and a sample of ten individuals is interviewed from population B. Both samples are randomly selected. For each sample, a measure of each person's age is obtained. These data are reported in Table 7.3. We begin by calculating the mean score for each sample. For Sample A it is 20.00, and for Sample B it is 30.00. We next calculate the standard deviation estimate for each sample, which turns out to be .82 for Sample A and 4.14 for Sample B. The respective estimates of the standard error of the mean are then calculated by dividing these standard deviation estimates by the square root of N. This yields values of .26 for Sample A and 1.31 for Sample B.

These analyses indicate that the mean of Sample A probably estimates the mean of population A better than the mean of Sample B estimates the mean of population B. Since the samples are the same size ($N = 10$), the difference in the estimated standard errors of the mean reflects the different sizes of the standard deviation estimates (for Sample A, $\hat{s} = .82$; for Sample B, $\hat{s} = 4.14$), which, in turn, presumably reflect differential variability of scores in the populations.

Study Exercise 7.2

Compute the estimated standard error of the mean for the scores in Study Exercise 7.1.

Answer   The estimated standard error of the mean is the standard deviation estimate divided by the square root of the sample size:

$$\hat{s}_{\overline{X}} = \frac{\hat{s}}{\sqrt{N}}$$

$$= \frac{2.36}{\sqrt{10}} = .75$$

TABLE 7.3   **Numerical Example for Estimating the Standard Error of the Mean**

	Sample A		Sample B	
	X	X^2	X	X^2
	19	361	26	676
	21	441	30	900
	20	400	34	1,156
	19	361	35	1,225
	20	400	25	625
	21	441	30	900
	21	441	24	576
	20	400	30	900
	20	400	30	900
	19	361	36	1,296

$\Sigma X = 200$ $\Sigma X^2 = 4{,}006$ $\Sigma X = 300$ $\Sigma X^2 = 9{,}154$
$\overline{X} = 20.00$ $\overline{X} = 30.00$

$$SS = \Sigma X^2 - \frac{(\Sigma X)^2}{N} \qquad\qquad SS = \Sigma X^2 - \frac{(\Sigma X)^2}{N}$$

$$= 4{,}006 - \frac{200^2}{10} = 6 \qquad\qquad = 9{,}154 - \frac{300^2}{10} = 154$$

$$\hat{s}^2 = \frac{SS}{N-1} = \frac{6}{9} = .67 \qquad\qquad \hat{s}^2 = \frac{SS}{N-1} = \frac{154}{9} = 17.1$$

$$\hat{s} = \sqrt{\hat{s}^2} = \sqrt{.67} = .82 \qquad\qquad \hat{s} = \sqrt{\hat{s}^2} = \sqrt{17.11} = 4.1\cdot$$

estimated standard error of the mean $\longrightarrow$
$$\hat{s}_{\overline{X}} = \frac{\hat{s}}{\sqrt{N}} = \frac{.82}{\sqrt{10}} = .26 \qquad\qquad \hat{s}_{\overline{X}} = \frac{\hat{s}}{\sqrt{N}} = \frac{4.14}{\sqrt{10}} = 1.31$$

$$120 - \frac{400}{4}$$

An Example of an Empirical Sampling Distribution of the Mean

We can demonstrate some of the above points by considering a contrived example with a small population, using sampling with replacement. Consider a population of four scores: 2, 4, 6, and 8. The mean of this population, μ, is $(2 + 4 + 6 + 8)/4 = 5.00$, and the standard deviation, σ, is 2.24.

We can construct an empirical sampling distribution of the mean for samples of size two by listing all possible samples of this size and their corresponding means. This has been done in columns 1 and 2 of Table 7.4. The mean of the 16 sample means is 5.00. Thus, the mean of the sampling distribution equals μ. The standard deviation of the 16 sample means can be derived with the usual procedures for a standard deviation. As shown in Table 7.4, the standard deviation of the sample means is 1.58. This is the standard error of the mean, $\sigma_{\overline{X}}$, and indicates how far, on the average, the sample means deviate from μ. This same value would result if we computed $\sigma_{\overline{X}}$ using Equation 7.4 because, in this example, $\sigma_{\overline{X}} = \sigma/\sqrt{N} = 2.24/\sqrt{2} = 1.58$.

Suppose that instead of a sample size of two, we construct a sampling

TABLE 7.4 **All Possible Random Samples and Sample Means for Samples of Size Two (Based on Minium, 1970)**

Population: 2, 4, 6, 8		
Sample	Sample mean	(Sample mean)^2
2, 2	2.00	4.00
2, 4	3.00	9.00
2, 6	4.00	16.00
2, 8	5.00	25.00
4, 2	3.00	9.00
4, 4	4.00	16.00
4, 6	5.00	25.00
4, 8	6.00	36.00
6, 2	4.00	16.00
6, 4	5.00	25.00
6, 6	6.00	36.00
6, 8	7.00	49.00
8, 2	5.00	25.00
8, 4	6.00	36.00
8, 6	7.00	49.00
8, 8	8.00	64.00
	Sum = 80.00	Sum = 440.00

$$\text{Mean} = \frac{80.00}{16} = 5.00$$

$$\text{Sum of squares of sample means} = 440.00 - \frac{80.00^2}{16} = 40.00$$

$$\text{Variance of sample means} = \frac{40.00}{16} = 2.50$$

$$\text{Standard deviation of sample means} = \sqrt{2.50} = 1.58 = \sigma_{\bar{X}}$$

distribution for a sample size of three. Table 7.5 lists all possible samples of size three and their corresponding means. The mean of these sample means is, as before, equal to 5.00, the value of μ. Due to space limitations, we have not calculated the standard deviation of the 64 sample means. If we were to do so, we would find that it equals 1.29, the same value we would obtain if we used the formula $\sigma_{\bar{X}} = \sigma/\sqrt{N} = 2.24/\sqrt{3} = 1.29$. Note that the standard error of the mean for samples of size three ($\sigma_{\bar{X}} = 1.29$) is smaller than the standard error of the mean for samples of size two ($\sigma_{\bar{X}} = 1.58$). This is because larger sample sizes will generally lead to more accurate mean estimates, everything else being equal.

The Shape of the Sampling Distribution of the Mean

A third implication of the central limit theorem is that the sampling distribution of the mean approximates a normal distribution when the sample size on which

TABLE 7.5   **All Possible Random Samples and Sample Means for Samples of Size Three (Based on Minium, 1970)**

			Population: 2, 4, 6, 8		
Sample	Sample mean	Sample	Sample mean	Sample	Sample mean
2, 2, 2	2.00	4, 4, 6	4.67	6, 8, 2	5.33
2, 2, 4	2.67	4, 4, 8	5.33	6, 8, 4	6.00
2, 2, 6	3.33	4, 6, 2	4.00	6, 8, 6	6.67
2, 2, 8	4.00	4, 6, 4	4.67	6, 8, 8	7.33
2, 4, 2	2.67	4, 6, 6	5.33	8, 2, 2	4.00
2, 4, 4	3.33	4, 6, 8	6.00	8, 2, 4	4.67
2, 4, 6	4.00	4, 8, 2	4.67	8, 2, 6	5.33
2, 4, 8	4.67	4, 8, 4	5.33	8, 2, 8	6.00
2, 6, 2	3.33	4, 8, 6	6.00	8, 4, 2	4.67
2, 6, 4	4.00	4, 8, 8	6.67	8, 4, 4	5.33
2, 6, 6	4.67	6, 2, 2	3.33	8, 4, 6	6.00
2, 6, 8	5.33	6, 2, 4	4.00	8, 4, 8	6.67
2, 8, 2	4.00	6, 2, 6	4.67	8, 6, 2	5.33
2, 8, 4	4.67	6, 2, 8	5.33	8, 6, 4	6.00
2, 8, 6	5.33	6, 4, 2	4.00	8, 6, 6	6.67
2, 8, 8	6.00	6, 4, 4	4.67	8, 6, 8	7.33
4, 2, 2	2.67	6, 4, 6	5.33	8, 8, 2	6.00
4, 2, 4	3.33	6, 4, 8	6.00	8, 8, 4	6.67
4, 2, 6	4.00	6, 6, 2	4.67	8, 8, 6	7.33
4, 2, 8	4.67	6, 6, 4	5.33	8, 8, 8	8.00
4, 4, 2	3.33	6, 6, 6	6.00		Sum = 320.00
4, 4, 4	4.00	6, 6, 8	6.67		

$$\text{Mean} = \frac{320.00}{64} = 5.00$$

it is based is sufficiently large. Of crucial importance is the fact that this holds *regardless of the shape of the underlying population.* When the sample size is greater than around 30, the normal approximation is quite good. For sample sizes of 30 or less, the approximation is less exact, although in many instances, particularly if the population is not highly skewed, the fit will be reasonable. The relationship of the sampling distribution of the mean to the normal distribution is very important in statistical theory and will be referred to extensively in later chapters.

The above points are illustrated in Figure 7.1, which presents graphs of frequency distributions for three different populations and corresponding frequency graphs of empirical sampling distributions of the mean for each of two sample sizes, $N = 10$ and $N = 30$. Note that in all cases, the sampling distribution is bell-shaped and symmetrical. The most frequently occurring sample mean is the true population mean, with very deviant sample means (from the true population mean) being infrequent. This is true even for population C even though all scores in the population occur with equal frequency. The frequency of highly deviant sample means is much less when the sample size is larger (for each population, compare the graphs for $N = 10$ and $N = 30$) and when the population variability is lower (for each sample size, compare the graphs for population A, which has a relatively small standard deviation, and population B, which has a relatively large standard deviation). This is consistent with principles developed earlier in the chapter.

FIGURE 7.1   **Frequency Distributions for Three Populations and Corresponding Sampling Distributions for Two Sample Sizes (Adapted from Johnson & Liebert, 1977)**

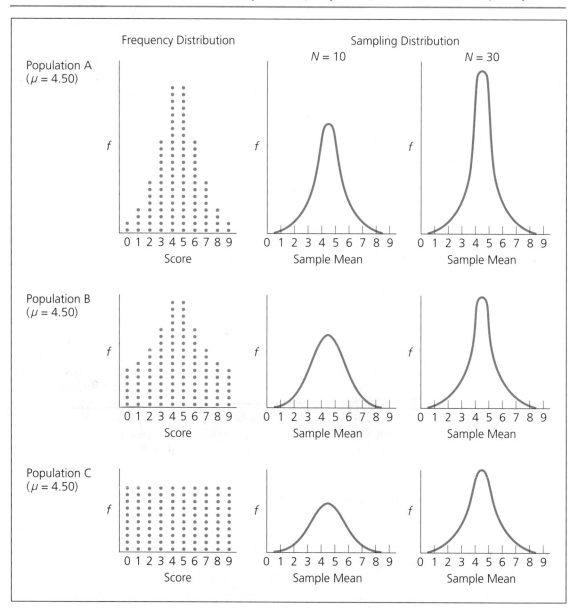

The Sampling Distribution of the Mean Summarized

The major points about the sampling distribution of the mean can be summarized as follows:

1. There is a different sampling distribution for every sample size. For instance, a sampling distribution of the mean for random samples of size ten is different from a sampling distribution of the mean for random samples of size 90.

2. The mean of a sampling distribution of the mean equals the population mean (of the raw scores).

3. The standard deviation of a sampling distribution of the mean is called the standard error of the mean.

4. The sample mean, on average, is a more accurate estimator of the population mean when the standard error of the mean is small (tends toward 0) than when the standard error of the mean is large, everything else being equal.

5. The standard error of the mean gets smaller as the sample size increases and as the variability of scores in the population decreases, everything else being equal.

6. The standard error of the mean can be estimated by dividing the standard deviation estimate by the square root of the sample size.

7. The sampling distribution of the mean approximates a normal distribution given a sufficiently large sample size. This is true regardless of the shape of the underlying population.

7.6 Types of Sampling Distributions

Although we have focused on the sampling distribution of the mean, it is possible to think of sampling distributions of other statistics, such as the mode, the median, or the variance. In the case of the median, for example, the sampling distribution reflects the medians for all possible random samples of size N that can be drawn from a population. Given the same population, the sampling distribution of the mean will show less variability (that is, it will have a smaller standard error) than either the sampling distribution of the median or the sampling distribution of the mode. It is for this reason that the mean is usually preferred by statisticians as a measure of central tendency.

**Applications
to the Analysis
of a Social Problem**

Using the parent–teen data discussed in Chapter 1, we can gain an appreciation of sampling error for selected sample values using the concepts developed in this chapter. Recall that the study design involved the selection of a random sample of inner-city African American youth in Philadelphia (between the ages of 14 and 17). The population to which we can generalize with a fair degree of confidence is African American youth (in the same age range) in the inner city of Philadelphia. To the extent that such youth are representative of inner city adolescents in other major urban areas, we can also generalize to them.

If parental communication about sex and birth control is going to be of use to teens, it is best if parents talk to them *before* the teens start to have sex. Our analyses in Chapter 2 indicated that 15.0% of the teens in the sample (excluding instances of sexual abuse) had intercourse by age 12. This means that discussion of sexual behavior should occur well before this if the parent is going to effectively discourage sexual intercourse or encourage the use of birth control for protecting against pregnancy prior to first intercourse.

As part of our study, we asked mothers how old they thought children should be when their parents should talk to them about sex and, in a separate question, how old they thought children should be when their parents start to talk to them about birth control. Tables 7.6 and 7.7 present computer outputs reporting frequency distributions, measures of central tendency, estimated standard errors of the mean, standard deviation estimates, and variance estimates for the two questions. The mean age at which mothers thought parents should talk with children about sex was 10.46, with an estimated population standard deviation of 2.83. This indicates that the ages suggested by mothers deviates, on average, by about 3 years from the mean preferred age (see Table 7.6). The estimated standard error of the mean of .11 (indicated by "Std err" on the printout) is relatively small and suggests that the sample mean is relatively close to the population mean. The mean age at which mothers thought parents should talk about birth control with children

tended to be older than that for sex (see Table 7.7). The mean was 12.09, with an estimated population standard deviation of 2.29. The estimated standard error was .09, which suggests a relatively small amount of sampling error.

Although these means, estimated standard deviations, and estimated standard errors are informative, they are not sufficient to gain an appreciation for the distributions of preferred ages provided by the mothers. For example, what are the most frequently given ages by the mothers? Are the distributions skewed? What percentage of mothers believe that ages 13 and above are best for talking about sex and birth control? This percentage is important because the analyses in Chapter 2 suggested that this age may be too late for a substantial minority of teens in relation to their first intercourse. We can gain insight into these questions only by examining other measures of central tendency and variability, and the frequency distributions, as described in earlier chapters.

Figures 7.2 and 7.3 present frequency histograms for the two age questions, with a normal distribution superimposed on each. There appears to be little skewness in either distribution. Despite the fact that the mean, median, and mode all have comparable values within a distribution (see Tables 7.6 and 7.7), both distributions are decidedly nonnormal because of the rather sharp drop in frequency for age 11 as compared to ages 10 and 12. It is interesting to note that had the decision been made to present these distributions using grouped intervals (e.g., ages 4–5, 6–7, and so on), this nonnormality would be masked.

In Chapter 3, we analyzed an important dichotomous variable: whether or not the teen (or the teen's partner) used birth control at his or her first intercourse. We noted that when such variables are scored as 1 or 0, the mean is simply the proportion of individuals who were assigned a score of 1. Table 7.8 presents computer output for the mean and estimated standard error of the mean for this dichotomous variable, where a score of 1 was assigned if the teen reported using birth control and a score of 0 was assigned if the teen reported not using birth control. The mean is

TABLE 7.6   Computer Output for Frequency Distribution and
             Summary Statistics for Age to Talk About Sex

AGESEX   Age when should talk about sex

Value Label	Value	Frequency	Percent	Valid Percent	Cum Percent
	4	16	2.3	2.3	2.3
	5	30	4.3	4.3	6.5
	6	30	4.3	4.3	10.8
	7	34	4.8	4.8	15.6
	8	51	7.2	7.2	22.8
	9	46	6.5	6.5	29.4
	10	162	23.0	23.0	52.3
	11	45	6.4	6.4	58.7
	12	137	19.4	19.4	78.2
	13	83	11.8	11.8	89.9
	14	23	3.3	3.3	93.2
	15	19	2.7	2.7	95.9
	16	22	3.1	3.1	99.0
	17	1	.1	.1	99.1
	18	5	.7	.7	99.9
	19	1	.1	.1	100.0
	Total	705	100.0	100.0	

Mean	10.464	Std err	.107	Median	10.000
Mode	10.000	Std dev	2.829	Variance	8.005
Range	15.000	Minimum	4.000	Maximum	19.000

.42, and the estimated standard error of the mean is .02. It turns out that this estimated standard error is an estimate of the standard deviation of a sampling distribution of a proportion. Thus, it estimates how far, on average, the proportions yielded by samples of a given size (in this case, $N = 425$) deviate from the true population proportion. The mean and estimated standard error can both be multiplied by 100 to convert them to percentage units.* Thus, we estimate that 42% of the population of teens used birth control at their first intercourse. The estimated standard error for this percentage is 2%, suggesting that the estimate of 42% is probably reasonably close to the true population percentage.

---

* This property of dichotomous variables can be used to calculate an estimated standard error for any percentage using computer programs that calculate means and estimated standard errors of means. One need only create a dichotomous variable that maps onto the percentage in question. For example, we could create a dichotomous variable where mothers who thought the preferred age to talk to their teen was 13 or greater receive a score of 1 and all other mothers receive a score of 0. Other formulas for estimating the standard error of a percentage are presented in Guilford (1965).

TABLE 7.7 **Computer Output for Frequency Distribution and Summary Statistics for Age to Talk About Birth Control**

AGEBC Age when should talk about birth control

Value Label	Value	Frequency	Percent	Valid Percent	Cum Percent
	4	1	.1	.1	.1
	5	3	.4	.4	.6
	6	5	.7	.7	1.3
	7	7	.9	1.0	2.2
	8	23	3.1	3.2	5.4
	9	26	3.5	3.6	9.1
	10	129	17.2	18.0	27.1
	11	41	5.5	5.7	32.8
	12	201	26.8	28.1	60.9
	13	131	17.4	18.3	79.2
	14	48	6.4	6.7	85.9
	15	39	5.2	5.4	91.3
	16	42	5.6	5.9	97.2
	17	6	.8	.8	98.0
	18	12	1.6	1.7	99.7
	19	1	.1	.1	99.9
	20	1	.1	.1	100.0
	.	35	4.7	Missing	
	Total	751	100.0	100.0	

Mean	12.092	Std err	.086	Median	12.000
Mode	12.000	Std dev	2.293	Variance	5.259
Range	16.000	Minimum	4.000	Maximum	20.000

TABLE 7.8 **Computer Output for Summary Statistics for Use of Birth Control**

BCFIRST Whether used birth control at first intercourse

Variable	Mean	S.E. Mean	Std Dev	Variance	Minimum	Maximum	N
BCFIRST	.42	.02	.49	.24	0	1	425

**FIGURE 7.2   Frequency Histogram of Age at Which Parent
Should Talk with Children About Sex**

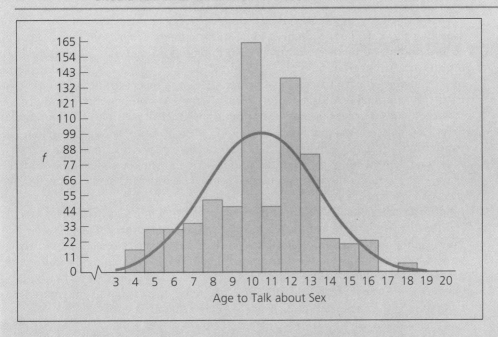

Age to Talk about Sex

**FIGURE 7.3   Frequency Histogram of Age at Which Parent
Should Talk with Children About Birth Control**

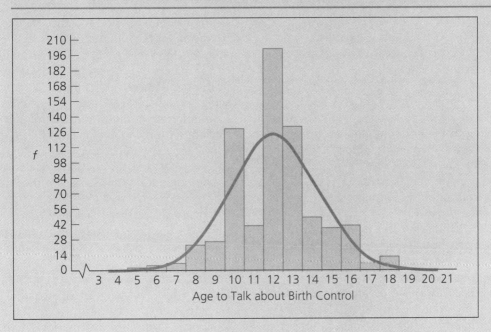

Age to Talk about Birth Control

Summary

Whenever behavioral scientists study samples, they must deal with the problem of sampling error, or the fact that a sample statistic might differ from the value of its corresponding population parameter. An unbiased estimator is a statistic whose average (mean) over all possible random samples of a given size equals the value of the population parameter being estimated. A key concept for understanding the nature of sampling error is sampling distributions. A sampling distribution of the mean refers to a distribution of mean scores based on all possible random samples of a given size. As stated in the central limit theorem, the mean of a sampling distribution of the mean will always be equal to the true population mean.

The central limit theorem also addresses the standard deviation and the shape of a sampling distribution of the mean. The standard deviation, called the standard error of the mean, indicates the average amount by which the means of random samples of size N deviate from the true population mean. The size of the standard error of the mean is influenced by two factors: the sample size (N) and the variability of scores in the population (σ). The shape of a sampling distribution of the mean approaches a normal distribution as N approaches infinity, regardless of the shape of the underlying population. The fit is particularly good when N is greater than around 30. These properties are crucial for inferential statistics and will be referred to extensively in later chapters.

Exercises

Answers to asterisked (*) exercises appear at the back of the book. Answers to exercises with two asterisks are also worked out step by step in the Study Guide.

1. Under what circumstance is sampling from a finite population analogous to sampling from an infinite population?

*2. What is sampling error? How can we represent the amount of sampling error that is present in a statistic?

*3. What is an unbiased estimator? What is a biased estimator?

4. Why, in the absence of any other information, is the observed sample mean one's "best estimate" of the value of the population mean?

5. Why is it necessary to divide the sum of squares by $N - 1$ rather than N when computing the variance estimate?

**6. Compute the variance and the standard deviation for the following scores: 2, 3, 3, 4, 4, 4, 5, 5, 5, 5, 6, 6, 6, 7, 7, 8. Estimate the variance and the standard deviation in the population, and compare the two sets of results.

7. Compute the variance and the standard deviation for the following scores: 4, 4, 4, 4, 4, 4, 5, 5, 5, 5, 6, 6, 6, 6, 6, 6. Estimate the variance and the standard deviation in the population, and compare the two sets of results.

*8. What are degrees of freedom? Why are there $N - 1$ degrees of freedom associated with a sum of squares around a sample mean?

9. What is a mean square? What is the relationship between a mean square and a variance estimate?

*10. What is a sampling distribution of the

mean? How is a sampling distribution of the mean different from a frequency distribution as discussed in Chapter 2?

*11. What three characteristics of the sampling distribution of the mean are addressed by the central limit theorem?

12. In general terms, what value will the mean of a sampling distribution of the mean always equal? Why?

*13. Given a population with three scores, 2, 4, and 6, specify all possible samples of size two that one could obtain from this population using sampling with replacement. (*Hint:* There are nine of them.) Compute the mean for each sample. Next compute the mean across the nine sample means. How does this result compare with the population mean, μ? What principle does this illustrate?

14. What is a standard error of the mean? What information does it convey?

*15. Distinguish a standard error of the mean from a standard deviation of a set of raw scores.

16. What two factors influence the size of the standard error of the mean?

*17. If a sampling distribution of the mean has an associated standard error of the mean of 0, what does this indicate about the means of the samples drawn from the relevant population?

*18. If a sampling distribution of the mean has an associated standard error of the mean of 0, what does this indicate about the variability of scores in the population (σ)?

*19. A random sample of size 30 is drawn from each of two populations. For population A, $\mu = 10.00$ and $\sigma = 5.00$. For population B, $\mu = 8.00$ and $\sigma = 7.00$. Which sample mean is probably a better estimate of its population mean? Why?

20. For each of two populations, $\mu = 10.00$ and $\sigma = 6.00$. A random sample of size 20 is drawn from population A, and a random sample of size 40 is drawn from population B. Which sample mean is probably a better estimate of its population mean? Why?

**21. Compute the mean and the estimated standard error of the mean for the data in Exercise 6.

22. Compute the mean and the estimated standard error of the mean for the data in Exercise 7.

23. Compare the estimated standard error of the mean calculated in Exercise 21 with that calculated in Exercise 22. Which sample mean is probably a better estimate of its population mean? Why?

24. Match each symbol in the first column with the appropriate term in the second column.

Symbol	Term
1. $\hat{s}_{\bar{X}}$	a. standard error of the mean
2. s	b. sample standard deviation
3. $\sigma_{\bar{X}}$	c. sample variance
4. σ^2	d. population variance
5. s^2	e. population standard
6. σ	deviation
7. $\hat{s}^2$	f. variance estimate
8. $\hat{s}$	g. estimated standard error
	of the mean
	h. standard deviation
	estimate

25. Discuss the relationship between the shape of the sampling distribution of the mean, the sample size, and the shape of the underlying population.

*26. Why do statisticians usually prefer the mean to the mode and the median as a measure of central tendency?

Multiple-Choice Questions

*27. A sampling distribution of the mean is always normal in shape.
   a. true   b. false

28. The sample mean is a biased estimator of the population mean.
    a. true   b. false

29. The symbol σ represents a
    a. sample standard deviation
    b. standard deviation estimate
    c. population standard deviation
    d. population variance

*30. Which of the following is the best measure of the amount of sampling error associated with a sample mean?
    a. population standard deviation
    b. sample standard deviation
    c. standard deviation estimate
    d. standard error of the mean

*31. As the degrees of freedom associated with an estimate of a population value increase, the accuracy of the estimate tends to
    a. decrease
    b. increase
    c. stay the same
    d. sometimes decrease and sometimes increase

32. Sampling error refers to the extent to which mistakes have been made in the collection and analysis of data.
    a. true   b. false

33. The standard deviation of a sampling distribution of the mean is called the
    a. standard deviation estimate
    b. population standard deviation
    c. standard error of the mean
    d. none of the above

*34. The standard error of the mean can never be directly calculated from a set of sample data.
    a. true   b. false

35. The sample variance is a(n) _____ estimator of the population variance because it _____ the population variance across all possible random samples of a given size.

    a. biased; underestimates
    b. biased; overestimates
    c. unbiased; equals
    d. unbiased; underestimates

*36. Behavioral science research is conceptualized as sampling from _____ populations _____ replacement.
    a. finite; with
    b. finite; without
    c. infinite; with
    d. infinite; without

37. The symbol $\hat{s}^2$ represents a(n)
    a. population variance
    b. sample variance
    c. estimated standard error of the mean
    d. variance estimate

Use the following information to complete Exercises 38–40:

Suppose we select a random sample of 100 individuals from a population and find them to have a mean self-esteem score of 82.00. Also suppose that self-esteem scores in the population are slightly negatively skewed, with a mean of 78.00 and a standard deviation of 25.00.

*38. The mean of the sampling distribution of the mean based on $N = 100$ will equal
    a. 77.00   c. 80.00
    b. 78.00   d. 82.00

*39. The standard deviation of the sampling distribution of the mean based on $N = 100$ will equal
    a. .25
    b. 2.50
    c. 25.00
    d. This cannot be determined

40. The sampling distribution of the mean based on $N = 100$ will approximate a _____ distribution.
    a. negatively skewed
    b. positively skewed
    c. normal
    d. leptokurtic

CHAPTER 8

Hypothesis Testing:
Inferences About
a Single Mean

In this chapter we consider basic principles of hypothesis testing. This material is central to all of inferential statistics and should be studied carefully. We rely on several concepts developed in previous chapters, and you should make sure that you are familiar with these concepts before covering the present material:

  probability of an event (see page 19)
  standard score (see page 105)
  normal distribution (see page 109)
  z score (see page 110)
  sampling distribution of the mean (see page 188)
  central limit theorem (see page 189)

You may wish to return to previous chapters and review these concepts.

8.1 A Simple Analogy for Principles of Hypothesis Testing

Some of the basic steps of hypothesis testing in behavioral science research can be characterized by a simple analogy. Suppose you were given a coin and asked to determine whether it was a fair coin or biased toward heads or tails. You might respond by conducting a simple experiment. If you assume that the coin is fair, then you would expect that flipping the coin a large number of times would result in heads about one-half the time because the probability of a head is .50. So you flip the coin 100 times and count the number of heads that occur. Suppose the tosses resulted in heads 52 times. This does not correspond *exactly* to what you would expect based on a probability of .50, but it certainly is close (52 out of 100 compared with 50 out of 100). Because the result is not very discrepant from the expected result, and because you know that your observed result could have deviated somewhat from the expected result just by chance, you might conclude that the coin is not biased. But suppose the result of 100 flips had yielded 65 heads? Or 95 heads? At some point, the discrepancy from the expected result becomes too great to attribute it to chance and, at this point, you would reject your original assumption that the coin is fair.

In behavioral science research, the process of **hypothesis testing** is very similar to this experiment. The investigator begins by stating a proposal, or *hypothesis,* that is assumed to be true (the coin is fair). Based on this assumption, an expected result is specified (we should obtain 50 heads out of 100 flips). The data are collected, and the observed result is compared with the expected result (52 heads versus an expectation of 50 heads). If the observed result is so discrepant from the expected result that the difference cannot be attributed to chance, then the original hypothesis is rejected. Otherwise, it is not rejected.

8.2 Statistical Inference and the Normal Distribution: The One-Sample z Test

Suppose an investigator is interested in the intelligence of the type of student who attends Victor University. Specifically, she wants to know whether the mean intelligence level of these students is different from the typical intelligence level of college students in general. Suppose that past research on a particular intelligence test has shown that a score of 100 represents the performance of the typical college student. The question of interest, then, is whether the mean intelligence test score for the population of students who attend Victor University is different from 100.

Competing Hypotheses

We can state this issue more precisely in terms of two competing hypotheses about the population mean. First, the mean intelligence test score for students who attend Victor University may, in fact, equal the value of interest, 100. This possibility is stated in a **null hypothesis:**

$$H_0: \quad \mu = 100$$

where H_0 is the symbol for a null hypothesis and μ represents the true population mean for the students who attend Victor University. We will provide a technical definition of the null hypothesis shortly. Informally, the null hypothesis can be thought of as the hypothesis of "no difference": If Victor University students are no different from the national average, then their mean must equal the national average; hence, the null hypothesis $\mu = 100$. Second, the true population mean for these students may not equal 100 but rather may be higher or lower than 100. We state this in an **alternative hypothesis:**

$$H_1: \quad \mu \neq 100$$

where H_1 is the symbol for an alternative hypothesis. The task at hand is to choose between these two competing hypotheses.

The investigator decides to do this by collecting some data. A random sample of 50 students who currently attend Victor University are administered the intelligence test. The mean score for the sample is found to be 105.00. These data would seem to support the second hypothesis stated above because the sample mean is not equal to 100, but rather is equal to 105.00. However, we know from Chapter 7 that a sample mean may not be an accurate descriptor of the population mean because of sampling error. Maybe the true population mean is 100 and we observed a sample mean of 105.00 because of sampling error. We need to test the viability of this possibility. Our knowledge of sampling distributions and the normal distribution helps us in this regard. More specifically, when the variable under study is quantitative in nature and measured on a level that at least approximates interval characteristics, hypotheses about the value of a population mean can be tested using the **one-sample z test.**

Analysis of Sampling Distributions

In the coin flipping example, we tested whether the coin was fair by assuming that such was the case and, based on this assumption, specifying an expected result that we then compared with the observed result. We will do the same for the problem with intelligence test scores. Assume that the true population mean for Victor University students is equal to 100. If we select a random sample of 50 students from the population, we would expect the mean of the sample to be near 100. We would not expect it to be exactly 100 because of sampling error (just as we would not expect flipping a fair coin 100 times to yield exactly 50 heads and 50 tails). How much sampling error can we reasonably expect to have if μ equals 100? If the sample mean were equal to 101.00, could this be due to sampling error? What about 110.00?

We can be quite specific on this matter by making reference to a sampling distribution of means based on all possible random samples of size 50. Recall that the mean of a sampling distribution of the mean equals the true population mean. We have assumed, for purpose of our statistical test, that this equals 100. Also recall that the standard deviation of the sampling distribution of the mean is called the standard error of the mean ($\sigma_{\bar{X}}$) and, per Equation 7.4, is equal to the population standard deviation divided by the square root of the sample size. In the present context, $\sigma_{\bar{X}}$ represents how much, on the average, sample means based on $N = 50$ deviate from the true population mean. Suppose we knew that the value of the population standard deviation was 17.68 and, thus, the value of the standard error of the mean was*

$$\sigma_{\bar{X}} = \frac{\sigma}{\sqrt{N}}$$

$$= \frac{17.68}{\sqrt{50}} = 2.50$$

This indicates that, on the average, sample means based on $N = 50$ deviate 2.50 units from the true population mean.

If the true population mean is 100 and the standard error of the mean is 2.50, how much sampling error can we reasonably expect in our data? Recall from Chapter 7 that a sampling distribution of means based on a relatively large N is approximately normally distributed. Also recall from Chapter 4 that 68.26% of all scores in a normal distribution occur between one standard deviation below the mean and one standard deviation above the mean. About 68% of all means in a sampling distribution of means will therefore occur between one standard error (standard deviation) below μ and one standard error above μ. Between what two scores will 95% of the sample means in a sampling distribution occur? If we consult column 2 of the z table in Appendix B, we find that 95% of the scores in a normal distribution fall between 1.96 standard deviations below the mean and 1.96 standard deviations above the mean. If 95% of all sample means fall within ± 1.96 standard errors of the mean, the probability that we would observe a sample mean *outside* this range is less than .05 (since less than 5% of the sample means are outside this range). This is a highly unlikely event. The mean of our sample *is* outside this range. It is $105.00 - 100 = 5.00$ units, or 2.00 standard errors (since the standard error is 2.50 and $5.00/2.50 = 2.00$), above the hypothesized population mean. Given this, we might reasonably conclude that the observed sample mean being as large as it is cannot be attributed to sampling error and that the assumption $\mu = 100$ is untenable. We would therefore reject the null hypothesis and, given that the sample mean is 105.00, conclude that the mean intelligence test score for the population of students who attend Victor University is higher than the typical score of 100 for college students in general.†

---

* In practice, the standard error of the mean is usually not known and must be estimated using the procedures discussed in Chapter 7. We will consider principles of hypothesis testing when estimates of the standard error are used in Section 8.9.

† The logic of using an extreme result to reject the null hypothesis has been questioned by Cohen (1994). He argues that a statistical test provides information about the probability of observing the data (or more extreme data) given that the null hypothesis is true, whereas what behavioral scientists are really interested in is the probability that the null hypothesis is true given the data. He correctly notes that these two conditional probabilities will not necessarily have the same values and that the logic of the test outlined here requires that the values be the same or similar. For the sake of exposition, we make this assumption, but we recognize that it may not always be true in practice. The mate-

We can formally summarize the above steps as follows:

1. Translate the research question into two competing hypotheses: a null hypothesis and an alternative hypothesis. The null hypothesis is the hypothesis of "no difference." More technically, it is the hypothesis that we assume to be true for purpose of conducting a statistical test.

In our example, the null hypothesis is H_0: $\mu = 100$, and the alternative hypothesis is H_1: $\mu \neq 100$.

2. Assuming the null hypothesis is true (i.e., the population mean is, in fact, 100), state an expected result in the form of a range of values within which the sample mean would be expected to fall. This is expressed in terms of the standard (*z*) scores, called **critical values,** that determine the endpoints of this range. The set of all standard scores more extreme than the critical values (that is, less than the negative critical value or greater than the positive critical value) is called a **rejection region** and constitutes an unexpected result.

In our example, the critical values are −1.96 and +1.96; thus, an expected result is defined by standard scores in the range −1.96 to +1.96, and an unexpected result (the rejection region) is defined by all standard scores less than −1.96 or greater than +1.96.

3. Characterize the mean and standard deviation of the sampling distribution of the mean, assuming the null hypothesis is true.

If we assume that the null hypothesis is true in the present example, then μ equals 100. It follows that the mean of the sampling distribution also equals 100 because the mean of a sampling distribution of the mean is always equal to the population mean. The standard deviation of the sampling distribution is the standard error of the mean. This was previously calculated to equal 2.50 in our example.

4. Convert the observed sample mean to a *z* value to determine how many standard errors it is from μ, *assuming the null hypothesis is true.* This is accomplished using the following formula for the one-sample *z* test:

$$z = \frac{\overline{X} - \mu}{\sigma_{\overline{X}}} \tag{8.1}$$

where $\overline{X}$ is the observed sample mean, μ is the population mean assuming the null hypothesis is true, and $\sigma_{\overline{X}}$ is the standard error of the mean. The *z* value in Equation 8.1 is called a **test statistic** to distinguish it from an ordinary descriptive statistic (such as the *z* score discussed in Chapter 4).

---

rial in this and later chapters represents the dominant practice in the behavioral sciences and, as such, is important to understand. However, we urge the reader to go beyond this text and consider alternative viewpoints, such as those expressed by Cohen. We will discuss this issue in more depth in Chapter 18.

In our example,

$$z = \frac{105.00 - 100}{2.50} = 2.00$$

5. Compare the observed result with the expected result. If the observed result falls within the rejection region, reject the null hypothesis. Otherwise, do not reject the null hypothesis. If the null hypothesis is rejected, compare the observed sample mean $(\overline{X})$ with the value of μ stated in the null hypothesis. If the observed sample mean is *greater* than the stated μ, conclude that the actual population mean is greater than the stated population mean. If the observed sample mean is *less* than the stated μ, conclude that the actual population mean is less than the stated population mean.*

In the present example, the observed z value of 2.00 exceeds the critical value of +1.96. This suggests that the observed difference between the sample mean and the hypothesized population mean is too large to be attributed to sampling error, and the null hypothesis is therefore rejected. Since the observed sample mean of 105.00 is greater than the stated μ of 100, the appropriate conclusion is that the mean intelligence test score of the population of students who attend Victor University is higher than 100.

8.3 Defining Expected and Unexpected Results

A major step in the preceding hypothesis testing procedure is the specification of what constitutes expected and unexpected results under the assumption that the null hypothesis is true. As discussed above, this is stated in terms of positive and negative critical values and a corresponding rejection region. Rejection regions are determined with reference to a probability value known as an **alpha level.** For example, when the alpha level is .05, a result is defined as being "unexpected" if the probability of obtaining that result, assuming the null hypothesis is true, is less than .05.

As noted previously, 95% of all sample means in the example pertaining to intelligence test scores at Victor University will fall between −1.96 and +1.96 standard errors of the true population mean, which, for purpose of hypothesis testing, is assumed to equal 100. Thus, there is less than a .05 probability of observing a sample mean outside this range. For an alpha level of .05, then, an unexpected result includes any sample mean that is more than 1.96 standard errors below or above the value of μ represented in the null hypothesis.

Although alpha can be set at levels other than .05, it is traditional to adopt an alpha level of .05 in behavioral science research. We discuss the issue of setting the alpha level in more detail shortly.

---

* Since the observed value of z will be positive if the sample mean is greater than the stated value of μ and negative if the sample mean is less than the stated value of μ, an alternative approach is to examine the sign of the observed z score.

Study Exercise 8.1

Suppose an investigator is interested in whether the mean score on an aptitude test for students who attend rural elementary schools is different from the typical score of elementary school students in general. The typical score for elementary school students in general is known to be 50. A random sample of 25 rural elementary school students is obtained, the aptitude test administered, and the mean score for the sample found to be 56.00. The population standard deviation, σ, is known to equal 10.00. Conduct a one-sample z test to test the viability of the hypothesis that $\mu = 50$.

Answer   We begin by explicitly stating the null and alternative hypotheses:

H_0: $\mu = 50$
H_1: $\mu \neq 50$

Next, we state an expected result under the assumption that the null hypothesis is true. For an alpha level of .05, this includes any z score between -1.96 and $+1.96$. An unexpected result thus consists of all z scores less than -1.96 or greater than $+1.96$.

Since $\sigma = 10.00$ and $N = 25$, the standard error of the mean can be calculated using Equation 7.4:

$$\sigma_{\overline{X}} = \frac{\sigma}{\sqrt{N}} = \frac{10.00}{\sqrt{25}} = 2.00$$

According to Equation 8.1, the observed z score is thus

$$z = \frac{\overline{X} - \mu}{\sigma_{\overline{X}}} = \frac{56.00 - 50}{2.00} = 3.00$$

Since 3.00 is greater than $+1.96$, we reject the null hypothesis and, based on a comparison of the observed sample mean (56.00) with the value of μ stated in the null hypothesis (50), we conclude that the actual population mean for rural elementary school students is higher than the typical score of 50 for elementary school students in general.

8.4   Failing to Reject Versus Accepting the Null Hypothesis

In the Victor University example, the observed sample mean was inconsistent with a sampling error interpretation of the data and the null hypothesis was therefore rejected. However, if the observed sample mean had been close to 100, say 100.50, would we have accepted the null hypothesis? The answer is no.

When a researcher obtains a result that is consistent with the null hypothesis (that is, when it falls within the range defined by the critical values rather than in the rejection region), he or she does not *accept* the null hypothesis as being true. Rather, the researcher *fails to reject* the null hypothesis. There is a subtle distinction here that is very important. In principle, *we can never accept the null hypothesis as being true from our statistical methods; we can only reject it as being untenable.*

Consider the null and alternative hypotheses from the Victor University example:

$$H_0: \quad \mu = 100$$
$$H_1: \quad \mu \neq 100$$

Note that the null hypothesis is stated such that the true population mean must equal one and only one value (100), whereas the alternative hypothesis is stated such that the true population mean could potentially equal any of an infinite number of values (for example, 100.51, 101.08, 97.63, or 112.30—anything but 100). When we observe a highly discrepant sample mean and reject the null hypothesis, we are saying that it is unlikely that the true population mean equals 100 and, in this sense, we "accept" the alternative hypothesis. In contrast, because of sampling error, we can never unambiguously conclude that the true population mean is equal to any one specific value based on sample data. If the observed sample mean is not extremely discrepant from the hypothesized population value of 100, we can say only that the sample mean is too close for us to confidently conclude that the true population mean does *not* equal 100. We cannot say that μ is equal to 100, but we also cannot confidently say that it is not. Even an observed sample mean of exactly 100 does not prove that the *population* mean is equal to 100 because sampling error could produce a sample mean of 100 even when the population mean is a much different value. Thus, when the observed value of z falls within the range defined by the critical values, we *fail to reject* the null hypothesis.

8.5   Type I and Type II Errors

When an investigator has drawn a conclusion with respect to the null hypothesis, that conclusion can be either correct or in error. Two types of errors are possible: (1) rejection of the null hypothesis when it is true (called a **Type I error**), and (2) failure to reject the null hypothesis when it is false (called a **Type II error**). These are illustrated in Table 8.1. The probability of making a Type I error is equal to the alpha level—in most cases, .05. As can be seen, alpha is symbolized by α, the lowercase Greek *a*. The probability of making a Type II error is traditionally called **beta** and is represented by β, the lowercase Greek *b*.

TABLE 8.1   **Two Types of Errors in Hypothesis Testing**

		True State of Affairs	
		H_0 is true	H_0 is false
Decision from Statistical Analysis	Reject H_0	Type I error (Probability = α)	Correct decision (Probability = $1 - \beta$)
	Fail to reject H_0	Correct decision (Probability = $1 - \alpha$)	Type II error (Probability = β)

The nature of decision errors can be explained with reference to Figure 8.1. Consider a population of scores in which the null hypothesis μ = 100 is true. Distribution A is the sampling distribution of the mean based on N = 50. The point marked "100" represents the population mean. A Type I error would occur if we selected a sample from this population and concluded, based on the sample mean, that μ did not equal 100. If the alpha level was .05, this would occur, on the average, only 5 times out of 100 because only 5% of the sample means would occur outside the range −1.96 to +1.96 standard errors. Thus, the probability of a Type I error equals the alpha level and is indicated by the shaded areas of Figure 8.1. The probability, given a true null hypothesis, of *not* making a Type I error—that is, of failing to reject the null hypothesis—is defined by the area labeled 1 − α .

Now consider the probability of a Type II error. Suppose the mean of a population of scores is 105 and we are testing the null hypothesis μ = 100. Distribution B in Figure 8.1 is the sampling distribution of the mean based on N = 50. The point marked "105" represents the true population mean. Note that if the observed sample mean occurred anywhere within the crosshatched area, we would fail to reject the null hypothesis when it is, in fact, false. This crosshatched area represents the probability of a **Type II error, or β**. The area labeled 1 − β defines the probability that an investigator will correctly reject the null hypothesis when it is false, and this probability is called the **power** of the statistical test.

These concepts can be illustrated intuitively with an electronics example. Suppose you are listening through a set of earphones and trying to decide whether you hear a particular signal. The static on the earphones makes this difficult for you. You have been told that you should hear the signal within 30 seconds. One type of error you could make is to say that you heard the signal when, in fact, it did not occur. This is analogous to a Type I error. Suppose that making such an error would lead to negative consequences. You would want to be very sure of yourself. Only if you are certain you heard the signal would you say you heard it. This is similar to setting a low alpha level (for example, .05) in an investigation.

FIGURE 8.1   **Illustration of Type I and Type II Errors (Based on McCall, 1980)**

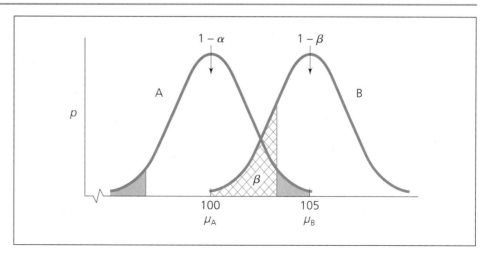

On the other hand, there is another type of error you could make—saying you did not hear the signal when, in fact, it was there. This corresponds to a Type II error. The ability not to miss the signal corresponds to the power of a statistical test. If you have a very sensitive ear, you will be likely to detect the signal when it occurs (high power). However, if you do not have a sensitive ear, you will be more likely to miss the signal (low power).

Notice that the value of the alpha level directly affects the power of the statistical test. If you are very conservative about saying you heard the signal (setting a very low alpha level), then this decreases the likelihood that you will say the signal is there when it is indeed present (that is, decreases the power).

8.6 Effects of Alpha and Sample Size on the Power of Statistical Tests

As previously noted, the alpha level in an investigation reflects the probability of making a Type I error. The tradition of adopting a low, or *conservative,* alpha level in behavioral science research evolved from experimental settings where a certain kind of error was very important and had to be avoided. An example of such an experiment is testing a new drug for medical purposes, with the aim of ensuring that the drug is safe for the general adult population. In this case, deciding that a drug is safe when, in fact, it tends to produce adverse reactions in a large proportion of adults is an error that is certainly to be avoided. Under these circumstances, the hypothesis that "the medicine is unsafe," or its statistical equivalent, would be cast as the null hypothesis and a low alpha level selected so as to avoid making the costly error. With a conservative alpha level, the medical researcher takes little risk of concluding that the drug is safe (H_1) when actually it is not (H_0). Thus, the practice of setting conservative alpha levels evolved from situations where one kind of error was extremely important and had to be avoided if possible. By casting such errors as Type I errors in the hypothesis testing framework and then setting a low alpha level, the risk of committing the error could be minimized.

Several researchers (Cohen, 1977; Greenwald, 1975) have argued that behavioral scientists have been preoccupied with Type I errors at the expense of Type II errors. The alpha level directly affects the power of the statistical test (and hence the probability of making a Type II error), with more conservative alpha levels yielding less powerful tests, everything else being equal. The argument is that for some behavioral science research, it is hard to justify that a Type I error should have the drastic character implied by a low alpha level. It is not necessarily worse, the argument goes, to conclude falsely that there is a difference between a mean and a hypothesized value (that is, to make a Type I error) than it is to conclude falsely that there is not a difference (that is, to make a Type II error). The issue concerns the balance between the risk of placing a false finding in the body of scientific knowledge versus the risk of letting an existing difference go undetected and thus unreported. By setting alpha at a less conservative level, we reduce the risk of the latter type of error, albeit at the expense of the former, everything else being equal. The issue of setting an alpha level enjoys much controversy. Interested readers are referred to Kirk (1972) for a detailed discussion of this issue.

In terms of the power of a statistical test, not only is the alpha level important, but so too is the sample size: The larger the sample size, the more powerful the statistical test will be, everything else being equal. This can be seen with reference to the formula for the one-sample z test (Equation 8.1):

$$z = \frac{\overline{X} - \mu}{\sigma_{\overline{X}}}$$

which can also be written as

$$z = \frac{\overline{X} - \mu}{\sigma/\sqrt{N}}$$

In this equation, as N becomes larger, $\sigma_{\overline{X}}$ becomes smaller, so z becomes more extreme, other things being equal. As z becomes more extreme, it is more likely that we will reject the null hypothesis when it is false, thereby increasing the power of the statistical test.

In summary, a researcher typically has control over the alpha level as well as the sample size to be used in the investigation. The power of one's test can be increased by selecting larger sample sizes and higher alpha levels. Selecting a larger sample size must be evaluated relative to the practical concerns of the increased costs of obtaining more research participants. In addition, increasing the alpha level must be evaluated relative to the importance of making a Type I versus a Type II error. There comes a point when increasing the power of a test is of diminishing value because the test is sufficiently powerful to draw a conclusion with an appropriate degree of confidence. As a rough guide, investigators generally attempt to achieve statistical power (the probability of correctly rejecting the null hypothesis when it is false) in the range of .80 to .95, depending on the nature of the proposition being investigated.

The power of a statistical test can be estimated from special tables developed for this purpose. In future chapters, we will encounter such tables, which are based on principles of power analysis discussed in Cohen (1977). These tables can also be used to estimate the sample sizes necessary to achieve desired levels of power, given the value of alpha. In most instances, researchers adopt the traditional alpha level of .05. We will follow this convention in the remainder of this book.

8.7   Statistical and Real-World Significance

If the null hypothesis is rejected, the results of a statistical test are commonly said to be *significant*. If the null hypothesis is not rejected, the phrase *nonsignificant* is often used instead. It is important to realize that these terms are meant to apply only to the *statistical* outcome. We have seen many instances in the popular press where studies are quoted as reporting "significant" findings when the researchers simply intended to convey that the null hypothesis was rejected. A statistically significant result (meaning the null hypothesis was rejected) may or may not have important practical implications.

Consider the following example. The poverty index in the United States for a family of four was defined in 1991 as an annual income of $14,120. Suppose that a researcher is interested in whether the mean 1991 income of a certain ethnic group differed from the official poverty level. The researcher is able to examine the issue using data from a large national survey that included a sample of 500,000 individuals from the ethnic group of interest. Suppose the population standard deviation is known, the observed sample mean for the ethnic group is found to be $14,300.23, and application of the one-sample z test leads to rejection of the null hypothesis. The researcher concludes that the mean income for the ethnic group population is "statistically significantly greater than the official poverty index." Such a conclusion says nothing about *how much* greater the mean income is than the official poverty index, nor does it say anything about the practical implications of the discrepancy. All the z test conveys is that the mean for the ethnic group population is not the same as the official poverty index. This general point is important to keep in mind when interpreting the results of a hypothesis test.

Why would such a small discrepancy from the official poverty index be found to be statistically significant? In this case, the sample mean is probably a very accurate estimator of the population mean because the sample size is extremely large. Because of this, when we observe a sample mean that differs at all from the official poverty index, we can be almost certain that the population mean is different from the poverty index as well. Stated more formally, the one-sample z test in this case has very high statistical power and is capable of detecting even minor departures from the official poverty index.

We believe that science would be better served by eliminating the jargon of "significance." However, this language is firmly entrenched in the behavioral sciences at this time. To help avoid possible confusion, we recommend that researchers use the term **statistically significant** in place of "significant" and **statistically nonsignificant** in place of "nonsignificant" in order to emphasize the statistical nature of the conclusion.

8.8 Directional Versus Nondirectional Tests

In each of the preceding examples, the alternative hypothesis has been *nondirectional*. It has stated that the true population mean is *either* higher *or* lower than the value specified in the null hypothesis. A *directional* alternative hypothesis, in contrast, specifies that a population mean is different from a given value and also indicates the direction of that difference. For instance, the alternative hypothesis

H_1: $\mu > 100$

states that the true population mean is *greater* than 100.

A **directional test** is designed to detect differences from a hypothesized population mean score (for example, $\mu = 100$) in one direction only. Suppose, for example, that a counselor is concerned with whether freshman students at her college have adequate reading skills. She decides to administer a reading test to a sample of incoming freshmen; if they score statistically significantly lower than the national test average of 112, she will consider instituting a remedial program. In this instance, the investigator would state a directional alternative hypothesis about the population mean of incoming freshmen:

H_1: $\mu < 112$

That is, the counselor wants the null hypothesis, H_0: $\mu = 112$, to be rejected only if the mean reading test score for the population of incoming freshmen is less than the national average of 112 because that will indicate a deficiency in reading skills. The concern in this case is only with detecting a mean *lower* than the national average.

Figure 8.2a presents a rejection region associated with only the lower end of a sampling distribution. Any score that falls in the rejection region leads to the rejection of the null hypothesis. Assuming an alpha level of .05, the rejection region is defined by z scores of less than -1.645, as indicated in column 3 of Appendix B.* Since they focus on only one tail of the distribution, directional tests are often referred to as **one-tailed tests**.

Figure 8.2b presents a rejection region for a **nondirectional test**—one that is designed to detect differences either above or below the hypothesized population mean. In this case, the .05 alpha level is "split" such that .025 of the scores occur at the upper end of the distribution and .025 of the scores occur at the lower end of the distribution. Thus, nondirectional tests are also called **two-tailed tests**. The z scores that define the rejection region are now -1.96 and $+1.96$. This can be verified from column 4 of Appendix B. Note in Figure 8.2 that the directional test is more sensitive than the nondirectional test to freshman scores being lower than the national average: If the null hypothesis is false because scores are lower than the national average, then the observed sample mean is more likely to fall in the rejection region for the directional test (z scores less than -1.645) than in the corresponding rejection region for the nondirectional test (z scores less than -1.96), since it is broader. Thus, we will be more likely to correctly reject the null hypothesis with the directional test than with the nondirectional test. In other words, in this example, the directional test is *more powerful* than the nondirectional test.

In general, a directional test will be more powerful than a corresponding nondirectional test if the actual population mean and the hypothesized population mean differ in the specified direction. However, if the actual population mean differs from the hypothesized population mean in the opposite direction from that stated in the alternative hypothesis, then a nondirectional test will be more powerful than its directional counterpart. This is because the null hypothesis cannot

---

* Specifically, Appendix B indicates that .0505 of scores in a normal distribution are less than or equal to a z score of -1.64 and that .0495 of scores in a normal distribution are less than or equal to a z score of -1.65. Therefore, .05, or 5%, of scores in a normal distribution must be less than or equal to a z score of $[-1.64 + (-1.65)]/2 = -1.645$.

FIGURE 8.2   **Rejection Regions for (a) Directional and (b) Nondirectional Tests (From Witte, 1980)**

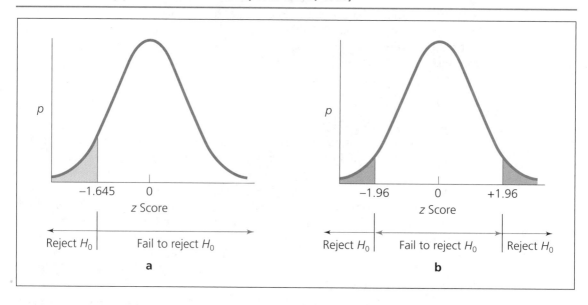

be rejected if the observed *z* score falls in the tail that does not contain the critical value. For instance, if the mean reading test score for the population of freshmen in the present example were greater than 112, perhaps yielding a sample mean corresponding to a *z* score of 2.36, the null hypothesis would be rejected if a nondirectional test were used (since 2.36 exceeds the critical value of +1.96) but not if a directional test were used (since 2.36 does not fall in the rejection region bounded by −1.645).

When there is *exclusive* concern that the population mean differs from a value in a specified direction, a directional test should be used because it will be more powerful than a nondirectional test. This concern must be stated before the data are analyzed. Never compute a statistical test and then, based on the results, decide that a directional hypothesis should be used. This defeats the logic of the hypothesis testing procedures. If concern is not with a specific direction of difference, then a nondirectional test should be used. Such tests ensure an equal likelihood of rejecting null hypotheses that are false because the actual population mean is less than the hypothesized value of μ and null hypotheses that are false because the actual population mean is greater than the hypothesized value of μ. Again, this concern must be stated before the data are analyzed.

8.9 Statistical Inference Using Estimated Standard Errors: The One-Sample *t* Test

An important property of the one-sample *z* test we developed in Section 8.2 was that the standard error of the mean, $\sigma_{\bar{X}}$, was known. It is more common that the standard error of the mean is not known and that it must be estimated from sample data.

Suppose an investigator administered to a random sample of 100 college students an attitude scale designed to measure attitudes toward living in dormitories. Scores on the scale can range from 1 to 7, with higher values representing greater favorability toward living in dormitories. A score of 1 indicates a very unfavorable attitude; a score of 4, a neutral attitude; and a score of 7, a very favorable attitude. The investigator is interested in whether the mean attitude score for the population is different from 4, the score that represents a neutral feeling. The null and alternative hypotheses are

$$H_0: \quad \mu = 4$$

$$H_1: \quad \mu \neq 4$$

Note that the alternative hypothesis is nondirectional because the investigator is interested in whether the mean attitude is either unfavorable (less than 4) or favorable (greater than 4).

The data collected from the sample of 100 students yielded a mean of 4.51 with a standard deviation estimate of 1.94. Although the sample mean is consistent with the alternative hypothesis, we must test for the possibility that the observed value is due to sampling error. But, unlike previous problems, we do not know the value of the standard error of the mean, $\sigma_{\overline{X}}$. However, we can estimate $\sigma_{\overline{X}}$ from the sample data using Equation 7.5:

$$\hat{s}_{\overline{X}} = \frac{\hat{s}}{\sqrt{N}}$$

$$= \frac{1.94}{\sqrt{100}} = .19$$

Given this estimate, you might reason that we can modify Equation 8.1 from $z = (\overline{X} - \mu)/\sigma_{\overline{X}}$ to $z = (\overline{X} - \mu)/\hat{s}_{\overline{X}}$ by simply substituting the estimated standard error, $\hat{s}_{\overline{X}}$, for the actual standard error, $\sigma_{\overline{X}}$. Unfortunately, some complications would result from this. Since $\hat{s}_{\overline{X}}$ is calculated from sample data, it is subject to sampling error, whereas $\sigma_{\overline{X}}$ is not. Stated differently, the estimate of the standard error will tend to vary from sample to sample; one could select several different samples from the same population and obtain several different values of $\hat{s}_{\overline{X}}$. This is true even for samples that have the same mean. Since the value yielded by the ratio $(\overline{X} - \mu)/\hat{s}_{\overline{X}}$ is influenced by the size of $\hat{s}_{\overline{X}}$, it would thus be possible to observe different z scores for the same value of $\overline{X}$.

Statisticians have developed procedures for dealing with this problem. Although the mathematics are complex, the idea can be stated in general terms: When $\sigma_{\overline{X}}$ is known and we calculate a z score for the sample mean, we are calculating the *exact* number of standard errors that the sample mean is from the mean of the sampling distribution, μ. If we substitute $\hat{s}_{\overline{X}}$ for $\sigma_{\overline{X}}$ and calculate a z score, we are calculating the *estimated* number of standard errors that the sample mean is from μ. The probability that a sample mean will be a certain number of *actual* standard errors from μ is not the same as the probability that a sample mean will be a certain number of *estimated* standard errors from μ. The sampling distribution in this latter instance does not follow a normal distribution, but, rather, approximates a well-known theoretical distribution called the *t* **distribution**, given certain assumptions that are discussed below. The *t* distribution can thus be used to determine the probability that a result would occur by chance, given a true null hypothesis.

The *t* distribution is similar to the normal distribution in that it is bell-shaped, unimodal, and symmetrical. In addition, as is the case with a distribution of *z* scores, the mean of the *t* distribution is 0.* As with the normal distribution, there is not a single *t* distribution but rather a family of *t* distributions. However, unlike with the normal distribution, the shape of the *t* distribution is influenced by the number of degrees of freedom that are associated with it, which, as we will see shortly, are related to the number of scores in the sample. Thus, there is a different *t* distribution for every sample size. Figure 8.3 presents example *t* and normal distributions for sample sizes of 5, 15, and 45. When N > 40, the normal and *t* distributions are quite similar. However, when the sample size is 40 or less, the shapes of the two distributions can differ somewhat. Roughly speaking, the *t* distribution tends to be a little "fatter" in the extreme regions (tails) and slightly "flatter" in the central region, with these differences becoming more pronounced as N decreases. Thus, given an interval at the end of a distribution, the proportion of scores in that interval will be greater in the *t* distribution than in the normal distribution. Stated in terms of probability, this means that the probability associated with an interval of scores at the end of the distribution will be greater in the *t* distribution than in the normal distribution. The smaller the sample size, the greater this discrepancy will be, everything else being equal.

A *t* **value**, then, is analogous to a *z* score except that it represents the num-

FIGURE 8.3 **Comparisons of *t* and Normal Distributions for Sample Sizes of 5, 15, and 45**

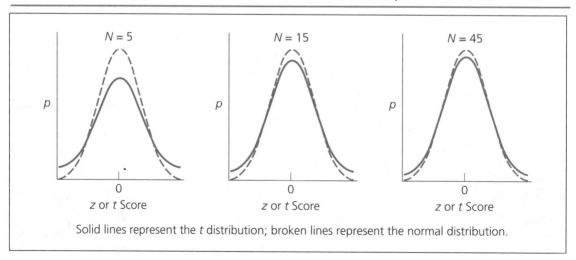

Solid lines represent the *t* distribution; broken lines represent the normal distribution.

---

* Technically, the mean of the *t* distribution is 0 only when the associated degrees of freedom, N − 1, are greater than 1.

ber of *estimated* standard errors a sample mean is from μ. The formula for calculating a *t* value is

$$t = \frac{\overline{X} - \mu}{\hat{s}_{\overline{X}}}$$ [8.2]

which is the same as the formula for calculating a *z* value but with $\hat{s}_{\overline{X}}$ substituted for $\sigma_{\overline{X}}$ in the denominator. This is formally known as the **one-sample *t* test**. Applying this test to the example concerning attitudes toward dormitories, we find

$$t = \frac{4.51 - 4}{.19} = 2.68$$

Just as statisticians have studied the normal distribution extensively, a considerable amount of information is also known about the *t* distribution. For example, it is possible to determine the probability of obtaining a *t* value greater than or equal to any specified *t* score. This is done in a similar manner as is done with *z* scores using knowledge of the normal distribution. As noted earlier, there is a different *t* distribution for each sample size. Technically, there is a different *t* distribution depending on the degrees of freedom associated with the *t* statistic, which, in the one-sample case that we are considering, equal $N - 1$.

Because of the complexity of the *t* distribution, Appendix D presents a table of critical *t* values (that is, values of *t* that define rejection regions) corresponding to only selected alpha levels. Instructions for using this table are presented in Appendix D. In the current example, the critical values of *t* that define the rejection region for an alpha level of .05, nondirectional test, and $N - 1 = 100 - 1 = 99$ degrees of freedom are approximately −1.987 and +1.987.* Thus, observed *t* values less than −1.987 or greater than +1.987 would lead us to reject the null hypothesis. The observed value of *t* in our problem is 2.68, and we therefore reject the null hypothesis. Given that the observed sample mean of 4.51 is greater than the hypothesized population mean of 4, we conclude that attitudes toward living in dormitories, on the average, are favorable.

Study Exercise 8.2

It is generally recommended that adults exercise at least twice a week for a minimum of 15 to 20 minutes. A researcher was interested in whether the mean num-

---

* Since there are no entries in Appendix D for df = 99, we must **interpolate**—that is, estimate the relevant critical values from the critical values associated with the next lowest and the next highest degrees of freedom that are listed in the *t* table. In the present example, the closest listed degrees of freedom below and above 99 are 60 and 120. For an alpha level of .05, nondirectional test, the respective critical values are ±2.000 and ±1.980, a difference of .020 unit. Since 99 is 39/60 = .65 of the way between 60 and 120, the positive critical value for df = 99 can be interpolated as 2.000 − (.65)(.020) = +1.987, thus yielding an estimated negative critical value of −1.987. The same general strategy can be applied whenever the degrees of freedom of interest are not included in a statistical table.

ber of workouts undertaken by 35- to 40-year-olds differs from the two-workouts-a-week recommendation. The mean number of times a random sample of 30 individuals of this age exercised in a particular week was found to be 1.84, with a standard deviation estimate of 1.68. Conduct a one-sample t test to test the viability of the hypothesis $\mu = 2$.

Answer   The null and alternative hypotheses are

$$H_0: \quad \mu = 2$$
$$H_1: \quad \mu \neq 2$$

For an alpha level of .05, nondirectional test, and $N - 1 = 30 - 1 = 29$ degrees of freedom, the critical values of t from Appendix D are ±2.045. Thus, if the observed t is less than −2.045 or greater than +2.045, we will reject the null hypothesis.
    Since $\hat{s} = 1.68$ and $N = 30$,

$$\hat{s}_{\bar{X}} = \frac{\hat{s}}{\sqrt{N}}$$

$$= \frac{1.68}{\sqrt{30}} = .31$$

Using Equation 8.2, we compute the observed t value:

$$t = \frac{\bar{X} - \mu}{\hat{s}_{\bar{X}}}$$

$$= \frac{1.84 - 2}{.31} = -.52$$

Since −.52 is not less than the negative critical value of −2.045, we fail to reject the null hypothesis that the mean number of workouts undertaken by the population of 35- to 40-year-olds is two per week.

Assumptions of the t Test

The one-sample t test is appropriate when the variable being studied is quantitative in nature and measured on a level that at least approximates interval characteristics. In addition, the test is based on the following assumptions:

1. The sample is independently and randomly selected from the population of interest. In most applications, independence is achieved by ensuring that the scores on the variable are provided by different individuals.

2. The scores on the variable are normally distributed in the population. This is known as the **normality assumption.**

    These assumptions are important because they assure that the sampling distribution of $(\bar{X} - \mu)/\hat{s}_{\bar{X}}$ reasonably approximates the theoretical t distribution. This, in turn, assures that the incidence of Type I errors will be equal to alpha and that the incidence of Type II errors will be equal to beta, as previously discussed.

This has obvious implications for the accuracy of the inferences to be drawn from the test.

Although it is essential that the assumption of independent and random selection be met for the test to be valid, under some conditions the one-sample t test is **robust** to violations of the normality assumption. When we say that a test is *robust* to violations of a distributional assumption, we mean that the frequency of Type I and Type II errors and, thus, the accuracy of our conclusions are relatively unaffected compared with when the assumption is met. As you might expect, the robustness of a test is influenced by several factors, including sample size (in general, robustness increases as sample size increases), the *degree* of violation (in general, robustness decreases as violations become more severe), and the *form* of the violation. In the present context, for instance, a population might be positively skewed, negatively skewed, bimodal, leptokurtic, platykurtic, and so forth.

When a test is robust to violations of an assumption, it can be appropriately applied even when that assumption is violated. The difficulty lies in establishing that the test is indeed robust for the specific circumstances under study. We will discuss procedures for assessing violations of distributional assumptions and their inherent problems in Chapter 9, when we return to the issue of the robustness of the one-sample t test.

Numerical Example

The legal highway speed limit near populated areas in many parts of the United States is 55 miles per hour. To determine whether people exceed the 55-mile-per-hour limit, suppose a state performed an investigation in which the speeds of 25 cars were monitored at selected highway locations. The observed speeds are listed in Table 8.2. As can be seen, the mean of the sample is 58.00, with a standard deviation estimate of 3.34.

The population from which this sample was drawn was considered all people who drive in this particular state. Can we conclude from these data that the average speed of this population is, in fact, higher than the 55-mile-per-hour limit? We begin by specifying a null hypothesis and an alternative hypothesis:

$$H_0: \quad \mu = 55$$

$$H_1: \quad \mu > 55$$

The problem dictates a directional test because we are interested only in whether people drive faster, on the average, than 55 miles per hour. For an alpha level of .05, directional test, and $N - 1 = 25 - 1 = 24$ degrees of freedom, the critical t value, taken from Appendix D, is 1.711. If the observed value of t is greater than 1.711, then the null hypothesis will be rejected.

Since $\hat{s} = 3.34$ and $N = 25$,

$$\hat{s}_{\overline{X}} = \frac{\hat{s}}{\sqrt{N}}$$

$$= \frac{3.34}{\sqrt{25}} = .67$$

TABLE 8.2   **Data and Calculations for Driving Speed Investigation**

X	X^2
55	3,025
60	3,600
60	3,600
55	3,025
57	3,249
60	3,600
55	3,025
58	3,364
63	3,969
54	2,916
65	4,225
56	3,136
61	3,721
58	3,364
55	3,025
57	3,249
59	3,481
53	2,809
59	3,481
65	4,225
56	3,136
61	3,721
55	3,025
54	2,916
59	3,481
$\Sigma X = 1,450$	$\Sigma X^2 = 84,368$
$\overline{X} = 58.00$	

$$SS = \Sigma X^2 - \frac{(\Sigma X)^2}{N}$$

$$= 84,368 - \frac{1,450^2}{25} = 268$$

$$\hat{s}^2 = \frac{SS}{N-1} = \frac{268}{25-1} = 11.17$$

$$\hat{s} = \sqrt{\hat{s}^2} = \sqrt{11.17} = 3.34$$

Thus,

$$t = \frac{\overline{X} - \mu}{\hat{s}_{\overline{X}}}$$

$$= \frac{58.00 - 55}{.67} = 4.48$$

Since the observed t of 4.48 is greater than 1.711, we reject the null hypothesis and conclude that people who drive in this state, on average, drive faster than the 55-mile-per-hour speed limit.

Although the statistical results are consistent with the conclusion that drivers in this particular state drive faster than 55 miles per hour, the conclusion is not definitive. Beyond the statistical analysis, one must consider the research design in order to draw an appropriate conclusion. For example, if the speeds were measured during one day only, perhaps there was something unique about that particular day relative to other days that caused drivers to speed. Driving speeds are affected by weather conditions such as temperature and wind current. Or maybe the day on which the measurements were taken was a holiday, so an unusual number of out-of-state drivers were passing through the state on vacation. Appropriate interpretation of one's data requires consideration of both the results of statistical analyses *and* the research design used to collect the data. Factors related to this issue will be considered in detail in the next chapter.

8.10 Confidence Intervals

Suppose we are trying to estimate the population mean score on an intelligence test for students at a large university. We do so by selecting a random sample of 100 students and administering the test to them. Suppose the mean score for the sample is 107.00. If we wanted to estimate the true population mean with one value, our best estimate would be 107.00. But we would also not be very confident that the true population mean is exactly 107.00, because of sampling error.

Another approach to estimating the population mean would be to specify a range of values that we are relatively confident the population mean is within (for example, between 100 and 110). The larger the range of values we specify, the more confident we are that the true population mean will be contained within it. Statisticians have developed a procedure for specifying such a range of values based on sampling distributions and probability theory.

The interval to be constructed is called a **confidence interval.** The values that define the boundaries of the interval are called the **confidence limits.** The degree of confidence we have that the population mean is contained within the confidence interval is stated in terms of a probability or a percentage. The confidence interval most commonly used by researchers in the behavioral sciences is the *95% confidence interval.*

The construction of confidence intervals differs somewhat, depending on whether the standard error of the mean is known or has to be estimated from sample data. As discussed below, this difference involves the use of $\sigma_{\bar{X}}$ and z versus $\hat{s}_{\bar{X}}$ and t in the formula for computing confidence intervals, according to whether $\sigma_{\bar{X}}$ is known or unknown.

Confidence Intervals When $\sigma_{\bar{X}}$ Is Known

Confidence intervals are conceptualized with respect to sampling distributions. In this example, the relevant sampling distribution is the sampling distribution of the mean for samples of size 100. The central limit theorem tells us that this distribution will be approximately normally distributed. When $\sigma_{\overline{X}}$ is known, we can invoke our knowledge of the area under the normal curve to determine the desired confidence interval. To illustrate the procedure, we continue with the intelligence test example.

Suppose $\sigma_{\overline{X}}$ is known to equal 2.00. Given that the sampling distribution is approximately normal in shape, we know that approximately 68% of all sample means based on $N = 100$ will fall between one standard error below μ and one standard error above μ or, in terms of raw score units, between $(1.00)(\sigma_{\overline{X}}) = (1.00)(2.00) = 2.00$ units below μ and 2.00 units above μ. Similarly, we know that 95% of all scores in a normal distribution fall between -1.96 and $+1.96$ standard deviations (in this case, standard errors) from the mean. Thus, 95% of all sample means based on $N = 100$ will fall between $(-1.96)(\sigma_{\overline{X}}) = (-1.96)(2.00) = -3.92$ and $(1.96)(\sigma_{\overline{X}}) = (1.96)(2.00) = +3.92$ raw score units of μ.

In practice, we will not know the value of μ and, in fact, μ is what we are trying to estimate based on data from our sample. Consequently, confidence intervals are calculated around the observed sample mean, $\overline{X}$, rather than μ. Establishing confidence intervals around $\overline{X}$ as opposed to μ makes sense when it is remembered that in the absence of additional information, $\overline{X}$ is one's "best estimate" of the value of μ. The observed sample mean in our example is 107.00, so the 95% confidence limits are

$$107.00 - (1.96)(2.00) = 107.00 - 3.92 = 103.08$$

and

$$107.00 + (1.96)(2.00) = 107.00 + 3.92 = 110.92$$

The range of 103.08 to 110.92 constitutes the confidence interval.

The preceding calculations reflect the general formula

$$\text{CI} = \overline{X} - (z)(\sigma_{\overline{X}}) \text{ to } \overline{X} + (z)(\sigma_{\overline{X}}) \tag{8.3}$$

where CI is the abbreviation for "confidence interval," $\overline{X} - (z)(\sigma_{\overline{X}})$ is the *lower confidence limit,* and $\overline{X} + (z)(\sigma_{\overline{X}})$ is the *upper confidence limit.* In this equation, z is the z score that corresponds to 1.00 minus the confidence level. When you use Appendix B to find the value of the desired z, always look at the nondirectional column (column 4) because the goal is to establish equivalent ranges below and above $\overline{X}$. In the example just given, we were concerned with the 95% confidence interval, so 1.00 minus the confidence level was $1.00 - .95 = .05$ and the appropriate value of z was thus 1.96.

The use of a single sample mean in the construction of a confidence interval raises an important interpretational issue. Suppose we draw all possible random samples of size N from a population and for each sample mean, compute the 95% confidence interval using Equation 8.3. We would find that 95% of these confidence intervals would contain the value of μ, and 5% of the confidence intervals would not. *A confidence claim reflects the long-term performance of an extended number of confidence intervals across all possible random samples of a given size.* In practice, only one confidence interval is constructed and that one interval either contains the population mean or does not contain the population mean. We never know for sure whether a particular confidence interval contains μ. However, the percentage associated with the confidence interval (e.g., 95%) gives us an appreciation of the degree of confidence we can have that the interval contains μ.

Sometimes investigators construct 99% confidence intervals (in which case $z = 2.575$) rather than 95% confidence intervals. The confidence level selected should depend on how important it is not to have a "false" interval (that is, one that does not contain μ). The lower the confidence level (for example, 80% as opposed to 95%), the more likely it is that false intervals will be observed. On the other hand, everything else being equal, the higher the degree of confidence (for example, 99% as opposed to 95%), the wider the confidence interval and the more likely it will contain μ. The problem with wide confidence intervals is that as the width of the confidence interval increases, the larger is the range of values that might contain μ. At some point, the range of the confidence interval becomes so large that the utility of the confidence interval is diminished. In our intelligence example, the 95% interval is 103.08 to 110.92, whereas the 99% interval is

$$CI = \overline{X} - (z)(\sigma_{\overline{X}}) \text{ to } \overline{X} + (z)(\sigma_{\overline{X}})$$

$$= 107.00 - (2.575)(2.00) \text{ to } 107.00 + (2.575)(2.00)$$

$$= 101.85 \text{ to } 112.15$$

Because the width of a confidence interval is influenced by $\sigma_{\overline{X}}$ in addition to the confidence level, it follows that the population standard deviation, σ, and the sample size, N, will also affect how wide the interval is. As σ becomes larger, so does $\sigma_{\overline{X}}$, thereby increasing the confidence interval, other things being equal. As N becomes larger, $\sigma_{\overline{X}}$ becomes smaller, so the width of the confidence interval decreases, other things being equal.

Confidence Intervals When $\sigma_{\overline{X}}$ Is Unknown

The structure of the formula for confidence intervals when $\sigma_{\overline{X}}$ is unknown is the same as when $\sigma_{\overline{X}}$ is known, except that the t distribution is used in place of the z distribution and $\hat{s}_{\overline{X}}$ is used in place of $\sigma_{\overline{X}}$. This formula can be represented as

$$CI = \overline{X} - (t)(\hat{s}_{\overline{X}}) \text{ to } \overline{X} + (t)(\hat{s}_{\overline{X}}) \tag{8.4}$$

where t is the nondirectional t value corresponding to 1.00 minus the confidence level and all other terms are as previously defined. When we determine the value of t to be used in Equation 8.4, the appropriate degrees of freedom are, as usual, $N - 1$.

For the driving speed example presented in Section 8.9, where $N = 25$, $\overline{X} = 58.00$, and $\hat{s}_{\overline{X}} = .67$, the 95% confidence interval is

$$CI = \overline{X} - (t)(\hat{s}_{\overline{X}}) \text{ to } \overline{X} + (t)(\hat{s}_{\overline{X}})$$

$$= 58.00 - (2.064)(.67) \text{ to } 58.00 + (2.064)(.67)$$

$$= 56.62 \text{ to } 59.38$$

We have a relatively high level of confidence (95%) that the true value of the population mean is within the range 56.62 to 59.38.

The calculation of confidence intervals bears a relationship to aspects of hypothesis testing. Consider the following null hypothesis:

$$H_0: \quad \mu = 100$$

Suppose that this hypothesis is tested for $N = 25$, $\overline{X} = 110.00$, and $\hat{s}_{\overline{X}} = 2.00$ at an alpha level of .05, nondirectional test, in which case the null hypothesis would be rejected. The 95% confidence interval about the mean is

$$CI = 110.00 - (2.064)(2.00) \text{ to } 110.00 + (2.064)(2.00)$$

$$= 105.87 \text{ to } 114.13$$

Note that the value of μ stated in the null hypothesis (100) is *not* contained within this interval. As it turns out, any null hypothesis that specifies a value of a population mean outside the given interval would be rejected based on the sample data. Furthermore, any null hypothesis that specifies a population mean within the interval would not be rejected. Because of this, confidence intervals provide the researcher with more information than the formal hypothesis testing procedures discussed earlier in this chapter. Specifically, the interval indicates to the investigator all null hypotheses that would or would not be rejected with a nondirectional test. Consequently, some behavioral scientists have advocated reporting confidence intervals rather than the results of formal hypothesis tests about specific values of μ when providing statistical results. Nevertheless, the large majority of research reports that you will encounter use the formal hypothesis testing procedures presented earlier rather than the present strategy, formally known as **interval estimation.**

Before concluding our discussion of confidence intervals, we should note that confidence intervals can be constructed for many parameters other than the mean, although these are rarely encountered in practice.

Study Exercise 8.3

An investigator administered a reading test to a sample of 30 students and found a mean of 83.00 with a standard deviation estimate of 17.35. Calculate the 95% and 99% confidence intervals.

Answer Since the population standard deviation is unknown, the standard error of the mean must be estimated using Equation 7.5:

$$\hat{s}_{\overline{X}} = \frac{\hat{s}}{\sqrt{N}} = \frac{17.35}{\sqrt{30}} = 3.17$$

Since df $= N - 1 = 30 - 1 = 29$, the t value used in determining the 95% confidence interval in this case is 2.045. Applying Equation 8.4, we find that the 95% confidence interval is

$$CI = \overline{X} - (t)(\hat{s}_{\overline{X}}) \text{ to } \overline{X} + (t)(\hat{s}_{\overline{X}})$$

$$= 83.00 - (2.045)(3.17) \text{ to } 83.00 + (2.045)(3.17)$$

$$= 76.52 \text{ to } 89.48$$

The t value used in determining the 99% confidence interval is 2.756, so the 99% confidence interval is

CI = 83.00 − (2.756)(3.17) to 83.00 + (2.756)(3.17)

= 74.26 to 91.74

8.11 Method of Presentation

The *Publication Manual of the American Psychological Association* (American Psychological Association, 1994) states that when presenting the results of a statistical test, "include information about the obtained magnitude or value of the test, the degrees of freedom, the probability level, and the direction of the effect. Be sure to include descriptive statistics (e.g., means or medians); where means are reported, always include an associated measure of variability" (p. 15). In addition, the alpha level should be explicitly stated either in a general statement or when providing the results of individual tests. We will illustrate the typical method of presentation that you will encounter for the one-sample t test by focusing on the driving speed example discussed in Section 8.9. The results for this problem might be reported as follows:

Results

A one-sample t test was performed comparing the sample mean against a hypothesized population mean of 55. An alpha level of .05 was used for the statistical analysis. The sample mean of 58.00 (SD = 3.34) was found to be statistically significantly different from this value, t(24) = 4.48, p < .0005, one-tailed, suggesting that the mean driving speed in the state is greater than 55 miles per hour.

The heading "Results" identifies this as a Results section. Other sections of a research report are typically headed "Method" and "Discussion." A fourth main section, the introduction, is not formally labeled because it always appears at the very beginning of a research article.

The first sentence states the type of test that was conducted and the value of μ specified in the null hypothesis. It should be noted that the null and alternative hypotheses are not formally written out when reporting the results of a statistical test. This is because the hypothesis testing steps discussed earlier in this chapter are implied whenever a statistical test is reported, so writing out each step would be unnecessary and, consequently, a poor use of journal space.

The second sentence specifies the alpha level that was used. As discussed earlier in this chapter, alpha is typically set equal to .05 by convention. If a different value of alpha is used, justification for this should be provided in the report.

The third sentence begins by specifying the value of the observed sample mean and the standard deviation estimate (reported as "SD" in parentheses) and then presents selected aspects of the statistical analysis, per American Psychological Association requirements. The symbol t indicates that a *t* test was performed. This is followed, in parentheses, by the degrees of freedom associated with the relevant *t* distribution. Next comes the observed value of *t* computed using Equation 8.2. The statement "p < .0005" indicates that the probability of obtaining a *t* value as extreme as the one observed in the study, assuming the null hypothesis is true, is less than .0005. The value associated with *p* is referred to as a **significance level.** The *p* value is also commonly called a *probability level.* A *p* value or significance level is distinct from the alpha level. An alpha level reflects the researcher's decision about how extreme the results of a statistical test should be before the null hypothesis is rejected, whereas a significance level represents the probability of obtaining a result as extreme as the one that was observed, given a true null hypothesis. The terminology *one-tailed* is used to indicate that a directional test was used. If this is not explicitly stated, a nondirectional test is assumed.

In the present example, the observed *t* value of 4.48 not only exceeds the critical value of 1.711 defining the (directional) .05 rejection region, but also exceeds the critical values of 2.064, 2.492, 2.797, and 3.745, respectively defining the .025, .01, .005, and .0005 rejection regions, as can be seen in Appendix D. The significance level is thus .0005. The advantage of reporting significance levels is that different readers might have different ideas about where alpha should be set, and if the probability associated with the observed result is specified, readers can immediately determine whether the null hypothesis would have been rejected had a more conservative (lower) alpha level been adopted.

Most statistical computer programs provide the exact significance level associated with a statistical result. If you wish, you may report this exact value. For instance, if the probability associated with a particular test result was .037, this could be presented as *p* = .037 rather than *p* < .05.

A Results section takes the same general form when the null hypothesis is not rejected as when it is. As before, the alpha level, the observed sample mean, the standard deviation estimate, the degrees of freedom, and the observed *t* value are reported. However, the statement of the significance level identifies the largest probability value from Appendix D that is exceeded by the test result (e.g., *p* > .10). Alternatively, the exact probability associated with the statistical result can be reported (e.g., *p* = .164).

8.12 Examples from the Literature

Accuracy of Subjective Life Expectancies

A subjective life expectancy is an individual's estimate of what age he or she expects to live to. One interesting question is how subjective life expectancies compare with actual life expectancies as represented by actuarial predictions. In a study of this issue, Robbins (1988) asked 49 female and 27 male college students to indicate their expected age in response to this question: Approximately how long do you expect to live? These estimates were then compared with actuarial predictions from the National Center for Health Statistics. Results showed that the mean subjective life expectancy for females (77.2 years) did not significantly differ from the actuarial prediction of 79.2 years, $t(48) = -.85, p > .20$. The mean subjective life expectancy of 77.6 years for males, on the other hand, was significantly greater than the actuarial prediction of 72.4 years, $t(26) = 2.49, p < .02$, thus indicating that males tend to overestimate their life spans.

Validation of a Priming Procedure

According to a cognitive principle known as the *availability heuristic,* the more readily instances of a class of objects come to mind, the larger that class of objects is judged to be. For instance, it has been found that people estimate that more words begin with the letter *K* than have *K* in the third position when, in actuality, the reverse is true. This bias presumably results from the fact that it is easier to think of words that start with *K* than words that have *K* as the third letter.

One factor that has been hypothesized to influence the ease with which instances of a class of objects come to mind and subsequent frequency judgments is the recency with which that class has been cognitively activated, or *primed,* through prior exposure; the more recent the exposure, the more available that information should be in memory and the greater the estimate of the class size should be. According to this perspective, this should be true even if the prior exposure occurs on an unconscious, or subliminal, level. In a test of this proposition, Gabrielcik and Fazio (1984) asked 15 college students to judge the frequency with which the letter *T* appears in the English language. Participants had previously been assigned to conditions where they were either subliminally exposed to a series of 40 words containing the letter *T* (primed condition) or subliminally exposed to strings of asterisks (control condition). If priming has the predicted effect on frequency judgments, students in the primed condition should estimate the letter *T* to be more common than should students in the control condition.

Crucial to this experiment was the establishment that the presentation of the 40 *T* words was indeed subliminal. This was accomplished in a preliminary study by asking eight undergraduate students to view a series of 40 words flashed before them for 1/500 second each. Unbeknownst to the students, each of these words contained one or more *T*s. After each presentation, students were given a slip of paper that contained four words. They were instructed to circle the word they had just viewed and to make a guess if uncertain. If the presentation of the words was indeed subliminal, students' responses to the recognition test should have been no better than chance. For a 40-item test with four response options, a chance result is $(40)(.25) = 10.00$ correct responses. The students' actual mean of 9.13 correct responses was compared with this value using a one-sample t test. This test was found to be statistically nonsignificant, $t(7) < 1$ (the specific value of t was not reported), $p > .20$, thus suggesting that word exposure of 1/500 second might be sufficiently short to be subliminal. With this established, Gabrielcik and Fazio were able to proceed with the main experiment. Consistent with their hypothesis, participants primed with the subliminal *T* words estimated the letter *T* to appear significantly more frequently than did the control students.

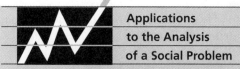

**Applications
to the Analysis
of a Social Problem**

We can apply the concepts in this chapter to the parent–teen data described in Chapter 1, although there is no conceptual question of interest that relates to the one-sample t test. We highlight here the application of confidence intervals. In Chapter 7, we calculated two means that were of practical interest: the mean age at which mothers think parents should talk with children about sex and the mean age at which mothers think parents should talk with children about birth control. We found that the former was 10.46 and the latter was 12.09. We can form 95% confidence intervals for these means. Table 8.3 presents the relevant computer printout. As indicated by the label "95% CI for Mean," the lower and upper confidence limits for the mean age to talk about sex are 10.25 and 10.67, respectively. The lower and upper confidence limits for the mean age to talk about birth control are 11.92 and 12.26, respectively. The confidence intervals are relatively narrow, which implies small amounts of sampling error.

**TABLE 8.3   Computer Output for Confidence Intervals for
Ages to Talk About Sex and Birth Control**

```
AGESEX      Age when should talk about sex

Valid cases: 705     Missing cases: 46     Percent missing: 6.1


Mean        10.4638    Std Err       .1066    Min        4.0000
Median      10.0000    Variance     8.0047    Max       19.0000
Std Dev      2.8293    Range       15.0000    IQR        3.0000
95% CI for Mean (10.2546, 10.6730)



AGEBC       Age when should talk about birth control

Valid cases: 716     Missing cases: 35     Percent missing: 4.7


Mean        12.0922    Std Err       .0857    Min        4.0000
Median      12.0000    Variance     5.2586    Max       20.0000
Std Dev      2.2932    Range       16.0000    IQR        3.0000
95% CI for Mean (11.9239, 12.2604)
```

Summary

In this chapter, we considered the basic logic underlying hypothesis testing. In doing so, we introduced two statistical tests for testing a hypothesized mean value. These tests are used whenever an investigator wants to test the viability of an assertion that a given population has a specific mean on some variable. The one-sample z test is used when the standard error of the mean is known, whereas the one-sample t test is used when the standard error of the mean must be estimated. Both tests require that the variable under study be quantitative in nature and measured on a level that at least approximates interval characteristics.

The logic of hypothesis testing begins with specifying a null hypothesis (a hypothesis that is tentatively accepted as being true for purpose of the statistical test) and an alternative hypothesis. The alternative hypothesis can be either directional or nondirectional, depending on the nature of the question being asked. Next, an alpha level, usually .05, is specified and, based on this (and in the case of the t test, the degrees of freedom), a rejection region is defined. The data are analyzed by converting the observed sample mean into a z or t value, as appropriate. This observed z or t value is compared with the critical z or t values that define the rejection region, and a conclusion is drawn. The null hypothesis is never accepted; rather, we either reject it or fail to reject it.

When we draw a conclusion from a statistical test, there is a possibility of error. The two types of errors we can make are (1) rejecting a true null hypothesis (Type I error) and (2) failing to reject a false null hypothesis (Type II error). The probability of making a Type I error is indicated by alpha (α), and the probability of making a Type II error is indicated by beta (β). The probability that an investigator will correctly reject the null hypothesis is called the power of the test and is indicated by $1 - \beta$.

One approach to estimating the value of a population parameter is to specify a range of values that has a high probability of containing the true population score. Such a range of values is called a confidence interval. When the standard error of the mean is known, confidence intervals about the mean can be determined from our knowledge of the area under the normal curve. When the standard error of the mean is estimated from sample data, the t distribution is used instead.

Exercises

Answers to asterisked (*) exercises appear at the back of the book. Answers to exercises with two asterisks are also worked out step by step in the Study Guide.

Exercises to Review Concepts

1. Define each of the following:
   a. null hypothesis
   b. alternative hypothesis
   c. critical values
   d. rejection region
   e. test statistic
   f. alpha level

*2. Why is it necessary to assume the null hypothesis is true in the context of hypothesis testing?

3. Summarize the five steps involved in hypothesis testing for the one-sample z test. Summarize the same steps for the one-sample t test.

*4. Why can we never accept the null hypothesis with traditional statistical tests?

*5. An economist is interested in whether high school students in a certain geographic area save more or less than $100 per month toward their college education. Translate this question into a null hypothesis and an alternative hypothesis.

*6. Given $H_0: \mu = 4$, $H_1: \mu \neq 4$, $\overline{X} = 7.31$, $\sigma = 14.78$, and $N = 64$, calculate the observed value of z.

*7. Test the viability of the null hypothesis for the problem in Exercise 6 and draw a conclusion about the actual value of μ.

8. Given $H_0: \mu = 10$, $H_1: \mu \neq 10$, $\overline{X} = 3.80$, $\sigma = 2.77$, and $N = 81$, calculate the observed value of z.

9. Test the viability of the null hypothesis for the problem in Exercise 8 and draw a conclusion about the actual value of μ.

10. Define each of the following:
    a. Type I error   d. beta
    b. Type II error   e. power
    c. alpha

*11. What is the relationship between alpha and the probability of a Type I error? What is the reason for this relationship?

*12. What is the relationship between power and the probability of a Type II error? What is the reason for this relationship?

*13. What is the relationship between alpha and power? What is the reason for this relationship?

14. What effect does sample size have on the power of a statistical test?

*15. Why is it important to use the phrases "statistically significant" and "statistically nonsignificant" rather than the words "significant" and "nonsignificant" when discussing the results of a statistical test?

16. Under what circumstance should a directional rather than a nondirectional test be used? Why? Under what circumstance should a nondirectional rather than a directional test be used? Why?

17. When is the t distribution used instead of the normal distribution to test hypotheses about population means?

*18. State the critical value(s) of t for a one-sample t test for an alpha level of .05 under each of the following conditions:
    a. $H_0: \mu = 3$, $H_1: \mu \neq 3$, $N = 20$
    b. $H_0: \mu = 3$, $H_1: \mu > 3$, $N = 20$
    c. $H_0: \mu = 3$, $H_1: \mu < 3$, $N = 20$
    d. $H_0: \mu = 3$, $H_1: \mu \neq 3$, $N = 10$
    e. $H_0: \mu = 3$, $H_1: \mu > 3$, $N = 10$
    f. $H_0: \mu = 3$, $H_1: \mu < 3$, $N = 10$

*19. Test the viability of the hypothesis that $\mu = 100$ using a nondirectional one-sample t test under each of the following conditions:
    a. $\overline{X} = 101.00$, $\hat{s} = 10.00$, $N = 10,000$
    b. $\overline{X} = 101.00$, $\hat{s} = 10.00$, $N = 100$
    c. $\overline{X} = 101.00$, $\hat{s} = 2.00$, $N = 100$
    In each of these problems, the difference between the sample mean ($\overline{X} = 101.00$) and the hypothesized population mean ($\mu = 100$) is the same. Why is the result in part **a** different from the result in part **b**? Why is the result in part **c** different from the result in part **b**?

20. A researcher administered a measure of life satisfaction to a sample of 30 individuals and found a mean of 121.00 with a standard deviation estimate of 10.77. Test the viability of the hypothesis that the true population mean equals 110 using a nondirectional one-sample t test.

21. What are the assumptions underlying the one-sample t test? Why are these assumptions important?

22. What is a confidence interval?

*23. A researcher administered a test of mathematical ability to a sample of 169 students. The mean of the sample was 74.40. The standard deviation for the population, σ, was known to be 13.00. Compute the 95% and 99% confidence intervals.

*24. Assuming the sample size for Exercise 23 was 200 instead of 169, compute the 95% and 99% confidence intervals. What is the effect of increasing N on the width of the intervals?

25. Assuming the population standard deviation for Exercise 23 was 7.00 instead of 13.00, compute the 95% and 99% confidence intervals. What is the effect of decreasing σ on the width of the intervals?

**26. Compute the 95% and 99% confidence intervals for the problem in Exercise 20.

27. Explain the relationship between interval estimation and formal hypothesis testing procedures.

28. What is a significance level? How does it differ from the alpha level?

Multiple-Choice Questions

29. Directional tests are also called two-tailed tests.
    a. true   b. false

*30. The alpha level is traditionally set equal to _____ in behavioral science research.
    a. .001   c. .05
    b. .01   d. .10

*31. Statistical power is the probability of
    a. making a Type I error
    b. making a Type II error
    c. correctly rejecting the null hypothesis
    d. none of the above

32. Beta is the probability of
    a. making a Type I error
    b. making a Type II error
    c. correctly rejecting the null hypothesis
    d. none of the above

*33. Alpha is the probability of
    a. making a Type I error
    b. making a Type II error
    c. correctly rejecting the null hypothesis
    d. none of the above

34. A statistically significant result for a statistical test means that
    a. the null hypothesis was not rejected
    b. the null hypothesis was rejected

c. a Type I error was made
d. the result has important practical implications

35. The alpha level and the p value represent the same information.
    a. true   b. false

*36. What is meant by $p < .01$?
    a. The probability of obtaining a statistical result as extreme as the one that was observed, given a true null hypothesis, is less than 1%.
    b. The probability of obtaining a statistical result as extreme as the one that was observed, given a true null hypothesis, is greater than 1%.
    c. Alpha was set equal to .01.
    d. Alpha was set at less than .01.

*37. Which of the following statements regarding confidence intervals is *not* true?
    a. The boundaries of confidence intervals are known as confidence limits.
    b. Any null hypothesis that specifies a population mean outside of the confidence interval will be rejected at a specified level of alpha.
    c. 95% of all 95% confidence intervals will contain the true population mean.
    d. The greater the probability that a confidence interval contains the true population mean, the narrower the confidence interval will be.

38. If $N = 26$, $\overline{X} = 16.00$, and $\hat{s}_{\overline{X}} = 1.00$, what is the 95% confidence interval?
    a. 13.94 to 18.06
    b. 14.04 to 17.96
    c. 14.29 to 17.71
    d. 15.60 to 16.40

*39. What are the degrees of freedom for a one-sample t test?
    a. $N + 1$
    b. $N/2$
    c. $N - 1$
    d. $N - 2$

40. When the sample size is greater than 40, the normal and t distributions are quite similar to each other.
    a. true   b. false

*41. When a statistical test is robust to violations of an assumption, it can be appropriately applied even when that assumption is violated.
   a. true   b. false

Exercises to Apply Concepts

**42. Recent population trends in the United States have shown a decrease in the fertility rate. Presently, the fertility rate is below the zero population growth level. The fertility rate that corresponds to zero population growth is 2.11 (that is, if couples have an average of 2.11 children, the size of the population will remain stable). One factor that has been shown to be related to family size is religion. Suppose a researcher interested in whether Catholics in the United States are having children at a rate consistent with zero population growth obtained the following hypothetical data for the numbers of children in a sample of Catholic families. Test the viability of the hypothesis that Catholics are reproducing at a zero population growth rate using a nondirectional one-sample t test, draw a conclusion, and report your results using the principles developed in the Method of Presentation section.

Number of children				
4	6	2	1	2
5	3	4	3	1
2	3	0	4	8
3	2	3	5	2
4	2	3	0	2

43. A large number of studies have investigated the effects of marijuana on human physiology and behavior. It is commonly believed by laypersons that marijuana affects human physiological processes. One of the more common beliefs is that smoking marijuana makes one hungry (gives one the "munchies"). Although the precise physiological mechanism that causes hunger is not well understood by psychologists, one factor that is often associated with hunger is blood sugar level. Several theorists have suggested that this mechanism may be the cause of marijuana-induced hunger. Weil, Zinberg, and Nelson (1968) examined this issue empirically. Ten 21- to 26-year-old men who smoked tobacco cigarettes regularly but who had never tried marijuana participated in the study. Each received a large dose of marijuana by smoking a potent marijuana cigarette. The level of blood sugar was measured for each man before he smoked marijuana and again 15 minutes after he smoked marijuana. A "change score" was then computed for each man by subtracting the amount of sugar in the blood after smoking marijuana from the amount of sugar in the blood before smoking marijuana. Hypothetical data (measured in mg/100 ml) representative of the results of the study are presented here. If marijuana has no effect on blood sugar level, the mean change score in the population would be 0. Test the viability of the hypothesis that marijuana has no effect on blood sugar level using a nondirectional one-sample t test, draw a conclusion, and report your results using the principles developed in the Method of Presentation section.

Blood sugar change score (before − after)	
14	−2
−2	−6
6	−2
−2	−18
−2	−6

44. One method researchers use to determine the accuracy with which people perceive the passage of time is *verbal estimation*. In this technique, research participants are asked to estimate how much time has passed during a given time interval. Hypothetical data representative of the estimation of 12 seconds are presented here. If people accurately perceive the passage of time, the mean time estimate in the population will equal the amount that has actually passed. Test the viability of the hypothesis that the mean population estimate is 12 seconds using a nondirectional one-sample t test, draw a conclusion, and report your results using the principles developed in the Method of Presentation section.

Estimated Time			
12	13	12	16
13	16	11	14
10	16	15	12
15	14	13	14

The Analysis of Bivariate Relationships

CHAPTER **9**

Research Design and Statistical Preliminaries for Analyzing Bivariate Relationships

Part 2 of this book focuses on the analysis of relationships between two variables, or **bivariate relationships.** In this chapter, we review some general issues of research design and test selection. In Chapters 10–16, we will discuss specific statistical techniques used in the bivariate case. We will make extensive use throughout of the basic statistical concepts outlined in Part 1.

9.1 Principles of Research Design: Statistical Implications

In order to facilitate an understanding of the use of statistics in interpreting research, it is instructive to review the general principles that guide research design. As noted in Chapter 1, statistics and research design are highly interwoven, and a consideration of design principles leads to a better understanding of the statistics considered in later chapters.

Two Strategies of Research

When studying the relationship between two variables, the investigator is essentially interested in determining how the values of one variable are associated with the values of another variable. For instance, if an investigator were interested in studying the relationship between gender (male versus female) and mathematical ability, the concern would be with whether the different values on the variable of gender are associated with different values on the variable of mathematical ability (that is, whether one gender has greater mathematical ability than the other).

Behavioral scientists use two general strategies for assessing the relationship between variables. First, they may use what is referred to as an **experimental strategy,** where a set of procedures or manipulations is performed to *create* different values of the independent variable for the research participants. The relationship of the different values of the independent variable to the dependent variable is then examined. For example, if the independent variable is test anxiety, the researcher might create three different values on this variable by telling one-third of the research participants that their performance on a test is very important and will reveal many aspects of their personal competencies (thus creating high test anxiety), telling another third that the test is unimportant and will not reflect on them personally (thus creating low test anxiety), and not addressing the issue of the test's importance with the final third. In this instance, the variable of test anxiety has three values, or *levels,* and each research participant can be distinguished with reference to a certain value.

Notice that in this example, the third group of participants was not actually exposed to the independent variable. A group of this type is formally known as a **control group.** The advantage of including a control group in an experimental strategy is that it provides a *baseline* for evaluating the effects of the experimental manipulation. Suppose, for instance, that we found that high test-anxious people scored higher than low test-anxious people on the relevant dependent variable. If a control group were not incorporated into the design, we would be unable to determine whether this was due primarily to high test anxiety *increasing* scores on the dependent variable, low test anxiety *decreasing* scores on the dependent variable, or some combination of the two. However, by including a control group, we can compare the dependent variable scores of each experimental group with those that naturally occur in the absence of the manipulation and thus determine the extent to which each value of the independent variable influences performance on the behavior of interest.

In contrast to an experimental strategy, an **observational strategy,** also called a **nonexperimental strategy,** does not involve actively creating values on an independent variable, but rather involves measuring differences in values that naturally exist in the research participants. For instance, a person's gender might be measured in a questionnaire based on response to the question: What is your gender, male or female? In this instance, the researcher is not using a set of manipulations to create values on a variable, but rather is measuring the values that naturally exist.

Experimental and observational strategies are often used together. For example, a study might investigate the effects of a person's gender *and* test anxiety on test performance. Gender would be indexed by an observational strategy and test anxiety could be indexed by the manipulations noted above. The effects of both of these variables on test performance might then be examined.

A dependent variable is always measured in the "observational" sense. This is because we are trying to determine how the dependent measure responds to the manipulation of the independent variable or varies with the naturally existing values of the independent variable. In this context, it is important to note that many of the statistical techniques that we will discuss in subsequent chapters are equally applicable to the experimental and observational situations.

Random Assignment to Experimental Groups

A major goal of research design is to control for alternative explanations. Consider an experiment in which an investigator is interested in the effect of alcohol on reaction time. Two mixed-gender groups of college students serve as participants. One-half respond to a reaction time task while under the influence of alcohol, and the other half respond to the same reaction task while not under the influence of alcohol. The task involves pressing a button when a certain slide appears in a series of slides shown sequentially. The investigator computes the mean reaction time in each group. Suppose the mean reaction time in the alcohol condition was 2.45 seconds, whereas in the no-alcohol group the mean reaction time was .98 second. It appears that alcohol has had an effect. However, there are alternative explanations. For instance, it may be the case that the alcohol did *not* affect reaction time, and that the difference between means is simply the result of students in the alcohol condition having slower reaction times than the students in the no-alcohol condition, independent of alcohol. If the study had been conducted without giving alcohol to any of the students, perhaps the results would still have yielded means of 2.45 and .98.

To control for differences in participants from one group to another, investigators typically assign individuals to groups using procedures similar to those used for selecting random samples (for instance, random number tables). If the condition in which a given person participates is determined on a completely random basis, then it is not more likely that students assigned to the alcohol condition will have slower reaction times than those assigned to the no-alcohol condition, independent of alcohol. Thus, **random assignment** helps to control for alternative explanations of results. Of course, random assignment is feasible only when an investigator is using a manipulative experimental strategy to "create" values on an independent variable. If gender is the independent variable, we cannot randomly assign participants to the conditions "male" and "female." By definition, men are males and women are females. Thus, random assignment is not possible with an observational strategy.

It is important to note that random assignment *does not guarantee* that the research groups will not differ beforehand on the dependent variable. Rather, it is *unlikely* that they will. There is always the chance that even with random assignment, the various groups will differ on the dependent variable.

Reducing Sampling Error

A second approach to controlling for alternative explanations focuses on sampling error. Consider the experiment on alcohol and reaction time. The investigator considered the two groups of participants as random samples from two populations: (1) a population of individuals who are under the influence of alcohol and (2) a population of similar individuals who are not under the influence of alcohol. The two populations are assumed to be similar in all respects except one—the presence or absence of alcohol. This is a reasonable assumption given the random assignment of participants to experimental conditions. If the alcohol has no effect on reaction time, then the two populations are, for all intents and purposes, similar in *all* respects, and we would expect the mean reaction time scores for the two populations to be the same. If the alcohol *does* have an effect on reaction time, then we would expect the population means to be different.

Suppose the mean reaction time in the alcohol condition was found to be 2.45 and the mean reaction time in the no-alcohol condition was found to be .98. This would appear to be consistent with the notion that the means for the two populations are different. However, we know from Chapter 7 that a sample mean does not usually equal the corresponding population mean because of sampling error. Perhaps the two population means are equal and the results of the experiment are nothing more than the result of sampling error. Ideally, we would like to minimize sampling error so as to rule out this interpretation as an alternative explanation. We now consider procedures that can be used to accomplish this.

Recall from Chapter 7 that two factors that influence the accuracy of a sample mean as an estimate of μ are the size of the sample and the variability of scores in the population. One way an investigator can reduce sampling error is to increase the sample sizes for the various groups. In general, the larger the sample size in each group, the smaller the sampling error. Obviously, there are practical limitations to the number of participants in a research study. Research is expensive, and the time and effort involved in collecting data can be extensive. As such, researchers will sometimes have to settle for relatively small sample sizes.

A second procedure for reducing sampling error is to define the groups such that the variances of scores in the populations (σ^2) will be relatively small. Consider the alcohol and reaction time example. There is some evidence to indicate that men, in general, have slightly faster reaction times than women on tasks similar to that used in the experiment. The population of individuals in the alcohol condition consists of both men and women, as does the population of individuals in the no-alcohol condition. The presence of both men and women within the populations yields more variability in reaction time scores than if either gender were considered separately. This can be illustrated with the following hypothetical sets of reaction time scores for a particular condition:

Men	Women	Men and women combined	
1.10	1.30	1.10	1.30
1.20	1.40	1.20	1.40
1.30	1.50	1.30	1.50
$\mu = 1.20$	$\mu = 1.40$	$\mu = 1.30$	
$\sigma = .007$	$\sigma = .007$	$\sigma = .017$	

Note that the reaction time for men is slightly faster, on the average, than the reaction time for women. Note also that the variance for the combined groups (.017) is greater than the variance for either group considered separately (.007 in both cases).

Sampling error could be reduced by *holding gender constant* by restricting the experiment to women. This would have the effect of yielding smaller population variances within both the experimental and the control group relative to a study using both men and women. Unfortunately, to reduce σ^2, we have then restricted the generalizability of the results of the study to women only. Behavioral scientists frequently find themselves in such trade-off situations. Researchers must weigh the theoretical and applied benefits of reducing sampling error against the cost of reducing the generalizability of their results.

Control of Confounding and Disturbance Variables

Much of behavioral science research is designed to study relationships between independent and dependent variables. In order to draw unambiguous inferences about such relationships, it is necessary to control other variables in the research setting. In the previous sections, we implicitly considered two basic types of variables a researcher must control. In this section, we make the nature of these variables explicit.

One type of variable that a researcher seeks to control is a *confounding variable*. Suppose a researcher wanted to test for the existence of gender discrimination and did so by examining the relationship between the gender of a job applicant and the likelihood of that individual being hired. An experiment might be designed in which the resumes of 50 applicants (25 men and 25 women) are gathered from personnel files. A group of personnel directors is then asked to read each resume and rate on a 20-point scale the likelihood that they would hire each applicant. Suppose the appropriate statistical analysis indicated there was, in fact, a relationship between the applicant's gender and the likelihood-of-hiring ratings, with men being more likely to be hired than women. Would this be evidence for gender discrimination? Not necessarily. The strongest evidence for gender discrimination would be if men are chosen over women who are equally or more qualified. In the above experiment, women may have been less qualified than the men. Rather than the judgments being a function of the applicants' gender, they may have simply reflected the applicants' qualifications, which just happened to be related to their gender. If this were the case, the test for gender discrimination would be ambiguous because the result could be attributed either to gender discrimination *or* to differences in the quality of applicants. In this experiment, the qualifications of the applicants represent a **confounding variable.** *A confounding variable is one that is related to the independent variable (the presumed influence) and that affects the dependent variable (the presumed effect), rendering a relational inference between the independent variable and the dependent variable ambiguous.* In the study on the effect of alcohol on reaction time, the failure to randomly assign participants to conditions could result in different reaction times for the two groups due to individual differences. Random assignment is one way to control confounding variables defined by individual differences.

A second type of variable that must be controlled in research aimed at drawing relational inferences is a **disturbance variable.** *A disturbance variable is one that is unrelated to the independent variable (and hence not confounded with it) but that affects the dependent variable.* As a result, a disturbance variable increases sampling error by increasing the variability within groups. In the alcohol and reaction time study, the gender of the participants is a disturbance variable. Disturbance variables obscure or mask a relationship that exists between the independent and dependent variables. To use an electronics analogy, they create "noise" in a system where we are trying to detect a "signal," and the more "noise" there is, the harder it is to detect the "signal."

Behavioral scientists use several procedures to control for confounding and disturbance variables. Three of the most common strategies are (1) holding a variable constant, (2) matching, and (3) random assignment to experimental groups. The first of these is applicable to both confounding and disturbance variables, whereas the last two can be applied only to confounding variables.

Consider the following example. In the sociological literature, a relationship has been established between a woman's religion and how many children she wants to have in her completed family (that is, her ideal family size). Generally speaking, Catholics want more children than Protestants, who in turn want more children than Jews. Sociologists have interpreted this in terms of the effects of the religious doctrine to which these women are exposed. Another interpretation is possible, however. Catholics tend to come from larger families than Protestants, who tend to come from larger families than Jews. It may not be religion that influences family size desires but rather that those people who are raised in large families prefer large families and those people who are raised in small families prefer small families. Thus, religion and the size of the family raised in might be confounded with each another. How might family size background be controlled?

We saw earlier with the alcohol and reaction time example how the disturbance variable of gender could be controlled by being held constant. **Holding a variable constant** can also be used to control for a confounding variable, such as family size background. For example, a study might be undertaken with Catholics, Protestants, and Jews who all come from families with two children. In this case, family size background and religion are *not* related because family size background has been held constant. Then differences in ideal family size among the three religious groups cannot be attributed to differences in family size background because everyone in the study comes from the same-sized family. The variables are no longer confounded.

As noted earlier, the major disadvantage of holding a variable constant is that it may restrict the generalizability of the results. If the above study is conducted on only individuals who come from families with two children, would the results generalize to individuals who come from families with five children? Perhaps religion influences family size desires when one comes from a relatively small family (two children) but not when one comes from a relatively large family (five children). When we hold a variable constant, we have no way of knowing the extent to which the results will generalize across the different levels of the variable that is held constant. One way to circumvent this problem would be to design an investigation in which one studied the effect of religion at each of several levels of family size background (for example, one child, two children, three children, and so on). A design of this type is called a *factorial design* and will be discussed in Chapter 17.

A second strategy that is used to control for confounding variables is called **matching.** In this approach, an individual in one group is "matched" with an individual in each of the other groups such that all of these individuals have the same value on the confounding variable. This strategy is different from holding a variable constant because within a group, the confounding variable can vary considerably. However, for each individual in one group, there is a comparable individual in each of the other groups who has the same value on the confounding variable. As an example, 15 women (5 in each group) who have the following family size backgrounds might be selected for inclusion in the religion and ideal family size study:

Catholics	Jews	Protestants
3	3	3
4	4	4
2	2	2
3	3	3
1	1	1
$\overline{X} = 2.60$	$\overline{X} = 2.60$	$\overline{X} = 2.60$

Note that the average family size background is identical in the three groups. Thus, any differences in the mean *ideal* family size among the three groups cannot be attributed to differences in family size background. Again, religion and family size background are now unconfounded. Furthermore, the problem of restricted generalizability of results that is encountered when holding a variable constant is not applicable because a range of family size backgrounds is included. Unfortunately, however, in practice it is often difficult to identify appropriate variables to serve as a basis for matching and, once identified, to readily complete the matching process.

A final strategy for dealing with confounding variables is **random assignment** to experimental groups. As discussed above, when used in this context, random assignment helps to control for confounding variables due to individuals' backgrounds. However, since random assignment cannot be done with observational groups, observational independent variables will always be confounded with all other variables that are naturally related to them (except, of course, those variables that can be controlled by being held constant or matching). For further discussion of confounding and disturbance variables, see Box 9.1.

An Electronics Analogy

Many of the concepts discussed thus far can be illustrated intuitively using the electronics analogy from Chapter 8. Suppose you are listening through a set of earphones and trying to decide whether you hear a particular signal. There is a good

BOX 9.1

Confounding and Disturbance Variables

The identification of confounding and disturbance variables is critical for evaluating any research design. Huck and Sandler (1979) have presented an interesting and very readable collection of 100 studies that have received attention in the popular press or professional forums. For each one, they provide a description and elaborate on some confounding and disturbance variables that could affect the interpretation of the results. An example from their book illustrates their approach and underscores the importance of controlling such variables.

Problem

The following story appeared about an advertisement in a weekly news magazine as well as in the local newspapers—you may have seen it yourself. It seems that the Pepsi-Cola Company decided that Coke's three-to-one lead in Dallas was no longer acceptable, so they commissioned a taste-preference study. The participants were chosen from Coke drinkers in the Dallas area and asked to express a preference for a glass of Coke or a glass of Pepsi. The glasses were not labeled "Coke" and "Pepsi" because of the obvious bias that might be associated with a cola's brand name. Rather, in an attempt to administer the two treatments (the two beverages) in a blind fashion, the Coke glass was simply marked with a "Q" and the Pepsi glass with an "M." Results indicated that more than half chose Pepsi over Coke. Besides a possible difference in taste, can you think of any other possible explanation for the observed preference of Pepsi over Coke? (p.11)

Solution

After seeing the results of the Pepsi experiment, the Coca-Cola Company conducted the same study, except that Coke was put in both glasses. Participants preferred the letter "M" over the letter "Q," thus creating the plausible rival hypothesis that letter preference rather than taste preference could easily have accounted for the original results. [In other words, the type of beverage might have been confounded with the letter used to label the glasses.] Since no statistical tests were given, another plausible rival hypothesis is that of instability; that is, we don't know whether "more than half" means 51% or 99% or how much confidence we should place in the finding. Flipping a coin 100 times is almost sure to result in either heads or tails occurring more than half the time.

Strangest of all was the fact that the same design error of using one letter exclusively for each brand was repeated in a second study conducted by Pepsi. In a feeble attempt to demonstrate that their initial results were not biased by the use of an "M" or a "Q," Pepsi duplicated their first study, this time using an "L" for Pepsi and an "S" for Coke! Clearly, these three studies indicate that there is sometimes more in advertisements than meets the eye (or the taste buds). (p. 158)

deal of static on the earphones. In research design, the static corresponds to disturbance variables, and your goal is to eliminate it to the extent that you can. Suppose you do some repairs and eliminate a large portion of the static. This is analogous to controlling for disturbance variables. There is still another problem, however: There are two other signals very much like the one you must detect. If you hear them, you will think your signal has occurred when, in fact, it has not. These other signals represent confounding variables. A mechanical device you hook up completely eliminates one of these signals and turns the other one into static (that is, a disturbance variable). The additional static is relatively minor, so you decide that you have done everything possible to assure accurate identification of the target signal.

However, you are confronted with yet another problem, as was discussed in Chapter 8. There are two types of errors you can make: (1) claiming you heard the signal when, in fact, it did not occur, and (2) claiming you did not hear the signal when, in fact, it did occur. These correspond to Type I and Type II errors. Suppose you determine that falsely saying the signal was present would be very detrimental, and that falsely saying it did not occur would be of little importance. In this case, you would want to ensure that you minimize the first type of error. This would correspond to setting a low alpha level in a study. With this in mind, you proceed to listen.

Problems with Inferring Causation

The statistics developed in this text are designed to indicate to a researcher whether there is a relationship between variables in the context of the research conducted. It must be emphasized that these statistics say nothing about whether two variables are *causally* related. It is entirely possible for two variables to be related to each other but for no causal relationship to exist between them. A good example of this is height and hair length. A random sample of adults in the United States would typically reveal a moderate relationship between length of hair and how tall someone is: People with shorter hair tend to be taller than people with longer hair. Is there a causal relationship between these variables? If you cut your hair, will you grow taller? Certainly not. It turns out that this relationship is due to a confounding variable, gender. Women tend to wear their hair longer than men. Women also tend to be shorter than men. These gender differences conspire to produce a relationship between hair length and height. If we were to remove the influence of gender, there would be no relationship between hair length and height.

Most statistics and research design books emphasize this point about causation only when considering correlational analysis. However, the issue holds for *all* statistics that assess the relationship between variables. *The ability to make a causal inference between two variables is a function of one's research design, not the statistical technique used to analyze the data that are yielded by that research design.* Since observational independent variables will always be confounded with all other variables that are naturally related to them, causal inferences are typically not possible when an observational research strategy is used. When an experimental research strategy is used, inferences of causation can be made only when confounding variables are controlled. The fact that two variables being related does not necessarily imply causality is important to keep in mind when interpreting statistics in the context of behavioral science research.

Between- Versus Within-Subjects Designs

Consider an experiment where the investigator is interested in the relationship between two variables: type of drug and learning. Specifically, the investigator wants to know whether two drugs, A and B, differentially affect performance on a learning task. Fifty participants are randomly assigned to one of two conditions. In the first condition, 25 participants are administered drug A and then read a list of 15 words. They are subsequently asked to recall as many of the words as possible. A learning score is derived by counting the number of words correctly recalled (hence, scores can range from 0 to 15). In the second condition, a different 25 participants read the same list of 15 words and respond to the same recall task after being administered drug B. The relative effects of the drugs on learning are determined by comparing the responses of the two groups.

In this experiment, the investigator is studying the relationship between two variables: (1) type of drug and (2) learning as measured on a recall task. Type of drug is the independent variable and the learning measure is the dependent variable. The independent variable is set up so that participants who received drug A did *not* receive drug B and those who received drug B did *not* receive drug A; that is, the two groups included different individuals. A variable of this type is known as a *between-subjects* variable because the values of the variable are "split up" between participants instead of occurring completely within the same individuals. Research designs that involve between-subjects independent variables are referred to as **between-subjects designs** or **independent groups designs.**

Now consider a similar experiment that is conducted in a slightly different fashion. A group of 25 participants are administered drug A and then given the learning task. One month later, the same 25 people return to the experiment and are given the learning task after being administered drug B. The performance of these participants under the influence of drug B is then compared with their earlier performance under the influence of drug A. Note that in this experiment the 25 subjects who received drug A also received drug B; that is, the same individuals participated in both conditions. A variable of this type is known as a *within-subjects* variable. Research designs that involve within-subjects independent variables are referred to as **within-subjects designs, correlated groups designs,** or **repeated measures designs.**

As illustrated by the above example, between-subjects designs and within-subjects designs are both viable strategies when the independent variable is experimental in nature. However, as we will see in later chapters, the two approaches require different statistical procedures. For many observational independent variables, only between-subjects designs are applicable. Consider the variable of gender, which has two levels: male and female. This variable is, by definition, a between-subjects variable. Individuals who are in the group called "male" by definition cannot also be in the group called "female."

The relative advantages and disadvantages of designing an investigation using a between-subjects versus a within-subjects design can be illustrated in the context of the example relating type of drug (A versus B) to learning. One advantage of the within-subjects approach is that it is more economical in terms of participants. For instance, in our example, half the number of participants were required to achieve the same per-condition sample size (25) when a within-subjects design was used as when a between-subjects design was used (25 versus 50). Participant economy is particularly important when a large amount of time, effort, or expense is necessary to recruit and train research participants.

A second advantage of within-subjects designs is the control of confounding variables that they provide. In the ideal experiment, individuals in the two conditions (drug A versus drug B) would be identical in all respects except one—the type of drug they are given. If the two groups differ in learning, then there would be one and only one logical explanation: The difference in drugs caused the difference in learning. In the between-subjects design, individuals are randomly assigned to one of the two conditions. The random assignment constitutes an attempt to "equalize" the two groups on all variables except the drug. If individuals are randomly assigned, it is unlikely that the individuals in one condition will, for example, be more intelligent (on the average) than the individuals in the other condition. But the key word here is *unlikely*. Although it is *unlikely* that a between-group difference in intelligence will occur, it *could* happen due to chance factors. When this does occur, the differential intelligence in the two conditions could make one drug appear superior when it is not. Or, it might offset the effect of the superior drug, making it appear as if there is no difference between the drugs when there actually is.

In contrast, this cannot happen with within-subjects designs. The same individuals who receive drug A also receive drug B. Since the same individuals are involved, intelligence must be the same in both conditions (unless it changed in the one month that separated the administration of the drugs). This is true not only of intelligence but of all individual differences that might otherwise render interpretation ambiguous. Thus, the within-subjects design can offer considerably more experimental control than the between-subjects design.

This last statement must be qualified by additional considerations. One potential problem with within-subjects designs is that the treatment in the first condition may have **carry-over effects** that influence performance in the second condition. For example, the effect of drug A may not have worn off completely when drug B is administered. This could make interpretation of the experiment ambiguous. Carry-over effects are not necessarily restricted to the independent variable of interest. Performance on the dependent measure during the second condition of our drug and learning experiment might be better than in the first condition. Instead of reflecting a difference in drugs, this may simply reflect the fact that the participants are taking the recall test for the second time and are more familiar with it. This increased familiarity rather than the type of drug could produce the difference in learning.

When an investigator is confident that no carry-over effects will occur, a within-subjects research design is usually superior to a between-subjects research design. When carry-over effects are possible, a between-subjects design may be more appropriate. We will return to the issue of within-subjects versus between-subjects designs in later chapters.

There is another type of research design, called a **matched-subjects design**, in which different individuals are included in the different conditions but are treated as if a within-subjects design is in force. In this design, an individual in one condition is "matched" with an individual in each of the other conditions who has similar characteristics. The data are then treated as if these different individuals represent the same participant. As noted earlier, matching is difficult to achieve in practice. For this and other reasons (see Thorndike, 1942), there are some serious problems with this strategy, and its use is recommended only under restricted circumstances.

Study Exercise 9.1

For each of the following studies, indicate whether the independent variable is between-subjects or within-subjects in nature, and whether an experimental or an observational research strategy is involved.

Study I

A researcher was interested in the effect of television viewing on aggressive behavior in children. Fifty children were identified as watching television less than 5 hours per week (low viewers), 50 children were identified as watching television between 5 and 10 hours per week (moderate viewers), and 50 children were identified as watching more than 10 hours per week (high viewers). For each child, a measure of aggressiveness was determined by interviewing the child's teacher and classmates. Children were subsequently rated as being low in aggressiveness, moderate in aggressiveness, or high in aggressiveness.

Answer   The independent variable, amount of television viewing, is between-subjects in nature because the 50 children who are low viewers are not the same children as the 50 who are moderate viewers, who, in turn, are not the same children as the 50 who are high viewers. Since the children characteristically watch low, moderate, or high amounts of television in their daily lives, this study involves an observational research strategy.

Study II

An investigator was interested in the effect of music on problem-solving performance. Two hypotheses are possible: (1) music helps to relax people and should therefore facilitate problem-solving performance, or (2) music serves to distract people and, hence, should interfere with problem-solving performance. One hundred individuals each tried to solve ten problems with soft background music playing. Three weeks later, the same individuals returned and tried to solve ten similar problems; this time, however, there was no background music. The number of problems correctly solved by each individual in each of the two conditions was computed.

Answer   The independent variable, music status, is within-subjects in nature because the same 100 individuals participated in both the background music and no-music conditions. Since the presence or absence of background music was manipulated by the investigator, this study involves an experimental research strategy.

9.2   Selecting the Appropriate Statistical Test to Analyze a Relationship: A Preview

Parametric Versus Nonparametric Statistics

Many of the statistical techniques that we discuss involve the analysis of means, variances, and sums of squares. Such techniques are called **parametric statistics.** Parametric statistics require quantitative dependent variables and are usually applied when these variables are measured on a level that at least approximates interval characteristics. Parametric statistics require assumptions about the distribution of scores within the populations of interest. These assumptions will be made explicit as the techniques are introduced. In contrast, there is a class of statistics called **nonparametric statistics** that focus on differences between *distributions of scores* and that can be used to analyze quantitative variables that are measured on an ordinal level. In addition, they do not require many of the assumptions about distributional properties of scores that parametric statistics rely on.

There is currently some controversy among behavioral scientists concerning the use of parametric as opposed to nonparametric statistics. Some argue that nonparametric analyses should be widely used because measures in the behavioral sciences often depart radically from interval level characteristics. Furthermore, the application of parametric analyses, they argue, is inappropriate when distributional assumptions are not met. Those who argue for the widespread use of parametric analyses note that parametric-based analyses are more refined and powerful than current nonparametric methods. They argue that many of these techniques are *robust* to violations of the distributional assumptions required of them. The term **robust** means that even though the assumptions of a technique are violated, the frequencies of Type I and Type II errors and thus the accuracy of one's conclusions are relatively unaffected compared with conditions under which the assumptions are met. Because of this, some statisticians, such as Bohrnstedt and Carter (1971), have concluded that "when one has a variable which is measured at the ordinal level, parametric statistics not only can be, but should be, applied" (p. 322).

A complete discussion of this issue is beyond the scope of this book. Interested readers are referred to Bohrnstedt and Carter (1971), Boneau (1960), Lord (1953), Stevens (1951), and an excellent collection of readings in Kirk (1972). We will discuss the issue in more depth in Chapter 16.

Robustness of Statistical Tests

The concept of robustness is important in inferential statistics. As noted above, robustness refers to the extent to which conclusions drawn on the basis of a statistical test (for example, rejection of the null hypothesis) are unaffected by violations of the assumptions underlying the test. Mathematicians who develop inferential tests use assumptions for several reasons. Sometimes an assumption is used to simplify mathematical derivations so as to make the statistics more *manageable*. Other times an assumption is used because it characterizes what is likely to be the case in the real world; that is, the assumption is *credible*. As an example, a statistical test that assumes every intelligence test score occurs with equal frequency would have little applicability to most real-world problems. The assumption of a normal distribution, with a large proportion of central scores and few extreme scores, is much more credible.

The most common distributional assumptions that we will encounter in future chapters are *normality* and *homogeneity of variance*. As discussed in Chapter 8, the normality assumption requires that scores on the variable of interest be normally distributed in the population from which they are drawn. The homogeneity of variance assumption is applicable when two or more groups of scores are considered in the research design. It requires that the variances of the scores be equal, or *homogeneous*, in the populations underlying each of the samples. Note that these assumptions relate to the populations from which the samples were drawn rather than to the samples themselves. Thus, the determination of violations of distributional assumptions requires sophisticated statistical procedures.

A number of methods have been proposed for assessing violations of the normality and homogeneity of variance assumptions. Unfortunately, each of these tests has problems associated with it. For instance, the most commonly cited tests for homogeneity of variance (Bartlett's test, Cochran's test, and Hartley's *F* max test) have been found to be unsatisfactory when the population data are not normally distributed.

When an inferential test is robust to violations of an assumption, it can be appropriately applied even when that assumption is violated. This is because the frequencies of Type I and Type II errors will be similar to what they would be under conditions where no violations occurred. Thus, the accuracy of the test's conclusions will be relatively unaffected.

Statisticians determine the effects of violating assumptions in two ways. They may be able to determine the consequences of assumption violation using mathematical logic. If it is not possible to logically determine this, a Monte Carlo study is performed instead. A **Monte Carlo study** involves a computer simulation in which the statistician generates scores for hypothetical populations that he or she knows violate a distributional assumption in some way. The statistical test in question is then applied hundreds of times to random samples selected from these populations, and the number of times that a wrong conclusion is made is determined. If a test consistently yields the correct conclusion even though its formal statistical assumptions are violated, then the test is said to be robust.

As you might expect, the results of studies investigating the robustness of inferential tests are quite complex. This is because, as noted in Chapter 8, robustness is influenced by several factors, including sample size, the degree of violation, and the form of the violation. For instance, there are many ways in which the homogeneity of variance assumption can be violated in the three-group situation. Focusing on the *form* of the violation, we may find that two of the population variances are the same but different from the third or that all three population variances differ from one another. In terms of the *degree* of violation, differences in population variances can range in magnitude from slight to large.

Although a given statistical test might be quite robust under one set of circumstances, its robustness might substantially decrease under somewhat different conditions. For example, statisticians have found that in many instances, even marked violations of the normality assumption will not seriously affect the validity of the one-sample *t* test as long as the sample size is larger than around ten (Pearson & Please, 1978). Although the test continues to be robust to various forms of minor to moderate violations when the sample size decreases, marked violations under these circumstances can seriously affect the frequency of both Type I and Type II errors.

Given the numerous ways that factors can combine to influence robustness, it is impossible to discuss robustness with any degree of precision when we con-

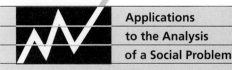

Applications

to the Analysis

of a Social Problem

Based on the material covered in this chapter, we can discuss additional features of the study design for the parent–teen data described in Chapter 1. First, the study uses an observational or nonexperimental strategy because no variables were formally manipulated. A major focus of the study was to explore adolescent gender and age differences in relation to parent–teen relationships and sexual behavior. Numerous hypotheses were formulated about potential gender effects. For example, one theory states that girls tend to be more "other oriented" than boys in their social relationships and therefore may be more influenced by peer pressure. This theory predicts, among other things, that sexually active girls are more likely to have sexually active friends, whereas this may not be true of boys. Suppose that statistical analyses found this to, in fact, be the case. Then the outcome of the analysis is consistent with the theory. However, other explanations for the result could also be in-

voked, based on the identification of confounding variables. For example, girls might be more likely than boys to seek out friends who are similar to them. The fact that sexually active girls tend to have sexually active friends may have nothing to do with girls succumbing to peer pressure; rather, it may reflect friendship selection processes unrelated to peer pressure. In order to differentiate between these two competing explanations, more fine-grained analyses or an additional study would need to be conducted.

In terms of the age variable, the study used a **cross-sectional design.** A cross-sectional design is one in which age serves as a between-subjects independent variable. The 14-year-olds in our study were not the same individuals as the 15-year-olds, who in turn were not the same individuals as the 16-year-olds, who were not the same individuals as the 17-year-olds. By contrast, in a **longitudinal design**, age serves as a within-subjects

sider the robustness of specific tests in future chapters. Our strategy will therefore be to provide general characterizations of the robustness of the tests that are presented and to cite a few articles that consider the revelant issues in more depth. If a researcher has reason to believe that a distributional assumption has been violated to the extent that a parametric test is no longer robust, one possible solution is to use a nonparametric alternative. These will be discussed in Chapter 16.

Selection of a Statistical Test

It is impossible to state any precise rules for determining the statistical test to apply in a given situation because it would always be possible to find exceptions where a different form of analysis might be more appropriate. Nevertheless, we can specify the most common research designs and the types of analyses typically used in the context of these designs. In Chapters 10–16, we will consider the most common statistical tests for analyzing bivariate relationships and, in each chapter, specify the issues to be considered in deciding to apply the test. As we will see, the essence of the proposed framework rests on distinguishing between qualitative and quantitative variables and between within-subjects and between-subjects designs. The required steps are: (1) identify the independent and dependent variables, (2) classify each as being qualitative or quantitative, (3) classify the independent variable as being between-subjects or within-subjects in nature, and independent variable. For example, a group of 14-year-olds might be interviewed every year for 4 years, yielding data for 14-year-olds, 15-year-olds, 16-year-olds, and 17-year-olds. Longitudinal designs are generally more expensive and time consuming than cross-sectional designs, and it was primarily because of these practical concerns that we decided to use a cross-sectional design. In so doing, we had to forego the ability to study how specific individuals changed over time and, instead, relied on comparisons of means, medians, and correlations across age groups to make developmental inferences.

Cross-sectional designs are efficient but have the potential problem of **cohort effects**. Cohort effects refer to historical variables that are confounded with age and that therefore serve as alternative explanations to developmental processes. Suppose, for example, that we theorized that because teens mature cognitively and morally throughout the adolescent years, they will become increasingly more responsible with age. This might lead us to predict that sexually active 17-year-old adolescents will be more consistent in their use of birth control than 16-year-olds, who in turn will be more consistent in their use of birth control than 15-year-olds, who will be more consistent in their use of birth control than 14-year-olds. Suppose furthermore that statistical analyses revealed this age trend in the consistency of birth control use. This result is consistent with the theory. However, it is also consistent with a number of alternative explanations, one of which involves a cohort effect. During the years surrounding our study, the problem of AIDS was receiving considerable attention in the media. The 17-year-old adolescents have been exposed to, on average, one more year of AIDS information campaigns than the 16-year-old adolescents, 2 more years of AIDS information than the 15-year-old adolescents, and 3 more years of AIDs information than the 14-year-old adolescents. Perhaps the age differences in the consistency of birth control use have little to do with maturational changes in cognitive and moral development, but instead merely reflect the differential exposure to AIDS information (and its subsequent effect on increased condom use) as is germane to the social history of this particular generation. Such confounds must be considered when interpreting data.

(4) note the number of levels that each variable has. Chapter 18 will formally discuss procedures for choosing a statistical test to analyze one's data.

Our discussion of statistics relevant to the analysis of bivariate relationships will consistently focus on three questions:

1. Given sample data, can we infer that a relationship exists between two variables in the population?

2. If so, what is the *strength* of the relationship?

3. If so, what is the *nature* of the relationship?

Summary

The results of a statistical analysis must be interpreted in the context of the research design used to generate the data. Behavioral scientists use two general types of research design. An experimental strategy involves performing a set of manipulations in order to create different values of the independent variable for the research participants. In contrast, an observational strategy does not involve actively creating values on the independent variable, but rather involves measuring differences in values that naturally exist in the research participants.

A major goal of research is to control for alternative explanations. This involves the control of confounding variables and disturbance variables. A confounding variable is one that is related to the independent variable and that influences the dependent variable, rendering a relational inference between the independent variable and the dependent variable ambiguous. A disturbance variable is one that is unrelated to the independent variable but that affects the dependent variable. Techniques for controlling confounding and disturbance variables include random assignment to experimental groups, holding a variable constant, matching, and the use of between-subjects versus within-subjects designs.

In Chapters 10–16, statistical tests for analyzing bivariate relationships will be examined. An important feature of these tests is the assumptions they make about the distribution of scores within the populations of interest. To the extent that conclusions drawn from a statistical test are relatively unaffected by the violation of these assumptions, the test is said to be robust. The issue of robustness will be examined for each test considered in future chapters.

Exercises

Answers to asterisked (*) exercises appear at the back of the book.

1. Differentiate between an experimental research strategy and an observational research strategy.

For each of the studies described in Exercises 2–5, indicate whether an experimental or an observational research strategy is involved.

*2. Morrow and Davidson (1976) studied the effect of race on family-size decisions. These investigators interviewed a total of 300 people; 100 were African American, 100 were Hispanic, and 100 were white. The participants were asked the number of children they wanted to have in their completed family. The average numbers of desired children were compared for the three groups.

*3. Harvath (1943) was interested in the effect of noise on problem-solving performance. One-hundred and fifty individuals participated in the study. They were randomly assigned to two conditions. Seventy-five participants tried to solve 30 problems while a steady "buzz" was present in the background. The other 75 tried to solve the same 30 problems with no background noise. The number of correct solutions was computed for each person and the average numbers of correct solutions compared for individuals in the noise versus the no-noise conditions.

4. Steiner (1972) reviewed studies on the effects of the presence of others on problem-solving performance. In one study, 100 female volunteers served as participants. At the first session, each woman was seated alone in a room and given a math problem to solve. The amount of time it took to solve the problem was measured. Two weeks later, the woman returned and solved another problem, but this time there was an observer present watching her. The amount of time it took to solve the problem was again measured. For each of the two conditions—observer present versus observer absent—the mean problem-solving time was computed across the 100 women. These means were then compared.

5. Sears (1969) reviewed studies on the relationship between gender and political party preference. In one investigation, 75 men and 75 women were interviewed. Respondents were asked whether they considered themselves Democrats, Republicans, or Independents. The frequencies with which men identified with the three classifications were compared with the corresponding frequencies for women.

*6. What is a control group? What is the advantage of including a control group in the research design?

7. For each of the studies described in Exercises 2–5, indicate whether a control group was used and, if so, identify the nature of this group.

8. What is random assignment? Why is it important?

*9. What are the limitations of random assignment?

*10. What procedures are available for reducing sampling error?

11. What are confounding variables? How can they be controlled?

*12. What are disturbance variables? How can they be controlled?

*13. What are the advantages of a within-subjects design compared with a between-subjects design? What is a potential problem?

*14. Indicate whether the independent variables in the studies described in Exercises 2–5 are between-subjects or within-subjects in nature.

15. How do parametric and nonparametric statistics differ?

*16. What is robustness? Why is it important?

17. What is the normality assumption? What is the homogeneity of variance assumption?

18. What are the two ways in which statisticians determine the effects of violating distributional assumptions?

*19. Identify three factors that influence the robustness of a statistical test.

20. What are the advantages of a cross-sectional design compared with a longitudinal design? What is the disadvantage?

Multiple-Choice Questions

21. A matched-subjects design is one in which
   a. the same individuals are included in all conditions
   b. the same individuals are included in all conditions but are treated as if a between-subjects design is in force
   c. different individuals are included in different conditions
   d. different individuals are included in different conditions but are treated as if a within-subjects design is in force

22. An experimental strategy does not involve the explicit manipulation of an independent variable.
   a. true   b. false

*23. A confounding variable
   a. is related to the independent variable and affects the dependent variable
   b. is related to the independent variable but does not affect the dependent variable
   c. affects the dependent variable but is not related to the independent variable
   d. is not related to either the independent variable or the dependent variable

24. A disturbance variable
   a. is related to the independent variable and affects the dependent variable
   b. is related to the independent variable but does not affect the dependent variable
   c. affects the dependent variable but is not related to the independent variable
   d. is not related to either the independent variable or the dependent variable

25. Whether or not two variables are *causally* related is an issue only when correlational analysis is used.
   a. true   b. false

*26. By definition, ethnic background is a within-subjects variable.
    a. true   b. false

*27. A statistical test that is robust
    a. can be used to analyze both qualitative and quantitative variables
    b. can be used to analyze both between-subjects and within-subjects variables
    c. will yield conclusions that are relatively unaffected by violations of underlying assumptions
    d. all of the above

28. Between-subjects and within-subjects designs are both viable strategies when the independent variable is experimental in nature.
    a. true   b. false

*29. Random assignment to groups guarantees that there will be no confounding variables.
    a. true   b. false

30. Holding a variable constant can be used to control for
    a. confounding variables
    b. disturbance variables
    c. both confounding variables and disturbance variables
    d. neither confounding variables nor disturbance variables

*31. The ability to make a causal inference between two variables is a function of
    a. the research design that is used
    b. the statistical techniques used to analyze the data
    c. the size of the samples
    d. whether parametric or nonparametric statistics are applied

32. The one-sample t test tends to be robust to violations of the normality assumption when the sample size is larger than about ten.
    a. true   b. false

*33. Distributional asssumptions relate both to the populations from which samples are drawn and to the samples themselves.
    a. true   b. false

34. Which of the following is *not* a step to be considered when selecting a stastistical test for analyzing bivariate relationships?
    a. Determine whether confounding variables or disturbance variables are more problematic.
    b. Note the number of levels that occur for both the independent and the dependent variables.
    c. Classify both the independent and the dependent variables as being qualitative or quantitative.
    d. Classify the independent variable as being between-subjects or within-subjects.

*35. Matching can be used to control for
    a. confounding variables
    b. disturbance variables
    c. both confounding variables and disturbance variables
    d. neither confounding variables nor disturbance variables

CHAPTER 10

Independent Groups
t Test

10.1 Use of the Independent Groups *t* Test

The statistical technique developed in this chapter is called the **independent groups**
t test. It is typically used to analyze the relationship between two variables under
these conditions:

1. The dependent variable is quantitative in nature and is measured on a level that
   at least approximates interval characteristics.

2. The independent variable is *between-subjects* in nature (it can be either qualitative or quantitative).*

3. The independent variable has two and only two levels.

Let us consider an example of an experiment that meets these conditions. When a
friend describes a stranger to you, you may form an impression as to whether you
would like that individual. One question of interest to behavioral scientists has

---

* Matched-subjects designs are analyzed as if the independent variable is within-subjects
in nature, using the procedures in Chapter 11.

been whether the order in which information is provided about a stranger influences the kind of impression formed about him. Consider the following procedures: Twenty college students at the same university read a verbal description of a stranger and then rate on a scale from 1 to 7 the extent to which they like or dislike that person. On the scale, the higher the score, the more the stranger is liked. The stranger is described by six adjectives: intelligent, sincere, honest, conceited, rude, and nervous. The first three characteristics are positive, whereas the last three characteristics are negative. Ten of the participants are randomly selected and given the description in the order listed above—that is, first the positive traits and then the negative traits. The other ten participants are given the same description, but in reverse order—that is, first the negative traits and then the positive traits. Thus, we have two groups reflecting different orders of presentation:

Group 1 (pro–con): intelligent, sincere, honest, conceited, rude, nervous
Group 2 (con–pro): nervous, rude, conceited, honest, sincere, intelligent

The likability ratings for the two conditions are presented in columns 1 and 3 of Table 10.1. If the order of information has no effect on the type of impression formed, we would expect that, on the average, the two sets of ratings would not differ. However, if the order of information does matter, then the average likability ratings should differ between the conditions.

TABLE 10.1   **Data and Calculations for Order of Information and Likability Experiment**

Pro–Con		Con–Pro	
X	X²	X	X²
7	49	1	1
4	16	5	25
3	9	2	4
5	25	5	25
5	25	3	9
6	36	3	9
5	25	3	9
5	25	2	4
4	16	4	16
6	36	2	4
$\Sigma X_1 = 50$	$\Sigma X_1^2 = 262$	$\Sigma X_2 = 30$	$\Sigma X_2^2 = 106$
$\overline{X}_1 = 5.00$		$\overline{X}_2 = 3.00$	

$$SS_1 = \Sigma X_1^2 - \frac{(\Sigma X_1)^2}{n_1}$$

$$= 262 - \frac{50^2}{10} = 12$$

$$\hat{s}_1^2 = \frac{SS_1}{n_1 - 1} \cdot \frac{12}{9} = 1.33$$

$$SS_2 = \Sigma X_2^2 - \frac{(\Sigma X_2)^2}{n_2}$$

$$= 106 - \frac{30^2}{10} = 16$$

$$\hat{s}_2^2 = \frac{SS_2}{n_2 - 1} \cdot \frac{16}{9} = 1.78$$

In this experiment, the order in which trait information is presented is the independent variable, and the likability of the stranger is the dependent variable. The independent variable has two levels (pro–con versus con–pro) and is between-subjects in nature. The dependent variable is quantitative in nature and is measured on a level that at least approximates interval characteristics. Given these conditions, the independent groups *t* test would typically be used to analyze the relationship between the variables. We now turn to our three central questions: (1) Is there a relationship between the variables? (2) If so, what is the strength of the relationship? (3) If so, what is the nature of the relationship?

10.2 Inference of a Relationship Using the Independent Groups t Test

Null and Alternative Hypotheses

The first question to be addressed is whether a relationship exists between the independent variable and the dependent variable. We begin by stating this question in terms of null and alternative hypotheses. This is done with reference to population means. Because we are interested in generalizing the results of our study beyond just those people who participated in it, we think of the participants as representing random samples from very large populations of similar individuals. In the experiment, there are two populations of interest: (1) individuals who read a verbal description of a stranger in which traits are presented in an order from pro to con, and (2) individuals who read a verbal description of a stranger in which traits are presented in an order from con to pro. Given that individuals were randomly assigned to groups, and if we assume that the order of information does *not* matter, then we would expect the population means for the two groups to be equal. The null hypothesis thus takes the form

$$H_0: \quad \mu_1 = \mu_2$$

where μ_1 is the population mean for the first group and μ_2 is the population mean for the second group. The null hypothesis posits that there is no relationship between the independent variable (order of information) and the dependent variable (likability) as measured by mean scores on the dependent variable. It states that it does not matter what the value of the independent variable is because the mean score on the dependent variable is the same at both levels. The alternative hypothesis states that there *is* a relationship between the two variables; the value of the independent variable *does* influence the average score on the dependent variable:

$$H_1: \quad \mu_1 \neq \mu_2$$

We have now restated the question in the context of two competing hypotheses: a null hypothesis ("there is no relationship between the variables") and an alternative hypothesis ("there is a relationship between the variables"). The next step is to choose between these two hypotheses. Examine the sample means in Table 10.1. The mean likability score for the pro–con group is 5.00, and the mean likability score for the con–pro group is 3.00. The two means are not equal, which appears to be consistent with the alternative hypothesis. However, we know from Chapter 7 that a sample mean may not reflect the true value of its population mean due to sampling error. Thus, the observed difference between the two sample means may not reflect the influence of the order of information on likability, but rather may reflect sampling error. Our task is to determine whether this is a reasonable interpretation of the observed difference between sample means.

Sampling Distribution of the Difference Between Two Independent Means

In order to test the sampling error interpretation, we use logic directly analogous to that developed in Chapter 8 for the one-sample *t* test. As an initial step, we must develop the concept of a **sampling distribution of the difference between two independent means.** Consider two large populations whose mean scores on a variable are equal (i.e., $\mu_1 = \mu_2$). Now suppose we select a random sample of size ten from each population and compute the mean of each sample as well as the difference between the means. We might find a result such as that illustrated in the first row of Table 10.2. In this table, $\overline{X}_1$ represents the sample mean for population 1, $\overline{X}_2$ represents the sample mean for population 2, and $\overline{X}_1 - \overline{X}_2$ represents the difference between them. Suppose we repeat this process again to get a second difference between sample means, as shown in Table 10.2. In principle, we could do this for all possible random samples of size ten to get a distribution of mean differences.* Table 10.2 presents 22 differences that might be observed. The distribution of the differences between two means represents a sampling distribution of the difference between two independent means. It is conceptually similar to a sampling distribution of the mean. However, now the concern is with a distribution of scores representing differences between two means. As we did in Chapter 7, we can compute the mean and standard deviation of this sampling distribution. If we were to do so, we would find that many of the properties of a sampling distribution of the mean also hold for a sampling distribution of the difference between two independent means. For instance, just as the mean of a sampling distribution of the mean is always equal to the population mean, *the mean of a sampling distribution of the difference between two independent means is always equal to the difference between the population means.* Consider the case where two population means are equal and hence their difference is 0 (that is, $\mu_1 - \mu_2 = 0$). If we were to generate a sampling distribution of the difference between means for these populations, the mean of the differences would equal 0. The underlying principle is much the same as that developed in Chapter 7. When we select all possible samples and compute the difference between means, some of the differences will overestimate the true mean difference while others will underestimate it. When we average all of these, the underestimations will cancel the overestimations, with the result being the true population difference between means.

---

* In practice, we would never actually do this. It is unnecessary because, as discussed below, the important information about a distribution of mean differences can be derived mathematically.

TABLE 10.2   **Illustrative Sample Means and Mean Differences from a Sampling Distribution of the Difference Between Two Independent Means**

$\overline{X}_1$	$\overline{X}_2$	$\overline{X}_1 - \overline{X}_2$
5.00	3.00	2.00
4.30	3.60	.70
5.60	6.40	−.80
4.80	5.40	−.60
4.70	4.70	.00
5.20	5.90	−.70
6.00	5.70	.30
4.30	3.90	.40
5.20	6.50	−1.30
6.10	7.50	−1.40
5.00	3.90	1.10
5.20	5.90	−.70
4.70	4.10	.60
4.90	4.40	.50
5.00	5.90	−.90
6.20	6.40	−.20
4.30	2.50	1.80
5.10	6.40	−1.30
5.70	6.60	−.90
5.10	4.10	1.00
4.90	5.40	−.50
4.50	5.00	−.50

The standard deviation of the sampling distribution of the difference between two independent means is called the *standard error of the difference between two independent means* or, more simply, the **standard error of the difference.** Like the standard error of the mean, the standard error of the difference indicates how much sampling error occurs, on the average. Statisticians have developed a formula that allows us to compute the standard error of the difference from the population standard deviations:

$$\sigma_{\overline{X}_1 - \overline{X}_2} = \sqrt{\frac{\sigma_1^2}{n_1} + \frac{\sigma_2^2}{n_2}} \qquad [10.1]$$

where $\sigma_{\overline{X}_1 - \overline{X}_2}$ is the standard error of the difference, σ_1^2 is the population variance for group 1, σ_2^2 is the population variance for group 2, n_1 is the sample size for group 1, and n_2 is the sample size for group 2.* Note the similarity of this formula to the one for the standard error of the mean (Equation 7.4):

---

* As we noted in Section 3.9, when more than one group is involved, *n* refers to the sample size for a particular group, whereas *N* refers to the total number of participants in the study. In the case of two groups, $N = n_1 + n_2$. The independent groups *t* test does not require that n_1 equal n_2.

$$\sigma_{\overline{X}} = \frac{\sigma}{\sqrt{N}} = \sqrt{\frac{\sigma^2}{N}}$$

Analogous to the standard error of the mean, the size of the standard error of the difference is influenced by two factors: (1) the sample sizes (n_1 and n_2) and (2) the variability of scores in the populations (σ_1^2 and σ_2^2). From the logic outlined in Section 7.5, the standard error of the difference becomes smaller as the sample sizes increase and the variability of scores in the populations decreases.

Pooled Variance Estimate

If we know the values of σ_1^2, σ_2^2, n_1, and n_2, we can compute the value of the standard error per Equation 10.1. In practice, we typically know the values of n_1 and n_2, but we do not know the values of σ_1^2 and σ_2^2. It is therefore necessary to estimate them from sample data. The independent groups *t* test assumes that the two population variances are equal, or *homogeneous*. This assumption, known as the assumption of **homogeneity of variance**, redefines the estimation problem. Instead of estimating σ_1^2 and σ_2^2 separately, our goal is to estimate σ^2, the variance of both populations. In other words, it is assumed that $\sigma_1^2 = \sigma_2^2 = \sigma^2$. The quantity σ^2 can best be estimated by combining, or *pooling*, the variance estimates from the two samples to obtain a **pooled variance estimate.** By pooling the variance estimates from two independent samples, we increase the degrees of freedom on which the estimate of σ^2 is based and thereby obtain a better estimate. The simplest method of pooling the two variance estimates is to compute their (unweighted) mean. This is, in fact, what is done when n_1 and n_2 are equal. However, if one of the groups has a larger sample size than the other, it makes sense to give the variance estimate from the group with the larger *n* more "weight" in determining the pooled variance estimate because the larger the sample size, the greater the degrees of freedom and, thus, the better the estimate, as discussed in Chapter 7. This is accomplished by the following equation:

$$\hat{s}_{\text{pooled}}^2 = \frac{(n_1 - 1)\hat{s}_1^2 + (n_2 - 1)\hat{s}_2^2}{n_1 + n_2 - 2} \qquad [10.2]$$

where $\hat{s}_{\text{pooled}}^2$ represents the pooled estimate of σ^2 (that is, the pooled variance estimate), $\hat{s}_1^2$ is the variance estimate for group 1, $\hat{s}_2^2$ is the variance estimate for group 2, and n_1 and n_2 are the sample sizes.* Examine the right-hand side of Equation 10.2. We have multiplied the variance estimate for each group by its degrees of freedom, summed the products for the two groups, and then divided the resulting quantity by the total number of degrees of freedom, $(n_1 - 1) + (n_2 - 1) = n_1 + n_2 - 2$.† This has been done so that the contributions of $\hat{s}_1^2$ and $\hat{s}_2^2$ to $\hat{s}_{\text{pooled}}^2$ are proportional to their degrees of freedom. For example, if $\hat{s}_1^2$ has twice as many degrees of

---

* When $n_1 = n_2$, Equation 10.2 reduces to $\hat{s}_{\text{pooled}}^2 = (\hat{s}_1^2 + \hat{s}_2^2)/2$.

† The same general procedure can be used with more complex designs to pool the variance estimates of more than two groups.

freedom as $\hat{s}_2^2$, it will contribute twice as much to $\hat{s}_{pooled}^2$. The meaning of Equation 10.2 becomes clearer if we rephrase it as follows:

$$\hat{s}_{pooled}^2 = \frac{(df_1)(\hat{s}_1^2) + (df_2)(\hat{s}_2^2)}{df_{TOTAL}}$$ [10.3]

where df_1 represents the degrees of freedom for the variance estimate for sample 1, df_2 represents the degrees of freedom for the variance estimate for sample 2, and df_{TOTAL} represents the total degrees of freedom, or $df_1 + df_2$. In short, Equations 10.2 and 10.3 characterize $\hat{s}_{pooled}^2$ as the mean of $\hat{s}_1^2$ and $\hat{s}_2^2$ after these have been adjusted by their respective degrees of freedom.

We demonstrate the application of Equation 10.2 by returning to the example on the effect of order of information on likability ratings discussed at the beginning of this chapter. Looking at Table 10.1, we find that $n_1 = 10$, $n_2 = 10$, $\hat{s}_1^2 = 1.33$, and $\hat{s}_2^2 = 1.78$. Thus,

$$\hat{s}_{pooled}^2 = \frac{(10 - 1)(1.33) + (10 - 1)(1.78)}{10 + 10 - 2} = 1.56$$

Estimated Standard Error of the Difference

Under the assumption of homogeneity of variance, $\hat{s}_{pooled}^2$ can be substituted for σ_1^2 and σ_2^2 in Equation 10.1 to yield the following formula for estimating the standard error of the difference:

$$\hat{s}_{\overline{X}_1 - \overline{X}_2} = \sqrt{\frac{\hat{s}_{pooled}^2}{n_1} + \frac{\hat{s}_{pooled}^2}{n_2}}$$ [10.4]

where $\hat{s}_{\overline{X}_1 - \overline{X}_2}$ represents the estimated standard error of the difference between two independent means and all other terms are as previously defined. For the example on the effect of order of information on likability ratings,

$$\hat{s}_{\overline{X}_1 - \overline{X}_2} = \sqrt{\frac{1.56}{10} + \frac{1.56}{10}} = .56$$

Thus, the mean differences that constitute the sampling distribution of the difference between two independent means are estimated to deviate an average of .56 unit from the true difference between the two population means.

If we wished, we could combine the above two steps into one by integrating Equations 10.2 and 10.4 as follows:

$$\hat{s}_{\overline{X}_1 - \overline{X}_2} = \sqrt{\frac{\hat{s}_{pooled}^2}{n_1} + \frac{\hat{s}_{pooled}^2}{n_2}}$$

$$= \sqrt{\hat{s}_{pooled}^2 \left(\frac{1}{n_1} + \frac{1}{n_2} \right)}$$

$$= \sqrt{\left[\frac{(n_1 - 1)\hat{s}_1^2 + (n_2 - 1)\hat{s}_2^2}{n_1 + n_2 - 2} \right] \left(\frac{1}{n_1} + \frac{1}{n_2} \right)}$$ [10.5]

Since a variance estimate is equivalent to the corresponding sum of squares divided by the sample size minus 1, this can also be expressed as

$$\hat{s}_{\bar{X}_1 - \bar{X}_2} = \sqrt{\left[\frac{(n_1 - 1)\left(\dfrac{SS_1}{n_1 - 1}\right) + (n_2 - 1)\left(\dfrac{SS_2}{n_2 - 1}\right)}{n_1 + n_2 - 2}\right]\left(\frac{1}{n_1} + \frac{1}{n_2}\right)}$$

$$= \sqrt{\left(\frac{SS_1 + SS_2}{n_1 + n_2 - 2}\right)\left(\frac{1}{n_1} + \frac{1}{n_2}\right)} \qquad [10.6]$$

where SS_1 is the sum of squares for the first group and SS_2 is the sum of squares for the second group. The decision to calculate $\hat{s}_{\bar{X}_1 - \bar{X}_2}$ using Equation 10.5 (or alternatively, Equations 10.2 and 10.4) or Equation 10.6 merely depends on whether one is dealing with variance estimates or sums of squares.

The *t* Test

We now have the background to test formally whether the observed difference in likability ratings between sample means (5.00 versus 3.00) can be attributed to sampling error or whether it reflects a true relationship between order of information and likability. To do so, we adapt the hypothesis testing steps outlined in Chapter 8 as follows:

1. Translate the research question into a null hypothesis and an alternative hypothesis.

In the example, the null hypothesis states that there is no relationship between order of information and likability. The alternative hypothesis states that there is a relationship between the two variables. Expressed in terms of population means, the null and alternative hypotheses are, respectively

H_0: $\mu_1 = \mu_2$

H_1: $\mu_1 \neq \mu_2$

Although the alternative hypothesis is nondirectional in this instance, directional alternative hypotheses are also possible. These would be phrased in terms of the mean of one population being larger than the mean of the other. As with the one-sample *t* test, directional tests should be used only when there is exclusive concern with a specific direction of difference and the researcher is uninterested in the effect should it be in the direction opposite to that stated in the directional alternative hypothesis. This is rarely the case.

Researchers often use subscripts based on the first letter of a condition's name rather than the more general "1" and "2" notation to identify specific conditions. For instance, we could state the null and alternative hypotheses for the order of information and likability experiment as

$H_0: \quad \mu_P = \mu_C$

$H_1: \quad \mu_P \neq \mu_C$

where the subscript "P" denotes the pro–con ordering and the subscript "C" denotes the con–pro ordering. The advantage of this approach is that the particular condition being represented is immediately identifiable. We will adopt this more precise notational system where appropriate in the remainder of this book.*

2. Assuming the null hypothesis is true, state an expected result in the form of a range of values within which the difference between sample means is expected to fall. This is expressed in terms of the critical values of *t* that define the endpoints of this range. The set of all *t* values more extreme than the critical values constitutes an unexpected result or, more formally, the rejection region.

The critical values of *t* are determined by reference to the appropriate *t* distribution in Appendix D. The degrees of freedom for the independent groups *t* test are equal to $n_1 + n_2 - 2$, reflecting that the degrees of freedom associated with the first sample are $n_1 - 1$, the degrees of freedom associated with the second sample are $n_2 - 1$, and $(n_1 - 1) + (n_2 - 1) = n_1 + n_2 - 2$.† Following the logic outlined in Chapter 8, by convention alpha is typically set equal to .05.

The *t* distribution in the present example has $n_1 + n_2 - 2 = 10 + 10 - 2 = 18$ degrees of freedom. For an alpha level of .05, nondirectional test, and 18 degrees of freedom, Appendix D defines the critical values of *t* as ±2.101. An expected result is therefore defined by *t* scores in the range −2.101 to +2.101, and an unexpected result is defined by all *t* scores less than −2.101 or greater than +2.101.

3. Characterize the mean and standard deviation of the sampling distribution of the difference between two independent means, assuming the null hypothesis is true.

If we assume that the null hypothesis is true in the present example, then μ_1 and μ_2 are equal, so $\mu_1 - \mu_2 = 0$. It follows that the mean of the sampling distribution also equals 0 because, as discussed previously, the mean of a sampling distribution of the difference between two independent means is always equal to the difference between the population means.

---

* So that students can become familiar with the various statistical formulas, these are consistently presented with the more general numerical subscripts.

† It will be remembered that $n_1 + n_2 - 2$ is also the degrees of freedom associated with the pooled variance estimate. This reflects that (as we will see shortly) the value of the *t* statistic in the independent groups case is dependent on the value of $\hat{s}_{\bar{X}_1 - \bar{X}_2}$, which, in turn, is dependent on the value of the pooled variance estimate.

The standard deviation of the sampling distribution is the standard error of the difference. This was estimated previously to equal .56 in our example.

4. Convert the observed difference between sample means into a *t* value to determine how many estimated standard errors it is from $\mu_1 - \mu_2$, *assuming the null hypothesis is true.* This is accomplished by the formula

$$t = \frac{(\overline{X}_1 - \overline{X}_2) - (\mu_1 - \mu_2)}{\hat{s}_{\overline{X}_1 - \overline{X}_2}} \qquad [10.7]$$

where $\overline{X}_1 - \overline{X}_2$ is the observed difference between sample means, $\mu_1 - \mu_2$ is the hypothesized difference between population means, and $\hat{s}_{\overline{X}_1 - \overline{X}_2}$ is the estimated standard error of the difference.

In most instances, the null hypothesis states that the population means are identical, so $\mu_1 - \mu_2 = 0$. However, other null hypotheses are also possible. For instance, a given null hypothesis might state that the mean of the first population is 10 units higher than the mean of the second population. This could be tested by setting $\mu_1 - \mu_2 = 10$ in Equation 10.7. Any other hypothesized difference between two population means can be similarly assessed.

When $\mu_1 - \mu_2$ is hypothesized to equal 0, Equation 10.7 reduces to

$$t = \frac{\overline{X}_1 - \overline{X}_2}{\hat{s}_{\overline{X}_1 - \overline{X}_2}} \qquad [10.8]$$

This is the most commonly used version of the independent groups *t* test. However, because it is conceptually clearer, we use the more general formula represented by Equation 10.7 whenever an independent groups *t* test is called for in the remainder of this chapter.

In our example, $\overline{X}_1 - \overline{X}_2 = 5.00 - 3.00 = 2.00$, $\hat{s}_{\overline{X}_1 - \overline{X}_2} = .56$, and $\mu_1 - \mu_2$ is hypothesized to equal 0. Hence,

$$t = \frac{(\overline{X}_1 - \overline{X}_2) - (\mu_1 - \mu_2)}{\hat{s}_{\overline{X}_1 - \overline{X}_2}}$$

$$= \frac{2.00 - 0}{.56} = 3.57$$

5. Compare the observed result with the expected result. If the observed result falls within the rejection region, reject the null hypothesis.* Otherwise, do not reject the null hypothesis.

In the present example, the observed *t* value of 3.57 exceeds the positive critical value of +2.101. This suggests that the observed mean difference is too large to be attributed to sampling error, and the null hypothesis is therefore

---

* If the null hypothesis is rejected, the nature of the relationship between the independent and dependent variables can be determined using the strategy outlined in Section 10.4.

rejected. Order of information is related to likability, and we have answered the first of the three questions for the analysis of bivariate relationships.

Study Exercise 10.1

In order to test for gender discrimination by women against other women, a researcher had 13 women read an essay that was presented as having been written by "John McKay" and then rate the essay for the quality of the writing style. The ratings were made on a scale from 1 to 10, with higher scores indicating greater perceived quality. An additional 13 women read and rated the identical essay but were led to believe that the author was "Joan McKay." The following summary data were observed. Use an independent groups t test to test for a relationship between the supposed gender of the author and the perceived quality of the essay.

Male author	Female author
$n_M = 13$	$n_F = 13$
$\overline{X}_M = 7.20$	$\overline{X}_F = 6.10$
$SS_M = 15$	$SS_F = 18$

Answer   The null and alternative hypotheses are

$$H_0: \quad \mu_M = \mu_F$$

$$H_1: \quad \mu_M \neq \mu_F$$

where the subscripts "M" and "F" represent the male author and the female author conditions, respectively. For an alpha level of .05, nondirectional test, and $n_1 + n_2 - 2 = 13 + 13 - 2 = 24$ degrees of freedom, the critical values of t from Appendix D are ± 2.064. Thus, if the observed t value is less than -2.064 or greater than $+2.064$, we will reject the null hypothesis.

The standard error of the difference is estimated using Equation 10.6:

$$\hat{s}_{\overline{X}_1 - \overline{X}_2} = \sqrt{\left(\frac{SS_1 + SS_2}{n_1 + n_2 - 2} \right)\left(\frac{1}{n_1} + \frac{1}{n_2} \right)}$$

$$= \sqrt{\left(\frac{15 + 18}{13 + 13 - 2} \right)\left(\frac{1}{13} + \frac{1}{13} \right)} = .46$$

The observed difference between sample means is $\overline{X}_1 - \overline{X}_2 = 7.20 - 6.10 = 1.10$, and the hypothesized difference between population means is $\mu_1 - \mu_2 = 0$. Using Equation 10.7, we compute the observed value of t:

$$t = \frac{(\overline{X}_1 - \overline{X}_2) - (\mu_1 - \mu_2)}{\hat{s}_{\overline{X}_1 - \overline{X}_2}}$$

$$= \frac{1.10 - 0}{.46} = 2.39$$

Since 2.39 is greater than $+2.064$, we reject the null hypothesis and conclude that there is a relationship between the supposed gender of the author and the perceived quality of the essay.

Assumptions of the *t* Test

As noted at the beginning of this chapter, the independent groups *t* test is appropriate when the independent variable is between-subjects in nature and the dependent variable is quantitative in nature and measured on a level that at least approximates interval characteristics. The *t* statistic approximately follows a *t* distribution when the following assumptions are met:

1. The samples are independently and randomly selected from their respective populations.

2. The scores in each population are normally distributed.

3. The scores in the two populations have equal variances; that is, $\sigma_1^2 = \sigma_2^2$. As discussed previously, this is called the assumption of homogeneity of variance.

   For the test to be valid, it is important that the assumption of independent and random selection be met. However, under certain conditions, the independent groups *t* test is robust to violations of the normality and homogeneity of variance assumptions. If the sample sizes in the two groups are each greater than 40 and roughly comparable, then the test is robust to rather severe departures from the normality assumption. Indeed, the test is fairly robust to normality violations for sample sizes as small as 15 per group, if not somewhat smaller.

   When the sample sizes are equal, the *t* test is quite robust to violations of the assumption of homogeneity of variance (Posten, 1978). For example, when the sample size is 15 in each group, the *t* test performs satisfactorily even when the population variance for one group is 4 times larger than the population variance for the other group. This trend is also evident for unequal *n* values in the two groups, as long as the sample sizes are comparable. With fairly discrepant sample sizes, there is a tendency for the test to be conservative (i.e., for there to be fewer Type I errors than the alpha level dictates) when the group with the larger sample size has the larger variance. By the same token, the *t* test is liberal (i.e., more Type I errors occur than the alpha level dictates) when the larger sample size is paired with the smaller variance (for further discussion of this issue, see Sawilosky & Hillman, 1992).

Relationships and Parameters

We have examined whether a relationship exists between two variables by conducting a test of the null hypothesis that the population means in different groups are equal. It is possible to define and test for a relationship between two variables using population parameters other than the mean. For example, one can examine the null hypothesis of equivalent population *medians*. If the groups defined by the independent variable have different population medians on the dependent variable, then one would conclude that a relationship exists between the independent and dependent variables (at least in terms of medians). The term *relationship* is general and must be qualified by the particular population parameter being examined. Two variables may exhibit a relationship in terms of one parameter (e.g., means) but not another (e.g., medians). The parameters used for characterizing relationships in this book are primarily means and correlations, although other parameters also are considered (see Chapters 15 and 16). Nevertheless, it is important to keep in mind that a relationship between two variables can be examined in terms of any of a number of parameters describing a set of scores and that the effect of an independent variable on a dependent variable might differ depending on the parameter under consideration.

10.3 Strength of the Relationship

If we reject the null hypothesis and conclude that a relationship exists between the independent and dependent variables, it becomes meaningful to ask how strong the relationship is. One approach is simply to examine the size of the mean difference between the two groups. For example, a mean discrepancy of $4,000 in annual starting salaries for entry-level secretaries who are white as opposed to African American is a sizeable difference in income. The difference is meaningful to us because we have implicit knowledge about the distribution of income in general and we also know what $1, $20, $100, or $1,000 will buy. However, sometimes we are not so informed about the dependent variable, as in the experiment on likability. Is a mean difference of 2.00 small or large? What are the implications of such a difference?

To help us gain perspective on the strength of the relationship, statisticians have developed a wide range of indexes. We develop the general logic of these approaches using an index known as *eta-squared* and then discuss the advantages and disadvantages of different measures. We initially derive eta-squared using procedures that are computationally inefficient but that best illustrate the concept. A computational formula is then presented. We use the experiment on the relationship between order of information and likability to develop the logic of the approach.

The second and third columns of Table 10.3 present the condition in which each of the 20 individuals participated and their corresponding scores on the dependent variable. Looking at column 3, we see that there is variability in the likability ratings; some individuals said they like the hypothetical stranger more than did other individuals. Our goal is to analyze this variability and determine what proportion of it is associated with the independent variable. Because of a desirable statistical property to be discussed shortly, we use the sum of squares as our measure of variability.

Sum of Squares Total

As a first step, we need to derive a numerical index of the amount of variability in the likability ratings. This involves computing the sum of squares for the dependent variable across all individuals in the experiment, which can be accomplished using the standard formula for a sum of squares. This has been done in columns 3 and 4 of Table 10.3. This sum of squares is called the **sum of squares total**

TABLE 10.3   **Computation of SS_{TOTAL} for Order of Information and Likability Experiment**

Individual	Condition	X	X^2
1	Pro–Con	7	49
2	Pro–Con	4	16
3	Pro–Con	3	9
4	Pro–Con	5	25
5	Pro–Con	5	25
6	Pro–Con	6	36
7	Pro–Con	5	25
8	Pro–Con	5	25
9	Pro–Con	4	16
10	Pro–Con	6	36
11	Con–Pro	1	1
12	Con–Pro	5	25
13	Con–Pro	2	4
14	Con–Pro	5	25
15	Con–Pro	3	9
16	Con–Pro	3	9
17	Con–Pro	3	9
18	Con–Pro	2	4
19	Con–Pro	4	16
20	Con–Pro	2	4
		$\Sigma X = 80$	$\Sigma X^2 = 368$

$$SS_{TOTAL} = \Sigma X^2 - \frac{(\Sigma X)^2}{N}$$

$$= 368 - \frac{80^2}{20} = 48$$

(symbolized SS_{TOTAL}) because it represents the total amount of variability that exists in the data. In this instance, $SS_{TOTAL} = 48$.

Treatment Effects and Variance Extraction

The next step in the analysis is to determine what effect the independent variable had on the dependent variable. This is done by comparing the mean score in each condition with the mean score for both conditions combined. Since it is based on all individuals in the study, the overall mean is known as the *grand mean* (symbolized *G*). As indicated in column 2 of Table 10.4, the grand mean in this instance is 4.00. The mean likability rating for the pro–con condition was previously found to be 5.00. Thus, the effect of having the traits presented in the pro–con order was to raise the likability ratings, on the average, 1 unit above the grand mean. For the con–pro condition, the mean likability score was previously found to be 3.00. The effect of having the traits presented in the con–pro order was thus to lower the likability ratings, on the average, 1 unit below the grand mean. These effects are formally called *treatment effects* (symbolized T) and are defined as the difference between a given group mean and the grand mean:

$$T_1 = \overline{X}_1 - G = 5.00 - 4.00 = 1.00$$

$$T_2 = \overline{X}_2 - G = 3.00 - 4.00 = -1.00$$

The treatment effects for the two groups are listed in column 3 of Table 10.4.

We can use the treatment effects to derive the strength of the relationship between the independent and dependent variables. This requires first removing, or *nullifying*, the influence of the independent variable on the dependent variable. For individuals in the pro–con condition, the effect of the order of information was to raise scores, on the average, 1 unit *above* the grand mean. If we want to nullify this effect, we can *subtract* 1 unit from each of these individuals' original likability scores. Consider the first person in Table 10.4. The likability rating for this individual is 7. To nullify the effect of being in the pro–con condition, we subtract

TABLE 10.4   **Computation of SS_{ERROR} for Order of Information and Likability Experiment**

Condition	X	T	$X_n = X - T$	X_n^2
Pro–Con	7	1	6	36
Pro–Con	4	1	3	9
Pro–Con	3	1	2	4
Pro–Con	5	1	4	16
Pro–Con	5	1	4	16
Pro–Con	6	1	5	25
Pro–Con	5	1	4	16
Pro–Con	5	1	4	16
Pro–Con	4	1	3	9
Pro–Con	6	1	5	25
Con–Pro	1	−1	2	4
Con–Pro	5	−1	6	36
Con–Pro	2	−1	3	9
Con–Pro	5	−1	6	36
Con–Pro	3	−1	4	16
Con–Pro	3	−1	4	16
Con–Pro	3	−1	4	16
Con–Pro	2	−1	3	9
Con–Pro	4	−1	5	25
Con–Pro	2	−1	3	9

$$\Sigma X = 80 \qquad\qquad \Sigma X_n = 80 \qquad \Sigma X_n^2 = 348$$

$$\overline{X} = G = 4.00$$

$$SS_{ERROR} = \Sigma X_n^2 - \frac{(\Sigma X_n)^2}{N}$$

$$= 348 - \frac{80^2}{20} = 28$$

1 from his score to get a *nullified score* (symbolized X_n) of 6. This process has been repeated in column 4 of Table 10.4 for each person in the pro–con condition.

On the other hand, the effect of the order of information for individuals in the con–pro condition was to lower scores, on the average, 1 unit *below* the grand mean. If we want to nullify this effect, we simply *add* 1 unit (that is, *subtract* 1 *negative* unit) to each of these individuals' original likability scores. The first participant listed in Table 10.4 in the con–pro condition has a likability rating of 1. To nullify the effect of being in the con–pro condition, we add 1 to his score to get a nullified score of 2. This process has been repeated for each person in the con–pro condition in column 4 of Table 10.4.

When you examine the nullified dependent variable scores in column 4, note that there is less variability than there is in the original dependent variable scores in column 2. This is because we have removed the variability that was associated with the independent variable. Note, however, that there is still variability in the X_n scores due to factors other than the order of information. In statistical terminology, the remaining variability is called **unexplained variance** or **error variance** and reflects the influence of disturbance variables, as discussed in Chapter 9.

We can derive a numerical index of how much error variance remains after we have removed the influence of the independent variable by computing a sum of squares for the nullified scores. This is done by applying the standard formula for a sum of squares to the X_n scores, as has been done in columns 4 and 5 of Table 10.4. This sum of squares is formally referred to as the **sum of squares error** (symbolized SS_{ERROR}). In this instance, $SS_{ERROR} = 28$.

Whereas the sum of squares total indexes the *total* amount of variability in the dependent variable, the sum of squares error indexes the amount of *unexplained* variability in the dependent variable—that is, variability that remains after the effects of the independent variable have been removed. If we subtract the sum of squares error from the sum of squares total, we obtain an index of **explained variance**—that is, variability associated with (explained by) the independent variable. This index is referred to as the **sum of squares explained** (symbolized $SS_{EXPLAINED}$). In our example, $SS_{EXPLAINED} = 48 - 28 = 20$.

Since $SS_{TOTAL} - SS_{ERROR} = SS_{EXPLAINED}$, it is also the case that

$$SS_{TOTAL} = SS_{EXPLAINED} + SS_{ERROR} \qquad [10.9]$$

In other words, the total variability in the dependent variable, as represented by SS_{TOTAL}, can be split up, or *partitioned,* into two components: one ($SS_{EXPLAINED}$) reflecting the influence of the independent variable and one (SS_{ERROR}) reflecting the influence of disturbance variables. It is this property of sums of squares—the fact that they can be meaningfully added to and subtracted from one another—that makes the sum of squares the preferred measure of variability when analyzing the strength of the relationship between variables.*

---

* In contrast, neither standard deviation estimates nor variance estimates can be manipulated in this manner. This is because they are averages (involving deviation or squared deviation scores), and averages cannot be meaningfully added to or subtracted from one another.

Eta-Squared

The stronger the influence of the independent variable on the dependent variable, the larger the sum of squares explained should be relative to the sum of squares total. By dividing the former by the latter, we can calculate the proportion of variability in the dependent variable that is explained by the independent variable. This statistic, the proportion of explained variance, is called **eta-squared** and is defined by the formula

$$\text{eta}^2 = \frac{SS_{\text{EXPLAINED}}}{SS_{\text{TOTAL}}} \qquad\qquad [10.10]$$

where eta^2 stands for "eta-squared" and the other terms are as previously defined.* Eta-squared indexes the strength of the relationship between the independent and dependent variables because it represents the proportion of variability in the dependent variable that is associated with the independent variable. Eta-squared can range from 0 to 1.00. As eta-squared approaches 1.00, the relationship between the variables is stronger, and as eta-squared approaches 0, the relationship between the variables is weaker. In our example,

$$\text{eta}^2 = \frac{20}{48} = .42$$

Thus, 42% of the variability in the dependent variable (likability) is explained by the independent variable (order of information).

Since eta-squared represents the proportion of variability in the dependent variable that is associated with the independent variable, 1.00 minus eta-squared must represent the proportion of variability in the dependent variable that is *not* associated with the independent variable—that is, the proportion of variability that is due to disturbance variables. In our example, $1.00 - .42 = .58$, or 58%, of the variability in likability ratings is attributable to disturbance variables.

Standards differ considerably among researchers on the substantive interpretation of eta-squared. Much depends on the specific area of application. One of us (JJ) has studied the relationship between attitudes and behavior. In this research, an eta-squared less than .20 represents a "weak" relationship, an eta-squared between .20 and .50 represents a "moderate" relationship, and an eta-squared greater than .50 represents a "strong" relationship between variables. In other contexts, different interpretations might apply. The behavior of organisms is complex, and rarely do only a small number of variables determine that behavior. To the extent that a behavior is determined by a multitude of variables, the explanation of as little as 5% of the variability in the dependent measure is, in some respects, a considerable amount. Typically, research in the behavioral sciences involves relatively small values of eta-squared, but this should not lead one to conclude that such effects are necessarily trivial. Rosenthal (1995) has described several research situations where eta-squared values less than .01 had tremendous

---

* It should be noted that eta-squared corresponds to the square of a statistic called the *point-biserial correlation coefficient*.

practical implications that reflected the saving of thousands of lives in a medical context. In the remainder of this book, we will consider an eta-squared near .05 as a weak effect, an eta-squared near .10 as a moderate effect, and an eta-squared greater than .15 as a strong effect. However, these conventions must be viewed as being somewhat arbitrary and can be revised upward or downward, depending on the research context. The values of eta-squared that we report in our worked examples will generally be inflated relative to those observed in actual behavioral science research.

Computational Formula for Eta-Squared

Although it is possible to derive eta-squared with Equation 10.10 using the approach discussed above, a much simpler computational formula for eta-squared is used in practice:

$$\text{eta}^2 = \frac{t^2}{t^2 + df} \qquad\qquad [10.11]$$

where *t* is the observed *t* value and df is the corresponding degrees of freedom.

For the experiment on the order of information and likability,

$$\text{eta}^2 = \frac{3.57^2}{3.57^2 + 18} = .41$$

This value agrees, within rounding error, with our earlier result and represents a strong effect.

Alternative Measures of the Strength of the Relationship

We have chosen to use eta-squared as the measure of the strength of the relationship between two variables because of its relationship to certain statistics discussed in later chapters and because it possesses desirable statistical properties (Kennedy, 1970; Kesselman, 1975). However, it must be emphasized that eta-squared describes the strength of the relationship between two variables in a set of *sample* data. Estimating the strength of the relationship in the *population* requires a somewhat different approach. A number of different estimation procedures have been proposed, and their applications are controversial among statisticians. *Eta-squared is a biased estimator in that it tends to slightly overestimate the strength of the relationship in the population across random samples.* Hays (1981) has argued for an alternative estimator called *omega-squared*. However, omega-squared has some undesirable statistical properties. The method of estimation one should use is not agreed upon by statisticians. For a discussion of the statistical properties of various indexes of strength of association, see Carrol and Nordholm (1975), Fisher (1950), Glass and Hakstian (1969), Haggard (1958), and Kesselman (1975).

Any index of the strength of the relationship derived from sample data is subject to sampling error and must be interpreted accordingly. An observed sample eta-squared of .50 could result from a population with a value of eta-squared that is quite different. The confidence one has in the accuracy of a sample eta-squared as an estimate of the corresponding population parameter is a function, in part, of the overall sample size. The larger the N, the better the estimate. In fact, for Ns less than 30, the amount of sampling error can be rather sizeable (Carrol & Nordholm, 1975). For this reason, indexes of the strength of association must be interpreted with caution. In light of the above, we advocate the use of eta-squared only as a heuristic index that will aid the researcher in appreciating relationships contained within sample data. The measure should *not* be used as an estimate of the strength of the relationship in the population because, in our opinion, *point estimation procedures* (the identification of one specific value) are not very useful in this respect. Instead, an *interval estimation approach* (the identification of a range of values) seems more reasonable. Analogous to the construction of confidence intervals about the mean discussed in Chapter 8, it is possible to construct confidence intervals about eta-squared. These provide the investigator with ranges of values that, with a specified degree of confidence, contain the population value of eta-squared. The mathematics for calculating such intervals are complex and are discussed by Fleishman (1980).

Eta-Squared Following a Nonsignificant Test

The foregoing discussion clearly demonstrates the utility of examining eta-squared following a significant statistical test. Given that it indexes the strength of the relationship between two variables in a set of sample data, eta-squared is also informative even when the statistical test is nonsignificant. This relates to the fact that, as discussed in Chapter 8, power increases with sample size. When sample sizes are small, power will tend to be low and we will be relatively unlikely to reject a false null hypothesis. By examining eta-squared, we can gain additional insight into the situation. If the null hypothesis is not rejected and eta-squared is small, then the statistical decision is reinforced. However, if the null hypothesis is not rejected and eta-squared is relatively large, then this is a "flag" of potentially low statistical power. We might want to collect additional data because this is an indication that the two variables may be related but that the sample sizes were not large enough to yield a significant result with the desired degree of consistency. With this in mind, we recommend that eta-squared be calculated following nonsignificant as well as significant statistical tests, and we will follow this strategy throughout the book.

10.4 Nature of the Relationship

We have now considered the first two questions for the analysis of bivariate relationships: (1) Is there a relationship between the independent and dependent variables? (2) If so, what is the strength of the relationship? The final question concerns the nature of the relationship. Although this is determined by examining sample means, the conclusion is intended to apply to the population means. In the experiment on order of information and likability of a stranger, the mean likability scores were 5.00 in the pro–con condition and 3.00 in the con–pro condition. Since the mean for the pro–con condition is greater than the mean for the con–pro condition, we conclude that the nature of the relationship between order of information and likability is such that when people are presented information going from positive to negative, they will form more favorable impressions of a stranger (at least in terms of likability) than when the order of information goes from negative to positive.

10.5   Methodological Considerations

Several methodological features should be noted about the experiment on the order of information and likability. These place the results of the statistical test in the proper context. First, the investigator attempted to control for confounding variables by randomly assigning participants to the two experimental conditions. Without this random assignment, differences between the two groups could be attributed to individual difference factors rather than to the experimental treatment. For example, psychologists have shown that some people have a tendency to view everyone in a positive light, whereas others tend to view everyone in a negative light. Without random assignment, it is possible that the pro–con condition could have contained more of the former individuals and the con–pro condition more of the latter. With random assignment, this is unlikely.

Another issue concerns the extent to which disturbance variables were controlled in the experiment. These have the effect of creating sampling error, the amount of which is reflected in the standard error of the difference. In this experiment, this was estimated to be .56, which suggests that, on the average, the difference between any two sample means based on random samples of size ten will deviate .56 unit from the true difference between the population means. Given a potential range of 6 (since likability ratings were made on a 1-to-7 scale), this suggests a relatively small amount of sampling error.

A second perspective on sampling error is gained by examining eta-squared. In this experiment, 42% of the variability in the dependent variable is associated with the independent variable and 58% is due to disturbance variables. The picture that emerges is one of the disturbance variables (and the consequent sampling error) not overwhelming any "signal" produced by the independent variable.

The experiment has limitations from the standpoint of generalizability. First, it used college students from one university. Would the results generalize to other individuals as well? Second, the experiment used only one pro–con list and one con–pro list. Maybe something about the particular adjectives used produced the experimental outcome. Would similar results occur if different adjectives were used? These and other questions of generalizability can be addressed through additional experimentation. It is important that one always consider such issues when trying to draw conclusions from research results. Statistics and research design go hand in hand when you interpret the results of an investigation.

10.6   **Numerical Example**

Individuals have long sought ways to help them to relax. One such method is transcendental meditation (TM). An individual who practices TM meditates twice a day for 15–20 minutes. During meditation, the individual subvocally repeats a mantra, which is a meaningless, two-syllable sound. The mantra is different for each individual and is assigned to the meditator by a trained teacher of TM. Proponents of TM emphasize the importance of the mantra in the meditation process. However, it has been suggested that the mantra is not essential to achieve optimal levels of relaxation during meditation and that the same results could be achieved by using the word *One* instead. Let us consider a hypothetical experiment designed to test this assertion.

Twenty individuals participated in the experiment. Ten were randomly assigned to the mantra condition, and ten were randomly assigned to the "One" condition. Each person practiced meditation using the relevant technique for 2 weeks prior to the test session. During the test session, individuals participated in their regular 15-minute meditation period while a number of physiological indicators of relaxation were measured. One of these was heart rate. In general, the more relaxed an individual, the slower his or her heart rate will be. The heart rate measures for the 20 participants are listed in columns 1 and 3 of Table 10.5.

The null hypothesis is that individuals who practice traditional TM will not differ in their mean heart rate from individuals who meditate using the word *One*

TABLE 10.5   **Data and Calculations for Meditation Technique and Heart Rate Experiment**

Mantra		"One"	
X	X^2	X	X^2
58	3,364	62	3,844
62	3,844	57	3,249
54	2,916	67	4,489
52	2,704	66	4,356
64	4,096	58	3,364
58	3,364	60	3,600
52	2,704	64	4,096
64	4,096	62	3,844
56	3,136	57	3,249
60	3,600	67	4,489
$\Sigma X_M = 580$	$\Sigma X_M^2 = 33,824$	$\Sigma X_O = 620$	$\Sigma X_O^2 = 38,580$
$\overline{X}_M = 58.00$		$\overline{X}_O = 62.00$	

$$SS_M = \Sigma X_M^2 - \frac{(\Sigma X_M)^2}{n_M}$$

$$= 33,824 - \frac{580^2}{10} = 184$$

$$SS_O = \Sigma X_O^2 - \frac{(\Sigma X_O)^2}{n_O}$$

$$= 38,580 - \frac{620^2}{10} = 140$$

in place of a mantra. The alternative hypothesis is that they will differ. These hypotheses are formally stated as

$$H_0: \quad \mu_M = \mu_O$$

$$H_1: \quad \mu_M \neq \mu_O$$

where the subscript "M" denotes the mantra condition and the subscript "O" denotes the "One" condition. In other words, the null hypothesis states that there is no relationship between the type of meditation technique and the relaxation level (as indicated by heart rate). The alternative hypothesis states that there is a relationship between these two variables.

Since the null hypothesis states that the population means are equivalent, $\mu_1 - \mu_2 = 0$ for the purpose of hypothesis testing. Other values necessary for the statistical analysis can be calculated using the information presented in Table 10.5:

$$\overline{X}_1 - \overline{X}_2 = 58.00 - 62.00 = -4.00$$

$$\hat{s}_{\overline{X}_1 - \overline{X}_2} = \sqrt{\left(\frac{SS_1 + SS_2}{n_1 + n_2 - 2} \right)\left(\frac{1}{n_1} + \frac{1}{n_2} \right)}$$

$$= \sqrt{\left(\frac{184 + 140}{10 + 10 - 2} \right)\left(\frac{1}{10} + \frac{1}{10} \right)} = 1.90$$

The observed value of *t* is thus

$$t = \frac{(\overline{X}_1 - \overline{X}_2) - (\mu_1 - \mu_2)}{\hat{s}_{\overline{X}_1 - \overline{X}_2}}$$

$$= \frac{-4.00 - 0}{1.90} = -2.11$$

For an alpha level of .05, nondirectional test, and $n_1 + n_2 - 2 = 10 + 10 - 2 = 18$ degrees of freedom, the critical values of *t* are ±2.101. Since the observed *t* value of –2.11 is less than the negative critical value of –2.101, we might be tempted to reject the null hypothesis and conclude that there is a relationship between the type of meditation technique and the relaxation level (heart rate). However, since the two values are so similar, we should first repeat our calculations to three (or more) decimal places to increase the precision of our answer. If we do so, we obtain an observed *t* value of –2.109. Since this is less than the negative critical value of –2.101, the null hypothesis is indeed rejected.

The strength of the relationship in the sample is indexed by eta-squared. Using Equation 10.11, we find that

$$\text{eta}^2 = \frac{t^2}{t^2 + df}$$

$$= \frac{-2.11^2}{-2.11^2 + 18} = .20$$

The proportion of variability in heart rate that is associated with the type of meditation technique is .20. This represents a strong effect.

The nature of the relationship is indicated by the mean scores for the two conditions. Since the mean heart rate for the mantra group (58.00) is lower than the mean heart rate for the "One" group (62.00), we conclude that the use of a mantra produces more relaxation during meditation than does the repetition of the word *One*.

You should think about these results in terms of the research design questions noted previously. Are there any confounding variables that have not been controlled? What has been the role of disturbance variables? What procedures could be used to reduce sampling error? What are the limitations of the experiment in terms of generalizability?

10.7   Planning an Investigation Using the Independent Groups *t* Test

When designing a study involving two independent groups, the investigator is faced with a number of important decisions. One concerns the delineation of confounding and disturbance variables and how to control for these in the context of the study. Another decision concerns the number of participants that should be sampled for each group. One consideration that influences the latter decision is practicality. A researcher cannot sample more participants than resources permit.

Practical matters aside, statistical considerations must also be taken into account. One major issue concerns the desired power of the statistical test that will be used to analyze the data. Recall from Chapter 8 that the power of a statistical test refers to the probability of rejecting the null hypothesis when it is false. When we say the power of a test is .70, this means that the chances are 7 in 10 we will correctly reject the null hypothesis when it is false. The power of a statistical test is influenced by three major factors: (1) the strength of the relationship between the two variables *in the population*, (2) the sample sizes, and (3) the alpha level. Consider the first factor. With sample sizes typical in the behavioral sciences, if the strength of the relationship in the population is weak, then it is relatively unlikely that this relationship will manifest itself in a sample over and above the "noise" of sampling error. Hence, it is relatively unlikely that we will correctly reject the null hypothesis. The power of the test will be relatively low, everything else being equal. If the strength of the population relationship between the two variables is strong, however, then it is more likely that the relationship will manifest itself in a sample over and above the "noise" of sampling error. We probably will detect the relationship between the independent and dependent variables; that is, it is more likely that we will correctly reject the null hypothesis. The power of the test will be relatively high. Thus, as the strength of the relationship in the population increases, the power of the statistical test also increases.

Next, consider sample size. As indicated in Chapters 8 and 9, when sample sizes are relatively large, it is more likely that we will detect an existing difference between population means (that is, the existence of a relationship). Again, the power of the statistical test is affected, with larger sample sizes leading to more powerful tests.

Finally, consider the alpha level. When the alpha level is low, the likelihood that we will reject the null hypothesis is also low because we must be very certain that a relationship exists before we are willing to say it does. Consequently, we will be relatively likely to overlook a relationship that does exist, especially if it is a weak one and we have small sample sizes. As the alpha level becomes larger, the power of a statistical test will increase.

If a Type I error is serious, then the investigator will want to minimize its chance of occurrence by setting a low alpha level. This means that the major factor the researcher can use to minimize a Type II error is the sample sizes in the study. If a Type II error is also serious, then one will want to achieve high power. Larger sample sizes can accomplish this.

Statisticians have developed procedures for estimating the sample sizes necessary to obtain a desired level of power, given the strength of the relationship in the population and the alpha level. The required sample sizes also differ, depending on whether the test is directional or nondirectional. Appendix E.1 contains tables of the per-group sample sizes necessary to achieve various levels of power for the independent groups *t* test.[*] A portion of this appendix for an alpha level of .05, nondirectional test, is reproduced in Table 10.6 for the purpose of exposition. The first column of the table lists various values of power. The numbers at the top of the table are population values of eta-squared. The table entries are required sample sizes *per group* to achieve the corresponding level of power.[†] For exam-

TABLE 10.6 **Approximate Sample Sizes Necessary to Achieve Selected Levels of Power for Alpha = .05, Nondirectional Test, As a Function of Population Values of Eta-Squared**

Power	.01	.03	.05	.07	.10	.15	.20	.25	.30	.35
.25	84	28	17	12	8	6	5	3	3	3
.50	193	63	38	27	18	12	9	7	5	5
.60	246	80	48	34	23	15	11	8	7	6
.67	287	93	55	39	27	17	12	10	8	6
.70	310	101	60	42	29	18	13	10	8	7
.75	348	113	67	47	32	21	15	11	9	7
.80	393	128	76	53	36	23	17	13	10	8
.85	450	146	86	61	41	26	19	14	11	9
.90	526	171	101	71	48	31	22	17	13	11
.95	651	211	125	87	60	38	27	21	16	13
.99	920	298	176	123	84	53	38	29	22	18

Above the eta-squared column headers: **Population Eta-Squared**

[*] All of the tables in Appendix E are based on power estimates given by Cohen (1977).

[†] Given the imprecision of the procedures for estimating the necessary sample sizes, the values contained in this and the other power tables presented in this book are approximate.

ple, if the researcher suspects (on the basis of past research, pilot studies, intuition, or theory) that the strength of the relationship he or she is trying to detect in the population is relatively weak, corresponding to an eta-squared of .03, then the necessary sample size to achieve a power of .95 for an alpha level of .05, nondirectional test, is 211 per group. If the researcher suspects that the strength of the relationship is much stronger, corresponding to an eta-squared of .20, then the necessary sample size to achieve a power of .95 is 27 per group. Table 10.6 and Appendix E.1 give a general feel for the relationships among directionality, the alpha level, sample size, the strength of the relationship one is trying to detect in the population, and power for the independent groups t test.

It is not always practical to obtain high levels of power by increasing sample sizes, especially when the population relationship is judged to be weak. In the above example where eta-squared was .03, the required 211 participants per group could be costly, especially if the study involved the use of nonhuman primates or other expensive laboratory animals. The seriousness of the lack of power again depends on the seriousness of a Type II error. When testing for gender discrimination, what would be the consequences of saying discrimination does not exist when, in fact, it does? If the consequences are judged to be serious, one should insist upon high power. Investigators generally strive to achieve a power of .80 to .95, as a rough guideline. However, the power can be revised upward or downward, depending on the situation.

10.8   Method of Presentation

When researchers report the results of an independent groups t test, all three questions outlined in Section 10.1 should be addressed (whether there is a relationship between the two variables, the strength of the relationship, and the nature of the relationship). Of course, the last question is relevant only if the null hypothesis is rejected. In fact, if the null hypothesis is not rejected, then the strength of the relationship is also typically not discussed.

The results for the experiment on order of information and likability discussed earlier in this chapter might be presented as follows:

Results

An independent groups t test was performed comparing the mean likability rating for the pro-con condition (M = 5.00, SD = 1.15) with that for the con-pro condition (M = 3.00, SD = 1.33). The alpha level was .05. This test was found to be statistically significant, $t(18)$ = 3.57, p < .01, indicating that strangers will be evaluated more favorably when positive information about them is followed by negative information than when the reverse is true. The strength of the relationship between order of information and likability, as indexed by eta^2, was .41.

The first sentence states the statistical test that the researcher used to analyze the data. It also conveys the values of the sample means and the standard deviation estimates, using the symbols M and SD, respectively.* The second sentence states the alpha level that was used. The statement that the *t* test was statistically significant signifies that the null hypothesis was rejected at whatever alpha level was used (in this case, .05). If it had not been rejected, the terminology "statistically nonsignificant" would have been used instead. Paralleling the format for the one-sample *t* test, this is followed by the degrees of freedom associated with the relevant *t* distribution, the observed value of *t*, and the significance level. Since there is nothing to indicate otherwise, a nondirectional test is assumed. The remainder of this sentence specifies the nature of the relationship.

The final sentence relates the value of eta-squared. In practice, researchers often fail to present this or other information regarding the strength of the relationship. This can be problematic because sometimes investigators interpret their data as if a strong relationship exists when, in fact, it is so weak as to be almost trivial. Fortunately, however, eta-squared can be derived from the degrees of freedom and the observed value of *t* using Equation 10.11.

10.9   Examples from the Literature

Monetary Compensation and Enjoyableness of a Boring Task

Behavioral scientists have devoted considerable effort to identifying psychological factors that influence attitude change. In a classic experiment in this area (Festinger & Carlsmith, 1959), subjects individually came to a psychology laboratory and worked on two very boring tasks for 1 hour. After completing the tasks, they were informed that the experiment was over and that it had been a study of the effect of "expectancy" on performance. Supposedly, half of the participants had been led to believe that the tasks would be interesting and enjoyable, whereas the other half had not been told anything about them so that the influence of the expectancy manipulation on task performance could be studied. In actuality, this was not the true purpose of the experiment, and none of the participants had actually been informed that the tasks would be interesting and enjoyable. Rather, as discussed below, it was a study of a psychological process known as cognitive dissonance.

The participants were then told that the experimenter who usually informed waiting subjects that the tasks were interesting and enjoyable was sick, and they were offered either $1 or $20 to take his place. All participants complied with this request. Afterward, they were given a questionnaire about the original experiment and asked to rate, among other things, how interesting and enjoyable the tasks

---

* The variance estimates were calculated in Table 10.1 to be 1.33 for the pro–con condition and 1.78 for the con–pro condition. The square roots of these values are 1.15 and 1.33, respectively.

were on a scale from −5 ("extremely dull and boring") to +5 ("extremely interesting and enjoyable").

At least two different principles could operate in this situation to influence the participants' ratings of the tasks. The first is a reinforcement principle, which holds that participants who were paid $20 for telling another person that the tasks were interesting and enjoyable received greater reinforcement for doing so than those who were paid only $1. This greater reinforcement should generalize to their own perception of the tasks; hence, people in the $20 condition should rate the tasks more positively than those in the $1 condition.

An alternative explanation predicts just the opposite. According to this viewpoint, all participants performed a behavior that was in contradiction to what they truly believed: They told someone that boring tasks were interesting and enjoyable. Because of this contradiction, the participants should experience an unpleasant internal state known as *cognitive dissonance,* which they should then be motivated to reduce. People in the $20 condition could easily justify their counterattitudinal behavior: They did it for the money. This was not the case, however, for people in the $1 condition, where the reward was not very great. These people would have to reduce the dissonance in another way, and one possibility would be to rationalize that the tasks were really not all that boring. If this dissonance mechanism were operating, rather than rating the task less positively, participants in the $1 condition should rate the tasks more positively than those in the $20 condition.

The results of the experiment were analyzed using an independent groups t test. The independent variable was the amount of money received for telling the waiting person about the tasks ($1 versus $20) and the dependent variable was the participants' ratings of how interesting and enjoyable the tasks were. The mean rating in the $1 condition was 1.35, and the mean rating in the $20 condition was −.05. Consistent with a dissonance interpretation, these means were found to be significantly different, $t(38) = 2.22$, $p < .05$. Festinger and Carlsmith did not report any statistics regarding the strength of the relationship. However, using Equation 10.11, we find that eta-squared equals .11, which represents a moderate effect.

Students' Perceptions of Self-Monitoring of Good and Poor Teachers

An interesting question for teachers and students alike concerns the personality characteristics that distinguish good from poor teachers. One potentially relevant characteristic is *self-monitoring.* Among other things, self-monitoring concerns the extent to which individuals are sensitive to the behavior of others and able to modify their own social behavior accordingly. As self-monitoring increases, individuals become relatively more concerned with the situational appropriateness of their behavior and relatively less constrained by their own internal dispositions. Thus, for instance, high self-monitoring teachers might be expected to be more responsive to the needs of their students and more flexible in their classroom approach.

In an attempt to determine students' perception of the role teacher self-

**Applications
to the Analysis
of a Social Problem**

It is commonly believed that parents want to discourage premarital sexual intercourse on the part of their teens. However, research suggests that this is not necessarily the case. Some parents feel more adamant about sexual abstinence for teens than others. For example, some parents believe that if it is with someone who is special and whom the teen has known for an extended period of time, then some sexual intercourse is permissible, given proper protection against unintended pregnancy and sexually transmitted diseases.

So that we could learn more about parental attitudes toward adolescent sexuality, teens in the parent–teen communication study described in Chapter 1 were asked a series of questions about their mother's position on sexual activity. Specifically, they were presented the following five statements and asked to indicate their agreement with each, using a five-point scale having the response options strongly disagree, moderately disagree, neither agree nor disagree, moderately agree, and strongly agree:

1. My mother would disapprove of my having sex at this time in my life.

2. My mother has specifically told me not to have sex.

3. My mother thinks I definitely should not be sexually active (having sexual intercourse) at this time in my life.

4. If it were with someone who was special to me and whom I knew well, like a steady boyfriend(girlfriend), my mother would not mind if I had sexual intercourse at this time in my life.

5. My mother thinks it is fine for me to be sexually active (having sexual intercourse) at this time in my life.

The questions were "split up" and asked in different sections of the interview. The teens answered the questions on a piece of paper that only they could see, so that they would not have to reveal their answers face to face to an interviewer. The responses to the first three statements were scored from 1 (strongly disagree) to 5 (strongly agree), whereas the responses to the last two items were reverse scored (1 = strongly agree to 5 = strongly disagree). An overall index of the extent to which the teen perceived the mother as disapproving of the teen engaging in sexual intercourse was obtained by summing the responses to the five statements.* This index of perceived maternal disapproval could range from 5 to 25, with higher scores indicating greater perceived maternal disapproval of engaging in sex.

One question we were interested in was whether mothers differ in the extent to which they discourage sexual intercourse in adolescent girls as opposed to boys. We felt there was a possibility for a double standard in this regard. Traditionally, preg-

---

* We conducted additional analyses, not described here, to ensure that the items were reasonably intercorrelated and met adequate psychometric criteria for summing them into an overall index.

monitoring plays in the teaching process, Larkin (1987) asked 116 college undergraduates to think of either the best or the worst teacher they had ever had. Students in each group rated their "best" or "worst" teacher on a 13-item self-monitoring scale. A dependent variable score was derived for each student by summing his or her responses across the 13 items. As predicted, an independent groups *t* test showed that the good teachers ($\overline{X}$ = 49.56) were rated as significantly higher in self-monitoring than the poor teachers ($\overline{X}$ = 29.81), $t(114)$ = 13.18, p < .001. As indexed by eta-squared, the strength of the relationship was .60. This represents a strong effect. Are these results consistent with your own educational experience?

nancy and birth control often have been framed as "female" issues that adolescent girls must face. However, unintended pregnancy results from the behavior of *both* the boy and the girl. If parents are going to discourage premarital sexual activity, they should do so equally for boys and girls.

To test whether teens perceive a difference in their mother's orientation toward their sexual activity, we conducted an independent groups *t* test on the perceived maternal attitude, using gender as a between-subjects independent variable. The first step was to examine summary statistics and frequency distributions for the dependent variable for boys and girls separately. These analyses were undertaken to identify outliers and to examine potential problems with variance heterogeneity (i.e., nonhomogeneity) and nonnormality. Although the assumptions apply to populations, examination of sample data can be informative in this regard. Table 10.7 presents the computer output for the summary statistics for the two groups, and Figure 10.1 presents a grouped frequency histogram for each group, with a normal curve superimposed on each. The sample sizes for the two groups were large and nearly identical (for boys, $n = 362$, for girls, $n =$

360), which suggests that the analysis will be robust to assumption violations. The variance estimate for girls (19.95) was comparable to that for boys (23.19), which suggests that violation of homogeneity of variance is not an issue. As can be seen in the frequency histograms, the distributions of scores are not normal. However, the departures from normality are not sufficiently large to undermine the validity of the *t* test (Posten, 1978). No outliers were evident in the data.

The computer output for the independent groups *t* test is presented in Table 10.8. The first part of the output presents the sample size, the sample mean, the standard deviation estimate, and the estimated standard error of the mean for each group. Directly beneath this section is a statement of the difference between the two means (15.83 − 19.10 = −3.27), followed by the results of a formal statistical test of the homogeneity of variance assumption. This test is called the **Levene test,** and it evaluates a null hypothesis of equal variances in the populations against an alternative hypothesis of unequal variances in the populations. If the significance level (i.e., *p* value) associated with the test

(continued)

TABLE 10.7   Computer Output for Summary Statistics for Perceived Maternal Attitude Toward Male and Female Teens Engaging in Sex

---

GENDER: Males

MDISSEX   Teen perception of maternal disapproval of sex

Mean	15.829	Std err	.253	Median	15.000
Mode	15.000	Std dev	4.816	Variance	23.189
Range	20.000	Minimum	5.000	Maximum	25.000

Valid Cases   362   Missing cases   12

GENDER: Females

MDISSEX   Teen perception of maternal disapproval of sex

Mean	19.103	Std err	.235	Median	19.000
Mode	25.000	Std dev	4.466	Variance	19.948
Range	20.000	Minimum	5.000	Maximum	25.000

Valid Cases   360   Missing cases   13

**FIGURE 10.1   Grouped Frequency Histograms of Perceived Maternal Attitude Toward
(a) Male and (b) Female Teens Engaging in Sex**

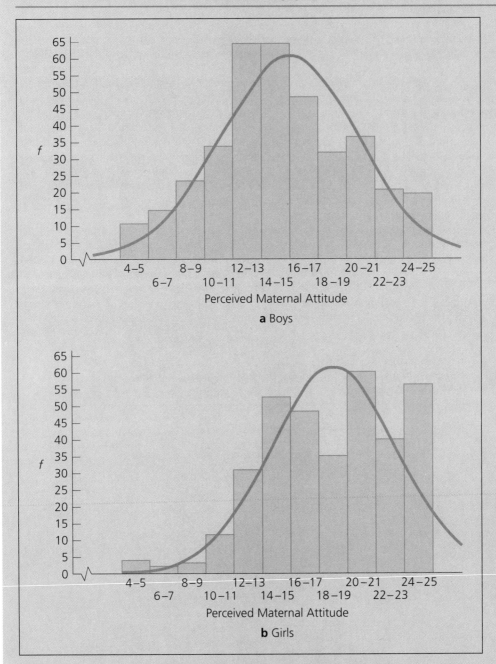

Perceived Maternal Attitude

a Boys

Perceived Maternal Attitude

b Girls

statistic (*F*) is less than .05, then there is evidence that the variances are not homogeneous. In this instance, the p value is .59, which is consistent with our observation earlier: The population variances do not appear to be highly discrepant..

Even if the Levene test yielded a statistically

TABLE 10.8   **Computer Output for Independent Groups *t* Test for Perceived Maternal Attitude Toward Teens Engaging in Sex**

```
t-test for Independent Samples of GENDER

MDISSEX     Teen perception of maternal disapproval of sex
```

Variable	Number of Cases	Mean	SD	SE of Mean
Male	362	15.8287	4.816	.253
Female	360	19.1028	4.466	.235

```
              Mean Difference = -3.2740

    Levene's Test for Equality of Variances: F= .291  P= .590

              t-test for Equality of Means
```

Variances	t-value	df	2-Tail Sig	SE of Diff
Equal	−9.47	720	.000	.346
Unequal	−9.47	716.53	.000	.346

significant value (i.e., $p < .05$), this would not necessarily be problematic because the independent groups *t* test is robust to variance heterogeneity. A *p* value less than .05 for the Levene test only allows one to conclude with a high degree of confidence that the variances in the populations are not equal. It does not speak to the *magnitude* of the discrepancy in the population variances, which must be substantial to undermine the *t* test. We will discuss the Levene test in more depth in Chapter 12. For now, we simply note that heterogeneity of variance is not a problem for the current analysis.

The bottom of Table 10.8 presents two versions of the *t* test, one in each row. The top row is the test we described in this chapter. It has a *t* value of −9.47 and 720 degrees of freedom. Although the *p* value for the test is reported as .000 in the column labeled "2-Tail Sig," its value is not actually .000. It appears as .000 because the *p* value is less than .0005, and the computer rounded to three decimal places. The column labeled "SE of Diff" presents the estimated standard error of the difference between two independent means, which is the denominator of the *t* test formula (see Equation 10.7). The estimated standard error of the difference is important because it provides a sense of the average amount of sampling error that occurs in sample mean differences. Recall that the mean sample difference between boys and girls was −3.27. The estimated standard error of the difference suggests that, on average, random samples of the size we used tend to yield sample mean differences that deviate from the true population mean difference by about .35 unit. The bottom row presents an alternative *t* test that would be used if the researcher determined that heterogeneity of population variances was a serious problem. This test does not use pooled variance estimates for calculating the estimated standard error of the difference, and it has a different number of degrees of freedom. The rationale and computational procedures for this test are described in Hays (1981).

From the printout we can state the following: Adolescent girls ($\overline{X} = 19.10$) tend to perceive *(continued)* their mothers as being more disapproving of sexual activity than adolescent boys do ($\bar{X} = 15.83$). As indicated by an independent groups *t* test, this difference was statistically significant, $t(720) = -9.47$, $p < .0005$. The strength of the relationship, as indexed by eta-squared, was .11 (as calculated by applying Equation 10.11). This represents a moderate effect.

Although this difference suggests a double standard on the part of mothers, there is a problem with making such an inference. The data focused on teen *perceptions* of maternal attitudes, not maternal attitudes per se. It may be that mothers of adolescent boys are just as disapproving of sexual intercourse by their teens as are mothers of adolescent girls, but that adolescent girls are more likely to perceive that disapproval compared with boys (or, alternatively, that adolescent girls are more likely to have that disapproval conveyed to them by their mothers than are adolescent boys). However, we also measured maternal attitudes directly by presenting mothers the same five items described above, but with appropriate wording changes (e.g., the item "My mother would disapprove of my having sex at this time in my life" was phrased "I disapprove of my teen having sex at this time in her/his life"). We summed the responses to the five items for mothers (with appropriate reverse scoring of the last two items) to yield an overall index of the actual maternal attitude toward her

teen engaging in sex. We then performed an independent groups *t* test on this measure as a function of the gender of the teen.

Preliminary analyses revealed no outliers. The distribution of scores for each group was not so discrepant from normality as to suggest that the *t* test would be undermined, and we found very similar variance estimates for the two groups. An independent groups *t* test showed that the maternal mean for adolescent girls ($\bar{X} = 20.99$) was statistically significantly greater than the mean for adolescent boys ($\bar{X} = 19.87$), $t(721) = 3.87$, $p < .001$. The strength of the relationship, as indexed by eta-squared, was .02. This represents a weak effect. Mothers are indeed more disapproving of sexual activity for adolescent girls than for adolescent boys.

Examination of the mean differences in the two analyses reveals an intriguing result. Figure 10.2 presents a graph of the means for the analyses. Note that the discrepancy in means between adolescent boys and girls is much greater for teen *perceptions* of attitudes than for actual maternal attitudes. This is evident by the fact that the bar for girls is relatively taller than the bar for boys for perceived maternal attitudes as compared with the difference for actual maternal attitudes. There is a statistical technique (not discussed in this book) that can formally compare these differences to determine whether the discrepancy for perceived at-

FIGURE 10.2   Graph of Means for Perceived and Actual Maternal Attitudes As a Function of Teen Gender

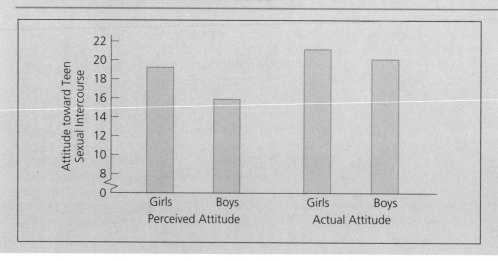

titudes between boys and girls is statistically significantly greater than the boy–girl discrepancy for actual maternal attitudes. We applied this technique to the data and found this to be the case. Thus, even though mothers are more disapproving of sexual activity for adolescent girls than for boys, the teens' perceived attitudes reflect a larger disparity. This could be because mothers communicate their attitudes more clearly or forcefully to their daughters than to their sons. Alternatively, there may be gender differences in the teens' ability to accurately perceive their mothers' attitudes.

Before concluding our analysis, we comment on one additional issue—namely, levels of measurement. The measures of our dependent variables were based on scales that involved summing responses to five agree–disagree questions that were all designed to measure the same construct. Most researchers would argue that the resulting "total" scores are probably ordinal in character and violate a strict assumption of intervalness. We would not argue with this assessment. However, this does not mean that the data cannot be effectively analyzed using the independent groups t test. The crucial issue is not whether a measure is ordinal or interval, but rather whether the measure approximates interval level characteristics to the degree that an independent groups t test can still be appropriately applied (see the discussion of measures versus scales in Section 1.4).

Several Monte Carlo studies have evaluated how severe the departure from intervalness can be for the independent groups t test without affecting the validity of the basic conclusions of the test. This research has shown the test to be quite robust to rather large departures from intervalness (e.g., see the discussion and review of Davison & Sharma, 1988).

Our experience with these particular measures of attitudes toward premarital sex in prior research has suggested that they are reasonable. They tend to be reliable, unrelated to measures of social desirability response tendencies, and to correlate with other constructs as we theoretically expect. Furthermore, the shapes of the frequency distributions of attitude scores are consistent with what one might expect on theoretical grounds. When the Monte Carlo studies of robustness are considered, we are reasonably confident that the psychometric properties of the measures are sufficient to effectively analyze the data using the independent groups t test.

Summary

The independent groups t test is typically used to analyze the relationship between two variables when (1) the dependent variable is quantitative in nature and is measured on a level that at least approximates interval characteristics, (2) the independent variable is between-subjects in nature, and (3) the independent variable has two and only two levels. The existence of a relationship between the two variables is tested by converting the observed difference between sample means into a t value that represents the number of estimated standard errors the difference is from the mean of the appropriate sampling distribution of the difference between two independent means (assuming the null hypothesis is true). This t value is compared with the critical value(s) of t, and the decision to reject or not to reject the null hypothesis is made accordingly. The strength of the relationship is measured using the eta-squared statistic, which represents the proportion of variability in the dependent variable that is associated with the independent variable. Finally, the nature of the relationship is determined by examining the mean scores for the two conditions.

Tables are available for determining the sample sizes necessary for the independent groups t test to achieve a desired level of power. The required sample sizes are dependent on the strength of the relationship between the independent variable and the dependent variable in the population, the alpha level, and whether the test is directional or nondirectional.

Exercises

Answers to asterisked (*) exercises appear at the back of the book. Answers to exercises with two asterisks are also worked out step by step in the Study Guide.

Exercises to Review Concepts

1. Under what conditions is the independent groups t test typically used to analyze a bivariate relationship?

*2. In general terms, what value will the mean of a sampling distribution of the difference between two independent means always equal?

*3. Explain the rationale that allows us to combine the variance estimates from two samples to obtain a pooled variance estimate.

**4. Compute the pooled variance estimate and the estimated standard error of the difference for $n_1 = 10$, $n_2 = 13$, $\hat{s}_1^2 = 6.48$, and $\hat{s}_2^2 = 4.73$.

*5. Consider the following information for two samples that were randomly selected from their respective populations:

Sample A	Sample B
$n_A = 49$	$n_B = 49$
$\overline{X}_A = 10.00$	$\overline{X}_B = 13.00$
$\hat{s}_A = 1.44$	$\hat{s}_B = 1.58$

a. Compute the estimated standard error of the mean for sample A.

b. Compute the estimated standard error of the mean for sample B.

c. Compute the estimated standard error of the difference between two independent means.

d. Your answer for part **c** should be larger than your answer for part **a** or **b**. In fact, the estimated standard error of the difference between two independent means will always be larger than the respective estimated standard errors of the mean. Why do you think this is the case?

6. Match each symbol in the first column with the appropriate term in the second column:

Symbol	Term
1. $\hat{s}_{\overline{X}_1 - \overline{X}_2}$	a. variance estimate
2. $\hat{s}_{pooled}^2$	b. null hypothesis
3. $\hat{s}_{\overline{X}}$	c. standard error of the mean
4. $\hat{s}^2$	
5. H_1	d. estimated standard error of the mean
6. H_0	
7. $\sigma_{\overline{X}}$	e. standard error of the difference between two independent means
8. $\sigma_{\overline{X}_1 - \overline{X}_2}^2$	
	f. alternative hypothesis
	g. estimated standard error of the difference between two independent means
	h. pooled variance estimate

7. Summarize the five steps involved in hypothesis testing for the independent groups t test.

*8. State the critical value(s) of t for an independent groups t test for an alpha level of .05 under each of the following conditions:

a. $H_0: \mu_1 = \mu_2$, $H_1: \mu_1 \neq \mu_2$, $n_1 = 10$, $n_2 = 10$

b. $H_0: \mu_1 = \mu_2$, $H_1: \mu_1 > \mu_2$,
   $n_1 = 10$, $n_2 = 10$
c. $H_0: \mu_1 = \mu_2$, $H_1: \mu_1 \neq \mu_2$,
   $n_1 = 16$, $n_2 = 14$
d. $H_0: \mu_1 = \mu_2$, $H_1: \mu_1 > \mu_2$,
   $n_1 = 16$, $n_2 = 14$
e. $H_0: \mu_1 = \mu_2$, $H_1: \mu_1 \neq \mu_2$,
   $n_1 = 23$, $n_2 = 19$
f. $H_0: \mu_1 = \mu_2$, $H_1: \mu_1 < \mu_2$,
   $n_1 = 23$, $n_2 = 19$

9. What are the assumptions underlying the independent groups t test?

Use the following information to complete Exercises 10–16.

An investigator tested the relationship between gender and discriminatory attitudes toward women. An attitude scale was administered to five men and five women. Scores could range from 1 to 10, with higher values indicating more discriminatory attitudes. The data are presented in the table:

Men	Women
7	4
7	3
8	4
7	5
6	4

*10. Test for a relationship between gender and discriminatory attitudes using a nondirectional independent groups t test.

**11. Compute the sum of squares total.

**12. Compute the treatment effect for men and the treatment effect for women.

**13. Based on your answers in Exercise 12, use the variance extraction procedures discussed in this chapter to nullify the effect of gender on discriminatory attitudes (that is, generate a set of scores on the dependent variable with the effect of gender removed).

**14. Compute the sum of squares error and the sum of squares explained for the data derived in Exercise 13.

*15. Based on your answers in Exercises 11 and 14, compute the value of eta-squared using Equation 10.10. Recalculate this index using Equation 10.11. Compare the two results. (*Note:* In practice, we would use only the approach of Equation 10.11.) Does the observed value represent a weak, moderate, or strong effect?

*16. Discern the nature of the relationship between gender and discriminatory attitudes.

Use the following information to complete Exercises 17–23.

An investigator tested the effect of alcohol on reaction time. Five participants were given alcohol to consume until a certain level of intoxication was achieved (as indexed by physiological measures). Another group was not given any alcohol but instead consumed a placebo. All then participated in a reaction time task. The reaction times (in seconds) for the two groups are listed in the table.

Alcohol	Placebo
2.00	1.00
2.50	.50
2.00	1.00
1.50	1.00
2.00	1.50

17. Test for a relationship between alcohol consumption and reaction time using a nondirectional independent groups t test.

18. Compute the sum of squares total.

19. Compute the treatment effect for alcohol and the treatment effect for the placebo.

20. Based on your answers in Exercise 19, use the variance extraction procedures

discussed in this chapter to nullify the effect of alcohol consumption on reaction time (that is, generate a set of scores on the dependent variable with the effect of alcohol consumption removed).

21. Compute the sum of squares error and the sum of squares explained for the data derived in Exercise 20.

22. Based on your answers in Exercises 18 and 21, compute the value of eta-squared using Equation 10.10. Recalculate this index using Equation 10.11. Compare the two results. (*Note:* In practice, we would use only the approach of Equation 10.11.) Does the observed value represent a weak, moderate, or strong effect?

23. Discern the nature of the relationship between alcohol consumption and reaction time.

*24. Explain the interrelationships among the sum of squares total, the sum of squares explained, and the sum of squares error.

25. What would the value of eta-squared be if an independent groups t test yielded an observed *t* value of −1.73 for $n_1 = 16$ and $n_2 = 18$? Does this represent a weak, moderate, or strong effect?

*26. Why is it inappropriate to estimate the strength of the relationship between two variables in the population from the sample value of eta-squared?

27. What is the rationale behind calculating eta-squared following nonsignificant statistical tests?

28. What are the three major factors that influence the power of a statistical test?

*29. If a researcher suspects that the strength of the relationship between two variables in the population is .07 as indexed by eta-squared, what sample size should she use per group in a study involving two independent groups and an alpha level of .05, nondirectional test, in order to achieve a power of .95?

30. Suppose an investigator conducted a study involving two independent groups

with $n = 23$ per group. If the value of eta-squared in the population was .10, what would the power of his statistical test be at an alpha level of .05, nondirectional test?

Multiple-Choice Questions

31. The null hypothesis for an independent groups *t* test states that
    a. $\overline{X}_1 = \overline{X}_2$ c. $\mu_1 = \mu_2$
    b. $\overline{X}_1 \neq \overline{X}_2$ d. $\mu_1 \neq \mu_2$

32. If a researcher suspects that the strength of the relationship between two variables in the population is .15 as indexed by eta-squared, what sample size should he use per group in a study involving two independent groups and an alpha level of .05, nondirectional test, in order to achieve a power of .85?
    a. $n = 21$ c. $n = 34$
    b. $n = 26$ d. $n = 38$

*33. A treatment effect is
    a. a nullified score
    b. the difference between the mean for one group and the mean for the second group
    c. the difference between the mean for a given group and the grand mean
    d. the mean score for both groups combined

34. If the mean score for men is 10.00, the mean score for women is 20.00, and the grand mean is 15.00, what is the "effect" of being a man?
    a. −10.00   c. 5.00
    b. −5.00   d. 10.0

35. Eta-squared should be calculated following both significant and nonsignificant statistical tests.
    a. true   b. false

*36. Which measure estimates the average amount by which the mean differences that constitute the sampling distribution of the difference between two independent means deviate from the true difference between the population means?

a. pooled variance estimate
b. *t*
c. eta-squared
d. estimated standard error of the difference

Use the following information to complete Exercises 37–46.

A researcher examined the effect of studying strategies on test performance. One mixed-gender group of ten individuals was instructed to study for an exam for 5 hours the night before the exam. A second group of ten individuals was told to study for 5 hours by studying only 1 hour per night on the five nights prior to the exam. Thus, the total amount of study time was the same, but in the first condition the study time was concentrated into one night whereas in the second condition the study time was spread out over five nights. The individuals then took an exam on which scores could range from 0 to 100, with higher values indicating better performance. In the "concentrated" group, the mean score on the exam was 65.00; in the "spread out" group, the mean score was 80.00. The estimated standard error of the difference was 5.00.

*37. If the value of eta-squared in the population is .30, what is the power of the statistical test at an alpha level of .05, nondirectional test?
   a. .30   c. .80
   b. .67   d. .85

*38. The independent variable is
   a. the amount of study time
   b. performance on the exam
   c. gender
   d. the type of studying strategy

39. The dependent variable is
   a. the amount of study time
   b. performance on the exam
   c. gender
   d. the type of studying strategy

*40. The observed value of *t* is
   a. −.60   c. −6.71
   b. −3.00   d. −15.00

*41. Based on this study, we can conclude that the independent variable is related to the dependent variable.
   a. true   b. false

*42. It is possible that the test results reflect a Type I error.
   a. true   b. false

43. It is possible that the test results reflect a Type II error.
   a. true   b. false

44. The strength of the relationship, as indexed by eta-squared, is
   a. .14   c. .43
   b. .33   d. .67

*45. The proportion of variability in the dependent variable that is due to disturbance variables is
   a. .45   c. .82
   b. .67   d. .86

46. What can we conclude about the effectiveness of the two studying strategies?
   a. The concentrated method is more effective.
   b. The spread out method is more effective.
   c. The two strategies are equally effective.
   d. The two strategies may or may not differ in their effectiveness.

Exercises to Apply Concepts

**47. Psychologists have studied extensively the effects of early experience on the development of individuals. It has long been recognized that a positive, challenging, and diverse environment (sometimes called an enriched environment) leads to the acquisition of more positive abilities and personality traits than an environment that is relatively impoverished and isolated. Bennett, Krech, and Rosenzweig (1964) suggested that the type of environment may even alter the physical characteristics of the brain, and they have reported a series of studies to investigate this possibility. In one study, laboratory rats from the same genetic strain were raised in one of two conditions. Half the rats were raised in an enriched environment, which involved being housed with 10 to 12 other animals in large cages that were equipped with a variety of "toys." Each day these rats were placed in a square field where they were allowed to explore a pattern of barriers that was changed daily. The other half of the rats were raised in an isolated environment. They were caged singly in a dimly lit room where they could not see or touch another animal (although they could hear and smell them). After 80 days, all animals were killed for the purpose of analyzing the structure of the brain as a function of the type of environment in which the animal was raised. One factor that was examined was the weight of the cortex (reported in milligrams). The hypothetical data presented in the table are representative of the results of the study. Analyze these data using a nondirectional test, draw a conclusion, and write up your results using the principles discussed in the Method of Presentation section.

Enriched environment	Isolated environment
685	660
690	642
675	640
660	626
645	612
630	610
635	592

48. McConnell (1966) reported a series of experiments that have been the subject of considerable controversy concerning the physiological bases of learning and memory. McConnell suggested that RNA and DNA protein molecules constitute the physiochemical substrate of learning and that it may be possible to transfer memory, biochemically, between organisms by transferring the relevant RNA and DNA molecules. McConnell's initial experiments were conducted with planaria (small, wormlike organisms). Classical conditioning procedures were used to teach a group of planaria to contract in size whenever they were exposed to a light. McConnell tried several different methods of transferring the relevant RNA and DNA molecules of these trained planaria to other planaria. The only practical method, however, was cannibalism. The trained planaria were chopped up into small pieces and fed to another group of planaria. The trained cannibals were the experimental group. A control group of untrained cannibals was fed chopped planaria that had not been taught to contract when exposed to the light. McConnell then exposed both the trained and the untrained cannibals to the light 25 times each and counted the number of times each planarian contracted. If the relevant RNA and DNA molecules had been transferred and had an impact on behavior, the trained cannibals should exhibit a greater number of contractions than the untrained cannibals. Such a demonstration would at least establish the *possibility* of transferring memory from one organism to another via biochemical means. The hypothetical data listed here are representative of the results of the study. Analyze these data using a nondirectional test, draw a conclusion, and write up your results using the principles discussed in the Method of Presentation section.

Trained cannibals					
14	8	10	13	22	7
6	15	17	10	8	15
21	9	15	4	11	
8	14	12	15	10	

Untrained cannibals					
6	4	19	16	10	6
1	7	6	7	10	5
3	6	1	4	4	
11	5	5	7	11	

49. One factor that has been proposed to affect creativity is how much choice individuals have in their approach to a task. In a study of this issue, Amabile and Gitomer (1974) varied the amount of choice children between 2 and 6 years of age had in selecting material with which to make collages. Children in the choice condition were presented with ten boxes containing collage materials and told to choose any five of these boxes for making their collages. Children in the no-choice condition had their five boxes selected for them by the experimenter. All sets of material were similar to one another. The children were given approximately 10 minutes to complete their collages, which were subsequently rated in terms of how creative they were by eight trained artists. Creativity ratings could range from 0 to 320, with higher scores indicating greater creativity. The hypothetical data in the table are representative of the results of the study. Analyze these data using a nondirectional test, draw a conclusion, and write up your results using the principles discussed in the Method of Presentation section.

Choice	No choice
207	130
203	142
180	137
167	149
212	146
192	150
172	128
200	154
170	142
164	166
178	145
	133
	161
	156

Correlated Groups
t Test

11.1 Use of the Correlated Groups *t* Test

The statistical technique developed in this chapter is called the **correlated groups *t* test.** It is typically used to analyze the relationship between two variables when these conditions are met:

1. The dependent variable is quantitative in nature and is measured on a level that at least approximates interval characteristics.

2. The independent variable is *within-subjects* in nature (it can be either qualitative or quantitative).

3. The independent variable has two and only two levels.

The major difference between the correlated groups *t* test and the independent groups *t* test is that the former is used when the independent variable is within-subjects in nature and the latter is used when the independent variable is between-subjects in nature.*

---

* Matched-subjects designs are analyzed as if the independent variable is within-subjects in nature.

An important advantage of the correlated groups *t* test over the independent groups *t* test relates to the control of disturbance variables. As discussed in Chapter 9, disturbance variables are variables that are unrelated to the independent variable but that influence the dependent variable. By thus contributing to variability in the dependent variable, disturbance variables create "noise" that makes it more difficult to detect a relationship between the independent and dependent variables.

One major source of "noise" is the difference in backgrounds and abilities of the individuals who participate in an investigation. For example, the fact that some individuals are more intelligent than others might be expected to influence performance on research tasks. If we know the influence of individual differences on the dependent variable, then their influence can be separated from the effects of the independent variable. However, if their influence cannot be isolated, then these differences in background and ability remain as uncontrolled sources of variability and increase the error variance. As will be demonstrated shortly, within-subjects designs (but not between-subjects designs) allow us to estimate this source of variability and extract it from the dependent variable. Consequently, the correlated groups *t* test provides a more sensitive test of the relationship between the independent and dependent variables than does the independent groups *t* test. A test's sensitivity is its ability to detect a relationship between variables when a relationship exists in the population.

As an example of the type of study that meets the conditions for the correlated groups *t* test, suppose an investigator is studying the relative effects of two drugs, A and B, on learning. Five people are administered drug A and then work on a learning task. One month later, the same five people are administered drug B and work on the same type of learning task as before. The investigator has introduced a 1-month time interval between the two sessions in order to control for potential carry-over effects. The number of errors made following the administration of each drug is tabulated and serves as the dependent variable. These data are presented in the second and third columns of Table 11.1. The experimental design involves a within-subjects independent variable with two levels (drug A versus drug B) and a quantitative dependent variable that is measured on a level that at least approximates interval characteristics. Hence, the correlated groups *t* test would typically be used to analyze the relationship between the variables.

11.2 Inference of a Relationship Using the Correlated Groups *t* Test

Null and Alternative Hypotheses

The first question to be considered is whether a relationship exists between the type of drug administered and learning. We begin by formally phrasing this in null and alternative hypotheses:

$$H_0: \quad \mu_A = \mu_B$$
$$H_1: \quad \mu_A \neq \mu_B$$

TABLE 11.1   Data and Calculations for Drug and Learning Experiment

Participant	X for drug A	X for drug B	Difference (D)	D^2
1	3	7	−4	16
2	1	5	−4	16
3	4	4	0	0
4	2	8	−6	36
5	5	6	−1	1

$$\Sigma X_A = 15 \qquad \Sigma X_B = 30 \qquad \Sigma D = -15 \qquad \Sigma D^2 = 69$$
$$\overline{X}_A = 3.00 \qquad \overline{X}_B = 6.00 \qquad \overline{D} = -3.00$$

$$SS_D = \Sigma D^2 - \frac{(\Sigma D)^2}{N} = 69 - \frac{(-15)^2}{5} = 24$$

$$\hat{s}_D^2 = \frac{SS_D}{N-1} = \frac{24}{4} = 6.00$$

$$\hat{s}_D = \sqrt{\hat{s}_D^2} = \sqrt{6.00} = 2.45$$

$$\hat{s}_{\overline{D}} = \frac{\hat{s}_D}{\sqrt{N}} = \frac{2.45}{\sqrt{5}} = 1.10$$

where the subscripts "*A*" and "*B*" respectively represent drug A and drug B. The null hypothesis states that there is no relationship between the type of drug and learning in terms of mean scores on the dependent variable. If the effects of the two drugs on learning are the same, then the population means should be equal. The alternative hypothesis states that there is a relationship between the type of drug and learning; that is, which drug is administered does matter, and the population means will, accordingly, not be equal.*

If the null hypothesis is true and the type of drug does not make a difference, then we would expect performance while under the influence of drug A to be the same as performance while under the influence of drug B. Another way of stating this is that the difference between a person's score in condition A and his or her score in condition B should be 0. Column 4 in Table 11.1 reports a difference score (*D*) for each individual in the experiment. The mean difference score across individuals ($\overline{D}$) is −3.00. Note that this equals the difference between the means for the two conditions (3.00 − 6.00 = −3.00). The question of interest is whether the mean difference of −3.00 is sufficiently different from 0 to reject the null hypothesis. To answer this question, it is necessary to specify the nature of the *sampling distribution of the mean of difference scores* for *N* = 5 difference scores.

---

* As with the tests discussed previously, if there is exclusive concern with a specific direction of mean differences, then the alternative hypothesis can be phrased directionally rather than nondirectionally.

Sampling Distribution of the Mean of Difference Scores and the *t* Test

A **sampling distribution of the mean of difference scores** is conceptually similar to a sampling distribution of the mean or a sampling distribution of the difference between two independent means. However, now the concern is with a distribution of scores that represent differences across individuals. More specifically, a sampling distribution of the mean of difference scores can be defined as a theoretical distribution consisting of mean difference scores across individuals for all possible random samples of a given size that could be selected from a population. A sampling distribution of the mean difference of scores is most readily understood by comparing it with a sampling distribution of the mean. As discussed in Chapters 7 and 8, this latter distribution is used for testing a sample mean against a hypothesized population mean; it has a mean that is equal to the true population mean (μ) and a standard deviation ($\sigma_{\bar{X}}$) that can be estimated using the formula $\hat{s}_{\bar{X}} = \hat{s} / \sqrt{N}$. Similarly, the mean of the sampling distribution of the mean of difference scores is equal to the mean difference score across individuals in the population. When the population means in the two conditions that define the difference scores are equivalent, the mean difference score across individuals in the population (represented by the symbol μ_D) will be equal to 0. The standard deviation of the sampling distribution of the mean of difference scores, referred to as the **standard error of the mean of difference scores** (represented by the symbol $\sigma_{\bar{D}}$), can be estimated as follows:

$$\hat{s}_{\bar{D}} = \frac{\hat{s}_D}{\sqrt{N}} \qquad [11.1]$$

where $\hat{s}_{\bar{D}}$ is the estimated standard error of the mean of difference scores, $\hat{s}_D$ is the estimated standard deviation of difference scores in the population as derived from the sample difference scores, and N is the number of difference scores.* The quantity $\hat{s}_D$ can be derived by applying the usual procedures for a standard deviation estimate to the difference scores. As shown in Table 11.1, $\hat{s}_D$ in the drug and learning example is equal to 2.45, which gives an estimated standard error of $2.45 / \sqrt{5} = 1.10$.

The existence of a relationship between an independent variable and a dependent variable in the correlated groups case is tested by converting the observed mean difference in the sample to a *t* value and comparing this with the critical *t* values that define the rejection region. Since we are dealing with difference scores, the relevant *t* distribution has $N - 1$ degrees of freedom associated with it. The observed value of *t* can be computed using the formula

$$t = \frac{\bar{D} - \mu_D}{\hat{s}_{\bar{D}}} \qquad [11.2]$$

When μ_D is hypothesized to equal 0 (that is, when the population means for the two conditions are hypothesized to be identical), Equation 11.2 reduces to

---

* It will always be the case that $n_1 = n_2 = N$ in the type of design considered in this chapter.

$$t = \frac{\overline{D}}{\hat{s}_{\overline{D}}}$$ [11.3]

This is the most commonly used version of the correlated groups *t* test because the null hypothesis usually states H_0: $\mu_1 = \mu_2$. However, for purpose of illustration, we use the more general Equation 11.2 whenever a correlated groups *t* test is called for in the remainder of this chapter.

In the drug and learning example,

$$t = \frac{-3.00 - 0}{1.10} = -2.73$$

For an alpha level of .05, nondirectional test, the critical values of *t* for df = $N - 1 = 5 - 1 = 4$ are ±2.776. Since −2.73 is not less than the negative critical value of −2.776, we fail to reject the null hypothesis of no relationship between the type of drug and learning.

Study Exercise 11.1

Eight individuals indicated their attitudes toward socialized medicine before and after listening to a pro–socialized medicine lecture. Attitudes were assessed on a scale from 1 to 7, with higher scores indicating more positive attitudes. The attitudes before and after listening to the lecture were as indicated in the second and third columns of the table. Use a correlated groups *t* test to test for a relationship between the time of assessment and attitudes toward socialized medicine.

Individual	Before speech	After speech
1	3	6
2	4	6
3	3	3
4	5	7
5	2	4
6	5	6
7	3	7
8	4	6

Answer The null and alternative hypotheses are

H_0: $\mu_B = \mu_A$

H_1: $\mu_B \neq \mu_A$

where the subscripts "*B*" and "*A*" denote the before- and after-speech conditions, respectively. For an alpha level of .05, nondirectional test, and $N - 1 = 8 - 1 = 7$ degrees of freedom, the critical values of *t* are ±2.365.

To obtain the observed value of *t*, we first calculate $\overline{D}$ and $\hat{s}_{\overline{D}}$ as follows:

Individual	Before speech	After speech	Difference (D)	D^2
1	3	6	−3	9
2	4	6	−2	4
3	3	3	0	0
4	5	7	−2	4
5	2	4	−2	4
6	5	6	−1	1
7	3	7	−4	16
8	4	6	−2	4

$$\Sigma X_B = 29 \qquad \Sigma X_A = 45 \qquad \Sigma D = -16 \qquad \Sigma D^2 = 42$$
$$\overline{X}_B = 3.62 \qquad \overline{X}_A = 5.62 \qquad \overline{D} = -2.00$$

$$SS_D = \Sigma D^2 - \frac{(\Sigma D)^2}{N} = 42 - \frac{(-16)^2}{8} = 10$$

$$\hat{s}_D^2 = \frac{SS_D}{N-1} = \frac{10}{7} = 1.43$$

$$\hat{s}_D = \sqrt{\hat{s}_D^2} = \sqrt{1.43} = 1.20$$

$$\hat{s}_{\overline{D}} = \frac{\hat{s}_D}{\sqrt{N}} = \frac{1.20}{\sqrt{8}} = .42$$

From Equation 11.2, the observed *t* is

$$t = \frac{\overline{D} - \mu_D}{\hat{s}_{\overline{D}}}$$

$$= \frac{-2.00 - 0}{.42} = -4.76$$

Since −4.76 is less than −2.365, we reject the null hypothesis and conclude that there is a relationship between the time of assessment and attitudes toward socialized medicine.

Assumptions of the *t* Test

The assumptions underlying the validity of the correlated groups *t* test parallel those for the one-sample *t* test, as discussed in Chapter 8. Specifically, these assumptions are made:

1. The sample is independently and randomly selected from the population of interest.

2. The population of difference scores is normally distributed.

In addition, the dependent variable should be quantitative in nature and measured on a level that at least approximates interval characteristics. The assumption of independent and random selection is an important one, as it is for all the statistical tests considered in this book. By contrast, the test is relatively robust to violations of the normality assumption. If the sample size is less than 15, the test may show more Type I errors for data that are markedly skewed (Posten, 1979).

However, for sample sizes of 40 or more, the test is remarkably robust, even for distributions that have considerable skewness.

11.3   Strength of the Relationship

The formula for computing eta-squared for the correlated groups *t* test is the same as that for the independent groups *t* test:

$$\text{eta}^2 = \frac{t^2}{t^2 + df} \qquad [11.4]$$

For the experiment on drugs and learning,

$$\text{eta}^2 = \frac{-2.73^2}{-2.73^2 + 4} = .65$$

However, whereas eta-squared represents the proportion of variability in the dependent variable that is associated with the independent variable in the independent groups case, in the correlated groups case eta-squared represents the proportion of variability in the dependent variable that is associated with the independent variable *after variability due to individual differences has been removed*. It follows that 1.00 minus eta-squared represents the proportion of variability in the dependent variable that is due to disturbance variables other than individual differences. In the present example, the proportion of variability in learning that is associated with the type of drug after the influence of individual differences has been removed is .65, which represents a strong effect. Thus, 1.00 − .65 = .35, or 35%, of the variability in learning is attributable to disturbance variables other than individual differences. This interpretation of eta-squared can be best demonstrated by using a variance extraction approach similar to that developed in Chapter 10.

Estimating and Extracting the Influence of Individual Differences

Consider an individual who tries to solve three math problems, each worth 5 points, on an exam. The individual gets a score of 5 on the first problem, a score of 1 on the second problem, and a score of 3 on the third problem. Another individual works on the same three problems and scores 5 on the first one, 5 on the second, and 2 on the third. Which individual has greater math ability? In the absence of any other information, the best approach to answering this question is to compare the average scores of the two individuals. The first individual's average score is (5 + 1 + 3)/3 = 3.00, and the second individual's average score is (5 + 5 + 2)/3 = 4.00. Using the average score across questions as an index of ability, you would conclude that the second individual has greater math ability than the first individual.

In within-subjects designs, estimates of the influence of individual difference variables are derived in a similar fashion. For each individual, the average score across conditions is computed. This has been done for the experiment on drugs and learning in column 4 of Table 11.2. For instance, the first person had a score of 3 with drug A and a score of 7 with drug B for a mean of $(7 + 3)/2 = 5.00$. We can see by inspection of column 4 that some participants had higher average scores than others. These differences reflect differences in the individuals' backgrounds (for example, intelligence, familiarity with the learning task, and so on).

We now develop the logic of extracting this source of variability. In order to do so, we must first compute a grand mean (G). This involves summing *all* of the scores in columns 2 and 3 of Table 11.2 and dividing by the number of summed scores. We find that $G = (3 + 1 + 4 + 2 + 5 + 7 + 5 + 4 + 8 + 6)/10 = 4.50$. Thus, the average score across all participants and across both experimental conditions is 4.50.

Consider the first participant. His average score across conditions is 5.00. The grand mean, which serves as a reference point for all participants, is 4.50. Because an individual's average score across conditions reflects the influence of his or her background, the effect of background in this instance was to raise the individual's score one-half a unit above the grand mean $(5.00 - 4.50 = .50)$. Using logic similar to that used to extract variability in scores in Chapter 10, we can nullify this effect by subtracting .50 unit from the scores of this individual. This has been done in columns 5 and 6 of Table 11.2. The score with drug A is 3, and this score is adjusted to 2.50 $(3 - .50 = 2.50)$. The score with drug B is 7, and this score is adjusted to 6.50 $(7 - .50 = 6.50)$. The new scores have had the effect of the individual's background removed, or nullified. This same approach is taken for each person. For the second participant, the average response across conditions was 3.00. This is 1.50 units below the grand mean $(3.00 - 4.50 = -1.50)$. The effect of this individual's background was to hold performance down 1.50 units from the overall average. To nullify the effect of this background, we simply add 1.50 to the two scores. The remaining participants' scores are nullified in a similar fashion.

We now have an adjusted data set in which the effects of individual differences in background have been removed. Compare the nullified data in Table 11.2 with the original data. Note that there is less variability in the nullified scores be-

TABLE 11.2   **Raw and Nullified Scores for Drug and Learning Experiment**

Participant i	X for drug A	X for drug B	$\overline{X}_i^a$	Nullified X for drug A	Nullified X for drug B
1	3	7	5.00	2.50	6.50
2	1	5	3.00	2.50	6.50
3	4	4	4.00	4.50	4.50
4	2	8	5.00	1.50	7.50
5	5	6	5.50	4.00	5.00
Mean =	3.00	6.00	4.50	3.00	6.00

$^a\overline{X}_i$ = Mean X score for participant i across conditions.

cause we have removed a source of variability—namely, individual abilities and background. If we were to compute the average nullified score across conditions for each participant, every person would have the same average score. Again, this is because we have extracted variability due to individual differences.

Since there is a set of nullified scores for drug A and a set of nullified scores for drug B, the adjusted data can be analyzed with an independent groups *t* test. This test tends to be more sensitive (and thus more powerful) than an independent groups *t* test applied to the raw scores because the "noise" created by individual differences has been eliminated. If we were to actually apply an independent groups *t* test to the adjusted data, we would get the same value of *t* (−2.73) as that obtained earlier when the computational formula for the correlated groups *t* test was applied to difference scores.* The equivalence of the two approaches clarifies the nature of *t* in the correlated groups case: A correlated groups *t* test is analogous to an independent groups *t* test with the effects of individual differences extracted from the dependent variable. It follows that eta-squared in the correlated groups case is analogous to eta-squared in the independent groups case, but with variability due to individual differences removed from the dependent variable.

11.4 Nature of the Relationship

The analysis of the nature of the relationship is identical to that for the independent groups *t* test and involves examining the mean scores for the two conditions. Because we failed to reject the null hypothesis, the question of the nature of the relationship is not meaningful for the present data. Had we rejected the null hypothesis, the nature of the relationship would be that drug B produces more errors in learning ($\overline{X}_B = 6.00$) than drug A ($\overline{X}_A = 3.00$).

11.5 Methodological Considerations

The experiment on drugs and learning raises a number of methodological issues. First, consider the potential role of confounding variables. In this study, the investigator attempted to control for carry-over effects, such as familiarity with the task or the persistence of the effects of drug A into the administration of drug B, by having a long time period between the administration of drug A and the administration of drug B. The introduction of the time interval may, in fact, reduce such carry-over effects, but an additional problem arises. Perhaps during the time interval between the administration of the two drugs, the participants experienced something that improved their performance on the learning task. For instance, if the participants are introductory psychology students, they might have learned about certain memory aids in their course work.

---

* Actually, a slight modification to the *t* test procedure is required, per Appendix 11.1.

A procedure that circumvents this problem and could have been used here is **counterbalancing.** With this technique, half the participants are given drug A first and then, at a later time when carry-over effects should be minimal, they are given drug B. The other half of the participants are given drug B first and then drug A. If this is done, then the intervening events and familiarity effects should no longer be related to the administration of the two drugs. Since they are now evenly distributed across conditions, any carry-over effects should influence learning performance under drug A (for those who were given drug A second) and drug B (for those who were given drug B second) to an equal extent.* In essence, these confounding variables are turned into disturbance variables, which now create "noise" in the experiment. It is possible to remove the disturbance influence of these variables using certain advanced statistical techniques. These are beyond the scope of this text, however, and are discussed in Winer (1971).

Another problem with this experiment is the lack of a control condition in which participants perform the learning task while not under the influence of any drugs. Although the effects of drug A and drug B relative to each other can be determined from the experiment, the effects of the drugs relative to baseline performance cannot. The addition of a control condition would be quite informative. Chapter 13 will introduce statistical techniques that could be used to analyze the study if a control condition were included.

In Chapter 10, we saw that the role of disturbance variables in the independent groups case can be clarified by examining the magnitude of eta-squared. A similar approach can be taken with within-subjects designs. In the experiment on drugs and learning, eta-squared was .65. An eta-squared of .65 indicates that, after variability due to individual differences has been removed, 65% of the variability in the dependent variable in the sample is associated with the independent variable and 35% is due to disturbance variables other than individual differences. This represents a strong relationship in the sample, yet the *t* test results were such that we could not conclude that a relationship exists in the population. Although this may appear contradictory, a consideration of the sample size helps to place this in proper perspective. The study was conducted with an extremely small sample size—namely, five individuals. With small sample sizes, very large values of eta-squared are necessary before we can conclude that there is a relationship between the independent and dependent variables. Although the size of eta-squared suggests that the role of disturbance variables is not very large, given the extremely small sample size, a relationship as strong as .65 in the sample is still not strong enough for us to be confident that sampling error is not producing the observed mean difference. The moral is to use large sample sizes when it is practical. As indicated in previous chapters, larger sample sizes will increase the power of the statistical test and increase the probability of detecting an existing relationship in the population.

---

* In general, carry-over effects will influence the dependent variable equally in the two conditions when the nature of such effects is the same for the two levels of the independent variable. This will not be the case when the nature of carry-over effects is different for the two levels of the independent variable.

11.6   Power of Correlated Groups Versus Independent Groups *t* Tests

Since variability due to individual differences is extracted from the dependent variable as part of the correlated groups *t* test procedure, a correlated groups *t* test will usually (but not always) be more powerful than a corresponding independent groups *t* test. We can illustrate this by reconsidering the formulas for the two *t* tests. Suppose we have two conditions for which we compute the sample means, $\overline{X}_1$ and $\overline{X}_2$. If the two conditions represent a between-subjects variable, then, as discussed in Chapter 10, the *t* statistic for the independent groups *t* test can be calculated using Equations 10.4 and 10.7:

$$t = \frac{(\overline{X}_1 - \overline{X}_2) - (\mu_1 - \mu_2)}{\sqrt{\dfrac{\hat{s}^2_{pooled}}{n_1} + \dfrac{\hat{s}^2_{pooled}}{n_2}}}$$

For the correlated groups *t* test, the formula (see Equation 11.2) is

$$t = \frac{\overline{D} - \mu_D}{\hat{s}_{\overline{D}}}$$

Simple algebra shows that the mean of the sample difference scores will always equal the sample mean for the first condition minus the sample mean for the second condition in a correlated groups design (i.e., $\overline{D}$ will always equal $\overline{X}_1 - \overline{X}_2$) and that this is also true of the population parameters (i.e., μ_D will always equal $\mu_1 - \mu_2$). Thus, the formula for the correlated groups *t* test can be rewritten as:

$$t = \frac{(\overline{X}_1 - \overline{X}_2) - (\mu_1 - \mu_2)}{\hat{s}_{\overline{D}}}$$

It is also the case that $\hat{s}_{\overline{D}}$ is equal to the following:

$$\hat{s}_{\overline{D}} = \sqrt{\frac{\hat{s}^2_1}{n_1} + \frac{\hat{s}^2_2}{n_2} - \frac{2r\hat{s}_1\hat{s}_2}{N}}$$

where $n_1 = n_2 = N$ and *r* represents the correlation between the scores in the first condition and the scores in the second condition. We can therefore reexpress the correlated groups *t* test formula as:

$$t = \frac{(\overline{X}_1 - \overline{X}_2) - (\mu_1 - \mu_2)}{\sqrt{\dfrac{\hat{s}^2_1}{n_1} + \dfrac{\hat{s}^2_2}{n_2} - \dfrac{2r_{12}\hat{s}_1\hat{s}_2}{N}}}$$

When *r* equals 0, the rightmost term in the denominator also equals 0 and the correlated groups *t* test formula reduces to

$$t = \frac{(\overline{X}_1 - \overline{X}_2) - (\mu_1 - \mu_2)}{\sqrt{\dfrac{\hat{s}^2_1}{n_1} + \dfrac{\hat{s}^2_2}{n_2}}}$$

which is strikingly similar to the formula for the independent groups t test (the only exception being the use of the pooled variance estimate in the independent groups t test as opposed to nonpooled estimates here). When the correlation between scores in the two conditions is positive, the estimated standard error for the correlated groups t test is reduced by a factor of $2r\hat{s}_1\hat{s}_2$. In most behavioral science research, r for the correlated groups t test is indeed positive, which results in an estimated standard error that is smaller than the estimated standard error for the independent groups t test. This, in turn, can yield a more powerful statistical test.

The exception to this statement occurs when r is so close to 0 that the magnitude of the estimated standard errors is comparable for the two tests. It turns out that the degrees of freedom for the correlated groups t test $(N - 1)$ are less than the degrees of freedom for the independent groups t test $(n_1 + n_2 - 2)$. For example, if there are two sets of ten scores, df $= 10 - 1 = 9$ for the correlated groups test and df $= 10 + 10 - 2 = 18$ for the independent groups test. Since the t distribution requires more extreme values of t in order to reject the null hypothesis as the degrees of freedom become less, the statistical advantage for the correlated groups t test of the smaller estimated standard error may be offset by the fewer degrees of freedom. In fact, under this circumstance, a correlated groups test might actually be less powerful than an independent groups test. Thus, it is important that a researcher who is considering a within-subjects design accurately anticipate the role of individual differences.

11.7 Numerical Example

Developmental psychologists have attempted to specify the age period when infants tend to show signs of fearing strangers. At very early ages (1 to 2 months), infants generally will show positive reactions when approached or held by any adult. At some point, however, infants begin to discriminate among adults and exhibit fear responses in the presence of a stranger, as opposed to a familiar person such as the mother or father. In one investigation, a researcher wanted to compare negative responses to a stranger, using infants at the age of 3 months and again at 6 months. Eight infants were involved one at a time in 10-minute interactions with a stranger in which the stranger attempted to engage them in playful behavior. The interactions were standardized as much as possible, and the number of minutes the infants spent crying was used as the measure of negative response. The scores on the dependent variable are presented in the second and third columns of Table 11.3.

The null hypothesis posits that there is no relationship between infants' ages and the severity of their negative responses to a stranger, whereas the alternative hypothesis states that these two variables are related. These hypotheses are formally stated as

$$H_0: \quad \mu_T = \mu_S$$

$$H_1: \quad \mu_T \neq \mu_S$$

TABLE 11.3   Data and Calculations for Age and Response to a Stranger Study

Individual	X for 3 months	X for 6 months	Difference (D)	D^2
1	0	3	−3	9
2	1	2	−1	1
3	2	4	−2	4
4	2	4	−2	4
5	2	4	−2	4
6	1	2	−1	1
7	0	3	−3	9
8	0	2	−2	4
	$\Sigma X_T = 8$	$\Sigma X_S = 24$	$\Sigma D = -16$	$\Sigma D^2 = 36$
	$\overline{X}_T = 1.00$	$\overline{X}_S = 3.00$	$\overline{D} = -2.00$	

$$SS_D = \Sigma D^2 - \frac{(\Sigma D)^2}{N} = 36 - \frac{(-16)^2}{8} = 4$$

$$\hat{s}_D^2 = \frac{SS_D}{N-1} = \frac{4}{7} = .57$$

$$\hat{s}_D = \sqrt{\hat{s}_D^2} = \sqrt{.57} = .75$$

$$\hat{s}_{\overline{D}} = \frac{\hat{s}_D}{\sqrt{N}} = \frac{.75}{\sqrt{8}} = .27$$

where the subscripts "T" and "S" denote the ages 3 months and 6 months, respectively.

The existence of a relationship is tested by converting the observed sample mean difference to a t value and comparing this with the values of t that define the rejection region. Since the null hypothesis states that the population means are equivalent, μ_D can be set equal to 0. Other intermediate values necessary for the calculation of t are given in Table 11.3. The observed t is found to equal

$$t = \frac{\overline{D} - \mu_D}{\hat{s}_{\overline{D}}}$$

$$= \frac{-2.00 - 0}{.27} = -7.41$$

For an alpha level of .05, nondirectional test, and $N - 1 = 8 - 1 = 7$ degrees of freedom, the critical values of t are ±2.365. Since −7.41 is less than −2.365, we reject the null hypothesis and conclude that a relationship exists between infants' ages and the severity of their negative responses to a stranger.

The strength of the relationship, as indexed by eta-squared, is

$$eta^2 = \frac{t^2}{t^2 + df}$$

$$= \frac{-7.41^2}{-7.41^2 + 7} = .89$$

The proportion of variability in the dependent variable (severity of negative response to a stranger) that is associated with the independent variable (age) after the influence of individual differences has been removed is .89. This represents a strong effect.

The nature of the relationship is such that infants who are 6 months old respond more negatively to a stranger than infants who are 3 months old. This conclusion stems from the fact that the mean crying times in the two conditions are 3.00 and 1.00 minutes, respectively.

You should think about these results in terms of the research design questions discussed in Chapter 9. Are there any potential confounding variables that have not been controlled? What has been the role of disturbance variables? What kinds of procedures could be used to reduce sampling error? What are the limitations of the study in terms of generalizability?

11.8   Planning an Investigation Using the Correlated Groups *t* Test

Appendix E.1 contains tables of the sample sizes necessary to achieve various values of power for the correlated groups *t* test. Table 11.4 reproduces a portion of

TABLE 11.4   **Approximate Sample Sizes Necessary to Achieve Selected Levels of Power for Alpha = .05, Nondirectional Test, As a Function of Population Values of Eta-Squared**

Power	Population Eta-Squared									
	.01	.03	.05	.07	.10	.15	.20	.25	.30	.35
.25	84	28	17	12	8	6	5	3	3	3
.50	193	63	38	27	18	12	9	7	5	5
.60	246	80	48	34	23	15	11	8	7	6
.67	287	93	55	39	27	17	12	10	8	6
.70	310	101	60	42	29	18	13	10	8	7
.75	348	113	67	47	32	21	15	11	9	7
.80	393	128	76	53	36	23	17	13	10	8
.85	450	146	86	61	41	26	19	14	11	9
.90	526	171	101	71	48	31	22	17	13	11
.95	651	211	125	87	60	38	27	21	16	13
.99	920	298	176	123	84	53	38	29	22	18

this appendix for an alpha level of .05, nondirectional test. As in Chapter 10, various values of power are listed in the first column, and various population values of eta-squared serve as column headings. These values of eta-squared are conceptualized as the proportion of variability in the dependent variable that is associated with the independent variable *after the effects of individual differences have been removed*. Thus, to the extent that the dependent variable is influenced by individual background, the population eta-squared will be greater in the correlated groups case than in the independent groups case.

To illustrate the use of the table, suppose the desired power is .80 and the investigator suspects that the strength of the relationship in the population corresponds to an eta-squared of .07. Then the number of participants he or she should use in a study for an alpha level of .05, nondirectional test, is 53. The power an investigator requires will, of course, depend on the seriousness of committing a Type II error. Examination of Table 11.4 and Appendix E.1 provides a general appreciation for the relationships among directionality, the alpha level, sample size, the strength of the relationship one is trying to detect in the population, and power for the correlated groups *t* test.

11.9 Method of Presentation

The method of presentation for the correlated groups *t* test is identical to that for the independent groups *t* test. The results should include statements of the alpha level that was used, the degrees of freedom, the observed value of *t*, the significance level, the sample means, the standard deviation estimates, and, if the analysis is statistically significant, the strength and nature of the relationship between the independent and dependent variables. For example, the results for the study of infants' ages and responses to a stranger discussed earlier in this chapter might be presented as follows:*

```
                        Results

     A correlated groups t test compared the mean crying
time for the infants when they were 3 months of age with
the mean crying time when they were 6 months of age. The
alpha level was .05. This test was found to be statisti-
cally significant, t(7) = -7.41, p < .001, suggesting that
infants react more negatively to strangers when they are
6 months old (M = 3.00, SD = .93) than when they are
3 months old (M = 1.00, SD = .93). The strength of the
```

---

* The standard deviation estimates were calculated by applying the usual formulas for the sum of squares, the variance estimate, and the standard deviation estimate to each of the age conditions in Table 11.3.

relationship between age and crying time was .89, as in-dexed by eta^2.

11.10 Examples from the Literature

Number of Social Contacts with Same- Versus Opposite-Gender Individuals

Psychologists have studied interpersonal relationships and friendship patterns in many contexts. One approach has been to document the daily patterns of people's social behavior through the use of diaries. For instance, Nezlek (1978) asked people to keep daily diaries of the social contact they had with other people over four 2-week periods. One of the variables studied was the number of people that individuals reported they had met. Specifically, Nezlek was interested in comparing the number of same-gender contacts with the number of opposite-gender contacts that people made. Sixty-three people who kept diaries indicated that they contacted, on the average, 1.54 same-gender individuals per day and 1.01 opposite-gender individuals per day. A correlated groups t test indicated that this difference was statistically significant, $t(62) = 2.10$, $p < .05$, such that people tend to make more same-gender than opposite-gender contacts. The strength of the relationship, as indexed by eta-squared, was .07. This represents a weak effect and indicates, as one would expect, that many factors other than a person's gender influence social contact.

Self-Schema Similarity Before and After Discussion of a Hypothetical Person

The term *self-schema* refers to an individual's conception of who and what one is. According to a model of self-schema development proposed by Deutsch and Mackesy (1985), during the course of conversations about other people, individuals become aware of the person-description dimensions used by the other discussants and come to adopt these dimensions in their descriptions of others and subsequently themselves. According to this model, as reflected in their self-descriptions, individuals' self-schemas should become more similar after they have had the opportunity to share their opinions of another individual.

In a test of this hypothesis, Deutsch and Mackesy first instructed experimental participants to list ten self-descriptive traits. The participants then independently read a description of a hypothetical person and discussed their impressions of this person with a randomly assigned partner. In a final phase, participants were again instructed to list ten self-descriptive traits.

The dependent variable was the number of overlapping self-traits reported by partners before versus after the discussion of the hypothetical person. Consistent with Deutsch and Mackesy's model, a correlated groups t test indicated that there was significantly more self-schema overlap after discussion ($\overline{X} = 2.05$) than before ($\overline{X} = 1.30$), $t(19) = 3.29$, $p < .01$. As indexed by eta-squared, the strength of the relationship was .36. This represents a strong effect.

**Applications
to the Analysis
of a Social Problem**

Past research suggests that the relationship between parent and teen is an important predictor of many teen problem behaviors: Teens who have relatively poor relationships with their parents are more likely to exhibit problem behavior, everything else being equal. In the parent–teen communication study described in Chapter 1, we obtained ratings of how satisfied the teens were with their relationship with their mother and, using a separate item, how satisfied the teens were with their relationship with their father.

One question we were interested in addressing was whether teens, on average, are more satisfied with their relationship with their mother or their father. We are developing educational interventions to help teach parents to communicate more effectively with their teens. Practical constraints are such that it is possible economically to engage only one parent in the intervention effort. If teens, on average, have a better relationship with one parent than the other, then this would be a consideration when deciding whether to develop the intervention around mothers or fathers. Our previous research suggests that information about adolescent problem behaviors is more likely to be exchanged between parent and child when a good relationship exists between them. If, for instance, teens tend to be more satisfied with their relationship with their mother, then it is probably more likely that information we provide to mothers (as opposed to fathers) about premarital pregnancy, sexually transmitted diseases, and birth control will be passed on to the teen, everything else being equal.

We asked teens to indicate how much they agreed or disagreed with the statements "I am satisfied with my relationship with my mother" and "I am satisfied with my relationship with my father." Responses were made on a five-point scale having the response options strongly disagree, moderately disagree, neither agree nor disagree, moderately agree, and strongly agree. Responses were scored from 1 to 5, with higher values indicating greater agreement with the statement. Although we relied on only a single item to measure overall satisfaction for each parent, prior research has shown that responses to the single item are highly correlated with more complex, multi-item measures of overall relationship satisfaction. In addition, although the measures are probably not strictly interval, our prior research suggests that they probably sufficiently approximate interval level characteristics so that we can use the correlated groups *t* test (per the discussion of this issue in Section 10.10).

To test whether teens' satisfaction with their maternal and paternal relationships differs, we conducted a correlated groups *t* test comparing the mean agreement rating assigned to mothers with the mean agreement rating assigned to fathers. Figure 11.1 presents a histogram of the difference scores, where the mother rating was subtracted from the father rating. Superimposed on this is a normal distribution. A score of −4 indicates strong agreement with the mother statement coupled with strong disagreement with the father rating. A score of +4 indicates the reverse. A score of 0 means the ratings for the mother and father were identical. Although the distribution of difference scores is nonnormal, the deviation from normality is not large enough to suggest that nonnormality in the population is great enough to undermine the *t* test, given the sample size that was used (see Posten, 1979).

The computer output for the statistical analysis is presented in Table 11.5. The first part of the output presents the sample size, designated "Number of pairs" (of scores), and the mean, standard deviation estimate, and estimated standard error of the mean for each level of the independent variable (SFATHER is the teen's satisfaction with the relationship with the father and SMOTHER is the teen's satisfaction with the relationship with the mother). The sample size (*N* = 446) was much smaller than the total sample of 751 adolescents who were interviewed. It turns out that more than 300 adolescents did not have a "father figure" currently living in their household (due to divorce, single parenthood, widowhood, and the like). This finding, in itself, is important because it suggests that our intervention should be directed at mothers, who are more likely to be a member of the teen's household. The estimated standard errors

FIGURE 11.1   **Frequency Histogram of Difference Scores Between Satisfaction with Paternal Relationship and Satisfaction with Maternal Relationship**

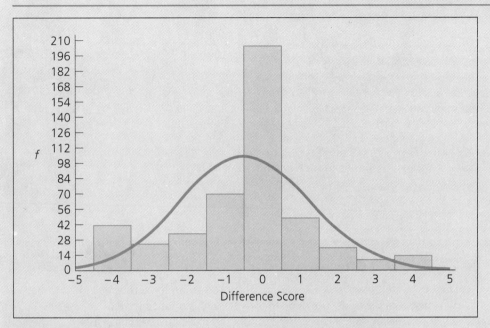

of the means were quite small, reflecting about .05 of one rating scale unit (on a 1 to 5 scale). This suggests that we can have reasonably high confidence in the accuracy of the sample means as estimators of the population means.

The bottom part of Table 11.5 presents the difference between the two means (labeled "Mean"), the estimated standard deviation of the population difference scores (labeled "SD"), the

estimated standard error of the mean of difference scores (labeled "SE of Mean"), the observed value of t, the degrees of freedom (labeled "df"), and the p value (labeled "2-tail Sig") for a nondirectional correlated groups t test. The mean difference was −.48, which corresponds to about half of a rating scale unit, and the estimated standard error of the difference was small, only .08. The t test was
(continued)

TABLE 11.5   **Computer Output for Correlated Groups t Test for Satisfaction with Paternal Versus Maternal Relationship**

Variable	Number of pairs	Mean	SD	SE of Mean
SFATHER		3.8991	1.379	.065
	446			
SMOTHER		4.3744	1.048	.050

Paired Differences

Mean	SD	SE of Mean	t-value	df	2-tail Sig
−.4753	1.702	.081	−5.90	445	.000

statistically significant, $t(445) = -5.90$, $p < .0005$, and the value of eta-squared was .07, which indicates a weak to moderate relationship. On average, teens tend to be more satisfied with their relationship with their mothers ($\bar{X} = 4.37$) than with their fathers ($\bar{X} = 3.90$).

We also examined whether these differences occurred for adolescent boys and girls considered separately. It is possible that adolescent girls tend to be more satisfied with their relationship with their mothers, but that adolescent boys tend to be more satisfied with their relationship with their fathers. This would follow from a formulation known as *identification theory,* which posits that boys identify more strongly with their fathers and girls with their mothers because of the common gender link. This stronger identification, in turn, leads to a more satisfying relationship. Although we do not report the specifics of the analyses here, we found that this was not the case. For both boys and girls considered separately, the mean rating for mothers tended to be about half a scale point higher than for fathers, and this difference was statistically significant in both cases. Overall, these data suggest that our intervention efforts be focused on the mother (although other issues, not considered here, must also be taken into account before making a decision).

Summary

The correlated groups *t* test is typically used to analyze the relationship between two variables when (1) the dependent variable is quantitative in nature and is measured at a level that at least approximates interval characteristics, (2) the independent variable is within-subjects in nature, and (3) the independent variable has two and only two levels. The existence of a relationship between the two variables is tested by converting the observed mean difference in the sample to a *t* value that represents the number of estimated standard errors the difference is from the mean of the appropriate sampling distribution of the mean of difference scores (assuming the null hypothesis is true). This *t* value is compared with the critical values of *t*, and the decision to reject or not reject the null hypothesis is made accordingly. An important advantage of the correlated groups *t* test over the independent groups *t* test is that it extracts the influence of individual differences from the dependent variable, thereby providing a more sensitive test of the relationship between the independent and dependent variables. The strength of the relationship is measured using the eta-squared statistic, which in this case represents the proportion of variability in the dependent variable that is associated with the independent variable after variability due to individual differences has been removed. Finally, the nature of the relationship is determined by examining the mean scores in the two conditions.

Appendix 11.1   Computational Procedures for the Nullified Score Approach

The logic and computational procedures for the nullified score approach to correlated groups designs are similar to those developed in Chapter 10 for the independent groups *t* test, with one important exception: As noted earlier in this chapter, the relevant *t* distribution is distributed with $N - 1$ rather than $n_1 + n_2 - 2$ degrees of freedom. This has implications not only for the critical values of t that are used in hypothesis testing, but also for the calculation of the estimated standard error of the difference.

We demonstrate the relevant procedures by referring to the experiment on drugs and learning discussed in this chapter. The nullified scores for this study from columns 5 and 6 of Table 11.2 are reproduced in the table below. Also appearing in this table are calculations for the means and sums of squares for the two conditions.

The formula for the independent groups t test (see Equation 10.7) is

$$t = \frac{(\overline{X}_1 - \overline{X}_2) - (\mu_1 - \mu_2)}{\hat{s}_{\overline{X}_1 - \overline{X}_2}}$$

In the present example, $\overline{X}_1 - \overline{X}_2 = 3.00 - 6.00 = -3.00$ and $\mu_1 - \mu_2 = 0$, because μ_1 and μ_2 are hypothesized to be equal. The standard error of the difference is estimated by the formula

$$\hat{s}_{\overline{X}_1 - \overline{X}_2} = \sqrt{\left(\frac{SS_1 + SS_2}{N-1}\right)\left(\frac{1}{n_1} + \frac{1}{n_2}\right)}$$

which is the same formula as that presented in Chapter 10 (see Equation 10.6) except that $N-1$ (the degrees of freedom for the correlated groups case) replaces $n_1 + n_2 - 2$ (the degrees of freedom for the independent groups case). In this equation, N is the number of pairs of scores, and n_1 and n_2 are the sample sizes in condition 1 and condition 2, respectively, such that $n_1 = n_2 = N$. In the present example,

$$\hat{s}_{\overline{X}_1 - \overline{X}_2} = \sqrt{\left(\frac{6+6}{5-1}\right)\left(\frac{1}{5} + \frac{1}{5}\right)} = 1.10$$

The observed t is thus

$$t = \frac{-3.00 - 0}{1.10} = -2.73$$

This is the same value that was obtained in Section 11.2 using the difference score approach.

Nullified Data for Drug A		Nullified Data for Drug B	
X	X^2	X	X^2
2.50	6.25	6.50	42.25
2.50	6.25	6.50	42.25
4.50	20.25	4.50	20.25
1.50	2.25	7.50	56.25
4.00	16.00	5.00	25.00
$\Sigma X_A = 15.00$	$\Sigma X_A^2 = 51.00$	$\Sigma X_B = 30.00$	$\Sigma X_B^2 = 186.00$
$\overline{X}_A = 3.00$		$\overline{X}_B = 6.00$	

$$SS_A = \Sigma X_A^2 - \frac{(\Sigma X_A)^2}{N} \qquad SS_B = \Sigma X_B^2 - \frac{(\Sigma X_B)^2}{N}$$

$$= 51.00 - \frac{15.00^2}{5} = 6 \qquad = 186.00 - \frac{30.00^2}{5} = 6$$

Exercises

Answers to asterisked (*) exercises appear at the back of the book. Answers to excercises with two asterisks are also worked out step by step in the Study Guide.

Exercises to Review Concepts

1. Under what conditions is the correlated groups t test typically used to analyze a bivariate relationship?

2. In general terms, what value will the mean of a sampling distribution of the mean of difference scores always equal?

*3. State the critical value(s) of *t* for a correlated groups *t* test for an alpha level of .05 under each of the following conditions:
   a. $H_0: \mu_1 = \mu_2$, $H_1: \mu_1 \neq \mu_2$, $N = 27$
   b. $H_0: \mu_1 = \mu_2$, $H_1: \mu_1 > \mu_2$, $N = 27$
   c. $H_0: \mu_1 = \mu_2$, $H_1: \mu_1 \neq \mu_2$, $N = 14$
   d. $H_0: \mu_1 = \mu_2$, $H_1: \mu_1 < \mu_2$, $N = 14$

4. What are the assumptions underlying the correlated groups *t* test?

5. What interpretation does eta-squared have in the context of the correlated groups *t* test? How does this interpretation differ from that for the independent groups *t* test?

*6. What is the rationale behind counterbalancing the sequence of conditions across participants in within-subjects designs?

*7. Why is a correlated groups *t* test usually more powerful than a corresponding independent groups *t* test? When will this not be the case? Why?

Use the following information to complete Exercises 8–11.

The following scores resulted on a learning test that was administered to a group of participants under quiet and noisy conditions:

Participant	Quiet	Noisy
1	16	10
2	5	3
3	12	10
4	9	5
5	23	15

*8. Test for a relationship between the amount of noise one is exposed to and learning scores using a nondirectional correlated groups *t* test.

*9. Compute the value of eta-squared. Does the observed value represent a weak, moderate, or strong effect?

*10. Discern the nature of the relationship between the amount of noise one is exposed to and learning scores.

*11. Analyze the data as if the independent variable were between-subjects in nature. That is, conduct a nondirectional independent groups *t* test, compute the value of eta-squared, and determine the nature of the relationship between the experimental condition one is exposed to and learning scores using the procedures developed in Chapter 10. Compare your findings with those in Exercises 8–10. How does this illustrate the advantage of within-subjects research designs?

Use the following information to complete Exercises 12–16.

The following self-esteem scores were provided by a group of individuals at two different times:

Individual	Time 1	Time 2
1	10	12
2	13	17
3	12	14
4	11	13
5	14	14

12. Test for a relationship between the time of assessment and self-esteem using a nondirectional correlated groups *t* test.

13. Compute the value of eta-squared. Does the observed value represent a weak, moderate, or strong effect?

14. Discern the nature of the relationship between the time of assessment and self-esteem.

**15. Extract the effects of individual differences (that is, generate a set of nullified scores). Compute the mean for time 1 and the mean for time 2 for the nullified scores. The respective mean values should be the same as for the original scores (for example, the mean for time 1 in the adjusted data set should equal the mean for time 1 in the original data set). Why do you think this is the case? (*Hint*: Remember that the individual differences whose effects have been extracted from the dependent variable serve as disturbance variables, as discussed in Chapter 9.)

**16. Using the procedures from Appendix 11.1, conduct a nondirectional independent groups t test on the nullified scores from Exercise 15. Compare the observed value of t with that obtained in Exercise 12.

For each of the studies described in Exercises 17–21, indicate whether an independent groups t *test or a correlated groups* t *test should be used to analyze the relationship between the independent and dependent variables. Assume that the underlying assumptions of the tests have been satisfied.*

*17. Frieze, Parsons, Johnson, Ruble, and Zellman (1978) examined the relationship between gender and mathematical ability as measured by the Graduate Record Exam. The mean scores on the quantitative test were compared for men and women who took the exam during 1972.

*18. Smith, Phillipus, and Guard (1968) studied the effect of a drug known as ethosuximide on the verbal skills of children who had learning problems. This research was based on previous research that had shown a similarity in EEG patterns between children who have learning problems and those who have a mild form of epilepsy, called petit mal. Smith and colleagues speculated that since ethosuximide is effective in treating petit mal, it might also serve as a "learning facilitator" for children with learning problems. A group of children participated in the study for 6 weeks. During the first 3 weeks, some of the children were injected with ethosuximide and others were given a placebo. During the second 3 weeks, the procedures were reversed (that is, children who had previously been given a placebo were given ethosuximide, and vice versa). Each child was given a verbal skills test after the first 3-week period and again after 6 weeks. For all children, the mean score on this test after they received the placebo was compared with the mean score after they received the ethosuximide.

19. A consumer psychologist was interested in the preference of consumers for two types of headache remedies. Five hundred individuals were surveyed and rated each brand on a ten-point scale. The mean ratings for the two brands were compared.

20. Jensen (1973) reviewed research on the relationship between scores on an intelligence test and race. In one study, the mean scores of African Americans and whites were compared.

21. Gallup (1976) studied attitudes toward nuclear energy over a period of 10 years. The same individuals were interviewed in both 1965 and 1975, and during each interview, the respondents indicated their attitudes toward nuclear energy on a five-point scale. The mean attitude scores were compared across interviews.

22. If a researcher suspects that the strength of the relationship between two variables in the population is .25 as indexed by eta-squared, what sample size should he use in a study involving two correlated groups and an alpha level of .05, directional test, in order to achieve a power of .90?

*23. Suppose an investigator conducted a study involving two correlated groups with $N = 7$. If the value of eta-squared in the population was .30, what would the power of her statistical test be at an alpha level of .05, nondirectional test?

Multiple-Choice Questions

24. In the context of a within-subjects design, an individual obtains scores of 5 in one condition and 7 in a second condition. If the grand mean is 5.00, what are the individual's scores with the effect of her background removed?
    a. 4.00 and 6.00   c. 5.00 and 7.00
    b. 5.00 and 5.00   d. 6.00 and 8.00

25. The correlated groups *t* test provides a less sensitive test of the relationship between the independent and dependent variables than does the independent groups *t* test.
    a. true   b. false

*26. In most behavioral science research, the correlation between individuals' scores in one condition of a study and the same individuals' scores in a second condition will be
    a. negative
    b. positive
    c. close to 0
    d. none of the above

*27. With small sample sizes, very large values of eta-squared are necessary before statistical tests allow us to conclude that there is a relationship between the independent and the dependent variables.
    a. true   b. false

Use the following information to complete Exercises 28–37.

A researcher wanted to tested the effect of the Three Mile Island nuclear accident on attitudes toward nuclear energy. As part of a larger survey, 48 individuals who lived near Three Mile Island were administered a scale to measure their attitudes toward nuclear energy 6 months before the accident occurred. Scores on this scale could range from 0 to 50, with higher values indicating a more favorable attitude. Three months after the accident, the attitude scale was administered to the 48 people for a second time. The mean attitude score was 45.26 prior to the accident and 35.71 after the accident. The estimated standard error of the mean of difference scores was 2.35.

28. If the value of eta-squared in the population is .10, what is the power of the statistical test at an alpha level of .05, nondirectional test?
    a. .10   c. .85
    b. .60   d. .90

29. The independent variable is
    a. the occurrence of the accident
    b. attitudes toward nuclear energy
    c. the time of attitude assessment
    d. the place of residence of the respondents

*30. The dependent variable is
    a. the occurrence of the accident
    b. attitudes toward nuclear energy
    c. the time of attitude assessment
    d. the place of residence of the respondents

*31. The observed value of *t* is
    a. 1.73   c. 4.06
    b. 2.02   d. 6.23

32. Based on this study, we can conclude that the independent variable is related to the dependent variable.
    a. true   b. false

33. It is possible that the test results reflect a Type I error.
    a. true   b. false

*34. It is possible that the test results reflect a Type II error.
    a. true   b. false

35. The strength of the relationship, as indexed by eta-squared, is
    a. .07   c. .26
    b. .08   d. .32

*36. The proportion of variability in the dependent variable that is due to disturbance variables other than individual differences is
    a. .74
    b. .86
    c. .93
    d. This cannot be determined.

*37. What can we conclude about the effect of the accident on attitudes toward nuclear energy?
   a. Attitudes became more positive.
   b. Attitudes became more negative.
   c. Attitudes stayed the same.
   d. The accident may or may not have affected attitudes toward nuclear energy.

Exercises to Apply Concepts

**38. The process of decision making has been studied extensively by psychologists. One area of inquiry has been the effect of making a decision on the subsequent evaluation of the decision alternatives. Brehm (1956) has suggested that when an individual is forced to choose between two equally attractive alternatives, the individual will justify his or her decision after the choice has been made by downgrading the unchosen alternative and upgrading the chosen one. In one experiment, participants rated the desirability of two household products. The products had been tested to ensure that they were approximately equally desirable. Participants then read marketing reports on each product and were asked to choose one as payment for participation in the study. Finally, participants were told that their first rating should be considered a first impression and that the experimenters wanted to get a second rating since the participant now had more time to think about the products. The ratings were made on an eight-point scale, where 1 indicated low desirability and 8 indicated high desirability. We will consider only the results for the unchosen alternative. The hypothetical data presented in the table are representative of the results of this study. Analyze these data using a nondirectional test, draw a conclusion, and write up your results using the principles developed in the Method of Presentation section.

Participant	Before choice	After choice
1	8	4
2	7	3
3	7	5
4	4	6
5	8	4
6	6	4
7	6	2
8	5	5
9	5	3
10	4	4

39. A common belief held by laypersons is that smoking marijuana affects a person's pupil size. To study this issue, Weil, Zinberg, and Nelson (1968) administered a high dose of marijuana to ten men by having them smoke a potent marijuana cigarette. These men were all 21 to 26 years of age and all smoked tobacco cigarettes regularly but had never tried marijuana. Pupil size was measured under constant illumination before and after a man smoked the marijuana cigarette. Measurements were taken with a millimeter ruler with the man's eyes focused on an object at a constant distance. The data are presented in the table. Analyze these data using a nondirectional test, draw a conclusion, and write up your results using the principles developed in the Method of Presentation section.

Participant	Before marijuana	After marijuana
1	5	7
2	7	5
3	6	8
4	7	5
5	6	6
6	5	7
7	3	9
8	3	5
9	5	9
10	3	9

40. Kleinke, Meeker, and Staneski (1986) identified categories of "opening lines" that individuals use when approaching persons of the opposite gender. Two such categories are direct and cute–flippant. The direct approach involves an overt statement of interest, such as "I hope you don't mind, but I'd really like to talk to you." The cute–flippant approach involves the use of humor, such as "Did anybody ever tell you that you look like a movie star?" In order to study the effectiveness of the two types of lines, suppose a researcher asks women to imagine that they are approached in a singles bar by an attractive man who uses one or the other of the two deliveries. Each woman rates her anticipated reaction to each of the two types of lines on a seven-point scale, where 1 indicates a very negative response and 7 indicates a very positive response. Hypothetical data for the study are presented in the table. Analyze these data using a nondirectional test, draw a conclusion, and write up your results using the principles developed in the Method of Presentation section.

Individual	Direct approach	Cute–flippant approach
1	5	4
2	6	3
3	6	4
4	4	5
5	5	3
6	6	5
7	5	4
8	6	3
9	6	4

One-Way Between-Subjects Analysis of Variance

12.1 Use of One-Way Between-Subjects Analysis of Variance

One-way between-subjects analysis of variance, more simply known as *one-way analysis of variance* (abbreviated as "one-way ANOVA"), is typically used to analyze the relationship between two variables when these conditions are met:

1. The dependent variable is quantitative in nature and is measured on a level that at least approximates interval characteristics.

2. The independent variable is *between-subjects* in nature (it can be either qualitative or quantitative).

3. The independent variable has three or more levels.

In short, one-way analysis of variance is used under the same circumstances as the independent groups *t* test except that the independent variable has more than two levels.* Let us consider an example of an investigation that meets the conditions for this technique.

---

* For ease of presentation, we restrict our discussion of one-way analysis of variance to the situation where sample sizes in the groups being considered are all equivalent. Although

(continued)

An investigator was interested in the relationship between individuals' religion and what they consider the ideal family size. Twenty-one people were interviewed—seven Catholics, seven Protestants, and seven Jews. Each was asked what he or she considered to be the ideal number of children to have in a family. Their responses are presented in Table 12.1.

In this investigation, religion is the independent variable. It is between-subjects in nature and has three levels. Ideal family size is the dependent variable. It is quantitative in nature and measured on a ratio level. Given these conditions, one-way between-subjects analysis of variance would typically be used to analyze the relationship between the variables.

12.2 Inference of a Relationship Using One-Way Between-Subjects Analysis of Variance

Null and Alternative Hypotheses

The first question to be addressed is whether a relationship exists between religion and ideal family size. The null hypothesis states that there is no relationship between religion and the ideal number of children. It is expressed in terms of population means, as we have done in previous chapters:

$$H_0: \quad \mu_C = \mu_P = \mu_J$$

where the subscript "C" represents the Catholic group, the subscript "P" represents the Protestant group, and the subscript "J" represents the Jewish group. The

TABLE 12.1   Data for Religion and Ideal Family Size Study

Catholic	Protestant	Jewish
4	1	1
3	2	1
2	0	2
3	2	0
3	4	2
4	3	0
2	2	1
$\Sigma X_C = 21$	$\Sigma X_P = 14$	$\Sigma X_J = 7$
$\overline{X}_C = 3.00$	$\overline{X}_P = 2.00$	$\overline{X}_J = 1.00$

the same general logic can be extended to unequal sample sizes, some of the computational procedures require minor modification. More generally, our discussion of analysis of variance in this and subsequent chapters will deal with only the type most commonly encountered in the behavioral sciences, referred to as *fixed-effects analysis*. For a discussion of other possibilities, see Hays (1981).

alternative hypothesis states that there is a relationship between religion and the ideal number of children; that is, that the three population means are not all equal to one another. Unlike previous tests, however, the alternative hypothesis for one-way analysis of variance cannot be summarized in a single mathematical statement. This is because there are four ways in which the three population means can pattern themselves so that they are not all equal to one another:

$$\mu_C = \mu_P \neq \mu_J$$

$$\mu_C = \mu_J \neq \mu_P$$

$$\mu_P = \mu_J \neq \mu_C$$

$$\mu_C \neq \mu_P \neq \mu_J$$

The alternative hypothesis does not distinguish among these different possibilities. It simply states that a relationship exists such that the three population means are *not* all equal to one another. The question of the exact patterning of means is addressed in the context of the nature of the relationship. Thus, the alternative hypothesis is

H_1:   The three population means are not all equal

The problem is to choose between the null and alternative hypotheses. If we look at the mean scores for the samples, we find that they are all different. The Catholic mean is 3.00, the Protestant mean is 2.00, and the Jewish mean is 1.00. This would seem to support the alternative hypothesis. However, we know that the non-equivalence of sample means could just reflect sampling error. We therefore want to test the viability of a sampling error interpretation.

The logic for testing this interpretation is similar to the logic discussed in previous chapters. First, we assume that the null hypothesis is true. We then state an expected result based on this assumption. Next, we compute a sample statistic (in this instance, the statistic is a *variance ratio,* which we discuss shortly). The statistic is treated as a score in a sampling distribution where the null hypothesis is assumed to be true. If the score falls within the rejection region (as defined by the alpha level), then we reject the null hypothesis and conclude that there is a relationship between the two variables. Otherwise, we fail to reject the null hypothesis.

Between- and Within-Group Variability

For the data in Table 12.1, we can distinguish two types of variability. The first is differences between the mean scores in the three groups or, in more formal terms, **between-group variability.** If the mean scores in the three groups were all equal, then there would be no between-group variability. The more different the three means are from one another, the more between-group variability there is.

Why do mean differences exist? Two factors contribute to between-group variability. The first is sampling error: Even if the population means were, in fact, equal, we might observe between-group variability in the sample data because of sampling error. A second factor that contributes to between-group variability is the effect of the independent variable on the dependent variable: If religion influences the ideal number of children, then this will tend to make the sample means different from one another. Thus, between-group variability for mean scores on the dependent variable reflects two things: (1) sampling error and (2) the effect of the independent variable on the dependent variable.

The second type of variability is the variability of scores *within* each of the three groups, or **within-group variability**. Examine the seven scores for Catholics in Table 12.1. Note that there is variability in the scores; some Catholics have a higher ideal number of children than other Catholics. The same is true for both Protestants and Jews. Since religion is held constant for each group, the variability in the scores cannot be attributed to religious affiliation. Other factors are operating to cause the variability in scores (such as how religious the person is, and other such disturbance variables). The fact that scores vary within a group suggests that there is variability in scores in the *population* represented by that group. Recall from Chapter 7 that when the variability of scores in a population is large, we expect more sampling error than when the population variability is small. As such, greater variability of scores within a group is indicative of greater variability of scores within the corresponding population and, thus, a greater amount of sampling error. In short, then, within-group variability reflects sampling error. Note that within-group variability is *not* influenced by the effect of the independent variable on the dependent variable. Since the variability of scores is considered within each group *separately*, the independent variable is held constant within each group and cannot be a source of within-group variability.

Since between-group variability reflects both sampling error *and* the effect of the independent variable, and within-group variability reflects just sampling error, the ratio of these two sources of variability can be used to test the sampling error interpretation discussed previously. This ratio, known as the **variance ratio**, can be represented as:

$$\frac{\text{Between-group variability}}{\text{Within-group variability}} = \frac{\text{sampling error} + \text{effect of the independent variable}}{\text{sampling error}} \qquad [12.1]$$

Consider the case where the null hypothesis is true. Then the independent variable has no effect on the dependent variable, so between-group variability reflects only sampling error. Since within-group variability also reflects sampling error, the ratio in Equation 12.1 is simply the division of one estimate of sampling error by another estimate of sampling error. The result is a value that, over the long run, will approach 1.00 (since any number divided by itself is 1.00). In contrast, if the null hypothesis is not true, then between-group variability reflects both sampling error and the effect of the independent variable. In this case, we would expect the variance ratio, over the long run, to be greater than 1.00.

Partitioning of Variability

To compute the variance ratio, it is necessary first to derive numerical estimates of between- and within-group variability. Although this is most readily done with the computational formulas presented later in this chapter, it is important to have a conceptual understanding of what these formulas entail. We demonstrate this by drawing a parallel between the way in which we can split up, or *partition*, an individual's score on the dependent variable into the mean score of the group of which he or she is a member and how far his or her score deviates from this mean, and the way in which we can partition the total variability in a set of scores into two similar components.

Consider the first individual in the Catholic group. Her score of 4 deviated $4 - 3 = 1$ unit from the group mean of 3.00. Her score can thus be represented as

$$4 = 3 + 1$$

We can write a general equation to reflect this relationship:

$$X = \overline{X}_j + d \tag{12.2}$$

where X represents an individual's score, $\overline{X}_j$ is the mean score of the group of which the individual is a member (group j), and d is the signed deviation of the individual's score from the group mean $(X - \overline{X}_j)$. It is possible to express the score of every individual in the study in this manner, as demonstrated in Table 12.2.

The column labeled X contains the score of each individual on the dependent variable. We can compute an index of how much variability there is in these scores by computing a sum of squares for this column of numbers using the

TABLE 12.2 **Breakdown of Scores on the Dependent Variable for Religion and Ideal Family Size Study**

Individual	Religion	X	$=$	$\overline{X}_j$	$+$	d[a]
1	C	4	=	3	+	1
2	C	3	=	3	+	0
3	C	2	=	3	+	−1
4	C	3	=	3	+	0
5	C	3	=	3	+	0
6	C	4	=	3	+	1
7	C	2	=	3	+	−1
8	P	1	=	2	+	−1
9	P	2	=	2	+	0
10	P	0	=	2	+	−2
11	P	2	=	2	+	0
12	P	4	=	2	+	2
13	P	3	=	2	+	1
14	P	2	=	2	+	0
15	J	1	=	1	+	0
16	J	1	=	1	+	0
17	J	2	=	1	+	1
18	J	0	=	1	+	−1
19	J	2	=	1	+	1
20	J	0	=	1	+	−1
21	J	1	=	1	+	0

[a] $d = X - \overline{X}_j$

standard formula for a sum of squares. This sum of squares is called the **sum of squares total** (symbolized SS_{TOTAL}) because it reflects the total variability in the dependent variable across all individuals. It is identical to the sum of squares total discussed in Chapter 10. If we were to perform the required calculations, we would find that $SS_{TOTAL} = 32$ in this instance.

Examine the scores in the column labeled $\overline{X}_j$ in Table 12.2. The source of variability in these scores is the differences between the group means. If the means of all three groups had been equal, every score in this column would be the same. The more the three means differ, the more dissimilar these scores will be from one another. A sum of squares for this column would therefore reflect between-group variability. This sum of squares, called the **sum of squares between** (symbolized $SS_{BETWEEN}$), is identical to the sum of squares explained discussed in Chapter 10 and can be calculated by applying the standard formula for a sum of squares to the $\overline{X}_j$ scores. In this instance, $SS_{BETWEEN} = 14$.

Examine the scores in the column labeled d. Each of these scores represents a deviation from the group mean, or how far an individual's score deviates from the average score *within the group*. If there were little variability within groups, all of these scores would tend to be close to 0 and to be very similar to one another. If there were considerable variability within groups, there would be considerable variability in these scores. A sum of squares for this column of numbers would therefore reflect how much within-group variability there is. This sum of squares is called the **sum of squares within** (symbolized SS_{WITHIN}) and can be calculated by applying the standard formula for a sum of squares to the d scores. In this instance, $SS_{WITHIN} = 18$. Another name given to the sum of squares within is the sum of squares error because it reflects the effects of disturbance variables and, hence, sampling error. This is the term we used when we first introduced this concept in Chapter 10.

In Chapter 10 we also noted that the total variability in the dependent variable can be partitioned into two components: one reflecting the influence of the independent variable and one reflecting the influence of disturbance variables. In the present context, the partitioning of variability can be represented as

$$SS_{TOTAL} = SS_{BETWEEN} + SS_{WITHIN} \qquad [12.3]$$

This partitioning is consistent with the idea of expressing each individual's score in terms of (1) the group mean and (2) the score's deviation from the group mean, as we did in Table 12.2. Thus, just as each score can be expressed in terms of these two components, the total *variability* in scores can be expressed in terms of (1) the *variability* between the group means (between-group variability) and (2) the *variability* of deviations from the group means (within-group variability). This is clearly demonstrated in our example (where $SS_{TOTAL} = 32$, $SS_{BETWEEN} = 14$, and $SS_{WITHIN} = 18$), as $32 = 14 + 18$.

Mean Squares and the *F* Ratio

The variance ratio of between-group variability divided by within-group variability that is computed to test the null hypothesis does not use sums of squares; rather, the variance ratio is based on measures of variance, or *mean squares*. As noted in Chapter 7, a mean square is simply a sum of squares divided by its corresponding degrees of freedom. The reason measures of mean squares rather than measures of sums of squares are used will be made explicit shortly. We first consider the computation of the relevant quantities:

$$MS_{BETWEEN} = \frac{SS_{BETWEEN}}{df_{BETWEEN}} \tag{12.4}$$

$$MS_{WITHIN} = \frac{SS_{WITHIN}}{df_{WITHIN}} \tag{12.5}$$

where $MS_{BETWEEN}$, referred to as the **mean square between,** represents the mean square for between-group variability; MS_{WITHIN}, referred to as the **mean square within,** represents the mean square for within-group variability; $df_{BETWEEN}$ is the degrees of freedom associated with the sum of squares between; and df_{WITHIN} is the degrees of freedom associated with the sum of squares within. These degrees of freedom are calculated as follows:

$$df_{BETWEEN} = k - 1 \tag{12.6}$$

$$df_{WITHIN} = N - k \tag{12.7}$$

where k is the number of levels of the independent variable (that is, the number of groups) and N is the total number of participants in the study.* As with the sums of squares, the degrees of freedom are additive; if we sum the degrees of freedom between and the degrees of freedom within, we get the degrees of freedom associated with the sum of squares total:

$$df_{TOTAL} = df_{BETWEEN} + df_{WITHIN} \tag{12.8}$$

Since $df_{BETWEEN} = k - 1$ and $df_{WITHIN} = N - k$, and $(k - 1) + (N - k) = N - 1$, the degrees of freedom total can be calculated directly as

$$df_{TOTAL} = N - 1 \tag{12.9}$$

In our example, $k = 3$ and $N = 21$, so

$$df_{BETWEEN} = 3 - 1 = 2$$

$$df_{WITHIN} = 21 - 3 = 18$$

$$df_{TOTAL} = 21 - 1 = 20$$

Using Equations 12.4 and 12.5, we obtain the mean squares:

$$MS_{BETWEEN} = \frac{14}{2} = 7.00$$

$$MS_{WITHIN} = \frac{18}{18} = 1.00$$

The mean square between reflects the variability between group means. As such, it bears a relationship to the quantity we would obtain if we treated the

---

* The rationale for the equivalence of the degrees of freedom to the indicated quantities is discussed in Appendix 12.1.

sample means like any other set of scores and applied the usual computations for a variance estimate. In fact, when sample sizes are equivalent, the mean square between is exactly equal to the variance estimate of the sample means multiplied by n to reflect the number of scores on which each mean is based. Recall from Chapter 7 that the formula for a variance estimate is $SS/(N - 1)$. In our example, the sample means are 3.00, 2.00, and 1.00. The sum of squares for these three "scores" is 2, so the variance estimate is $2/(3 - 1) = 1.00$. Multiplying this by 7, the per-group sample size, we obtain a value of 7.00, the same quantity obtained using Equation 12.4.

"Mean square within" is merely another name for the pooled variance estimate discussed in Chapter 10. It will be remembered that a pooled variance estimate is derived by multiplying the variance estimate for each group by its degrees of freedom, summing the products across groups, and then dividing the resulting quantity by the total number of degrees of freedom. In our example, there are three groups: Catholics, Protestants, and Jews. If we were to individually compute the variance estimate for each, we would find that the estimated variance for Catholics is .67; for Protestants, 1.67; and for Jews, .67. Since each sample contains seven people, each variance estimate has $n - 1 = 7 - 1 = 6$ degrees of freedom associated with it, and the total degrees of freedom are $6 + 6 + 6 = 18$. The pooled variance estimate is thus

$$\hat{s}^2_{pooled} = \frac{(6)(.67) + (6)(1.67) + (6)(.67)}{18} = 1.00$$

This is the same quantity obtained using Equation 12.5.

The variance ratio, formally referred to as the **F ratio** (in honor of the statistician Sir Ronald Fisher, who developed this approach), is

$$F = \frac{MS_{BETWEEN}}{MS_{WITHIN}} \qquad [12.10]$$

which in our example equals

$$F = \frac{7.00}{1.00} = 7.00$$

To recapitulate, the F ratio is a ratio of two variances. In the context of one-way analysis of variance, it is the mean square between (an index of between-group variability) divided by the mean square within (an index of within-group variability). The mean square between and the mean square within are derived by computing the sum of squares between and the sum of squares within and dividing each by its corresponding degrees of freedom. If the null hypothesis is true, we would expect the F ratio, over the long run, to approach 1.00. If the null hypothesis is not true, we would expect the F ratio, over the long run, to be greater than 1.00. Let us now consider this issue in more detail.

Sampling Distribution of the F Ratio

Consider three populations, A, B, and C, where $\mu_A = \mu_B = \mu_C$. Suppose we select a random sample of 30 individuals from population A, a random sample of 30 individuals from population B, and a random sample of 30 individuals from population C. Using the procedures described previously, we could compute an F ratio. Although we would expect the value of this ratio to be near 1.00 (because the null hypothesis is true), it might not equal exactly 1.00 because $MS_{BETWEEN}$ and MS_{WITHIN} are independent estimates of variability in the populations, based on sample data. Suppose the F ratio were found to equal 1.52. Furthermore, suppose that we repeated this procedure by randomly selecting new samples of the same sizes from the same populations and this time got an F ratio of 1.32. We could repeat this procedure for all possible random samples of size 30 for each population, and the result would be a large number of F ratios. These F ratios can be treated as "scores" in a distribution and constitute a **sampling distribution of the F ratio,** analogous to the sampling distributions we have discussed previously.

A sampling distribution of the F ratio, under certain conditions to be discussed shortly, is very similar to an important theoretical distribution called the **F distribution.** The reason measures of mean squares rather than measures of sums of squares are used to define the F ratio is that when certain conditions are met, measures of mean squares yield F ratios that have a sampling distribution that closely approximates an F distribution, whereas measures of sums of squares do not.

As illustrated in Figure 12.1, the F distribution takes on different shapes depending on the particular values of $df_{BETWEEN}$ and df_{WITHIN} associated with it. In all of the distributions, the lowest possible value for F is 0. Also, the median value for F in all of the distributions is 1.00. This should not be surprising because the F ratios are calculated from the case where the null hypothesis is true. Values of F become less frequent as they become increasingly greater than 1.00, and large values of F are highly unlikely. The question then becomes similar to the one posed in Chapter 8: How large must the F ratio computed from the sample information be before we reject the null hypothesis of no relationship?

Just as probability statements can be made with respect to scores in the normal and t distributions, so can probability statements be made with respect to scores in the F distribution. It is therefore possible to set an alpha level and define a critical value such that if the null hypothesis is true, the probability of obtaining an F ratio larger than that critical value is less than alpha. Since all departures from the null hypothesis are reflected in the upper tail of the F distribution (as defined by the critical value), the **F test** is, by its nature, nondirectional.

Before conducting the F test, we summarize all of our previous calculations in a **summary table:**

Source	SS	df	MS	F
Between	14.00	2	7.00	7.00
Within	18.00	18	1.00	
Total	32.00	20		

The first column of the summary table indicates the source of the information in the other columns. The second column reports the sums of squares between,

FIGURE 12.1   *F* **Distributions for Various Degrees of Freedom**

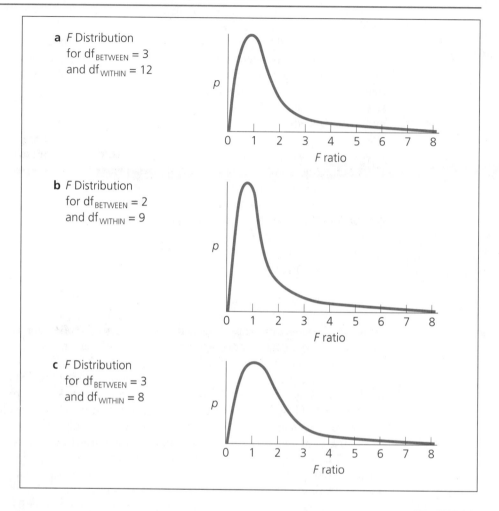

a *F* Distribution
for $df_{BETWEEN} = 3$
and $df_{WITHIN} = 12$

b *F* Distribution
for $df_{BETWEEN} = 2$
and $df_{WITHIN} = 9$

c *F* Distribution
for $df_{BETWEEN} = 3$
and $df_{WITHIN} = 8$

within, and total. The next column provides the associated degrees of freedom for each sum of squares, and column 4 presents the mean squares for the two sources of variability. The last column contains the *F* ratio.

The test of the null hypothesis involves comparing the observed value of *F* with the appropriate critical value of *F* from Appendix F. Instructions for using this table are also contained in the appendix. When an *F* value is reported, it is conventional to present the degrees of freedom between followed by the degrees of freedom within. For an alpha level of .05 and $df_{BETWEEN} = 2$ and $df_{WITHIN} = 18$, the critical value of *F* is 3.55. In order to be statistically significant, the observed value of *F* must exceed the critical value of *F*. Since the observed *F* value of 7.00 is greater than 3.55, we reject the null hypothesis and conclude that a relationship exists between religion and ideal family size. We have answered the first of the three questions for the analysis of bivariate relationships.

Computational Formulas

There are several different ways to derive the information necessary for the calculation of the F ratio. We have already seen how the sum of squares between and the sum of squares within can be thought of as variability between the group means and variability of deviations from the group means, respectively. We have also seen the equivalence between the mean square between and the variance estimate for the sample means scaled by per-group sample size, on the one hand, and the mean square within and the pooled variance estimate, on the other.

Although both of these approaches help to clarify the conceptual foundation of one-way analysis of variance, they are computationally inefficient. In practice, the F ratio is most commonly (and most efficiently) obtained by calculating the sum of squares between and the sum of squares within with computational formulas available for this purpose and then dividing these by their corresponding degrees of freedom. We now consider these formulas.

We begin by examining the computational formula for the sum of squares total. This is simply the standard computational formula for a sum of squares applied to the entire data set. More specifically, to obtain the sum of squares total using the computational formula, square each score in the data set, add these squared scores, and subtract the square of the sum of the scores divided by the total sample size. Symbolically,

$$SS_{TOTAL} = \sum X^2 - \frac{(\sum X)^2}{N} \qquad [12.11]$$

where $\sum X^2$ is the *sum of the squared scores,* $(\sum X)^2$ is the *square of the summed scores,* and N is the total sample size.

The data for the example on religion and ideal family size are reproduced in Table 12.3. To obtain $\sum X^2$ we must first square each score in the data set and then add these squared scores. For each group, the scores have been squared in the column labeled X^2. Adding all 21 of these squared scores together, we find that

$$X^2 = 16 + 9 + \cdots + 0 + 1 = 116$$

The notation "$\cdots$" is used to indicate that not all of the X^2 scores contributing to the observed sum of 116 are physically represented. Rather, as a means of saving space, we have provided only the first two and the last two X^2 scores. When the "$\cdots$" notation is used, the inclusion in the summation of *all* relevant scores falling between the written values is implicit.

The sum of the X scores is

$$\sum X = 4 + 3 + \cdots + 0 + 1 = 42$$

where the "$\cdots$" notation again indicates that *all* relevant scores (in this case, X scores) are included in the summation. Squaring the observed sum of 42 and dividing by N, we obtain

$$\frac{(\sum X)^2}{N} = \frac{42^2}{21} = 84$$

TABLE 12.3 **Data and Computation of Sums of Squares for Religion and Ideal Family Size Study**

Catholic		Protestant		Jewish	
X	X^2	X	X^2	X	X^2
4	16	1	1	1	1
3	9	2	4	1	1
2	4	0	0	2	4
3	9	2	4	0	0
3	9	4	16	2	4
4	16	3	9	0	0
2	4	2	4	1	1

$T_C = \Sigma X_C = 21$ $\Sigma X_C^2 = 67$ $T_P = \Sigma X_P = 14$ $\Sigma X_P^2 = 38$ $T_J = \Sigma X_J = 7$ $\Sigma X_J^2 = 11$

$T_C^2 = 441$ $T_P^2 = 196$ $T_J^2 = 49$

$$SS_C = 67 - \frac{21^2}{7} = 4 \qquad SS_P = 38 - \frac{14^2}{7} = 10 \qquad SS_J = 11 - \frac{7^2}{7} = 4$$

$$\Sigma X^2 = 16 + 9 + \cdots + 0 + 1 = 116$$

$$\Sigma X = 4 + 3 + \cdots + 0 + 1 = 42$$

$$\frac{(\Sigma X)^2}{N} = \frac{42^2}{21} = 84$$

$$\frac{\Sigma T_j^2}{n} = \frac{441 + 196 + 49}{7} = 98$$

$$SS_{WITHIN} = SS_C + SS_P + SS_J = 4 + 10 + 4 = 18$$

Thus, the sum of squares total in this example is

$$SS_{TOTAL} = \Sigma X^2 - \frac{(\Sigma X)^2}{N}$$

$$= 116 - 84 = 32$$

which is the same value as reported previously.

The sum of squares within is equal to the total of the sums of squares within the individual groups. For any given group, this is equal to the sum of that group's squared scores minus that group's squared sum after it has been divided by the relevant sample size. These procedures are mathematically equivalent to the computational formula for the sum of squares within:

$$SS_{WITHIN} = \Sigma X^2 - \frac{\Sigma T_j^2}{n} \qquad [12.12]$$

where ΣX^2 is as defined above, T_j^2 is the square of the sum of the scores in group j [in other words, $T_j^2 = (\Sigma X)^2$ for a given group], and n is the per-group sample size. In our example, T_j^2 equals 441 for Catholics, 196 for Protestants, and 49 for Jews. Thus,

$$\frac{\sum T_j^2}{n} = \frac{441 + 196 + 49}{7} = 98$$

so

$$SS_{WITHIN} = \sum X^2 - \frac{\sum T_j^2}{n}$$

$$= 116 - 98 = 18$$

This is the same value reported previously for the sum of squares within and is also equal to the total of the individual groups' sums of squares as reported in Table 12.3.

The computational formula for the sum of squares between is

$$SS_{BETWEEN} = \frac{\sum T_j^2}{n} - \frac{(\sum X)^2}{N} \qquad [12.13]$$

where all terms are as defined above. In our example,

$$SS_{BETWEEN} = 98 - 84 = 14$$

as previously reported.

Of course, given that the sum of squares total is equal to the sum of squares between plus the sum of squares within, if two of these quantities have already been calculated, the third can be determined through simple algebraic manipulation. For instance, we could have obtained the sum of squares between as follows:

$$SS_{BETWEEN} = SS_{TOTAL} - SS_{WITHIN}$$

$$= 32 - 18 = 14$$

However, it usually is a good idea to calculate all three sum of squares using the above formulas. This allows us to check for computational errors by determining whether the calculated values of the sum of squares between and the sum of squares within add up to the calculated value of the sum of squares total.

Study Exercise 12.1

An investigator was interested in the effects of various types of performance feedback on self-esteem. To examine the relationship between these two variables, she had 15 people take a general knowledge test. Five participants were randomly assigned to a positive feedback condition where, irrespective of actual performance, they were informed that they had scored at a very high level. Another five participants were randomly assigned to a negative feedback condition and informed that they had performed very poorly. A final group of five constituting a control condition were not provided with any feedback regarding their test scores. All participants then responded to a measure of self-esteem. Scores on this measure could range from 0 to 10, with higher values indicating greater self-esteem. The data for the experiment are listed in the table. Use a one-way analysis of variance to test for a relationship between the type of feedback and self-esteem.

Positive feedback X	Negative feedback X	Control X
8	5	2
7	6	4
9	7	5
10	4	3
6	3	6

Answer   The null and alternative hypotheses can be stated as follows:

H_0: $\mu_P = \mu_N = \mu_C$

H_1:   The three population means are not all equal

where the subscripts "P," "N," and "C" respectively denote the positive feedback, negative feedback, and control conditions.

The requisite calculations for analyzing the data are as follows:

Positive Feedback		Negative Feedback		Control	
X	X²	X	X²	X	X²
8	64	5	25	2	4
7	49	6	36	4	25
9	81	7	49	5	25
10	100	4	16	3	9
6	36	3	9	6	36
$T_P = 40$		$T_N = 25$		$T_C = 20$	
$\overline{X}_P = 8.00$		$\overline{X}_N = 5.00$		$\overline{X}_C = 4.00$	
$T_P^2 = 1{,}600$		$T_N^2 = 625$		$T_C^2 = 400$	

$$\sum X^2 = 64 + 49 + \cdots + 9 + 36 = 555$$

$$\sum X = 8 + 7 + \cdots + 3 + 6 = 85$$

$$\frac{(\sum X)^2}{N} = \frac{85^2}{15} = 481.67$$

$$\frac{\sum T_j^2}{n} = \frac{1{,}600 + 625 + 400}{5} = 525.00$$

Intermediate statistics necessary for calculating the sums of squares are given in the table. The sum of squares between is computed using Equation 12.13:

$$SS_{BETWEEN} = \frac{\sum T_j^2}{n} - \frac{(\sum X)^2}{N}$$

$$= 525.00 - 481.67 = 43.33$$

The sum of squares within is computed using Equation 12.12:

$$SS_{WITHIN} = \sum X^2 - \frac{\sum T_i^2}{n}$$

$$= 555 - 525.00 = 30.00$$

The sum of squares total is computed using Equation 12.11:

$$SS_{TOTAL} = \sum X^2 - \frac{(\sum X)^2}{N}$$

$$= 555 - 481.67 = 73.33$$

The degrees of freedom are computed using Equations 12.6, 12.7, and 12.9:

$$df_{BETWEEN} = k - 1 = 3 - 1 = 2$$

$$df_{WITHIN} = N - k = 15 - 3 = 12$$

$$df_{TOTAL} = N - 1 = 15 - 1 = 14$$

The mean square between and the mean square within are then computed by dividing the corresponding sums of squares by their degrees of freedom:

$$MS_{BETWEEN} = \frac{SS_{BETWEEN}}{df_{BETWEEN}} = \frac{43.33}{2} = 21.67$$

$$MS_{WITHIN} = \frac{SS_{WITHIN}}{df_{WITHIN}} = \frac{30.00}{12} = 2.50$$

Finally, the F ratio is derived by dividing $MS_{BETWEEN}$ by MS_{WITHIN}:

$$F = \frac{MS_{BETWEEN}}{MS_{WITHIN}} = \frac{21.67}{2.50} = 8.67$$

These calculations yield the following summary table:

Source	SS	df	MS	F
Between	43.33	2	21.67	8.67
Within	30.00	12	2.50	
Total	73.33	14		

The critical value of F from Appendix F for an alpha level of .05 and 2 and 12 degrees of freedom is 3.88. Since the observed F value of 8.67 is greater than 3.88, we reject the null hypothesis and conclude that a relationship exists between the type of feedback and self-esteem.

Assumptions of the F Test

The F test for one-way between-subjects analysis of variance is appropriate to use when the dependent variable is quantitative in nature and measured on a level that at least approximates interval characteristics. The sampling distribution of the F ratio tends to closely approximate an F distribution when the following assumptions are met:

1. The samples are independently and randomly selected from their respective populations.

2. The scores in each population are normally distributed.

3. The scores in each population have equal variances.

Note that these assumptions parallel those for the independent groups t test, as discussed in Chapter 10.

For the F test to be valid, it is important that the assumption of independent and random selection be met. However, under certain conditions, the F test is robust to violations of the normality and homogeneity of variance assumptions. This is particularly true when the sample sizes are moderate (greater than 20) to large and the same for all groups. Under these conditions, the F test is quite robust to even marked skewness and kurtosis. Heterogeneity of population variances can be problematic, more so as the number of groups increases (Wilcox, Charlin & Thompson, 1986). The F test is relatively robust when the population variance for one condition is as much as 2 to 3 times larger than the population variances for the other conditions (even with somewhat unequal sample sizes). However, the Type I error rate begins to become unacceptable when the population variance for one condition is 4 or more times larger than the population variances for the other conditions. For more details, see Milligan, Wong, and Thompson (1987) and Harwell, Rubinstein, and Hayes (1992).

12.3 Relationship of the F Test to the t Test

The F test that we have been considering is closely related to the independent groups t test presented in Chapter 10. The distinction between the two approaches is more a matter of historical than practical significance. For the two-group between-subjects situation, the F distribution bears a mathematical relationship to the t distribution:

$$F = t^2 \qquad\qquad [12.14]$$

One could apply the procedures developed in this chapter to the problems in Chapter 10 and get exactly the same conclusions. The values of F would be equal to the squares of the corresponding t scores.

12.4 Strength of the Relationship

Earlier, we noted that the sum of squares between is identical to the sum of squares explained discussed in Chapter 10. Accordingly, the strength of the relationship between the independent and dependent variables can be indexed by substituting $SS_{BETWEEN}$ for $SS_{EXPLAINED}$ in the defining formula for eta-squared (Equation 10.10):

$$\text{eta}^2 = \frac{SS_{\text{EXPLAINED}}}{SS_{\text{TOTAL}}}$$

This yields the following formula for eta-squared for one-way analysis of variance:

$$\text{eta}^2 = \frac{SS_{\text{BETWEEN}}}{SS_{\text{TOTAL}}} \qquad [12.15]$$

For the religion and ideal family size study,

$$\text{eta}^2 = \frac{14}{32} = .44$$

In this investigation, 44% of the variability in ideal family size is associated with religion. This represents a strong effect.

An alternative formula for eta-squared based on the degrees of freedom and the observed value of F is

$$\text{eta}^2 = \frac{(df_{\text{BETWEEN}})F}{(df_{\text{BETWEEN}})F + df_{\text{WITHIN}}} \qquad [12.16]$$

For the religion and ideal family size study,

$$\text{eta}^2 = \frac{(2)(7.00)}{(2)(7.00) + 18} = .44$$

which is the same result as obtained above. Equation 12.16 is useful because in some journal reports, researchers report the values of F and the degrees of freedom but not the values of eta-squared. Using Equation 12.16, you will be able to calculate eta-squared from the information typically provided.

12.5   Nature of the Relationship

As noted at the beginning of this chapter, the F test considers the null hypothesis against all possible alternatives. For example, the possible patterns of unequal population means for three groups are $\mu_1 = \mu_2 \neq \mu_3$, $\mu_1 = \mu_3 \neq \mu_2$, $\mu_2 = \mu_3 \neq \mu_1$, and $\mu_1 \neq \mu_2 \neq \mu_3$. If any one of the alternatives holds, the null hypothesis will be rejected, unless a Type II error occurs. However, the alternative hypothesis states only that the population means are not all equal. Given this state of affairs, it becomes necessary to conduct additional analyses to determine the exact nature of the relationship between the two variables when three or more groups are involved.

Which procedure to use for determining the nature of the relationship after the null hypothesis has been rejected is controversial among statisticians. Among the most common **multiple comparison procedures** are the Scheffé test, the Newman–Keuls test, Duncan's multiple range test, Tukey's honest significant difference (HSD) test, and Fisher's least significant difference (LSD) test. Each of these is discussed in Kirk (1968, 1995). The most general technique is the one proposed by Scheffé. However, the Scheffé procedure tends to produce a high incidence of Type II errors and, for this reason, is often not the test of choice for researchers. Because of its ease of presentation and desirable statistical properties (see Jaccard, Becker & Wood, 1984), we focus on the test proposed by Tukey. A recent competitor to Tukey's test that can be used effectively in a wide range of situations has been proposed by Hayter (1986).*

The **Tukey HSD test** discerns the nature of the relationship by testing a null hypothesis for each possible pair of group means. In the study on religion and ideal family size, there are three groups: Catholic, Protestant, and Jewish. The questions of interest for the HSD test are: (1) Is the population mean for Catholics different than the population mean for Protestants? (2) Is the population mean for Catholics different than the population mean for Jews? and (3) Is the population mean for Protestants different than the population mean for Jews? These three questions can be cast in three sets of null and alternative hypotheses:

$$H_0: \quad \mu_C = \mu_P$$
$$H_1: \quad \mu_C \neq \mu_P$$

$$H_0: \quad \mu_C = \mu_J$$
$$H_1: \quad \mu_C \neq \mu_J$$

$$H_0: \quad \mu_P = \mu_J$$
$$H_1: \quad \mu_P \neq \mu_J$$

These hypotheses are similar to the null and alternative hypotheses for the independent groups t test discussed in Chapter 10. Could we simply test each pair of group means by conducting three independent groups t tests? The answer is no. The problem with doing this is that multiple t tests increase the probability of making a Type I error for at least one of the tests beyond the probability specified by the alpha level for each individual analysis. Witte (1980) has presented a coin-tossing analogy that illustrates the problem:

> When a fair coin is tossed only once, the probability of heads equals .50—just as when a single t test is to be conducted at the .05 level of significance, the probability of a Type I error equals .05. When a fair coin is tossed three times, however, heads can appear not only on the first toss but on the second or third toss as well, and hence the probability of heads on at least one of the three tosses exceeds .50. By the same token, when a series of three t tests are conducted at the .05 level of significance, a Type I error can be committed not only on the first test but on the second or third test as well, and hence the probability of committing a Type I error on at least one of the three tests exceeds .05.

The Tukey HSD test circumvents this problem by maintaining the probability of making one or more Type I errors in the set of multiple comparisons at the specified alpha level. In most applications, researchers adopt an *overall alpha level* of .05. We will follow this practice in the remainder of the book. The logic underlying the HSD test is rather complex and is presented in Tukey (1953). We focus on the mechanics of applying the approach.

---

* Hayter's test has greater statistical power than Tukey's test, but it is not as general.

To facilitate our presentation, the summary table for the study on religion and ideal family size is reproduced here:

Source	SS	df	MS	F
Between	14.00	2	7.00	7.00
Within	18.00	18	1.00	
Total	32.00	20		

The HSD test involves the computation of a **critical difference**, which is defined as follows:

$$CD = q\sqrt{\frac{MS_{WITHIN}}{n}}$$ [12.17]

In this formula, CD represents the critical difference, MS_{WITHIN} is the mean square within from the summary table, n is the per-group sample size, and q is a *Studentized range value* obtainable from the *Studentized range table* in Appendix G. The value of q is determined with reference to the overall alpha level (in this case, .05), the degrees of freedom within from the summary table (in this case, 18), and the number of groups, k (in this case, 3). Using the instructions provided in Appendix G, we find that $q = 3.61$. We calculate the critical difference as follows:

$$CD = 3.61\sqrt{\frac{1.00}{7}} = 1.36$$

Consider the first set of hypotheses specified above:

H_0: $\mu_C = \mu_P$

H_1: $\mu_C \neq \mu_P$

We want to make a decision with respect to these two competing hypotheses. The rule for doing so for the HSD test is the following: If the absolute difference between sample means for the two groups involved in the comparison exceeds the critical difference, then reject the null hypothesis. Otherwise, fail to reject the null hypothesis. If the null hypothesis is rejected, conclude that the group with the larger sample mean also has a larger mean in the population. In our example, the absolute difference between the sample means is $|\overline{X}_C - \overline{X}_P| = |3.00 - 2.00| = 1.00$. Since 1.00 is not greater than 1.36, we fail to reject the null hypothesis stated above.

The same logic is applied to the other sets of hypotheses. We can summarize the results of the entire analysis in a table:

Null hypothesis tested	Absolute difference between sample means	Value of CD	Null hypothesis rejected?
$\mu_C = \mu_P$	$|3.00 - 2.00| = 1.00$	1.36	No
$\mu_C = \mu_J$	$|3.00 - 1.00| = 2.00$	1.36	Yes
$\mu_P = \mu_J$	$|2.00 - 1.00| = 1.00$	1.36	No

The nature of the relationship is now apparent: Catholics ($\overline{X}_C = 3.00$) have a larger ideal family size than Jews ($\overline{X}_J = 1.00$). However, we cannot confidently conclude that the ideal family size for either of these groups differs from the ideal family size for Protestants ($\overline{X}_P = 2.00$). The HSD test provides a clear picture of just how religion is related to a person's ideal number of children.

Study Exercise 12.2

Calculate eta-squared and use the HSD test to analyze the nature of the relationship between the two variables for the data in Study Exercise 12.1.

Answer Eta-squared is most easily calculated using Equation 12.15:

$$\text{eta}^2 = \frac{SS_{\text{BETWEEN}}}{SS_{\text{TOTAL}}} = \frac{43.33}{73.33} = .59$$

This represents a strong effect.

The first step in applying the HSD test is to calculate the critical difference. Referring to Appendix G, we find that for an overall alpha level of .05, $df_{\text{WITHIN}} = 12$, and $k = 3$, q is equal to 3.77. Thus,

$$CD = q\sqrt{\frac{MS_{\text{WITHIN}}}{n}} = 3.77\sqrt{\frac{2.50}{5}} = 2.67$$

The HSD procedure can now be applied as follows:

Null hypothesis tested	Absolute difference between sample means	Value of CD	Null hypothesis rejected?
$\mu_P = \mu_N$	$|8.00 - 5.00| = 3.00$	2.67	Yes
$\mu_P = \mu_C$	$|8.00 - 4.00| = 4.00$	2.67	Yes
$\mu_N = \mu_C$	$|5.00 - 4.00| = 1.00$	2.67	No

The nature of the relationship is such that self-esteem will be greater when people receive positive feedback ($\overline{X}_P = 8.00$) than when they receive negative feedback ($\overline{X}_N = 5.00$) or no feedback ($\overline{X}_C = 4.00$). However, we cannot confidently conclude that negative feedback affects self-esteem relative to no feedback.

The overall approach to analysis of variance that we have described can be characterized as (1) making an overall, or *omnibus*, test of whether a relationship exists between two variables (by means of the F test), (2) evaluating the strength of the relationship, and (3) analyzing pairwise mean differences using multiple comparisons to discern the nature of the relationship. The analysis of variance framework is sometimes modified and is a very flexible analytic method. For example, sometimes investigators are not interested in performing all possible pairwise comparisons when evaluating the nature of the relationship, but instead focus only on one or two contrasts that are of particular theoretical interest. In this case, the Tukey HSD test is not used, and an alternative analytic strategy is pursued. For a discussion of the many different approaches that adopt an analysis of variance framework, see Kirk (1995).

12.6   Methodological Considerations

Although the results of the study of religion and ideal family size suggest that a relationship exists between the two variables, the data must be interpreted in light of certain methodological constraints. Consider first the role of confounding variables. Because religion cannot be manipulated in an experiment, research participants could not be randomly assigned to groups. Consequently, all variables that are naturally related to religion are potential confounding variables. These include such factors as social class, size of the family in which one grew up, and education, to name a few. It is impossible to conclude unambiguously that a causal relationship exists between religion and ideal family size. Religion per se might have no effect on ideal family size, and the observed relationship might simply be a function of the causal influence of one or more confounding variables.

A large number of disturbance variables were uncontrolled in the study. For instance, research has shown that ideal family size is influenced by one's religiosity (that is, how religious one is). Individuals within a given religion in this study almost certainly differed in religiosity, and this would, in turn, create within-group variability. One minus eta-squared provides an index of the extent to which disturbance variables have influenced the dependent measure. Specifically, 1.00 minus eta-squared represents the proportion of variability in the dependent variable that can be attributed to disturbance variables. For the religion and ideal family size data, this equals $1.00 - .44 = .56$. Thus, more than half of the variability in ideal family size is due to disturbance variables. Certainly, this could have been reduced through additional control procedures.

The results of the study must also be considered in terms of their generalizability. No details were given about who the participants were. If they were all college students, this would suggest limitations on the generalizability of the findings. Research reports should provide reasonable descriptions of the nature of the samples studied.

12.7   Numerical Example

An important issue in court cases is the validity of eyewitness testimony. Behavioral scientists have suggested that eyewitness accounts can be influenced by many psychological factors. One of these is the way in which a question is phrased to an eyewitness. If subtle wording of questions can influence the answer an eyewitness gives, then such reports must be interpreted with considerable caution. Consider the following hypothetical experiment.

Twenty individuals watched a film of a car accident in which car A ran through a stop sign and hit car B. Car A was traveling at 20 miles per hour. The entire incident was filmed, including the arrival of a police officer and the citation of the motorist who was in the wrong. After watching the film, each individual was asked to estimate the speed of car A at the moment of impact (they did not know the actual speed of the car). The question was phrased in four different ways. The first five individuals were asked, "How fast was car A going at the time of the accident with car B?" The second five individuals were asked, "How fast was car A going when it hit car B?" An additional five individuals were asked, "How fast was car A going when it crashed into car B?" Finally, the last five individuals were asked, "How fast was car A going when it smashed into car B?" The issue of interest is whether estimates of the car's speed vary as a function of the specific wording used in asking about the accident. The speed estimates are provided in Table 12.4.

The null hypothesis is that the phrasing of the question does not influence the speed estimates. The alternative hypothesis is that the phrasing of the question does influence the speed estimates. These hypotheses are formally stated as

$$H_0: \quad \mu_A = \mu_H = \mu_C = \mu_S$$

$H_1:$ The four population means are not all equal

TABLE 12.4 **Data and Computation of Sums of Squares for Phraseology and Speed Estimation Experiment**

Accident		Hit		Crashed		Smashed	
X	X^2	X	X^2	X	X^2	X	X^2
18	324	23	529	25	625	29	841
20	400	20	400	27	729	28	784
17	289	22	484	26	676	30	900
19	361	19	361	23	529	27	729
21	441	21	441	24	576	31	961
$T_A = 95$		$T_H = 105$		$T_C = 125$		$T_S = 145$	
$\overline{X}_A = 19.00$		$\overline{X}_H = 21.00$		$\overline{X}_C = 25.00$		$\overline{X}_S = 29.00$	
$T_A^2 = 9,025$		$T_H^2 = 11,025$		$T_C^2 = 15,625$		$T_S^2 = 21,025$	

$$\sum X^2 = 324 + 400 + \cdots + 729 + 961 = 11,380$$

$$\sum X = 18 + 20 + \cdots + 27 + 31 = 470$$

$$\frac{(\sum X)^2}{N} = \frac{470^2}{20} = 11,045$$

$$\frac{\sum T_j^2}{n} = \frac{9,025 + 11,025 + 15,625 + 21,025}{5} = 11,340$$

where the subscript "A" represents the "accident" phraseology, the subscript "H" represents the "hit" phraseology, the subscript "C" represents the "crashed" phraseology, and the subscript "S" represents the "smashed" phraseology.

Intermediate statistics necessary for calculating the sums of squares are given in Table 12.4. The sum of squares between is

$$SS_{BETWEEN} = \frac{\sum T_j^2}{n} - \frac{(\sum X)^2}{N}$$

$$= 11{,}340 - 11{,}045 = 295$$

The sum of squares within is

$$SS_{WITHIN} = \sum X^2 - \frac{\sum T_j^2}{n}$$

$$= 11{,}380 - 11{,}340 = 40$$

Finally, the sum of squares total is

$$SS_{TOTAL} = \sum X^2 - \frac{(\sum X)^2}{N}$$

$$= 11{,}380 - 11{,}045 = 335$$

The degrees of freedom are

$$df_{BETWEEN} = k - 1 = 4 - 1 = 3$$

$$df_{WITHIN} = N - k = 20 - 4 = 16$$

$$df_{TOTAL} = N - 1 = 20 - 1 = 19$$

The relevant mean squares are

$$MS_{BETWEEN} = \frac{SS_{BETWEEN}}{df_{BETWEEN}} = \frac{295}{3} = 98.33$$

$$MS_{WITHIN} = \frac{SS_{WITHIN}}{df_{WITHIN}} = \frac{40}{16} = 2.50$$

Thus, the F ratio is

$$F = \frac{MS_{BETWEEN}}{MS_{WITHIN}} = \frac{98.33}{2.50} = 39.33$$

These calculations yield the following summary table:

Source	SS	df	MS	F
Between	295.00	3	98.33	39.33
Within	40.00	16	2.50	
Total	335.00	19		

The first question is whether there is a relationship between the type of wording used and the speed estimates. For an alpha level of .05 and 3 and 16 degrees of freedom, the critical value of F is 3.24. The observed value of F is 39.33. This exceeds the critical value, so we reject the null hypothesis and conclude that a relationship exists between the two variables.

The second question concerns the strength of the relationship. This is indexed by eta-squared and is computed as follows:

$$\text{eta}^2 = \frac{SS_{BETWEEN}}{SS_{TOTAL}} = \frac{295}{335} = .88$$

This represents a strong effect and indicates that 88% of the variability in speed estimates is associated with the way in which the question was phrased.

The final question concerns the nature of the relationship. This requires application of the HSD test. The critical difference is derived using Equation 12.17. From Appendix G, the value of q for an overall alpha level of .05, $df_{WITHIN} = 16$, and $k = 4$ is 4.05. Thus,

$$CD = q\sqrt{\frac{MS_{WITHIN}}{n}} = 4.05\sqrt{\frac{2.50}{5}} = 2.86$$

The HSD procedure can now be applied as follows:

Null hypothesis tested	Absolute difference between sample means	Value of CD	Null hypothesis rejected?
$\mu_A = \mu_H$	$\lvert 19.00 - 21.00 \rvert = 2.00$	2.86	No
$\mu_A = \mu_C$	$\lvert 19.00 - 25.00 \rvert = 6.00$	2.86	Yes
$\mu_A = \mu_S$	$\lvert 19.00 - 29.00 \rvert = 10.00$	2.86	Yes
$\mu_H = \mu_C$	$\lvert 21.00 - 25.00 \rvert = 4.00$	2.86	Yes
$\mu_H = \mu_S$	$\lvert 21.00 - 29.00 \rvert = 8.00$	2.86	Yes
$\mu_C = \mu_S$	$\lvert 25.00 - 29.00 \rvert = 4.00$	2.86	Yes

Inspection of the table suggests the following conclusions: Speed estimates obtained when the question is phrased in terms of car A smashing into car B ($\overline{X}_S = 29.00$) will be higher than speed estimates obtained when the question is phrased in terms of the two cars being involved in an accident ($\overline{X}_A = 19.00$), car A hitting car B ($\overline{X}_H = 21.00$), or car A crashing into car B ($\overline{X}_C = 25.00$). Furthermore, speed estimates obtained using the "crash" phraseology will be higher than such estimates obtained using either the "accident" or the "hit" phraseology. However, we cannot confidently conclude that the "accident" phraseology affects speed estimates relative to the "hit" phraseology.

You should think about the experiment in terms of basic research design questions: What confounding variables might be operating? What has been the role of disturbance variables? What kinds of procedures could be used to reduce sampling error? What are the potential limitations of the experiment in terms of generalizability? All these questions are critical to drawing appropriate conclusions from the study.

12.8 Planning an Investigation Using One-Way Between-Subjects Analysis of Variance

Appendix E.2 contains tables of the per-group sample sizes necessary to achieve various levels of power for one-way between-subjects analysis of variance. A set of tables is presented for each research design involving a different number of degrees of freedom between. Table 12.5 reproduces a portion of this appendix for $df_{BETWEEN} = 2$ (that is, for three groups) and an alpha level of .05. The first column presents various values of power, and the column headings are values of eta-squared in the population. To illustrate the use of this table, suppose the desired power level is .80 and the researcher suspects that the strength of the relationship in the population corresponds to an eta-squared of .15. Then the number of participants that should be sampled in *each group* is 19. Inspection of Table 12.5 and Appendix E.2 provides a general appreciation for the relationships among $df_{BETWEEN}$, the alpha level, sample size, the strength of the relationship one is trying to detect in the population, and power for one-way between-subjects analysis of variance.

12.9 Method of Presentation

Reports of a one-way between-subjects analysis of variance should include statements of the alpha level that was used, the degrees of freedom between, the degrees of freedom within, the observed value of F, the significance level, the sample means, and the standard deviation estimates. In addition, if the analysis is statistically significant, the strength and nature of the relationship between the independent and dependent variables should be addressed.

The results for the study on religion and ideal family size discussed earlier in this chapter might be presented as follows:

TABLE 12.5 Approximate Sample Sizes Necessary to Achieve Selected Levels of Power for $df_{BETWEEN} = 2$ and Alpha = .05 As a Function of Population Values of Eta-Squared

	Population Eta-Squared									
Power	.01	.03	.05	.07	.10	.15	.20	.25	.30	.35
.10	22	8	5	4	3	2	2	2	—	—
.50	165	55	32	23	16	10	8	6	5	4
.70	255	84	50	35	24	16	11	9	7	6
.80	319	105	62	44	30	19	14	11	9	7
.90	417	137	81	57	39	25	18	14	11	9
.95	511	168	99	69	47	30	22	16	13	11
.99	708	232	137	96	65	41	29	22	18	14

Results

A one-way analysis of variance compared the mean ideal family sizes of Catholics, Jews, and Protestants. The alpha level was .05. This test was found to be statistically significant, $\underline{F}$(2, 18) = 7.00, $\underline{p}$ < .01. The strength of the relationship, as indexed by eta^2, was .44. A Tukey HSD test indicated that the mean for Catholics ($\underline{M}$ = 3.00, $\underline{SD}$ = .82) was significantly greater than the mean for Jews ($\underline{M}$ = 1.00, $\underline{SD}$ = .82). The mean for Protestants ($\underline{M}$ = 2.00, $\underline{SD}$ = 1.29) did not significantly differ from the mean for either of these groups.

The first sentence identifies the statistical technique that was used to analyze the data. The term *between-subjects* is not used to describe the analysis of variance because, unless stated otherwise, it is assumed that a between-subjects analysis was performed.

The second sentence states the alpha level, and the third sentence states the results for the test of a relationship. The numbers within the parentheses are the degrees of freedom between and the degrees of freedom within, respectively. This is followed by the observed value of *F* and the significance level. Sometimes the researcher will also report a formal summary table as described earlier, although this is becoming rarer because of cost and limited journal space.

The fourth sentence indicates the strength of the relationship. In practice, this is rarely reported. Fortunately, however, eta-squared can be derived from the degrees of freedom and the observed value of *F* using Equation 12.16.

The last two sentences present the results of the HSD test, including the sample means, and the standard deviation estimates.* The intermediate statistics computed for the application of a multiple comparison procedure are not reported.

12.10 Examples from the Literature

The Effectiveness of Different Incentives for Learning

In the early 1900s, many educators and parents assumed that the threat of punishment or the actual application of punishment was the most effective means for motivating children to learn. More recently, the dominant educational philosophy seems to advocate rewards as an incentive for learning.

Hurlock (1925) reported a study relevant to this issue in which fourth- and sixth-grade students were divided into four groups. Each group took addition tests in class on five successive days. Students in the first group were separated from the other students and told to work on these tests as usual. This constituted the con-

---

* The variance estimates were identified previously to be .67 for Catholics and Jews and 1.67 for Protestants. The square roots of these values are .82 and 1.29, respectively.

trol condition. Students in another group were brought to the front of the room each day before the test was given and praised for their good work. Students in the third group were also brought to the front of the room, but these students were reproved (reprimanded) for their poor work. The students in the fourth group were ignored. Of interest to Hurlock was whether the different incentives had different effects on performance as indicated by scores on the final addition test. Since students were randomly assigned to the four conditions, group means were expected to reflect the effect of the type of incentive on learning.

Hurlock's investigation was conducted before the technique of analysis of variance had been developed. However, Kerlinger (1973) analyzed the Hurlock data using one-way analysis of variance and found the null hypothesis to be rejected, $F(3, 102) = 10.08$, $p < .001$, indicating that the type of incentive is related to performance. The strength of the relationship, as indexed by eta-squared, was .23. This represents a strong effect. An HSD test indicated that the mean for the praised group ($\overline{X} = 20.22$) was significantly greater than the mean for each of the other three groups ($\overline{X} = 11.35$ for the control group, $\overline{X} = 12.38$ for the ignored group, $\overline{X} = 14.19$ for the reproved group). The means for these latter groups, however, did not significantly differ. Thus, consistent with a reward philosophy, the most effective incentive in this experiment was praise.

Body Posture of Message Recipients and Susceptibility to Influence

Research in nonverbal behavior has shown that the body posture of a communicator can affect his or her ability to influence the attitudes of the message recipients. A related issue, but one that has received far less attention, concerns how susceptibility to influence is affected by the recipient's body posture. In one of the few studies on this topic, Petty, Wells, Heesacker, Brock, and Cacioppo (1983) had 78 college students listen to a persuasive message advocating a 20% tuition increase at their university while either standing, sitting, reclining on a cushioned table, or reclining on an uncushioned table. Following exposure to this message, the students responded to the question, "In general, to what extent do you agree that the tuition should be increased?" Responses were made on a scale of 1 to 12, with higher scores indicating greater agreement.

A one-way analysis of variance applied to these data was statistically significant, $F(3, 74) = 3.33$, $p < .025$. As indexed by eta-squared, the strength of the relationship was .12. This represents a moderate effect. Multiple comparisons showed that students who had heard the persuasive communication while reclining on the cushioned table ($\overline{X} = 7.60$) reported significantly greater positive attitudes than did students who had heard this message while standing ($\overline{X} = 5.63$). The means for the sitting ($\overline{X} = 6.00$) and uncushioned reclining ($\overline{X} = 6.95$) conditions did not significantly differ from each other or from the means reported above.

A follow-up study by Petty and colleagues suggests that "a reclining posture facilitates message-relevant thinking over a standing posture and thereby enhances the importance of message content in producing persuasion" (p. 219). According to this perspective, reclining individuals are better able to differentiate strong from

Applications

to the Analysis

of a Social Problem

As noted in Chapter 11, an important variable in predicting teen problem behaviors is the quality of the relationship between parent and child: The poorer the teen's relationship with the parent, the more likely the teen is to engage in problem behaviors, everything else being equal. One question of theoretical interest is how the quality of the relationship between parent and teen changes as a function of the adolescent's age. Some theorists argue that the relationship between parent and teen becomes increasingly strained as teens progress from the early to the later stages of adolescence. With each passing year, the teen is closer to adulthood. This transition to adulthood places strain on parents as children begin to assert their independence and develop their own lifestyles. The quality of the parent–teen relationship is thus negatively influenced by the teen's increasing rejection of the traditional parent–child roles.

An alternative point of view argues just the opposite: The quality of the relationship between parent and teen improves with age. According to this view, during the younger years of adolescence, teens experience stress and strain as they undergo puberty and the tremendous physical and hormonal changes that result from it. These stresses detract from the quality of the parent–teen relationship as the teen tries to cope with the outside world. In addition, it is at the younger ages when adolescents typically initiate independence and a "break" from the traditional "parent-as-all-knowing" role that has typified their upbringing. In later adolescence, the teen has already dealt with the harsh stresses of the adolescent growth spurt. In addition, parents and teens have already resolved their differing expectations and have effectively moved from the now inappropriate "parent–child" relationship to the "parent–emerg-

ing adult" relationship. Thus, the quality of the parent–teen relationship should improve as a function of age.

We evaluated the opposing predictions made by these theories by applying a one-way between-subjects analysis of variance to data collected in the parent–teen communication study described in Chapter 1. We first asked adolescents a series of questions designed to measure how satisfied they were with their relationship with their mother. Specifically, we asked them to indicate how much they agreed or disagreed with each of 11 statements about different facets of their maternal relationship (e.g., "I am satisfied with the way my mother and I communicate with each other," "I am satisfied with the way my mother and I resolve conflicts," "I am satisfied with the respect my mother shows me," "I am satisfied with the emotional support my mother gives me"). Each statement was rated on a five-point scale having the response options strongly disagree, moderately disagree, neither agree nor disagree, moderately agree, and strongly agree. Responses were scored from 1 to 5, with higher values indicating greater agreement with the statement. After performing item analyses to ensure it was appropriate psychometrically, we summed the responses across the 11 statements to yield an overall index of relationship satisfaction. Scores on this measure could range from 5 to 55, with higher values indicating greater satisfaction. This constituted our dependent variable. Although the measure may not be strictly interval, our previous research suggested that the approximation is sufficiently close that the use of analysis of variance would not be inappropriate (see Davison & Sharma, 1988).

The independent variable, age of the teen, had four levels: 14, 15, 16, and 17 years old. The

weak arguments than are standing individuals and are thus more susceptible to influence when persuasive messages are compelling but less susceptible to influence when persuasive messages are specious. The exact reasons for this, however, are unclear.

sample sizes in the four groups were 114, 270, 220, and 133, respectively. These represent large sample sizes per group, which indicates that the analysis will be relatively robust to assumption violations. The unequal sample sizes suggest that some caution be exercised, however, because the robustness of the F test can be weakened by unequal n values.

The next step in the analysis was to examine the response distributions for the satisfaction measure for each of the four age groups separately. Grouped frequency histograms are presented in Figure 12.2 with a normal distribution superimposed on each. In all cases, the satisfaction scores were strongly negatively skewed. Although the F test is probably robust even to the levels of skewness observed here (given our sample sizes), we must exercise some caution in our interpretation.

Because a few 14-year-olds seemed to have abnormally low scores, we double checked the coding of the data and the responses these individuals made to other questions. This revealed neither coding errors nor anything unique about these potential "outliers" relative to other teens. Because they represented only a few cases (and because additional, more advanced analyses we performed indicated the results were unaffected by their presence), we decided to proceed with the analysis. In addition, despite the skewness in the sample data, we felt that the sample deviations from normality were not marked enough to indicate population nonnormality that would undermine the analysis of variance (see Harwell, Rubinstein & Hayes, 1992; Milligan, Wong & Thompson, 1987). However, the strong skew also led us to supplement the present *(continued)*

FIGURE 12.2   Grouped Frequency Histograms of Satisfaction with Maternal Relationship for Four Age Groups

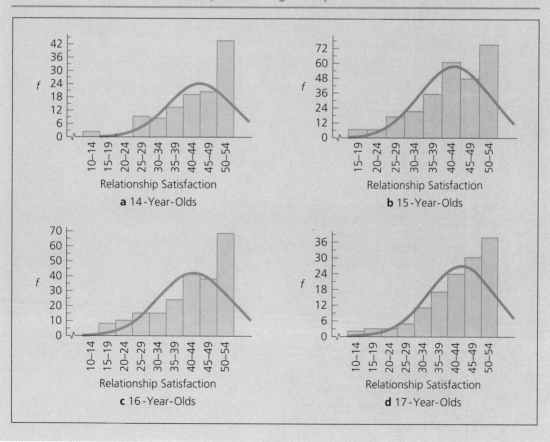

analyses with a nonparametric test. These results are discussed in Chapter 16.

A computer printout of the analysis of variance results is presented in Table 12.6. The first section contains the value of the F statistic, the degrees of freedom, and the significance level for the Levene test of the assumption of homogeneous variances discussed in Chapter 10. The test is performed by computing a one-way between-subjects analysis of variance on the absolute difference between each individual's score and the mean for that individual's group. Of particular interest is the p value yielded by the analysis (labeled "2-tail Sig"). If this value is less than .05, the null hypothesis of equal population variances across age groups will be rejected. In this instance, the p value was .358, so we fail to reject the null hypothesis of equal population variances.

Despite this nonsignificant result, we cannot accept the null hypothesis of equal population variances based on the Levene test (recall from Chapter 8 that one never "accepts" the null hypothesis based on a statistical test). The data are *consistent* with homogeneous population variances (with the sample differences in variance estimates being due to sampling error), but we cannot positively conclude that the population variances are homogeneous. Another problem with the Levene test is that it is somewhat sensitive to nonnormality in the data. Although it is much better than alternative tests of homogeneity of variance in this regard, it still must be interpreted cautiously given nonnormal population data. The test has a tendency to yield false significance when population distributions are highly skewed.

The standard deviation estimates for each of the four age groups are presented directly beneath the Levine test output in Table 12.6, along with group sizes (labeled "Count"), means, and estimated standard errors of the mean. The standard

TABLE 12.6   Computer Output for One-Way Between-Subjects Analysis of Variance for Satisfaction with Maternal Relationship

```
Variable SMOTHER
By Variable  AGE

Levene Test for Homogeneity of Variances

    Statistic     df1        df2          2-tail Sig.
     1.0771         3         733             .358

                                    Standard    Standard
Group           Count      Mean    Deviation     Error

Grp14            114     44.9298    9.4651        .8865
Grp15            270     43.2148    9.3370        .5682
Grp16            220     43.0773   10.3949        .7008
Grp17            133     43.7368    9.9285        .8609

Total            737     43.5332    9.7907        .3606
```

Analysis of Variance

Source	D.F.	Sum of Squares	Mean Squares	F Ratio	F Prob.
Between Groups	3	300.9804	100.3268	1.0468	.3711
Within Groups	733	70250.4552	95.8396		
Total	736	70551.4355			

deviation estimates (and thus the variance estimates) are fairly homogeneous; the largest standard deviation estimate is only 1.11 times larger than the smallest. Under these circumstances, and given the large sample sizes and the statistically nonsignificant Levene test, heterogeneity of variance is not problematic for the F test (see Milligan, Wong & Thompson, 1987). In general, all estimated standard errors were relatively small.

The summary table for the analysis of variance is presented in the last section. The F ratio was statistically nonsignificant, as indicated by the fact that the p value (labeled "F Prob") of .37 is higher than the .05 alpha level. We fail to reject the null hypothesis of no age differences in relationship satisfaction. The value of eta-squared was .004.

Given that we failed to reject the null hypothesis, there is the risk of a Type II error. To evaluate this possibility, we calculated the statistical power for a one-way between-subjects analysis of variance having an alpha level of .05, four groups, and the sample sizes in our study for what we considered to be a small effect size, a population eta-squared of .03. We used the procedure described in Cohen (1988) to calculate power, although a rough estimate can be obtained by using the tables in Appendix E.2.* The statistical power exceeds .95, suggesting that a Type II error is unlikely.

The data are not consistent with the predictions made by either of the theories described earlier. There is no evidence that age is related to teen satisfaction with the maternal relationship. It might be that *both* processes described by the two competing theories are operating and that the effects of each process cancel each other out, yielding the near-equal mean scores as a function of age. Future research or more fine-grained analyses can address this possibility.

---

* In Appendix E.2, the table for $df_{BETWEEN} = 3$ and an alpha level of .05 shows that the power for a population eta-squared of .03 will exceed .95 if there are 140 cases in each group. The average sample size in our example was $(114 + 270 + 220 + 133)/4 = 184.25$. This suggests that our power also exceeds .95, consistent with the conclusion based on Cohen's procedure.

Summary

One-way between-subjects analysis of variance is typically used to analyze the relationship between two variables when (1) the dependent variable is quantitative in nature and is measured on a level that at least approximates interval characteristics, (2) the independent variable is between-subjects in nature, and (3) the independent variable has three or more levels. The essence of one-way analysis of variance is the partitioning of variability into between-group and within-group components. Between-group variability reflects both sampling error and the effect of the independent variable on the dependent variable. Within-group variability reflects just sampling error. The ratio of variance measures (mean squares) based on these sources of variability is called the F ratio. When certain conditions are met, the F ratio has a sampling distribution that closely approximates an F distribution. This latter distribution is the basis for the F test used to test for a relationship between the independent and dependent variables. The strength of the relationship is measured using eta-squared, and the nature of the relationship is analyzed using the Tukey HSD test. The HSD test ensures that the probability of making one or more Type I errors in the set of multiple comparisons will be equal to alpha.

Appendix 12.1 Rationale for the Degrees of Freedom

Unlike the sum of squares discussed in Chapter 7, the sum of squares between does not have $N - 1$ degrees of freedom associated with it. This is because the sum of squares between is based on deviations of the k group means from the grand mean. Since the sum of signed deviations about a mean (in this case, the grand mean) will always equal 0 (see Chapter 3 for the logic underlying this property), if all but one of the group means are known, then the last one is not free to vary. Thus, the degrees of freedom between is $k - 1$.

The sum of squares within is based on the deviation of scores about their respective group means. Within each group, there will be $n - 1$ degrees of freedom. Given k groups, this yields $k(n - 1)$ degrees of freedom, which can be rewritten as $kn - k$. Since $kn = N$, the degrees of freedom within is $N - k$.

Exercises

Answers to asterisked (*) exercises appear at the back of the book. Answers to exercises with two asterisks are also worked out step by step in the Study Guide.

Exercises to Review Concepts

1. Under what conditions is one-way between-subjects analysis of variance typically used to analyze a bivariate relationship?
*2. What general form does the alternative hypothesis take for one-way between-subjects analysis of variance? Why can't it be summarized in a single mathematical statement?
*3. Distinguish between between-group variability and within-group variability.
4. What does between-group variability reflect? Why?
5. What does within-group variability reflect? Why?
*6. Under what circumstance will the F ratio, over the long run, approach 1.00? Under what circumstance will the F ratio, over the long run, be greater than 1.00?
7. Distinguish among the sum of squares total, the sum of squares between, and the sum of squares within. How are they interrelated?
*8. Consider the following scores in an experiment involving three conditions—A, B, and C:

A	B	C
3	5	7
3	5	7
3	5	7
3	5	7
3	5	7

Without actually computing the sum of squares within, what must its value be? Why?
*9. What is the relationship between the mean square between and the sum of squares between? What is the relationship between the mean square within and the sum of squares within?
10. What is the relationship between the mean square within and the pooled variance estimate?
*11. State the critical value of F for a one-way between-subjects analysis of variance for an alpha level of .05 under each of the following conditions:

a. $k = 3$, $n = 7$
b. $k = 4$, $n = 5$
c. $k = 3$, $n = 10$
d. $k = 5$, $n = 15$

12. What are the assumptions underlying one-way between-subjects analysis of variance?

*13. Insert the missing entries in the summary table for a one-way analysis of variance with three levels of the independent variable and $n = 20$.

Source	SS	df	MS	F
Between	—	—	—	—
Within	152.00	—	—	
Total	182.00	—		

14. Insert the missing entries in the summary table for a one-way analysis of variance with four levels of the independent variable:

Source	SS	df	MS	F
Between	—	—	18.00	3.60
Within	—	—	—	
Total	—	23		

15. What is the advantage, following rejection of the null hypothesis, of determining the nature of the relationship between an independent variable and a dependent variable with the Tukey HSD test rather than with multiple t tests?

**16. An investigator wanted to test the effect of marital status on attitudes toward divorce. A scale measuring attitudes on this issue was administered to ten single, ten married, and ten divorced individuals. Scores could range from 1 to 12, with higher values representing more positive attitudes. The means and summary table are given here:

$$\overline{X}_S = 6.00 \qquad \overline{X}_M = 8.00 \qquad \overline{X}_D = 10.00$$

Source	SS	df	MS	F
Between	80.00	2	40.00	8.00
Within	135.00	27	5.00	
Total	215.00	29		

Analyze the nature of the relationship between marital status and attitudes toward divorce using the Tukey HSD test.

Use the following information to complete Exercises 17–19.

Suppose one of "Nader's Raiders" was interested in comparing the performance of three types of cars, X, Y, and Z. Random samples of five owners were drawn from the list of owners of each model. These owners were asked how many times their cars had undergone major repairs in the past 2 years. The data follow:

Car X	Car Y	Car Z
2	5	9
1	4	6
2	3	3
3	4	7
2	4	5

*17. Test for a relationship between the type of car and repair records using a one-way analysis of variance.

*18. Compute the value of eta-squared using Equation 12.15. Recalculate eta-squared using Equation 12.16. Compare the two results. Does the observed value represent a weak, moderate, or strong effect?

*19. Analyze the nature of the relationship between the type of car and repair records using the Tukey HSD test.

Use the following information to complete Exercises 20–22.

An investigator tested the relationship between supposed task difficulty and task performance.

Twenty-four participants worked on the identical spatial ability task, but six of these individuals were led to believe that the task was of low difficulty, six were led to believe that the task was of moderate difficulty, six were led to believe that the task was of high difficulty, and six were not given any information about the task's difficulty. Scores could range from 0 to 10, with higher values indicating better task performance. The data follow:

Low	Moderate	High	No information
8	6	4	4
7	7	1	5
5	4	2	5
8	5	4	6
9	4	6	8
7	6	3	6

20. Test for a relationship between supposed task difficulty and task performance using a one-way analysis of variance.
21. Compute the value of eta-squared. Does the observed value represent a weak, moderate, or strong effect?
22. Analyze the nature of the relationship between supposed task difficulty and task performance using the Tukey HSD test.
*23. Consider the following scores for two groups of people who were tested under either condition A or condition B:

A	B
6	12
4	10
5	11
3	9
7	13

Test for a relationship between the independent and dependent variables using the one-way analysis of variance procedures developed in this chapter. (*Note:*

Even though one-way between-subjects analysis of variance is typically used when there are three or more groups, it can also be applied to the two-group situation.)

*24. Test for a relationship between the independent and dependent variables for the data in Exercise 23 using the procedures developed in Chapter 10 for a nondirectional independent groups t test. Square the observed value of t. Compare this with the observed value of F from Exercise 23. Square the critical values of t. Compare this with the critical value of F from Exercise 23. What does this indicate about the relationship between one-way between-subjects analysis of variance and the independent groups t test in the two-group case?

For each of the studies described in Exercises 25–27, indicate the appropriate statistical test for analyzing the relationship between the independent and dependent variables. Assume that the underlying assumptions of the tests have been satisfied.

*25. Barron (1965) administered an intelligence test to creative individuals in four different occupations: mathematicians, writers, psychologists, and architects. The mean intelligence scores were compared for the four groups.

26. A researcher studied the ability of people to process information presented to their right versus their left ear. Twelve individuals participated in the experiment. Headphones were used to present simultaneously one list of 20 words to participants' right ears and a different list of 20 words to participants' left ears. After the lists were presented, the participants were asked to recall as many words from each list as they could. The mean number of words recalled from the list presented to the right ear was compared with the mean number of words recalled from the list presented to the left ear.

27. An investigator examined the effect of practice on problem-solving performance. Study participants were randomly assigned to two conditions. Seventy-five participants tried to solve 30 problems after having a 10-minute practice session on a similar set of problems. A different 75 participants tried to solve the same 30 problems, but with no practice session. The average number of correct solutions was compared for people in the practice condition and those in the no-practice condition.

*28. If a researcher suspects that the strength of the relationship between two variables in the population is .07 as indexed by eta-squared, what sample size should she use per group in a study involving four independent groups and an alpha level of .05 in order to achieve a power of .80?

29. Suppose an investigator conducted a study involving five independent groups with $n = 5$ per group. If the value of eta-squared in the population was .25, what would the power of his statistical test be at an alpha level of .05?

Multiple-Choice Questions

*30. The alternative hypothesis for one-way between-subjects analysis of variance states that
   a. the population means are all equal to one another
   b. specific population means are greater than specific other population means
   c. the sample means are not all equal to one another
   d. the population means are not all equal to one another

31. Researchers most commonly adopt an overall alpha level of .01 when applying the Tukey HSD test.
   a. true
   b. false

32. The F ratio is a ratio of _____ variability divided by _____ variability.

   a. between-group; within-group
   b. within-group; between-group
   c. between-subjects; within-subjects
   d. within-subjects; between-subjects

33. How many comparisons will be included in the set of multiple comparisons if the Tukey HSD test is applied to the five-group situation?
   a. 5   c. 8
   b. 7   d. 10

*34. Given a research report that states $F(2, 30) = 5.00$, $p < .05$, what is the strength of the relationship, as indexed by eta-squared?
   a. .06   c. .50
   b. .25   d. .96

Use the following information to complete Exercises 35–45.

A researcher examined the effects of three presentation formats (audio only, visual only, audio and visual) on recall of a speech. Thirty participants, ten in each condition, were presented either a tape recording of a speech (audio only), a written text of a speech (visual only), or both (audio and visual). After presentation of the speech, the participants tried to recall the 15 major points covered. The mean number of points recalled was 2.00 in the audio-only condition, 4.00 in the visual-only condition, and 6.00 in the audio and visual condition. The summary table follows:

Source	SS	df	MS	F
Between	80.00	2	40.00	4.00
Within	270.00	27	10.00	
Total	350.00	29		

*35. If the value of eta-squared in the population is .15, what is the power of the statistical test at an alpha level of .05?
   a. .10   c. .50
   b. .15   d. .80

*36. The independent variable is
   a. the time interval between presentation of the speech and trying to recall the major points
   b. the type of presentation format
   c. the number of participants in each condition
   d. the number of points recalled

37. The dependent variable is
   a. the time interval between presentation of the speech and trying to recall the major points
   b. the type of presentation format
   c. the number of participants in each condition
   d. the number of points recalled

*38. Based on this study, we can conclude that the independent variable is related to the dependent variable.
   a. true   b. false

*39. It is possible that the test results reflect a Type I error.
   a. true   b. false

40. It is possible that the test results reflect a Type II error.
   a. true   b. false

*41. What is the strength of the relationship, as indexed by eta-squared?
   a. .05   c. .30
   b. .23   d. .77

42. What is the proportion of variability in the dependent variable that is due to disturbance variables?
   a. .70
   b. .77
   c. .93
   d. This cannot be determined.

*43. What can we conclude about recall for the audio-only versus the visual-only presentation formats?
   a. Recall is better with the audio-only presentation.
   b. Recall is better with the visual-only presentation.
   c. Recall is the same for the two presentation formats.

d. Recall may or may not differ for the two presentation formats.

44. What can we conclude about recall for the audio-only versus the audio and visual presentation formats?
   a. Recall is better with the audio-only presentation.
   b. Recall is better with the audio and visual presentation.
   c. Recall is the same for the two presentation formats.
   d. Recall may or may not differ for the two presentation formats.

45. What can we conclude about recall for the visual-only versus the audio and visual presentation formats?
   a. Recall is better with the visual-only presentation.
   b. Recall is better with the audio and visual presentation.
   c. Recall is the same for the two presentation formats.
   d. Recall may or may not differ for the two presentation formats.

Exercises to Apply Concepts

**46. One topic of interest to psychologists is jury decision making and factors that influence the jurors' judgments of guilt. Stephen (1975) reviewed a number of studies in this area on the effect of a defendant's race. In one experiment, participants were presented a transcript of a trial and asked to indicate the probability that the defendant was guilty. Judgments were made on a 0 to 10 scale, with higher scores representing a greater perceived probability of guilt. All participants read the same transcript. However, one-third of them were told that the defendant was white, another third were told that the defendant was African-American, and the last third were told that the defendant was Hispanic. The hypothetical data in the table are representative of the results of the experiment. Analyze these data, draw a conclusion, and write up your results using the principles discussed in the Method of Presentation section.

White defendant	African-American defendant	Hispanic defendant
6	10	10
7	10	6
2	9	10
3	4	5
5	4	10
0	10	5
1	10	2
0	10	10
6	3	2
0	10	10

47. Ainsworth, Blehar, Waters, and Wall (1978) studied how infants become psychologically attached to their mothers. They delineated three qualitatively different types of attachment, which Sroufe and Waters (1977) labeled "security," "avoidance," and "ambivalence." Infants who are securely attached generally seek to be near their mother and to have contact with her. These infants, when separated from their mother, may or may not exhibit distress and are generally not anxious about being left alone. Infants who are "avoidant" tend to resist proximity and contact with the mother. Finally, infants who are "ambivalent" display proximity- and contact-seeking behaviors as well as proximity- and contact-avoiding behaviors (hence the term *ambivalent*). When separated from the mother, these infants tend to exhibit anger or become conspicuously passive until the mother returns.

In one study, Ainsworth and colleagues were interested in examining the relationship between the type of attachment exhibited by infants and maternal behavior. Trained observers watched interactions between mothers and infants and rated these interactions on a number of dimensions. One of these was the extent to which the mother was sensitive to the infant's signals and communications. Ratings were made on a 1 to 9 scale, with higher scores indicating greater sensitivity. Hypothetical data representative of the results of Ainsworth and colleagues are presented in the table. Analyze these data, draw a conclusion, and write up your results using the principles discussed in the Method of Presentation section.

Security	Avoidance	Ambivalence
5	1	3
9	5	3
9	1	1
5	1	1
9	5	3
5	5	1

48. One way in which psychologists study aggression is through the use of the *teacher–learner paradigm*. With this approach, two individuals at a time take part in a study that is presented as an investigation of the effect of punishment on learning. Unbeknownst to the actual participant, the second person in each pair is actually an experimental accomplice. The experimental procedures are such that the participant is always assigned the role of teacher and the accomplice is always assigned the role of learner, whose task is to memorize a list of word pairs. During the test phase, the teacher reads the first word of each pair, along with four response alternatives. Whenever the learner makes a mistake,

the teacher is to punish him by delivering one or more electric shocks. The total number of shocks administered during the course of the experiment constitutes the measure of aggressive behavior. To ensure that all participants have the same opportunity to aggress, the learner makes errors in a predetermined sequence. In reality, no shocks are ever actually delivered; the teacher is only led to believe that they are.

In one study using the teacher–learner paradigm, suppose that a researcher examines the effect of verbal provocation on aggression by assigning participants to conditions where the learner acts in a noninsulting, mildly insulting, moderately insulting, or highly insulting manner. Hypothetical data for this study are presented in the table. Analyze these data, draw a conclusion, and write up your results using the principles discussed in the Method of Presentation section.

Noninsulting	Mildly insulting	Moderately insulting	Highly insulting
8	12	16	21
8	16	15	16
10	14	23	27
12	10	20	18
8	16	14	32
8	8	21	27
15	10	18	23
8	8	15	24
10	17	18	30
8	15	17	26

One-Way Repeated Measures Analysis of Variance

13.1 Use of One-Way Repeated Measures Analysis of Variance

One-way repeated measures analysis of variance (abbreviated as "one-way repeated measures ANOVA") is typically used to analyze the relationship between two variables when the following conditions are met:

1. The dependent variable is quantitative in nature and is measured on a level that at least approximates interval characteristics.

2. The independent variable is *within-subjects* in nature (it can be either qualitative or quantitative).

3. The independent variable has three or more levels.

In short, one-way repeated measures analysis of variance is used under the same circumstances as one-way between-subjects analysis of variance except that the independent variable is within-subjects in nature rather than between-subjects. Just as one-way between-subjects analysis of variance is an extension of the independent groups *t* test for instances when the independent variable has more than two levels, one-way repeated measures analysis of variance is an extension of the correlated groups *t* test. Let us consider an example of an investigation that meets the conditions for this technique.

Suppose a consumer psychologist is interested in the effect of label information on the perceived quality of wine. She designs an experiment with three conditions. In each condition, an individual tastes a wine and then rates its taste. The ratings are made on a scale of 1 to 20, with higher scores indicating that the wine tasted better. In all three conditions, the wine is identical; however, in one condition the label indicates that it is a French wine, in a second condition the label indicates that it is an Italian wine, and in a third condition the label indicates that it is an American wine. The experiment is conducted as a within-subjects design in that each of the six participants tastes and rates the wines in the three different conditions. In order to counteract possible confounding variables associated with carry-over effects, the label presented first, second, and third is randomized for each participant.* The taste ratings are presented in the second, third, and fourth columns of Table 13.1.

In this experiment, the independent variable is the type of label, and it has three levels. It is within-subjects in nature because the same individuals participate in all three conditions. The dependent variable is the perceived quality of the wine as reflected in the taste ratings. It is quantitative in nature and is measured on a level that at least approximates interval characteristics. Given these conditions, one-way repeated measures analysis of variance is the statistical technique that would typically be used to analyze the relationship between the variables.

The mean score across conditions for each participant is calculated in column 5 of Table 13.1. Note that there is variability in the mean scores; some participants, on the average, rated the wines higher than others. This reflects the influence of individual differences. If we could remove this influence (which is a disturbance variable), we could increase the sensitivity of the statistical test of the relationship between the independent and dependent variables.

When discussing the correlated groups *t* test in Chapter 11, we noted that

TABLE 13.1 Data for Wine Label and Perceived Quality Experiment

Participant	French	Italian	American	$\overline{X}_i$
1	14	10	9	11.00
2	16	12	12	13.33
3	17	13	14	14.67
4	16	14	16	15.33
5	15	12	10	12.33
6	12	11	8	10.33
	$\Sigma X_F = 90$	$\Sigma X_I = 72$	$\Sigma X_A = 69$	
	$\overline{X}_F = 15.00$	$\overline{X}_I = 12.00$	$\overline{X}_A = 11.50$	

* The rationale for randomly ordering the sequence of conditions across participants as a way to deal with confounding variables associated with carry-over effects is discussed in Section 13.5.

an advantage of within-subjects designs is their ability to systematically remove variability due to individual differences from the dependent variable, and we developed the logic of extracting this source of variability in the two-group case. The same procedures can be used to generate a set of nullified scores when a within-subjects independent variable has more than two levels. The nullified data can then be analyzed to determine whether a relationship exists between the independent variable and the dependent variable after the effects of individual background have been removed. The result of this analysis would be identical to the result that would be obtained using the more computationally efficient procedure for one-way repeated measures analysis of variance described in the following section. As with the correlated groups *t* test, the computational approach uses raw rather than nullified scores. For illustration, we return to the wine label and perceived quality experiment.

13.2 Inference of a Relationship Using One-Way Repeated Measures Analysis of Variance

Null and Alternative Hypotheses

The null hypothesis posits that the type of label does not influence the perceived quality of the wine. The alternative hypothesis states that there is a relationship between the two variables. These hypotheses are formally phrased as

H_0: $\mu_F = \mu_I = \mu_A$

H_1: The three population means are not all equal

where the subscripts "*F*," "*I*," and "*A*" denote the French, Italian, and American labels, respectively.

Partitioning of Variability

In Chapter 12, we demonstrated that the total variability in the dependent variable in a between-subjects design (the sum of squares total) can be partitioned into two components, one reflecting the influence of the independent variable (the sum of squares between) and one reflecting the influence of disturbance variables (the sum of squares within). We can also identify and partition the total variability when a repeated measures design is used. As before, the total variability in the dependent variable across all individuals and all conditions can be represented by the sum of squares total. This can be partitioned into three components: one reflecting the influence of the independent variable (called the **sum of squares IV**), one reflecting the influence of individual differences (called the **sum of squares across subjects**), and one reflecting the influence of disturbance variables other than individual differences (called the **sum of squares error**).* Symbolically,

---

* Some textbooks refer to the sum of squares across subjects as the *sum of squares between subjects*. However, we have found this latter terminology to be confusing to students.

$$SS_{TOTAL} = SS_{IV} + SS_{ACROSS\ SUBJECTS} + SS_{ERROR} \qquad [13.1]$$

The sum of squares IV is conceptually equivalent to the sum of squares between in the between-subjects design discussed in Chapter 12. The sum of squares error and the sum of squares within are also conceptually equivalent because both reflect only the influence of disturbance variables.

The sum of squares across subjects has no counterpart in a between-subjects analysis of variance design. In fact, it is this component that differentiates the two approaches. When a between-subjects design is used (as in Chapter 12), each score is provided by a different person, so it is not possible to estimate the effects of individual differences. Differences in background (including such things as affinity for wine) thus contribute to sampling error, as reflected in the sum of squares within. However, when a repeated measures design is used, the influence of individual differences can be estimated and statistically removed from the dependent variable. This is reflected in the sum of squares across subjects and explains why a repeated measures analysis of variance is usually a more sensitive test of the relationship between the independent and dependent variables than is a between-subjects analysis of variance: The sum of squares within in a between-subjects design includes the effects of individual differences, whereas the sum of squares error in a repeated measures design does not. Thus, given a common data set, the mean square error, which forms the denominator in the F test of a relationship between the independent and dependent variables in the repeated measures case, will tend to be smaller than the mean square within, which forms the denominator of the F test in the between-subjects case. To the extent that the mean square error is in fact smaller than the mean square within, a larger F ratio and a greater likelihood of rejecting the null hypothesis should result.*

Computation of the Sums of Squares

The formula for calculating the sum of squares total is

$$SS_{TOTAL} = \sum X^2 - \frac{(\sum X)^2}{kN} \qquad [13.2]$$

where $\sum X^2$ is the *sum of the squared scores* constituting the data set, $(\sum X)^2$ is the *square of the summed scores*, k is the number of levels of the independent variable (that is, the number of conditions), and N is the sample size.

The data for the wine label and perceived quality experiment have been reproduced in Table 13.2, where it can be seen that the sum of the 18 X^2 scores is

$$\sum X^2 = 196 + 256 + \cdots + 100 + 64 = 3,081$$

and the sum of the 18 X scores is

$$\sum X = 14 + 16 + \cdots + 10 + 8 = 231$$

---

* We return to the issue of the sensitivity of repeated measures versus between-subjects analysis of variance later in this chapter.

TABLE 13.2 Data and Computation of Sums of Squares for Wine Label and Perceived Quality Experiment

Participant	French		Italian		American		s_i	s_i^2
	X	X^2	X	X^2	X	X^2		
1	14	196	10	100	9	81	33	1,089
2	16	256	12	144	12	144	40	1,600
3	17	289	13	169	14	196	44	1,936
4	16	256	14	196	16	256	46	2,116
5	15	225	12	144	10	100	37	1,369
6	12	144	11	121	8	64	31	961

$$T_F = 90 \qquad T_I = 72 \qquad T_A = 69$$
$$T_F^2 = 8{,}100 \qquad T_I^2 = 5{,}184 \qquad T_A^2 = 4{,}761$$

$$\sum X^2 = 196 + 256 + \cdots + 100 + 64 = 3{,}081$$

$$\sum X = 14 + 16 + \cdots + 10 + 8 = 231$$

$$\frac{(\sum X)^2}{kN} = \frac{231^2}{(3)(6)} = 2{,}964.50$$

$$\frac{\sum T_j^2}{N} = \frac{8{,}100 + 5{,}184 + 4{,}761}{6} = 3{,}007.50$$

$$\frac{\sum s_i^2}{k} = \frac{1{,}089 + 1{,}600 + 1{,}936 + 2{,}116 + 1{,}369 + 961}{3} = 3{,}023.67$$

Squaring the latter value and dividing by kN, we obtain

$$\frac{(\sum X)^2}{kN} = \frac{231^2}{(3)(6)} = 2{,}964.50$$

Thus, the sum of squares total in this example is

$$SS_{TOTAL} = \sum X^2 - \frac{(\sum X)^2}{kN}$$

$$= 3{,}081 - 2{,}964.50 = 116.50$$

The formula for calculating the sum of squares IV is

$$SS_{IV} = \frac{\sum T_j^2}{N} - \frac{(\sum X)^2}{kN} \qquad [13.3]$$

where T_j^2 is the square of the sum of the scores in condition j and all other terms are as defined previously. In our example, T_j^2 equals 8,100 for the French label, 5,184 for the Italian label, and 4,761 for the American label. Thus,

$$\frac{\sum T_j^2}{N} = \frac{8{,}100 + 5{,}184 + 4{,}761}{6} = 3{,}007.50$$

so

$$SS_{IV} = \frac{\sum T_j^2}{N} - \frac{(\sum X)^2}{kN}$$

$$= 3{,}007.50 - 2{,}964.50 = 43.00$$

The formula for calculating the sum of squares across subjects is

$$SS_{ACROSS\ SUBJECTS} = \frac{\sum s_i^2}{k} - \frac{(\sum X)^2}{kN} \qquad [13.4]$$

where s_i^2 is the square of the sum of the scores of individual i and all other terms are as defined previously. For instance, the first participant in Table 13.2 obtained a score of 14 in the French condition, 10 in the Italian condition, and 9 in the American condition. The value of s_i^2 for this individual is thus $(14 + 10 + 9)^2 = 1{,}089$. The s_i^2 values for the other participants have been calculated in the last column of Table 13.2. We find that s_i^2 equals 1,600 for the second participant, 1,936 for the third, 2,116 for the fourth, 1,369 for the fifth, and 961 for the sixth. Thus,

$$\frac{\sum s_i^2}{k} = \frac{1{,}089 + 1{,}600 + 1{,}936 + 2{,}116 + 1{,}369 + 961}{3} = 3{,}023.67$$

so

$$SS_{ACROSS\ SUBJECTS} = \frac{\sum s_i^2}{k} - \frac{(\sum X)^2}{kN}$$

$$= 3{,}023.67 - 2{,}964.50 = 59.17$$

The formula for calculating the sum of squares error is

$$SS_{ERROR} = \sum X^2 + \frac{(\sum X)^2}{kN} - \frac{\sum T_j^2}{N} - \frac{\sum s_i^2}{k} \qquad [13.5]$$

where all terms are as defined above. In our example,

$$SS_{ERROR} = 3{,}081 + 2{,}964.50 - 3{,}007.50 - 3{,}023.67 = 14.33$$

Of course, given that the sum of squares total is equal to the sum of squares IV plus the sum of squares across subjects plus the sum of squares error, if three of these quantities have already been calculated, the fourth can be determined through simple algebraic manipulation. For instance, we could have obtained the sum of squares error as follows:

$$SS_{ERROR} = SS_{TOTAL} - SS_{IV} - SS_{ACROSS\ SUBJECTS}$$

$$= 116.50 - 43.00 - 59.17 = 14.33$$

This relationship reflects the fact that, as noted above, the sum of squares error represents the influence of disturbance variables other than individual differences—that is, variability in the dependent variable that remains after the effects of the independent variable and individual differences have been partitioned out. However, it usually is a good idea to calculate all four sums of squares using the above formulas. This allows us to check for computational errors by determining whether the calculated values of the sum of squares IV, the sum of squares across subjects, and the sum of squares error add up to the calculated value of the sum of squares total.

Derivation of the Summary Table

Each of the above sums of squares has a certain number of degrees of freedom associated with it. These are calculated as follows:*

$$df_{IV} = k - 1 \tag{13.6}$$

$$df_{ACROSS\ SUBJECTS} = N - 1 \tag{13.7}$$

$$df_{ERROR} = (k - 1)(N - 1) \tag{13.8}$$

As with the sums of squares, the degrees of freedom are additive:

$$df_{TOTAL} = df_{IV} + df_{ACROSS\ SUBJECTS} + df_{ERROR} \tag{13.9}$$

Since $df_{IV} = k - 1$, $df_{ACROSS\ SUBJECTS} = N - 1$, and $df_{ERROR} = (k - 1)(N - 1)$, and $(k - 1) + (N - 1) + [(k - 1)(N - 1)] = k - 1 + N - 1 + (kN - k - N + 1) = kN - 1$, the degrees of freedom total can be calculated directly as

$$df_{TOTAL} = kN - 1 \tag{13.10}$$

In our example,

$$df_{IV} = 3 - 1 = 2$$

$$df_{ACROSS\ SUBJECTS} = 6 - 1 = 5$$

$$df_{ERROR} = (3 - 1)(6 - 1) = 10$$

$$df_{TOTAL} = (3)(6) - 1 = 17$$

In order to test the null hypothesis of equivalent population means, it is necessary to calculate mean squares for the independent variable and error components. These are obtained by dividing the relevant sums of squares by their degrees of freedom:

$$MS_{IV} = \frac{SS_{IV}}{df_{IV}} \tag{13.11}$$

$$= \frac{43.00}{2} = 21.50$$

$$MS_{ERROR} = \frac{SS_{ERROR}}{df_{ERROR}} \tag{13.12}$$

$$= \frac{14.33}{10} = 1.43$$

---

* The rationale for the equivalence of the degrees of freedom to the indicated quantities is conceptually similar to that discussed in Appendix 12.1 for the degrees of freedom in the between-subjects case.

The mean square across subjects, on the other hand, does not directly figure in the F test that constitutes the hypothesis testing procedure and, thus, is not typically calculated. This is consistent with the fact that the sole function of the sum of squares across subjects is to remove variability due to individual differences from the dependent variable so that a more sensitive test of the relationship between the independent and dependent variables can be performed.

The F test for one-way repeated measures analysis of variance follows the same general logic as outlined in Chapter 12 for one-way between-subjects analysis of variance. First, we form an F ratio by dividing the mean square IV by the mean square error. If the null hypothesis is true, both of these components reflect only sampling error. Thus, we would expect the F ratio, over the long run, to approach 1.00. If the null hypothesis is not true, the mean square error again reflects sampling error but the mean square IV reflects both sampling error and the effect of the independent variable on the dependent variable. We would thus expect the F ratio, over the long run, to be greater than 1.00. Next, we compare the observed value of F with the appropriate critical value of F from Appendix F. If the observed value of F exceeds the critical value of F, we reject the null hypothesis and conclude that there is a relationship between the independent and dependent variables. Otherwise, we fail to reject the null hypothesis.

The F ratio for one-way repeated measures analysis of variance is*

$$F = \frac{MS_{IV}}{MS_{ERROR}} \qquad [13.13]$$

which in our example equals

$$F = \frac{21.50}{1.43} = 15.03$$

As we did in Chapter 12, we can summarize our calculations in a summary table:

Source	SS	df	MS	F
IV	43.00	2	21.50	15.03
Error	14.33	10	1.43	
Across subjects	59.17	5		
Total	116.50	17		

When an F value is reported, it is conventional to present the degrees of freedom IV followed by the degrees of freedom error. For an alpha level of .05 and

---

* Analogous to the between-subjects case, the F distribution in the two-group within-subjects situation bears a mathematical relationship to the t distribution such that the values of F obtained using Equation 13.13 will be equal to the squares of the t scores that would be obtained if a correlated groups t test were applied to the same data.

$df_{IV} = 2$ and $df_{ERROR} = 10$, the critical value of F from Appendix F is 4.10. Since the observed F value of 15.03 is greater than 4.10, we reject the null hypothesis and conclude that there is a relationship between the type of label and the perceived quality of the wine.

Study Exercise 13.1

An investigator was interested in the effect of exercise on psychological well-being. To examine the relationship between these two variables, she placed five volunteers on an aerobic exercise regimen. All of these individuals were of normal weight and health, and none was presently involved in a physical fitness program. Participants responded to a measure of psychological well-being on four occasions: before beginning the exercise regimen and 2, 4, and 6 weeks later. Scores on this measure could range from 1 to 70, with higher values indicating greater psychological well-being. The data for the study are listed in the accompanying table. Use a one-way repeated measures analysis of variance to test for a relationship between the amount of exercise and psychological well-being.

Answer   The null and alternative hypotheses can be stated as follows:

H_0: $\mu_Z = \mu_T = \mu_F = \mu_S$

H_1:   The four population means are not all equal

where the subscripts "Z," "T," "F," and "S" represent 0, 2, 4, and 6 weeks of exercise, respectively.

Intermediate statistics necessary to calculate the sums of squares are given in the table below. Applying Equations 13.2–13.5, we obtain the following values:

$$SS_{IV} = \frac{\sum T_j^2}{N} - \frac{(\sum X)^2}{kN}$$
$$= 70{,}726.00 - 70{,}567.20 = 158.80$$

$$SS_{ACROSS\ SUBJECTS} = \frac{\sum s_i^2}{k} - \frac{(\sum X)^2}{kN}$$
$$= 71{,}524.00 - 70{,}567.20 = 956.80$$

$$SS_{ERROR} = \sum X^2 + \frac{(\sum X)^2}{kN} - \frac{\sum T_j^2}{N} - \frac{\sum s_i^2}{k}$$
$$= 71{,}714 + 70{,}567.20 - 70{,}726.00 - 71{,}524.00 = 31.20$$

$$SS_{TOTAL} = \sum X^2 - \frac{(\sum X)^2}{kN}$$
$$= 71{,}714 - 70{,}567.20 = 1{,}146.80$$

The degrees of freedom are computed using Equations 13.6, 13.7, 13.8, and 13.10:

$$df_{IV} = k - 1 = 4 - 1 = 3$$

$$df_{ACROSS\ SUBJECTS} = N - 1 = 5 - 1 = 4$$

$$df_{ERROR} = (k - 1)(N - 1) = (4 - 1)(5 - 1) = 12$$

$$df_{TOTAL} = kN - 1 = (4)(5) - 1 = 19$$

The mean square IV and the mean square error are then computed by dividing the corresponding sums of squares by their degrees of freedom:

$$MS_{IV} = \frac{SS_{IV}}{df_{IV}} = \frac{158.80}{3} = 52.93$$

$$MS_{ERROR} = \frac{SS_{ERROR}}{df_{ERROR}} = \frac{31.20}{12} = 2.60$$

Finally, the F ratio is derived by dividing MS_{IV} by MS_{ERROR}:

$$F = \frac{MS_{IV}}{MS_{ERROR}} = \frac{52.93}{2.60} = 20.36$$

These calculations yield the following summary table:

Source	SS	df	MS	F
IV	158.80	3	52.93	20.36
Error	31.20	12	2.60	
Across subjects	956.80	4		
Total	1,146.80	19		

The critical value of F from Appendix F for an alpha level of .05 and 3 and 12 degrees of freedom is 3.49. Since the observed F value of 20.36 is greater than 3.49, we reject the null hypothesis and conclude that a relationship exists between the amount of exercise and psychological well-being.

	0 Weeks		2 Weeks		4 Weeks		6 Weeks			
Individual	X	X^2	X	X^2	X	X^2	X	X^2	s_i	s_i^2
1	60	3,600	59	3,481	63	3,969	68	4,624	250	62,500
2	52	2,704	53	2,809	58	3,364	61	3,721	224	50,176
3	61	3,721	67	4,489	69	4,761	69	4,761	266	70,756
4	44	1,936	46	2,116	50	2,500	50	2,500	190	36,100
5	63	3,969	62	3,844	66	4,356	67	4,489	258	66,564

$$T_Z = 280 \qquad T_T = 287 \qquad T_F = 306 \qquad T_S = 315$$
$$\overline{X}_Z = 56.00 \quad \overline{X}_T = 57.40 \quad \overline{X}_F = 61.20 \quad \overline{X}_S = 63.00$$
$$T_Z^2 = 78,400 \quad T_T^2 = 82,369 \quad T_F^2 = 93,636 \quad T_S^2 = 99,225$$

$$\sum X^2 = 3,600 + 2,704 + \cdots + 2,500 + 4,489 = 71,714$$

$$\sum X = 60 + 52 + \cdots + 50 + 67 = 1,188$$

$$\frac{(\sum X)^2}{kN} = \frac{1,188^2}{(4)(5)} = 70,567.20$$

$$\frac{\sum T_j^2}{N} = \frac{78,400 + 82,369 + 93,636 + 99,225}{5} = 70,726.00$$

$$\frac{\sum s_i^2}{k} = \frac{62,500 + 50,176 + 70,756 + 36,100 + 66,564}{4} = 71,524.00$$

Assumptions of the *F* Test

The *F* test for one-way repeated measures analysis of variance is appropriate to use when the dependent variable is quantitative in nature and measured on a level that at least approximates interval characteristics. Its validity rests on the following assumptions:

1. The sample is independently and randomly selected from the population of interest.
2. Each population of scores is normally distributed.
3. The variance of the population difference scores for any two conditions is the same as the variance of the population difference scores for any other two conditions. This is known as the **sphericity** assumption.

As an example of the sphericity assumption, the variance of the scores we would obtain across the population by subtracting individuals' scores in the French condition of the wine label and perceived quality experiment from individuals' scores in the Italian condition is assumed to be the same as the variance of the scores we would obtain by subtracting individuals' scores in the French or Italian condition from their scores in the American condition.

As with one-way between-subjects analysis of variance, the assumption of independent and random sampling is important for the validity of the *F* test. However, the *F* test is quite robust to violations of the normality assumption. For sample sizes larger than 30, the Type I error rate remains near the specified alpha level even in the face of marked nonnormality. By contrast, the *F* test is not robust to violations of sphericity. This is important because many applications in the behavioral sciences almost certainly violate this assumption, sometimes substantially. Because of this, many statisticians recommend that the traditional *F* test be modified unless one is confident that sphericity holds. This modification takes the form of multiplying the degrees of freedom IV and the degrees of freedom error by an *adjustment factor* to obtain new degrees of freedom to be used in assessing the significance of the observed *F* ratio. Two of the most frequently encountered adjustment factors are the *Huynh–Feldt epsilon* and the *Greenhouse–Geisser epsilon*. Both approaches result in adjusted degrees of freedom that are less than or equal to the usual degrees of freedom of $df_{IV} = k - 1$ and $df_{ERROR} = (k - 1)(N - 1)$. The use of these adjusted degrees of freedom decreases the Type I error rate below what it would otherwise be and, thus, increases the robustness of the statistical test to violations of the sphericity assumption. Unfortunately, the formulas for calculating the adjustment factors are complex and difficult to apply by hand. Most statistical computer programs provide both the Huynh–Feldt epsilon and the Greenhouse–Geisser epsilon as part of the statistical output for repeated measures analysis of variance, as we demonstrate later. In general, the Huynh–Feldt procedure is the preferred alternative for most applications in the behavioral sciences, although there are scenarios where other adjustments may be preferred (see Maxwell & Delaney, 1990).

The probable violation of sphericity in a repeated measures design can be evaluated using the **Mauchly test**. Unfortunately, studies have suggested that this test is of little diagnostic value in practical applications of repeated measures analysis of variance, in part because it is highly sensitive to nonnormality in the data (Kesselman, Rogan, Mendoza, & Breen, 1980). Unless one is confident on theoretical grounds that sphericity holds, it is probably best to apply the appropriate adjustment to the degrees of freedom IV and the degrees of freedom error using the Huynh–Feldt procedure.*

13.3   Strength of the Relationship

The formula for computing eta-squared for one-way repeated measures analysis of variance is

$$\text{eta}^2 = \frac{SS_{IV}}{SS_{IV} + SS_{ERROR}} \tag{13.14}$$

As with the correlated groups t test, eta-squared in the context of one-way repeated measures analysis of variance represents the proportion of variability in the dependent variable that is associated with the independent variable *after variability due to individual differences has been removed*. The denominator of Equation 13.14 is often called the *sum of squares within subjects*; that is, $SS_{\text{WITHIN SUBJECTS}} = SS_{IV} + SS_{ERROR}$. This should not be confused with the sum of squares within (SS_{WITHIN}) discussed in Chapter 12 that represents variability within *groups*.

The strength of the relationship for the wine label and perceived quality experiment is

$$\text{eta}^2 = \frac{43.00}{43.00 + 14.33} = .75$$

In this experiment, 75% of the variability in the dependent variable (the perceived quality of the wine) is associated with the independent variable (the type of label) after the influence of individual differences has been removed. This represents a strong effect.

An alternative formula for eta-squared based on the degrees of freedom and the observed value of F is

$$\text{eta}^2 = \frac{(df_{IV})F}{(df_{IV})F + df_{ERROR}} \tag{13.15}$$

For the wine label and perceived quality experiment,

$$\text{eta}^2 = \frac{(2)(15.03)}{(2)(15.03) + 10} = .75$$

which is the same result as obtained above.

---

* All of the examples in this chapter assume that the sphericity assumption is satisfied.

13.4 **Nature of the Relationship**

The nature of the relationship following a statistically significant one-way repeated measures analysis of variance is addressed differently, depending on whether or not the assumption of sphericity is satisfied. If it is, then the Tukey HSD test is applied in a similar manner as it was in Chapter 12. The only differences are that N and MS_{ERROR} replace n and MS_{WITHIN} in the formula for the critical difference, and that the value of q is determined with reference to the overall alpha level, the degrees of freedom error, and k (the number of conditions) rather than the overall alpha level, the degrees of freedom within, and k. Specifically, the critical difference for the Tukey HSD test in the repeated measure case is defined as

$$CD = q\sqrt{\frac{MS_{ERROR}}{N}}$$ [13.16]

Referring to Appendix G, we find that for an overall alpha level of .05, $df_{ERROR} = 10$, and $k = 3$, q is equal to 3.88. Thus, the critical difference for the wine label and perceived quality experiment is

$$CD = 3.88\sqrt{\frac{1.43}{6}} = 1.89$$

The HSD procedure can now be applied as follows:

Null hypothesis tested	Absolute difference between sample means	Value of CD	Null hypothesis rejected?
$\mu_F = \mu_I$	$|15.00 - 12.00| = 3.00$	1.89	Yes
$\mu_F = \mu_A$	$|15.00 - 11.50| = 3.50$	1.89	Yes
$\mu_I = \mu_A$	$|12.00 - 11.50| = .50$	1.89	No

The nature of the relationship is such that the wine will be rated as tasting better when it is labeled as French ($\overline{X}_F = 15.00$) than when it is labeled as either Italian ($\overline{X}_I = 12.00$) or American ($\overline{X}_A = 11.50$). However, we cannot confidently conclude that the wine will be rated differently as a function of being labeled Italian versus American.

If the sphericity assumption is not met, then the HSD test is not appropriate for evaluating the nature of the relationship. In this case, we use a **modified Bonferroni procedure**, which is described in Appendix 13.1.

Study Exercise 13.2

Calculate eta-squared and use the Tukey HSD test to analyze the nature of the relationship between the two variables for the data in Study Exercise 13.1.

Answer Eta-squared is most easily calculated using Equation 13.14:

$$eta^2 = \frac{SS_{IV}}{SS_{IV} + SS_{ERROR}}$$

$$= \frac{158.80}{158.80 + 31.20} = .84$$

This represents a strong effect.

The first step in applying the Tukey HSD test is to calculate the critical difference. Referring to Appendix G, we find that for an overall alpha level of .05, $df_{ERROR} = 12$, and $k = 4$, q is equal to 4.20. Thus,

$$CD = q\sqrt{\frac{MS_{ERROR}}{N}}$$

$$= 4.20\sqrt{\frac{2.60}{5}} = 3.03$$

The HSD procedure can now be applied as follows:

Null hypothesis tested	Absolute difference between sample means	Value of CD	Null hypothesis rejected?
$\mu_Z = \mu_T$	$|56.00 - 57.40| = 1.40$	3.03	No
$\mu_Z = \mu_F$	$|56.00 - 61.20| = 5.20$	3.03	Yes
$\mu_Z = \mu_S$	$|56.00 - 63.00| = 7.00$	3.03	Yes
$\mu_T = \mu_F$	$|57.40 - 61.20| = 3.80$	3.03	Yes
$\mu_T = \mu_S$	$|57.40 - 63.00| = 5.60$	3.03	Yes
$\mu_F = \mu_S$	$|61.20 - 63.00| = 1.80$	3.03	No

The nature of the relationship is such that psychological well-being will be greater after 4 ($\overline{X}_F = 61.20$) or 6 ($\overline{X}_S = 63.00$) weeks of aerobic exercise than before beginning the exercise regimen ($\overline{X}_Z = 56.00$) or after exercising for only 2 weeks ($\overline{X}_T = 57.40$). However, we cannot confidently conclude that 2 weeks of exercise affects psychological well-being relative to no exercise, or that 6 weeks of aerobic exercise affects psychological well-being relative to 4 weeks of exercise.

13.5 Methodological Considerations

The sum of squares across subjects was 59.17 in the wine label and perceived quality experiment. This constitutes a relatively large portion of the total sum of squares (116.50), which indicates that individual differences played an important role in the taste ratings. One way to examine the extent of this influence is to compare the observed F ratio with the F ratio that would have been observed if the experiment had been conducted using a between-subjects rather than a repeated measures design (that is, if each wine had been rated by six *different* individuals) and the same data hypothetically obtained.

As discussed earlier, since it is not possible to identify the effects of individual differences in a between-subjects analysis, variability due to individual differences contributes to the sum of squares within along with all other disturbance variables that are operating. Not surprisingly, then, it turns out that the sum of squares within from a one-way between-subjects analysis of variance is mathematically equal to the total of the sum of squares across subjects (which reflects the influence of individual differences) and the sum of squares error (which reflects the influence of disturbance variables other than individual differences) from a one-way repeated measures analysis of variance applied to the same scores. Furthermore, given a common data set, it also turns out that the degrees of freedom within in the between-subjects case is mathematically equal to the total of the degrees of freedom across subjects and the degrees of freedom error in the repeated measures case, and that the sum of squares and degrees of freedom between are mathematically equal to the sum of squares and degrees of freedom IV, respectively.

Thus, if the wine label and perceived quality data had been obtained using a between-subjects design, the summary table would take the following form:

Source	SS	df	MS	F
Between	43.00	2	21.50	4.39
Within	73.50	15	4.90	
Total	116.50	17		

Note that the F ratio is 4.39 compared with the F ratio of 15.03 computed earlier when a repeated measures analysis was used. Since the mean square IV and the mean square between are identical, the difference in F ratios must be due to a difference in the denominator of the F test. Indeed, a comparison of the two summary tables shows that the denominator of the F ratio increased from 1.43 (the mean square error) to 4.90 (the mean square within) when variability due to individual differences was not separately removed from the dependent variable. Although the F test in this instance is still statistically significant (since the critical value of F for an alpha level of .05 and 2 and 15 degrees of freedom is 3.68), there will be instances where the F ratio yielded by a repeated measures analysis of variance will be large enough to lead to rejection of a false null hypothesis, whereas the F ratio yielded by the corresponding between-subjects analysis will not be.

On the other hand, the degrees of freedom for the denominator of the F test will always be less in the repeated measures case than in the between-subjects case. Since the value of F required to reject the null hypothesis becomes more extreme as the degrees of freedom become smaller, repeated measures analysis of variance might actually be less powerful than between-subjects analysis of variance when individual differences have only a minimal influence on the dependent variable. This is parallel to the situation that exists with the independent and correlated groups t tests, as discussed in Section 11.6. Thus, as noted in that section, it is important that a researcher who is considering a within-subjects (repeated measures) design accurately anticipate the role of individual differences.

In Chapter 11 we saw that one way to deal with confounding variables associated with carry-over effects is *counterbalancing*. This involves exposing individuals to *predetermined* sequences of the independent variable such that carry-over effects are evenly distributed across conditions. This technique turns confounding variables associated with treatment order into disturbance variables. An alternative to counterbalancing as a means of controlling confounding variables is to *randomly* order the conditions for each participant. This is the approach that was taken in the wine label and perceived quality experiment. The rationale is that when the sequence of conditions across participants is randomly determined, chance will ensure that each condition occurs in each position an approximately equal number of times and, thus, that carry-over effects are evenly distributed across conditions. This approach is particularly valuable when the number of conditions is so great that counterbalancing is impractical. For instance, there are 24 different ways that four conditions can be ordered and 120 different ways that five conditions can be ordered!* Can you think of some potential confounding variables that might have been counteracted in the wine label and perceived quality experiment by using the randomization procedure?

13.6 Numerical Example

One area of interest to cognitive psychologists is memory. To examine the role that the amount of exposure to a stimulus plays in the memory process, suppose a cognitive psychologist conducts an experiment in which the same list of ten nonsense syllables (for instance, *blux*, *gonk*, and *delp*) is presented to participants four different times. The list is presented for 30 seconds on each occasion, and participants are given 60 seconds to recall as many of the syllables as they can after the list is removed. Test sessions are separated by 10 minutes, during which time participants work on a math task designed to occupy their thoughts so that they will not be able to practice the nonsense syllables. Hypothetical data for the seven experimental participants are presented in Table 13.3.

The null hypothesis is that there is no relationship between the amount of exposure and recall, whereas the alternative hypothesis is that there is a relationship. These hypotheses are formally stated as

H_0: $\mu_1 = \mu_2 = \mu_3 = \mu_4$

H_1: The four population means are not all equal

where the subscripts "1," "2," "3," and "4" represent the four test times.

Intermediate statistics necessary to calculate the sums of squares are given in Table 13.3. The sum of squares IV is

---

* This can be determined using the formula for permutations (Equation 6.10) presented in Chapter 6: $_nP_r = n!/(n - r)!$. For instance, the number of permutations of five conditions taken five at a time is $5!/(5 - 5)! = [(5)(4)(3)(2)(1)]/1 = 120$.

TABLE 13.3   Data and Computation of Sums of Squares for Amount of Exposure and Recall Experiment

Participant	Time 1 X	X^2	Time 2 X	X^2	Time 3 X	X^2	Time 4 X	X^2	s_i	s_i^2
1	5	25	6	36	6	36	5	25	22	484
2	7	49	6	36	7	49	8	64	28	784
3	8	64	9	81	9	81	10	100	36	1,296
4	3	9	4	16	4	16	6	36	17	289
5	9	81	8	64	9	81	7	49	33	1,089
6	5	25	4	16	6	36	6	36	21	441
7	7	49	10	100	8	64	9	81	34	1,156

$$T_1 = 44 \quad T_2 = 47 \quad T_3 = 49 \quad T_4 = 51$$
$$\overline{X}_1 = 6.29 \quad \overline{X}_2 = 6.71 \quad \overline{X}_3 = 7.00 \quad \overline{X}_4 = 7.29$$
$$T_1^2 = 1,936 \quad T_2^2 = 2,209 \quad T_3^2 = 2,401 \quad T_4^2 = 2,601$$

$$\sum X^2 = 25 + 49 + \cdots + 36 + 81 = 1,405$$

$$\sum X = 5 + 7 + \cdots + 6 + 9 = 191$$

$$\frac{(\sum X)^2}{kN} = \frac{191^2}{(4)(7)} = 1,302.89$$

$$\frac{\sum T_j^2}{N} = \frac{1,936 + 2,209 + 2,401 + 2,601}{7} = 1,306.71$$

$$\frac{\sum s_i^2}{k} = \frac{484 + 784 + 1,296 + 289 + 1,089 + 441 + 1,156}{4} = 1,384.75$$

$$SS_{IV} = \frac{\sum T_j^2}{N} - \frac{(\sum X)^2}{kN}$$
$$= 1,306.71 - 1,302.89 = 3.82$$

The sum of squares across subjects is

$$SS_{ACROSS\ SUBJECTS} = \frac{\sum s_i^2}{k} - \frac{(\sum X)^2}{kN}$$
$$= 1,384.75 - 1,302.89 = 81.86$$

The sum of squares error is

$$SS_{ERROR} = \sum X^2 + \frac{(\sum X)^2}{kN} - \frac{\sum T_j^2}{N} - \frac{\sum s_i^2}{k}$$
$$= 1,405 + 1,302.89 - 1,306.71 - 1,384.75 = 16.43$$

Finally, the sum of squares total is

$$SS_{TOTAL} = \sum X^2 - \frac{(\sum X)^2}{kN}$$

$$= 1,405 - 1,302.89 = 102.11$$

The degrees of freedom are

$$df_{IV} = k - 1 = 4 - 1 = 3$$

$$df_{ACROSS\ SUBJECTS} = N - 1 = 7 - 1 = 6$$

$$df_{ERROR} = (k - 1)(N - 1) = (4 - 1)(7 - 1) = 18$$

$$df_{TOTAL} = kN - 1 = (4)(7) - 1 = 27$$

The relevant mean squares are

$$MS_{IV} = \frac{SS_{IV}}{df_{IV}} = \frac{3.82}{3} = 1.27$$

$$MS_{ERROR} = \frac{SS_{ERROR}}{df_{ERROR}} = \frac{16.43}{18} = .91$$

Thus, the F ratio is

$$F = \frac{MS_{IV}}{MS_{ERROR}} = \frac{1.27}{.91} = 1.40$$

These calculations yield the following summary table:

Source	SS	df	MS	F
IV	3.82	3	1.27	1.40
Error	16.43	18	.91	
Across subjects	81.86	6		
Total	102.11	27		

The critical value of F for an alpha level of .05 and 3 and 18 degrees of freedom is 3.16. Since the observed F value of 1.40 does not exceed 3.16, we fail to reject the null hypothesis of equivalent recall during the four test sessions.

As discussed in Section 10.3, it is worthwhile to calculate eta-squared even when the inferential statistical test is nonsignificant. In our example,

$$eta^2 = \frac{SS_{IV}}{SS_{IV} + SS_{ERROR}}$$

$$= \frac{3.82}{3.82 + 16.43} = .19$$

Thus, the proportion of variability in recall scores that is associated with the amount of exposure to the nonsense syllables after the effects of individual differences have been removed is .19. Although this indicates a strong relationship between the independent variable and the dependent variable in the sample, we are unable to conclude that the two variables are related in the population. If the null hypothesis had been rejected, we would have determined the nature of the relationship using the Tukey HSD procedure discussed in Section 13.4. As it is, it is interesting to note the high variability due to individual differences extracted from the dependent variable by the sum of squares across subjects and to contemplate whether additional disturbance or confounding variables might have been operating in the experiment. Given that a sample eta-squared of .19 indicates a strong effect, one possibility is that the amount of exposure to nonsense syllables and recall are actually related in the population but that the sample size ($N = 7$) was not large enough to yield a statistically significant result with the desired degree of consistency.

13.7   Planning an Investigation Using One-Way Repeated Measures Analysis of Variance

Appendix E.3 contains tables of the sample sizes necessary to achieve various values of power for one-way repeated measures analysis of variance. A set of tables is presented for each research design involving a different number of degrees of freedom IV. Table 13.4 reproduces a portion of this appendix for $df_{IV} = 2$ (that is, for three conditions) and an alpha level of .05. The first column presents various values of power, and the column headings are values of eta-squared in the population. It should be noted that these values of eta-squared are conceptualized as the proportion of variability in the dependent variable that is associated with the independent variable *after the effects of individual differences have been removed*. Thus, to the extent that the dependent variable is influenced by individual background, the population eta-squared will be greater in the within-subjects case than in the between-subjects case.

To illustrate the use of the table, suppose the desired power level is .80 and the researcher suspects that the strength of the relationship in the population

TABLE 13.4   **Approximate Sample Sizes Necessary to Achieve Selected Levels of Power for $df_{IV} = 2$ and Alpha = .05 As a Function of Population Values of Eta-Squared**

Power	Population Eta-Squared									
	.01	.03	.05	.07	.10	.15	.20	.25	.30	.35
.10	32	11	7	5	4	3	2	2	2	2
.50	247	81	48	34	23	15	11	8	7	6
.70	382	125	74	52	36	23	16	13	10	8
.80	478	157	93	65	44	28	20	15	12	10
.90	627	206	121	85	58	37	26	20	16	13
.95	765	251	148	104	70	45	32	24	19	15
.99	1,060	347	204	143	97	62	44	33	26	21

corresponds to an eta-squared of .15. Then the number of participants that should be used in the study is 28. Inspection of Table 13.4 and Appendix E.3 provides a general appreciation for the relationships among df_{IV}, the alpha level, sample size, the strength of the relationship one is trying to detect in the population, and power for one-way repeated measures analysis of variance.

13.8 Method of Presentation

The method of presentation for a one-way repeated measures analysis of variance parallels that for a one-way between-subjects analysis of variance, except that the degrees of freedom IV and the degrees of freedom error are reported instead of the degrees of freedom between and the degrees of freedom within. For example, the results for the wine label and perceived quality experiment discussed earlier in this chapter might be reported as follows:*

Results

A one-way repeated measures analysis of variance was performed relating the type of label (French, Italian, or American) to the perceived quality of the wine. For an alpha of .05, the obtained F ratio was found to be statistically significant, $\underline{F}(2, 10) = 15.03$, $\underline{p} < .01$. The strength of the relationship, as indexed by eta^2, was .75. A Tukey HSD test revealed that the wine was rated significantly higher when it was labeled as French ($\underline{M} = 15.00$, $\underline{SD} = 1.79$) than when it was labeled as either Italian ($\underline{M} = 12.00$, $\underline{SD} = 1.41$) or American ($\underline{M} = 11.50$, $\underline{SD} = 3.08$), but that the means for the latter two conditions did not significantly differ.

13.9 Examples from the Literature

Age Regression and the Magnitude of the Poggendorff Illusion

The use of hypnosis to "return" individuals to an earlier chronological age is known as *age regression*. The technique is based on the assumption that individuals regressed to a particular age will behave as if they actually were that age. One type of age-regressed behavior that has been studied is perception. For instance, Parrish, Lundy, and Leibowitz (1969) investigated the effect of age regression on

---

* The standard deviation estimates were calculated by applying the usual formulas from Chapters 3 and 7 to each of the wine label conditions in Table 13.2.

the magnitude of the Poggendorff illusion. This illusion refers to the fact that the right half of a diagonal white bar running upward from left to right through a solid black bar appears to be higher than it actually is.

The magnitude of the Poggendorff illusion decreases with age from age 5 to approximately age 10 and then stabilizes. If age regression affects perceptual behavior, this same pattern of results should be observed with age-regressed individuals. In an attempt to determine whether this is indeed the case, Parrish and colleagues exposed ten college students to each of four conditions: no hypnosis, hypnosis but no age regression, hypnosis and regression to age 5, and hypnosis and regression to age 9. The order of these conditions was randomized such that students participated in one condition at a time at weekly intervals. The experimental task involved adjusting the right half of the diagonal bar on a metal Poggendorff figure to make it straight with the stationary left half. The magnitude of the illusion was measured as the difference in inches between where students set the right half of the bar and the point at which this portion was actually aligned with the left half. Six trials were undertaken during each experimental session, with the average of the six responses constituting the dependent variable.

A one-way repeated measures analysis of variance applied to these data yielded statistically significant results, $F(3, 27) = 4.10$, $p < .05$. The strength of the relationship, as indexed by eta-squared, was .31. This represents a strong effect. Multiple comparisons showed that the Poggendorff illusion was stronger for the hypnosis/regressed-to-age-5 condition ($\overline{X} = 1.41$ inches) than for either the no-hypnosis ($\overline{X} = .96$ inch) or the hypnosis/no-age regression ($\overline{X} = 1.03$ inches) condition. (The mean for the hypnosis/regressed-to-age-9 condition was 1.26 inches.) None of the other comparisons was statistically significant. When considered in conjunction with similar findings for a second illusion (the Ponzo illusion), these results suggest that age regression affects perceptual behavior in the expected manner.

Sex-Role Orientation and Perceived Sexual Attractiveness

Sex-role orientation has been defined in the psychological literature in terms of two broad classes of traits that an individual may exhibit: instrumental traits (for instance, aggression, risk taking, ambition) and expressive traits (for instance, warmth, sympathy, sensitivity). Several psychologists classify individuals into one of four sex-role groups based on their scores on psychological tests designed to measure these constructs: (1) *masculine* individuals, who score high on instrumental traits but low on expressive traits; (2) *feminine* individuals, who score high on expressive traits but low on instrumental traits; (3) *androgynous* individuals, who score high on both instrumental and expressive traits; and (4) *undifferentiated* individuals, who score low on both instrumental and expressive traits.

Among the findings of research on sex-role orientation is that an individual's sex-role standing affects the ways in which he or she is perceived by others. For instance, Becker and Gaeddert (1988) presented 18 male college students with descriptions of masculine, feminine, androgynous, and undifferentiated female targets. For each of the four descriptions, the students were asked to indicate how sexually attractive they thought the target person was on a 1 to 9 scale ranging

Applications

to the Analysis

of a Social Problem

It is important that parents take responsibility for ensuring that their teens are well informed about sex and birth control and place such conversations into a context of family values that the parents espouse. Nevertheless, many parents do not talk with their teens about these issues. This is unfortunate because research shows that many teens believe that their parents would be a useful source of information about these topics. Why don't parents talk with their teens about sex and birth control? In the parent–teen communication study described in Chapter 1, we asked mothers to indicate their agreement with a set of statements that focus on possible reasons. These statements included the following:

It would be difficult for me to explain things if I talked with my daughter(son) about sex and birth control.

It wouldn't do much good if I talked with my daughter(son) about sex and birth control.

My daughter(son) will think I do not trust her(him) if I try to talk to her(him) about sex and birth control.

It would be difficult to find a convenient time and place to talk to my daughter(son) about sex and birth control.

Talking about birth control with my daughter(son) will only encourage her(him) to have sex.

Each statement was responded to on a five-point scale having the response options strongly disagree, moderately disagree, neither agree nor disagree, moderately agree, and strongly agree. Responses were scored from 1 to 5, with higher values indicating greater agreement. We were interested in determining whether there are differences in the extent to which mothers endorse each of these reasons.

To accomplish this, we conducted a one-way repeated measures analysis of variance that had the type of reason as the independent variable and the amount of agreement as the dependent variable. We first examined the frequency histogram for each level of the independent variable (that is, for each of the five statements). These are presented in Figure 13.1, with a normal distribution superimposed on each histogram. All of the distributions were strongly positively skewed, with most mothers indicating that they strongly disagree that these are obstacles to discussion. Although the analysis of means will be informative, this dominance of "strongly disagree" responses must be kept in mind when drawing conclusions. The nonnormality observed in the sample raises questions about the normality of the population scores. However, given the large sample size (732 mothers), studies have suggested that the analysis will be relatively robust to nonnormality, even to the degree that we observe in this case.

Table 13.5 presents the computer printout for the analysis. The first section of the output presents the Mauchly test of the sphericity assumption. The W value (.91) is the value of the Mauchly statistic. If the p value associated with the Mauchly

from "not at all" to "very much." The order of presentation was randomized for each participant.

A one-way repeated measures analysis of variance comparing the mean sexual attractiveness ratings for the four targets was found to be statistically significant, $F(3, 51) = 29.42$, $p < .01$. As indexed by eta-squared, the strength of the relationship was .63. This represents a strong effect. Application of the Tukey HSD test showed that the feminine ($\overline{X} = 5.33$) and the androgynous ($\overline{X} = 6.17$) targets were rated as significantly more sexually attractive than were the masculine ($\overline{X} = 2.78$) and the undifferentiated ($\overline{X} = 2.22$) targets. The means for the feminine versus the androgynous target and the masculine versus the undifferentiated target did not significantly differ.

FIGURE 13.1   **Frequency Histograms for Maternal Agreement That (a) It Would Be Difficult to Explain Things, (b) It Wouldn't Do Much Good, (c) Teen Will Think I Don't Trust Her/Him, (d) It Would Be Difficult to Find a Time and Place, and (e) It Will Encourage Teen to Have Sex**

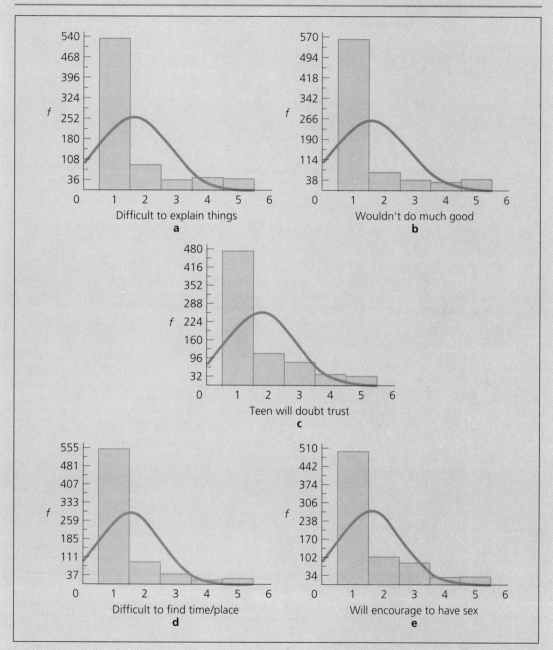

test (labeled "Significance") is less than .05, this implies that the sphericity assumption has been violated. However, as noted in Section 13.2, the Mauchly test is quite sensitive to nonnormality in the data and this can undermine the test.

(continued)

TABLE 13.5 **Computer Output for One-Way Repeated Measures Analysis of Variance for Agreement with Reasons for Not Talking About Sex and Birth Control**

```
Tests involving 'FACTOR1' Within-Subject Effect

  Mauchly sphericity test, W =    .91216
  Significance =                  .000


  Greenhouse-Geisser Epsilon =    .96141
  Huynh-Feldt Epsilon =           .96712
  Lower-bound Epsilon =           .25000
```

Source of Variation	SS	DF	MS	F	Sig of F
FACTOR1	21.23	4	5.31	8.08	.000
(Huynh-Feldt)		3.87		8.08	.000
WITHIN+RESIDUAL	1923.57	2928	.66		
(Huynh-Feldt)		2831.72			

Following the recommendation of Kesselman, Rogan, Mendoza, and Breen (1980), we suggest that a sphericity correction be adopted independent of the results of the Mauchly test.

The second part of the printout presents three different adjustments for the degrees of freedom: the Huynh–Feldt and Greenhouse–Geisser epsilons discussed earlier, as well as a "lower-bound" epsilon that is highly conservative (and generally not recommended). We chose to use the Huynh–Feldt correction, which equals .97. The degrees of freedom for the F ratio are multiplied by this value when assessing its significance.

The bottom portion of the printout in Table 13.5 provides a summary table for both the standard repeated measures analysis of variance and the Huynh–Feldt adjusted analysis. The source of variability due to error is called "WITHIN + RESIDUAL" in this computer output, and the source of variability due to the independent variable is called "FACTOR 1." This summary table does not report the sum of squares across subjects or the sum of squares total, but it contains all of the information necessary to complete our analysis. Note that the sums of squares, the mean squares, and the F ratio on the printout are the same for the standard repeated measures analysis of variance as for the Huynh–Feldt analysis. The only difference in the

two analyses is the degrees of freedom. In the standard F test, the F of 8.08 is evaluated with 4 and 2,928 degrees of freedom. With the Huynh–Feldt analysis, this F is evaluated with 3.87 and 2,831.72 degrees of freedom. In both cases, the p value associated with the F ratio was less than .0005, as indicated by the fact that its value is reported as .000 in the summary table.

The statistically significant F ratio suggests that the population means are not all equal. The strength of the relationship, as indexed by eta squared, was .01, which suggests that the effect, though statistically significant, is weak. We conducted follow-up tests to determine the nature of the relationship using the modified Bonferroni procedure discussed in Appendix 13.1.* The mean scores for each reason, rank ordered from highest to lowest, were:

Reason	Mean
Teen will doubt trust	1.72
Will encourage teen to have sex	1.63
Difficult to explain things	1.60
Wouldn't do much good	1.56
Difficult to find time and place	1.50

---

* We did not use the Tukey HSD test because it assumes that the condition of sphericity is met.

The modified Bonferroni analysis showed that the concern about the teen forming an impression about lack of trust was rated statistically significantly higher than all of the other reasons except thinking that engaging in a discussion will encourage the teen to have sex. None of the other pairwise comparisons was statistically significant.

It is evident that the means for the variables are quite similar. When this is coupled with the small effect size and the strong positive skewness in the distributions due to the preponderance of "strongly disagree" responses, substantive conclusions about the mean differences do not seem warranted. Mothers tend not to see these five reasons as problematic, and any mean differences between them appear to be trivial from a practical point of view.

Summary

One-way repeated measures analysis of variance is typically used to analyze the relationship between two variables when (1) the dependent variable is quantitative in nature and is measured on a level that at least approximates interval characteristics, (2) the independent variable is within-subjects in nature, and (3) the independent variable has three or more levels. One-way repeated measures analysis of variance differs from one-way between-subjects analysis of variance in that it removes the influence of individual differences from the dependent variable. The repeated measures analysis thus yields a more sensitive test of the relationship between the independent and dependent variables. Otherwise, the logic underlying the two techniques is identical. The strength of the relationship is measured using eta-squared, and the nature of the relationship is analyzed using the Tukey HSD test, unless the sphericity assumption is violated. In that case, a modified Bonferroni procedure is used.

Appendix 13.1 Determining the Nature of the Relationship Under Sphericity Violations

When the sphericity assumption is violated, it is not possible to use the mean square error from the overall analysis to calculate the critical difference as is done with the Tukey HSD test. Instead, we use a modified Bonferroni procedure. This involves performing all possible pairwise comparisons between group means using the traditional correlated groups t test. For the wine label and perceived quality experiment, there are three pairwise comparisons. Using the procedures from Chapter 11, we conduct a correlated groups t test for each pair of means involved in a comparison and determine a p value for each of the t values. This usually must be done with the aid of a computer. The absolute t values are ordered in a column from largest to smallest, and the corresponding p values are each compared against a "critical alpha." If the p value is less than the "critical alpha," the null hypothesis in question is rejected.

The value of the "critical alpha" changes for each successive t value as one proceeds from the top to the bottom of the column. The largest t value must yield a p value less than the overall alpha level that we adopt for the set of comparisons divided by the total number of comparisons (also called *contrasts*)—in this case, three—for the null hypothesis to be rejected.

The next largest t value must yield a p value less than the overall alpha level divided by the number of contrasts minus 1 for the null hypothesis to be rejected. The next largest t value must yield a p value less than the overall alpha level divided by the number of contrasts minus 2 for the null hypothesis to be rejected. And so on, subtracting an additional unit from the total number of comparisons with each successive contrast. If a contrast is statistically significant, the nature of the relationship is determined by examining the mean scores in the two conditions, per the usual t test procedure. As soon as the first statistically nonsignificant contrast is encountered, as one proceeds from the largest to the smallest absolute t value, all remaining comparisons are declared statistically nonsignificant. The rationale for this modified Bonferroni procedure is described in Holland and Copenhaver (1988).

The results for the modified Bonferroni procedure for the wine label and perceived quality experiment can be summarized in a table as follows. In deriving this table, we used an overall alpha level of .05.

Null hypothesis tested	Absolute value of t	p value	Critical alpha	Null hypothesis rejected?
$\mu_F = \mu_I$	5.81	.002	.05/3 = .017	Yes
$\mu_F = \mu_A$	4.58	.006	.05/2 = .025	Yes
$\mu_I = \mu_A$	.65	.542	.05/1 = .050	No

Exercises

Answers to asterisked (*) exercises appear at the back of the book. Answers to exercises with two asterisks are also worked out step by step in the Study Guide.

Exercises to Review Concepts

1. Under what conditions is one-way repeated measures analysis of variance typically used to analyze a bivariate relationship?

2. Distinguish among the sum of squares total, the sum of squares IV, the sum of squares across subjects, and the sum of squares error. How are they interrelated?

*3. Why is a repeated measures analysis of variance usually a more sensitive test of the relationship between the independent and dependent variables than a between-subjects analysis of variance?

*4. Insert the missing entries in the summary table for a one-way repeated measures analysis with five levels of the independent variable and $N = 12$.

Source	SS	df	MS	F
IV	20.00	—	—	—
Error	132.00	—	—	
Across subjects	—	—		
Total	198.00	—		

5. Insert the missing entries in the summary table for a one-way repeated measures analysis of variance with three levels of the independent variable.

Source	SS	df	MS	F
IV	—	—	20.00	10.00
Error	—	18	—	
Across subjects	—	9		
Total	106.00	—		

*6. For a repeated measures analysis of variance with $N = 21$ and five conditions, what would be the values of the degrees of freedom IV, degrees of freedom across subjects, degrees of freedom error, and degrees of freedom total?

*7. State the critical value of F for a one-way repeated measures analysis of variance for an alpha level of .05 under each of the following conditions:
   a. $k = 3, N = 20$ c. $k = 3, N = 15$
   b. $k = 4, N = 21$ d. $k = 5, N = 12$

8. What are the assumptions underlying one-way repeated measures analysis of variance?

*9. How do the Huynh–Feldt epsilon and the Greenhouse–Geisser epsilon increase the robustness of one-way repeated measures analysis of variance to violations of the sphericity assumption?

10. What interpretation does eta-squared have in the context of one-way repeated measures analysis of variance? How does this interpretation differ from that for one-way between-subjects analysis of variance?

*11. Under what circumstance might a one-way repeated measures analysis of variance be less powerful than a corresponding one-way between-subjects analysis of variance? Why?

12. What is the rationale behind randomly ordering the sequence of conditions across participants in repeated measures designs? How does this differ from counterbalancing?

Refer to the following scores on an anxiety test that was administered to five individuals at three different times to do Exercises 13–16:

Individual	Time 1	Time 2	Time 3
1	2	3	4
2	3	5	8
3	4	5	6
4	5	7	9
5	5	6	4

*13. Test for a relationship between the time of assessment and anxiety using a one-way repeated measures analysis of variance.

*14. Compute the value of eta-squared using Equation 13.14. Recalculate eta-squared using Equation 13.15. Compare the two results. Does the observed value represent a weak, moderate, or strong effect?

*15. Analyze the nature of the relationship between the time of assessment and anxiety using the Tukey HSD test.

*16. Analyze the data as if the independent variable were between-subjects in nature. That is, conduct a one-way between-subjects analysis of variance, compute the value of eta-squared, and use the Tukey HSD test to analyze the nature of the relationship. Compare your findings with those from Exercises 13–15. How does this illustrate the advantage of within-subjects research designs?

Refer to the following numbers of pages read in an assigned book during a test period by five students under three levels of distraction to do Exercises 17–19.

Subject	Low	Moderate	High
1	5	7	2
2	9	7	5
3	8	5	6
4	6	5	4
5	9	6	6

17. Test for a relationship between distraction and the number of pages read using a one-way repeated measures analysis of variance.

18. Compute the value of eta-squared. Does the observed value represent a weak, moderate, or strong effect?

19. Analyze the nature of the relationship between distraction and the number of pages read using the Tukey HSD test.

For each of the studies described in Exercises 20–23, indicate the appropriate statistical test for analyzing the relationship between the independent and dependent variables. Assume that the underlying assumptions of the tests have been satisfied.

*20. Kelman and Hovland (1953) studied the effect of the source of a persuasive communication on attitude change. Three groups of participants listened to a persuasive message on the treatment of juvenile delinquents. For one group of participants, the message was attributed to a trustworthy, well-informed source; for another group, it was attributed to an untrustworthy, poorly informed source; for a third group, it was attributed to a "neutral" source. The mean amounts of attitude change were compared for the three groups.

*21. A researcher tested the effect of age on infants' memory capacities. Forty infants were given a memory test at the age of 5 months and again at the age of 7 months. Scores on this test could range from 1 to 15. The mean test scores were compared for the two ages.

22. An investigation was undertaken to determine the effects of alcohol and marijuana on driving skills. Thirty participants took part in a driving simulation task under each of three conditions: (1) while under the influence of a small amount of alcohol, (2) while under the influence of a small amount of marijuana, and (3) while not under the influence of either drug. The simulation task yielded "driving scores" that could range from 1 to 100. The mean driving scores were compared for the three conditions.

23. Advertisements of medical products often include the results of surveys that assess the relative number of physicians who give a positive recommendation. A researcher was interested in whether the context in which such information is presented influences its impact on consumer attitudes. Two groups of individuals were presented with identical descriptions of a brand of aspirin. However, in one group, participants were told that "8 out of 10 doctors" recommend the aspirin, whereas the other group was told that "80 out of 100 doctors" recommend the aspirin. Although the proportions of doctors who recommended the aspirin were identical for the two groups, the *numbers* of doctors differed. After reading one of the two descriptions, participants indicated their attitude toward the aspirin on a 1 to 10 scale. The mean attitude scores were compared for the two groups.

24. If a researcher suspects that the strength of the relationship between two variables in the population is .10 as indexed by eta-squared, what sample size should he use in a study involving a within-subjects independent variable with five levels and an alpha level of .05 in order to achieve a power of .80?

*25. Suppose an investigator conducted a study involving a within-subjects independent variable having four levels with $N = 12$. If the value of eta-squared in the population was .15, what would the power of her statistical test be at an alpha level of .05?

Multiple-Choice Questions

*26. Which sum of squares in a between-subjects design is conceptually equivalent to the sum of squares error in a repeated measures design?
   a. sum of squares between
   b. sum of squares within
   c. sum of squares total
   d. none of the above

*27. Which sum of squares in a between-subjects design is conceptually equivalent to the sum of squares across subjects in a repeated measures design?

a. sum of squares between
b. sum of squares within
c. sum of squares total
d. none of the above

28. Which sum of squares in a between-subjects design is conceptually equivalent to the sum of squares IV in a repeated measures design?
    a. sum of squares between
    b. sum of squares within
    c. sum of squares total
    d. none of the above

29. The Mauchly test is of great diagnostic value in evaluating the probable violation of the sphericity assumption in practical applications of repeated measures analysis of variance.
    a. true   b. false

*30. If a researcher suspects that the strength of the relationship between two variables in the population is .05 as indexed by eta-squared, what sample size should he use in a study involving a within-subjects independent variable with three levels and an alpha level of .05 in order to achieve a power of .80?
    a. 62   c. 93
    b. 70   d. 133

31. Participants in a repeated measures design are exposed to
    a. all levels of the independent variable
    b. only one level of the independent variable
    c. a varying number of levels of the independent variable, depending on which condition they are assigned to
    d. a randomly determined number of levels of the independent variable

*32. The mean square error for a one-way repeated measures analysis of variance will tend to be _____ the mean square within for a one-way between-subjects analysis of variance applied to the same data set.
    a. larger than
    b. equal to
    c. smaller than
    d. half the size of

Use the following information to complete Exercises 33–43.

A developmental psychologist studied the number of close friends that a mixed-gender group of 21 children reported having at 8, 10, and 12 years of age. The mean number of friends was 3.48 at 8 years of age, 4.17 at 10 years of age, and 4.35 at 12 years of age. The summary table is given here:

Source	SS	df	MS	F
IV	8.86	2	4.43	3.57
Error	49.60	40	1.24	
Across subjects	96.49	20		
Total	154.95	62		

33. If the value of eta-squared in the population is .35, what is the power of the statistical test at an alpha level of .05?
    a. .35   c. .90
    b. .80   d. .99

34. The independent variable is
    a. the gender of the investigator
    b. the gender of the children
    c. the number of close friends
    d. age

*35. The dependent variable is
    a. the gender of the investigator
    b. the gender of the children
    c. the number of close friends
    d. age

36. Based on this study, we can conclude that the independent variable is related to the dependent variable.
    a. true   b. false

37. It is possible that the test results reflect a Type I error.
    a. true   b. false

*38. It is possible that the test results reflect a Type II error.
    a. true   b. false

39. What is the strength of the relationship, as indexed by eta-squared?
    a. .02   c. .35
    b. .15   d. .39

*40. What is the proportion of variability in the dependent variable that is due to disturbance variables other than individual differences?

a. .61   c. .85
b. .72   d. .92

*41. What can we conclude about the number of close friends one has at 8 versus 10 years of age?

a. The number of close friends is greater at age 8.
b. The number of close friends is greater at age 10.
c. The number of close friends is the same at the two ages.
d. The number of close friends may or may not differ at the two ages.

*42. What can we conclude about the number of close friends one has at 8 versus 12 years of age?

a. The number of close friends is greater at age 8.
b. The number of close friends is greater at age 12.
c. The number of close friends is the same at the two ages.
d. The number of close friends may or may not differ at the two ages.

43. What can we conclude about the number of close friends one has at 10 versus 12 years of age?

a. The number of close friends is greater at age 10.
b. The number of close friends is greater at age 12.
c. The number of close friends is the same at the two ages.
d. The number of close friends may or may not differ at the two ages.

Exercises to Apply Concepts

**44. Although numerous studies have examined the factors that women take into consideration when choosing a birth control method, relatively few studies have examined birth control from the standpoint of men. Male oral contraceptives will probably be available to the general public in the next 10 years. Jaccard (1980) conducted a study to discover which factors men would be most concerned about when evaluating male oral contraceptives. In this investigation, men rated how important each of four factors would be to them in deciding whether or not to use an oral contraceptive. The ratings were made on a 1 to 21 scale, with higher scores indicating greater importance. The hypothetical data presented in the table are representative of those obtained by Jaccard. Analyze these data, draw a conclusion, and write up your results using the principles given in the Method of Presentation section.

Individual	Health risks	Effectiveness	Cost	Convenience
1	16	16	12	12
2	19	15	11	11
3	18	14	10	10
4	20	12	8	8
5	17	13	9	9

45. Evidence suggests that many rape victims fail to report their victimization to the police or other public authorities. For instance, a survey conducted in five large metropolitan areas found that only about 50% of most crimes—including rape—are ever reported to the police. With this in mind, Feldman-Summers and Ashworth (1980) investigated the extent to which various factors influence intentions to report sexual victimization to various agencies and individuals. As part of this study, each participant was asked the likelihood that she would report a rape to her husband/boyfriend, the police, her parents, and a female friend. A separate probability judgment was obtained for each source. The judgments

were made on a 1-to-7 scale, with higher scores indicating a greater likelihood of reporting the crime. Hypothetical data representative of the results of Feldman-Summers and Ashworth are presented in the table. Analyze these data, draw a conclusion, and write up your results using the principles given in the Method of Presentation section.

Individual	Husband/ boyfriend	Police	Parents	Female friend
1	7	6	3	4
2	6	5	4	5
3	7	6	1	2
4	6	5	6	7
5	7	6	3	4
6	6	5	4	5

46. A researcher was interested in the effectiveness of a particular weight-loss program. Eight subscribers to this program were weighed before beginning the program, at the completion of the program, and, to determine the long-term effect of the program, again 6 months later (this constitutes the *follow-up* measure). The data for the study are presented in the table. Analyze these data, draw a conclusion, and write up your results using the principles given in the Method of Presentation section.

Individual	Before Program	After Program	Follow-up
1	194	186	193
2	246	227	241
3	211	195	209
4	185	172	188
5	207	204	207
6	239	228	236
7	188	175	180
8	226	211	230

Pearson Correlation and Regression: Inferential Aspects

14.1 Use of Pearson Correlation

In Chapter 5, we noted the conditions under which Pearson correlation is used to determine the extent of linear relationship between two variables. Formally stated, Pearson correlation is typically used to analyze the relationship between variables under these conditions:

1. Both variables are quantitative in nature and are measured on a level that at least approximates interval characteristics.

2. The two variables have been measured on the same individuals.

3. The observations on each variable are *between-subjects* in nature.

Our discussion of correlation in Chapter 5 focused on the *description* of the relationship between two variables. In practice, the most common use of correlation procedures is to make inferences about correlation coefficients in populations based on sample data. For instance, the correlation between traditionalism and ideal family size is .66 for the ten individuals discussed in Chapter 5. The question of interest is whether we can conclude that a correlation exists between these two variables in the *population,* given a correlation of this magnitude in the observed *sample.* This chapter considers the procedure for drawing inferences about population correlation coefficients from sample correlation coefficients.

To develop the logic of correlational inference, we restate the linear model developed in Chapter 5, but we now use population notation with Greek letters for the slope and intercept:

$$Y = \alpha + \beta X + \varepsilon \qquad\qquad [14.1]$$

The model states that a person's score on Y is a linear function of X, with α representing the intercept and β representing the slope. It is rare in the behavioral sciences to have a perfect linear relationship between Y and X, and this is why there is another term in Equation 5.1, ε (lowercase Greek e, called "epsilon"). This term is the **error score** and reflects all factors that are uncorrelated with X that influence Y. It is because of these other factors that the absolute correlation between Y and X is not always 1.00 (as well as the fact that relationships are sometimes nonlinear rather than linear). If everyone in the population has an ε score of 0, then the standard deviation of the ε scores across individuals is also 0, and Y is a perfect linear function of X. In this case, the absolute correlation between X and Y is 1.00. As ε scores deviate from 0, Y scores deviate from $\alpha + \beta X$ values, and the relationship between the variables is no longer perfectly linear. In this chapter, we discuss procedures for estimating α, β, and the standard deviation of ε, as well as the population correlation between X and Y, ρ (lowercase Greek r, called "rho").

14.2   Inference of a Relationship Using Pearson Correlation

As noted in previous chapters, even if a relationship is observed between two variables in sample data, this does not necessarily mean that a relationship exists between the variables in the corresponding population. A relationship might exist in a sample even though it does not exist in the population because of sampling error. The task at hand, then, is to test the viability of a sampling error interpretation. We demonstrate the procedures for doing this using the traditionalism and ideal family size example from Chapter 5.

Null and Alternative Hypotheses

Recall that a correlation coefficient can range from −1.00 to +1.00. A coefficient of 0 means that there is no linear relationship between the variables. In contrast, nonzero correlation coefficients indicate some approximation to a linear relationship. We can therefore state the null and alternative hypotheses as follows:

$$H_0: \quad \rho = 0$$

$$H_1: \quad \rho \neq 0$$

where ρ represents the true correlation in the population. The null hypothesis states that the population correlation between the two variables is 0. The alternative hypothesis states that the correlation between the two variables is not 0 in the population.

Based on the data collected in an investigation, we want to make a decision with respect to the two hypotheses. We use hypothesis testing logic analogous to that developed in Chapter 8. In the context of Pearson correlation, this involves converting the value of r in the sample to a t value under the assumption that the null hypothesis is true and then determining whether this value of t falls in the rejection region. If it does, then we reject the null hypothesis and conclude that there is some degree of a linear relationship between the two variables. Otherwise, we fail to reject the null hypothesis.

Sampling Distribution of the Correlation Coefficient

Consider a population in which the correlation coefficient between two variables is 0. From the population, we select a random sample of ten individuals and compute the correlation coefficient. We might find that $r = .15$. The fact that r does not equal 0 is because of sampling error, as discussed in previous chapters. Now suppose we randomly select another sample of ten individuals from the population. For this sample, the correlation might be .03. In principle, we could continue this process a large number of times until we have calculated r for all possible samples of size ten. The resulting distribution of correlation coefficients based on all random samples of size ten would constitute a **sampling distribution of the correlation coefficient** and would have many of the same properties as the sampling distributions we have discussed in earlier chapters.

The mean of a sampling distribution of the correlation coefficient is approximately ρ, the true population correlation. When $\rho = 0$ and the scores in the population are bivariate normally distributed (that is, the distribution of Y scores at any value of X is normal), then as N increases, the distribution of the sample correlation coefficients tends, somewhat slowly, toward a normal distribution. With sample sizes larger than 50, the standard error of this distribution is approximately equal to $1/\sqrt{N}$. When $\rho \neq 0$, the sampling distribution is skewed. Figure 14.1 illustrates sampling distributions for $\rho = -.80$, $\rho = 0$, and $\rho = +.80$ for $N = 10$.

Testing the Statistical Significance of a Correlation Coefficient

We can test the null hypothesis that $\rho = 0$ by converting the sample correlation coefficient to a statistic that has a sampling distribution that closely approximates the t distribution with $N - 2$ degrees of freedom.* The relevant formula is

$$t = \frac{r}{\sqrt{(1 - r^2)/(N - 2)}} \qquad [14.2]$$

---

* Null hypotheses other than $\rho = 0$ can be tested using the procedures discussed in Appendix 14.1.

FIGURE 14.1   **Sampling Distributions of *r* for Three Values of ρ for *N* = 10 (From Minium, 1970)**

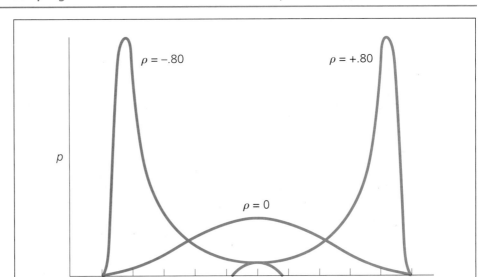

For the traditionalism and ideal family size study,

$$t = \frac{.66}{\sqrt{(1 - .66^2)/(10 - 2)}} = 2.44$$

For an alpha level of .05, nondirectional test, and $N - 2 = 10 - 2 = 8$ degrees of freedom, the critical values of *t* from Appendix D are ±2.306. Since the observed *t* value is greater than +2.306, we reject the null hypothesis. A sample correlation coefficient of .66 based on 8 degrees of freedom is too large to attribute to sampling error, assuming the null hypothesis is true. We therefore conclude that the population correlation is nonzero—that is, that there is some degree of a linear relationship between traditionalism and ideal family size.

Tabled Values of *r*

Because the calculation of *t* using Equation 14.2 depends on only the values of *r* and *N*, it is possible, through algebraic manipulation, to determine values of *r* that will lead to rejection of the null hypothesis that ρ = 0, given the sample size. This is done in Appendix H, which contains a table of critical values of *r* for selected degrees of freedom of $N - 2$. If the observed value of *r* is greater than the positive critical value from this table, or less than the corresponding negative critical value, we will reject the null hypothesis. Otherwise, we will fail to reject the null hypothesis. For instance, based on an alpha level of .05, nondirectional test, and $N - 2 = 10 - 2 = 8$ degrees of freedom, the critical values of *r* for the traditionalism and ideal family size example are ±.632. Since the observed correlation of .66 is greater than +.632, we reject the null hypothesis and conclude that there is some degree of a linear relationship between traditionalism and ideal family size. This is the same conclusion we reached using the t test procedure. Given its greater computational ease, we follow the strategy of referring to Appendix H when testing the significance of correlation coefficients in the remainder of this chapter.

Study Exercise 14.1

In Study Exercise 5.2, the correlation between voters' perceptions that a candidate supported labor unions and their willingness to vote for her was calculated to be .50 for a sample of nine individuals. Test the viability of the hypothesis that there is no linear relationship between the two variables in the population.

Answer   The null and alternative hypotheses are

$$H_0: \quad \rho = 0$$

$$H_1: \quad \rho \neq 0$$

For an alpha level of .05, nondirectional test, and $N - 2 = 9 - 2 = 7$ degrees of freedom, the critical values of r from Appendix H are $\pm.666$. Since .50 is neither greater than $+.666$ nor less than $-.666$, we fail to reject the null hypothesis of no linear relationship between the two variables.

Assumptions of Pearson Correlation

The test of the null hypothesis that $\rho = 0$ is based on the following assumptions:

1. The sample is independently and randomly selected from the population of interest.

2. The population distributions of X and Y are such that their joint distribution (that is, their scatterplot) represents a **bivariate normal distribution**. This means that the distribution of Y scores at any value of X is normal in the population

3. The Y scores at each value of X have equal variances in the population.

In addition, both variables should be quantitative in nature and measured on a level that at least approximates interval characteristics. Although it is important that the first assumption not be violated, Van den Brink (1988) has shown that the t test of the correlation coefficient is robust to large deviations from bivariate normality when the sample size is larger than 15. Violation of the assumption of equal Y-score variability at each value of X can be problematic, especially if one is going to pursue estimation of the population regression equation. Tests for evaluating variance heterogeneity of Y scores and remedial solutions are discussed in Judge, Griffiths, Hill, Lutkephol, and Lee (1985).

14.3 Strength of the Relationship

The strength of the relationship between two variables in a correlational analysis can be represented by eta-squared in the form of the ratio of explained variability ($SS_{EXPLAINED}$) to total variability (SS_{TOTAL}). However, it is not necessary to actually calculate $SS_{EXPLAINED}$ and SS_{TOTAL} because eta-squared bears a direct relationship to the correlation coefficient. Specifically,

$$r^2 = eta^2 = \frac{SS_{EXPLAINED}}{SS_{TOTAL}} \qquad [14.3]$$

Eta-squared and hence r^2, represent the proportion of variability in one variable that is associated with the other variable. This is sometimes conceptualized as the proportion of variability that is *shared* by the two variables.

In the example on traditionalism and ideal family size, r was found to equal .66. Thus, r^2, formally known as the **coefficient of determination,** is equal to

$$r^2 = .66^2 = .44$$

The proportion of variability in ideal family size that is associated with traditionalism is .44, as indexed by a linear model. This represents a strong effect.

Another useful index of the strength of the relationship between two variables is the slope of the regression line. This indicates the number of units Y is predicted to change given a one-unit change in X. From our calculations in Chapter 5, the slope for regressing desired family size onto traditionalism is .58. For every unit that traditionalism increases, the desired family size is predicted to increase by just over half a child (.58). If the slope had been larger, then the effect of traditionalism on desired family size would be greater (e.g., if the slope were 1.13, for every unit that traditionalism increases, desired family size would be predicted to increase by 1.13 children). The use of the slope as an index of effect size makes sense only when the correlation between Y and X is relatively high. Otherwise, a linear model is being used to characterize an effect when the linear model is either not applicable or is of little explanatory value.

14.4 Nature of the Relationship

The nature of the relationship between two correlated variables is determined by examining the sign of the correlation coefficient observed in the sample. If the null hypothesis is rejected and the sample correlation coefficient is positive, then the appropriate conclusion is that the population correlation coefficient is also positive—that is, that the variables approximate a direct linear relationship. On the other hand, if the null hypothesis is rejected and the sample correlation coefficient is negative, the appropriate conclusion is that the two variables approximate an inverse linear relationship in the population. Given an observed correlation of .66 in the traditionalism and ideal family size example, we conclude that the variables approximate a direct linear relationship.

14.5   Planning an Investigation Using Pearson Correlation

Appendix E.4 contains tables of the sample sizes necessary to achieve various values of power for Pearson correlation for tests of the null hypothesis that $\rho = 0$. Table 14.1 reproduces a portion of this appendix for an alpha level of .05, nondirectional test. The first column presents various values of power, and the column headings are population values of the correlation coefficient squared (that is, ρ^2). To illustrate the use of this table, suppose the desired level of power is .80 and the researcher suspects that the strength of the relationship in the population corresponds to a rho-squared of .15. Then the number of participants that should be sampled is 49. Inspection of Table 14.1 and Appendix E.4 provides a general appreciation for the relationships among directionality, the alpha level, sample size, the strength of the relationship one is trying to detect in the population, and power for Pearson correlation.

14.6   Method of Presentation for Pearson Correlation

Reports of a correlation analysis should present the alpha level that was used, the degrees of freedom, the observed value of r, and the significance level. It is conventional to also give the sample mean and the standard deviation estimate for each variable. It is not necessary to explicitly state the strength of the relationship between the variables because this is indicated by the square of the correlation coefficient. Although the nature of the relationship is indicated by the sign of the correlation coefficient, researchers often choose to make this explicit by stating it ver-

Table 14.1   **Approximate Sample Sizes Necessary to Achieve Selected Levels of Power for Alpha = .05, Nondirectional Test, As a Function of the Population Correlation Coefficient Squared**

	Population Correlation Coefficient Squared									
Power	.01	.03	.05	.07	.10	.15	.20	.25	.30	.35
.25	166	56	34	25	17	12	9	8	6	6
.50	384	127	76	54	38	25	19	15	12	10
.60	489	162	97	69	48	31	23	18	15	13
.67	570	188	112	80	55	36	27	21	17	14
.70	616	203	121	86	59	39	29	23	18	15
.75	692	228	136	96	67	43	32	25	20	17
.80	783	258	153	109	75	49	36	28	23	19
.85	895	294	175	124	85	56	41	32	26	21
.90	1,046	344	204	144	100	65	47	37	30	25
.95	1,308	429	255	180	124	80	58	46	37	30
.99	1,828	599	355	251	172	111	81	63	50	42

bally. The results for the investigation of traditionalism and ideal family size discussed earlier might be reported as follows:*

Results

A Pearson correlation addressed the relationship between traditionalism ($\underline{M}$ = 5.00, $\underline{SD}$ = 2.98) and ideal family size ($\underline{M}$ = 5.00, $\underline{SD}$ = 2.58). For an alpha level of .05, the correlation was found to be statistically significant, $\underline{r}$(8) = .66, $\underline{p}$ < .05, indicating that these two variables are positively related.

14.7   Examples from the Literature

Personality Characteristics and Creativity

Psychologists have studied extensively the process of creativity and what distinguishes creative individuals from noncreative individuals. For instance, Barron (1965) reported the results of research designed to illuminate personality differences between certain creative and noncreative persons. As part of this research, 44 female mathematicians were invited to attend a 3-day interview session at the Institute of Personality Assessment and Research (IPAR) at the University of California at Berkeley. These women were chosen by a "nomination" technique and included 16 women considered by a panel of mathematics experts to be "the most original and important women mathematicians in the United States and Canada" and 28 women who were nationally recognized but not distinctively creative in the mathematics field.

While at IPAR, the mathematicians interacted in both formal and informal settings with the staff of the institute, who were kept unaware of who the creative versus noncreative women were. The staff rated each woman on a number of personality characteristics, and these ratings were correlated with ratings of the women's creativity as previously generated by the above-mentioned panel of mathematics experts. The strongest correlations, all of which are statistically significant, are shown in the table at the top of page 400. In each case, these represent strong effects, as indexed by r^2.

Barron summarized these findings as follows: "The emphasis is upon genuine unconventionality, high intellectual ability, vividness or even flamboyance of character, moodiness and preoccupation, courage, and self-centeredness. These are people who stand out, and who probably are willing to strike out if impelled to do so."

---

* The reported sample means are as derived in Table 5.2. The standard deviation estimates were calculated by applying the usual procedures to the sums of squares of 80 for traditionalsim and 60 for ideal family size, as also identified in Table 5.2.

Positive Correlations			Negative Correlations		
Personality characteristic	r	r^2	Personality characteristic	r	r^2
Thinks and associates to ideas in unusual ways; has unconventional thought processes	.64	.41	Judges self and others in conventional terms like "popularity," the "correct thing to do," "social pressures," and		
Is an interesting, arresting person	.55	.30	so forth	−.62	.38
Tends to be rebellious and nonconforming	.51	.26	Is a genuinely dependable and responsible person	−.45	.20
Genuinely values intellectual and cognitive matters	.49	.24	Behaves in a sympathetic or considerate manner	−.43	.18
Appears to have a high degree of intellectual capacity	.46	.21	Favors conservative values in a variety of areas	−.40	.16
Is self-dramatizing; histrionic	.42	.18	Is moralistic	−.40	.16
Has fluctuating moods	.40	.16			

Desire to Control and Gambling Frequency

Among the factors that have been hypothesized to influence gambling behavior is the belief that one can exert personal control over outcomes that are largely determined by chance. This belief, referred to as *illusion of control,* is supposedly induced by elements of gambling situations that hint at potential control (for instance, the observation of horses during prerace warm-ups or the decision to hold or draw cards at a poker game).

Burger and Smith (1985) proposed that one's characteristic desire to control events in one's life should relate to gambling behavior on games that hold an element of illusion of control but not on games that do not. To test this hypothesis, they asked 18 members of Gamblers Anonymous to complete a scale assessing characteristic desire to control and to indicate the frequency with which they had bet on a number of gambling games. Some of these games (casino games, lotteries) were conceptualized as not inducing an illusion of control, whereas others (poker and card games, horse racing, sports events) were considered capable of inducing such an illusion. For each type of game, a total frequency score was reached by adding the appropriate frequency ratings. As predicted, desire-to-control scores were found to correlate significantly with gambling frequency on illusion-of-control games, $r(16) = .46$, $p < .05$, but not on the other games, $r(16) = .04$, $p > .10$. As indexed by r^2, the strengths of the relationships were .21 (representative of a strong effect) and .002 (representative of a weak effect), respectively.

14.8 Regression

If a test of Pearson correlation yields statistically significant results, researchers sometimes pursue regression analysis. In Chapter 5, we noted that a regression equation can be used to identify the value of Y that is predicted to be paired with an individual's score on X. Our discussion at that time focused on the scores of individuals who were members of the group on which the regression equation was based. An important characteristic of regression is that prediction procedures can be extended to individuals who were not included in the original data set. This is accomplished as follows: Scores on X and Y are determined for a sample of individuals. The procedures presented in Chapter 5 are then used to derive a regression equation, which takes the general form $\hat{Y} = a + bX$, based on the sample data. The values of a and b in this equation are estimates of α and β, respectively, in the linear model presented in Equation 14.1. The regression equation can then be applied to individuals outside the original sample to make predictions about their scores on Y from their scores on X. This is done by substituting an individual's X score into the regression equation. The resulting value of $\hat{Y}$ is that individual's predicted score on Y.

In the context of regression, prediction merely refers to the fact that we are making inferences about one variable from a second variable; it does not imply that the latter variable necessarily precedes the former. For instance, individuals might formulate their opinions about ideal family size (variable Y) several years before they develop their traditionalism orientation (variable X). Nevertheless, we can use regression procedures to get an idea of what a person's ideal family size might be given knowledge of her traditionalism score. The variable being predicted, Y, is formally known as the *dependent* or **criterion variable**. The variable from which predictions are made, X, is formally known as the *independent* or **predictor variable.***

Consider the following example: The personnel director of an insurance company has rated every underwriter who has worked for the company during the past 5 years on a scale of 1 to 10, indicating how successful the underwriter has been during his or her tenure. The higher the rating, the more successful the underwriter. Each underwriter has also completed an aptitude test designed to measure potential on the job. Scores on this test could range from 0 to 100, with higher values indicating greater potential. Suppose correlation analysis yielded a statistically significant (positive) Pearson correlation between test scores (X) and success ratings (Y). Given the significance of this relationship, the personnel director might construct a regression equation to help him screen future applicants for underwriting jobs by predicting who is likely to be successful, operationalized as a success rating of 8 or greater. Suppose this equation took the form

$$\hat{Y} = .50 + .10X$$

---

* Although the use of X to represent the independent (predictor) variable in the case of regression differs from the use of X to represent the dependent variable as presented in previous chapters, this notation is consistent with conventional practice in both instances.

Furthermore, suppose an applicant took the test and obtained a score of 85. If we substitute this score into the regression equation, we find a predicted success rating of

$$\hat{Y} = .50 + (.10)(85) = 9.00$$

A success rating of 9.00 exceeds the minimum score of 8; thus, based on this criterion, this applicant remains a viable candidate for employment. On the other hand, the predicted success rating for an applicant who obtained a test score of 40 is

$$\hat{Y} = .50 + (.10)(40) = 4.50$$

A success rating of 4.50 is lower than the minimum score of 8, so this individual will probably not be considered for a position with the company. The success ratings for other applicants can be similarly estimated by substituting their test scores into the regression equation.

The Estimated Standard Error of Estimate

Unless the two variables are perfectly correlated, the use of a regression equation to predict scores on Y from scores on X will have some degree of error associated with it. For instance, the regression equation for predicting ideal family size from traditionalism was found in Chapter 5 to be $\hat{Y} = 2.10 + .58X$. The predicted ideal family size for a traditionalism score of 8 according to this equation is $2.10 + (.58)(8) = 6.74$ or, rounded to the nearest whole number, 7. Surely, though, not every woman who gets a score of 8 on the traditionalism questionnaire desires a family this large. What is needed to gain insight into the predictive utility of a regression equation is an index of how much error will occur when predicting Y from X. Such an index is provided by the **estimated standard error of estimate.**

When we introduced the standard error of estimate in Chapter 5, it was defined in Equation 5.12 as

$$s_{YX} = \sqrt{\frac{\sum (Y - \hat{Y})^2}{N}}$$

where s_{YX} is the standard error of estimate, Y is an individual's actual score on Y, $\hat{Y}$ is an individual's predicted score on Y based on the regression equation, and N is the sample size. Although this formula represents the average amount of predictive error *for the sample data,* it is not an appropriate measure of how much error occurs when predicting scores on Y across the population. The population value of the standard error of estimate (σ_{YX}) can be estimated from sample data as follows:

$$\hat{s}_{YX} = \sqrt{\frac{\sum (Y - \hat{Y})^2}{N - 2}} \qquad [14.4]$$

where $\hat{s}_{YX}$ is the estimated standard error of estimate and all other terms are as previously defined. The only difference between this formula and the formula for the sample standard error of estimate from above is that the denominator reflects the associated degrees of freedom $(N - 2)$ rather than the sample size (N). The degrees of freedom are equal to $N - 2$ because both the slope and the intercept of the regression line must be estimated from sample data. The estimated standard error of estimate thus estimates the average error that will be made across individuals when predicting scores on Y from the regression equation. Indeed, it is an estimate of the standard deviation of the ε scores in the linear model presented in Equation 14.1.

Analogous to the computational formulas for the sample standard error of estimate presented in Chapter 5 (Equations 5.13 and 5.14), in practice the estimated standard error of estimate is most efficiently calculated using the equation

$$\hat{s}_{YX} = \hat{s}_Y \sqrt{\left(\frac{N-1}{N-2}\right)(1 - r^2)} \qquad [14.5]$$

if one is working with the standard deviation estimate for Y, and using the equation

$$\hat{s}_{YX} = \sqrt{\frac{SS_Y(1 - r^2)}{N - 2}} \qquad [14.6]$$

if one is working with the sum of squares for Y.

The interpretation of $\hat{s}_{YX}$ directly follows that for s_{YX}. First, its absolute magnitude provides a measure of the amount of predictive error that occurs. Second, $\hat{s}_{YX}$ can be compared with the estimated standard deviation of Y. The estimated standard deviation of Y estimates what the average error in prediction would be if one were to predict that everyone in the population had a Y score equal to the mean of Y. If X helps to predict Y, then the estimated standard error of estimate will be smaller than the estimated standard deviation of Y, and the better the predictor X is, the smaller the estimated standard error of estimate will be.

14.9 Numerical Example

Generally, a participant in a computer dating program completes a questionnaire indicating his or her interests and habits. Responses to the questions are then analyzed, and the person is matched with a member of the opposite gender who has similar qualities. Underlying the matching of individuals is an important assumption: People with similar interests will be attracted to one another.

To test the validity of this assumption, suppose a researcher asked 15 women who had been on a computer date to indicate how attracted they were to their assigned partners on a scale from 1 to 10. A score of 1 indicated no attraction and a score of 10 indicated very great attraction. Before being matched, participants had responded to a ten-item questionnaire about their interests and habits, as had their eventual partners. For purpose of the study, the similarity between assigned partners differed. Some individuals were randomly assigned partners who responded the same way on all ten questions, some were assigned partners who responded the same way on only nine of the questions, and so on, all the way down to one similar questionnaire response. Thus, similarity between partners could range from 1 to 10, with low scores indicating little similarity and high scores indicating a great deal of similarity. The data for the experiment are listed in columns 2 and 3 of Table 14.2. In this table, partner similarity is represented as variable X and attraction is represented as variable Y.

The null hypothesis of no linear relationship states that the population correlation between these two variables is 0. The alternative hypothesis states that the population correlation is nonzero. These hypotheses can be formally stated as

$$H_0: \quad \rho = 0$$

$$H_1: \quad \rho \neq 0$$

The correlation coefficient can be represented as (see Equation 5.4)

$$r = \frac{SCP}{\sqrt{SS_X SS_Y}}$$

and calculated as (see Equation 5.8)

$$r = \frac{\sum XY - \frac{(\sum X)(\sum Y)}{N}}{\sqrt{\left[\sum X^2 - \frac{(\sum X)^2}{N}\right]\left[\sum Y^2 - \frac{(\sum Y)^2}{N}\right]}}$$

TABLE 14.2   **Data and Calculation of Intermediate Statistics for Similarity and Attraction Experiment**

Individual	X	Y	X^2	Y^2	XY
1	10	8	100	64	80
2	8	6	64	36	48
3	6	4	36	16	24
4	4	2	16	4	8
5	2	3	4	9	6
6	10	6	100	36	60
7	8	8	64	64	64
8	6	5	36	25	30
9	7	9	49	81	63
10	4	5	16	25	20
11	1	3	1	9	3
12	3	1	9	1	3
13	5	3	25	9	15
14	7	5	49	25	35
15	9	7	81	49	63
	$\sum X = 90$	$\sum Y = 75$	$\sum X^2 = 650$	$\sum Y^2 = 453$	$\sum XY = 522$
	$\bar{X} = 6.00$	$\bar{Y} = 5.00$			

The intermediate statistics necessary for the application of this formula are computed in Table 14.2. From these we can derive the correlation coefficient:

$$r = \frac{522 - \dfrac{(90)(75)}{15}}{\sqrt{\left(650 - \dfrac{90^2}{15}\right)\left(453 - \dfrac{75^2}{15}\right)}}$$

$$= \frac{72}{\sqrt{(110)(78)}} = .78$$

For an alpha level of .05, nondirectional test, and $N - 2 = 15 - 2 = 13$ degrees of freedom, the critical values of r from Appendix H are ±.514. Since .78 is greater than +.514, we reject the null hypothesis and conclude that some degree of a linear relationship exists between similarity with one's partner and attraction.

The strength of the relationship is indicated by the square of the correlation coefficient:

$$r^2 = .78^2 = .61$$

The proportion of variability in attraction ratings that is associated with partner similarity is .61, as indexed by a linear model. This represents a strong effect.

The nature of the relationship is indicated by the sign of the correlation coefficient. In this case, the correlation is positive, indicating that as similarity between partners increases, so does attraction.

If the researcher was interested in identifying the regression equation for predicting attraction (Y) from similarity scores (X), he would first determine the slope using Equation 5.10:

$$b = \frac{SCP}{SS_X}$$

$$= \frac{72}{110} = .65$$

The obtained value of b can then be substituted into Equation 5.11 along with the means of the two variables to yield the intercept:

$$a = \overline{Y} - b\overline{X}$$

$$= 5.00 - (.65)(6.00) = 1.10$$

The regression equation is thus

$$\hat{Y} = a + bX$$

$$= 1.10 + .65X$$

This equation can be used to make predictions about the outcome of dating situations. For instance, a woman is unlikely to enjoy a computer date with a man who agrees with her on only three questionnaire items because the predicted attraction level for a similarity score of 3 is $1.10 + (.65)(3) = 3.05$. In contrast, the predicted attraction score is $1.10 + (.65)(10) = 7.60$ when the two individuals agree on all ten questions.

These predictions are, of course, subject to error. Specifically, the standard error of estimate in this example is estimated to be

$$\hat{s}_{YX} = \sqrt{\frac{SS_Y(1 - r^2)}{N - 2}}$$

$$= \sqrt{\frac{78(1 - .78^2)}{15 - 2}} = 1.53$$

which indicates that, on the average, predicted Y (attraction) scores are estimated to deviate from actual Y scores by 1.53 units. This represents a relatively small degree of error. To gain further insight into the predictive utility of the regression equation, we can compare the estimated standard error of estimate with the estimated standard deviation of Y. Since $SS_Y = 78$ and $N = 15$, $\hat{s}_Y = \sqrt{SS_Y /(N - 1)} = \sqrt{78 /(15 - 1)} = 2.36$. Thus, predicting scores on Y from scores on X leads to considerably less error (an estimated average prediction error of 1.53 units) than if all Y scores were predicted to be equal to $\bar{Y}$ (an estimated average prediction error of 2.36 units).

As noted in Chapter 9, the ability to make a causal inference between two variables is a function of one's research design, not the statistical technique used to analyze the data that are yielded by that research design. Thus, as discussed in Section 5.4, correlation does not necessarily imply causation. Note, however, that the present study used an experimental research strategy: Individuals were *randomly assigned* partners who ranged in similarity from 1 to 10. To the extent that confounding variables were controlled, we can thus infer that in this investigation similarity between partners *caused* attraction to vary as it did. As with all statistical techniques, causality can similarly be inferred from a statistically significant analysis anytime an experimental, as opposed to an observational, research strategy is used (assuming, of course, that confounding variables are controlled). We should note, however, that in practice most applications of correlation and regression use an observational approach.

The above results relate only to the linear relationship between the two variables. If we were interested in the extent of nonlinear relationship, we could use curvilinear regression procedures, as discussed in Chapter 5. The use of regression to predict attraction scores from similarity scores is predicated on the fact that the researcher explicitly manipulated similarity so that he could determine how it affects attraction. Thus, there would be no justification for deriving the regression equation for predicting similarity from attraction in this example. Since the range of similarity scores in the sample was from one to ten similar questionnaire responses, our conclusions apply only across these values. For instance, we are unable to infer the effect of *no* similarity (no similar questionnaire responses) on attraction ratings or what attraction ratings would be if assigned dating partners shared more than ten interests and habits.

Applications

to the Analysis

of a Social Problem

As noted in previous chapters, evidence suggests that the quality of the relationship between a teen and his or her mother has an impact on adolescent risk behavior. For example, data suggest that there is a negative correlation between satisfaction with the maternal relationship and sexual activity: Adolescents who have relatively poor relationships with their mother tend to be more sexually active. There are at least two interpretations of this correlation. The first holds that because teens are unhappy at

home, they reject parental values and seek out risky behaviors to assert their independence. Having distanced themselves from parents, such teens are more susceptible to peer pressure to engage in sexual behavior, with the result being increased sexual activity. This interpretation, in essence, holds that the poor relationship between parents and teens causes increased sexual activity.

A second interpretation reverses the causal
(continued)

14.10   Method of Presentation for Regression

The results of a regression analysis might be presented in several ways. One approach is to include the regression equation and the estimated standard error of estimate along with the information for the correlation. For the example on similarity and attraction, this might appear as follows:*

Results

Using an alpha level of .05, a Pearson correlation between the number of similar questionnaire responses ($\underline{M}$ = 6.00, $\underline{SD}$ = 2.80) and attraction to one's partner ($\underline{M}$ = 5.00, $\underline{SD}$ = 2.36) was found to be statistically significant, $\underline{r}$(13) = .78, $\underline{p}$ < .01. The regression equation for predicting attraction from partner similarity was found to be $\hat{\underline{Y}}$ = 1.10 + .65$\underline{X}$, and the estimated standard error of estimate was found to be 1.53.

On occasion, written reports of regression results are supplemented by a scatterplot that shows the observed data points and the regression line.

In practice, regression analyses that involve only one predictor variable are rarely encountered in the behavioral sciences. Far more common is the practice of predicting a criterion variable from two or more predictor variables. This technique, known as *multiple regression*, will be reviewed in Chapter 18.

---

* The standard deviation estimate for partner similarity was calculated from the sum of squares for this variable as follows: $\hat{s}_x = \sqrt{SS_x /(N - 1)} = \sqrt{110/(15 - 1)} = 2.80$.

direction between the two variables. According to this view, teens become sexually active for reasons unrelated to the parental relationship (e.g., because of peer pressure or hormonal influences). At some point, the parent discovers that the teen has become sexually active and this causes arguments between the parent and teen. According to this interpretation, increased sexual activity on the part of the teen causes a deterioration in the relationship between the parent and the teen because the parent learns that the teen is sexually promiscuous.

We can gain some perspective on these interpretations from the parent–teen communication study described in Chapter 1. As noted in previous chapters, the parent–teen data included a measure of the teen's satisfaction with the maternal relationship as well as self-reports of the number of times the teen engaged in sexual intercourse during the past 6 months. We also obtained a measure of whether the mother thought the teen had ever engaged in sexual intercourse. Suppose that we select for analysis only those adolescents whose mothers stated that their child has *never* engaged in sexual intercourse. If the second interpretation above is correct (i.e., the quality of the parent–teen relationship deteriorates because the parent finds out that the child is promiscuous), then the correlation between relationship quality and sexual activity for these adolescents should be near 0, as none of the mothers is aware of sexual activity on the part of the teen. If the first interpretation is correct (i.e., a poor relationship with the parent leads to increased sexual activity on the part of the teen), then a negative correlation between relationship satisfaction and sexual activity should exist even for adolescents whose parents do not think their child is sexually active.

Relationship satisfaction was assessed using the measure described in Chapter 12. Recall that we asked adolescents to indicate how much they agreed or disagreed with each of 11 statements about different facets of their maternal relationship (e.g., "I am satisfied with the way my mother and I communicate with each other," "I am satisfied with the way my mother and I resolve conflicts," "I am satisfied with the respect my mother shows me," "I am satisfied with the emotional support my mother gives me"). Each statement was rated on a five-point scale having strongly disagree, moderately disagree, neither agree nor disagree, moderately agree, and strongly agree as response options. Responses were scored from 1 to 5, with

higher values indicating greater agreement with the statement. After performing item analyses to ensure that each was appropriate psychometrically, we summed the responses across the 11 statements to yield an overall index of relationship satisfaction. Scores on this measure could range from 5 to 55, with higher values indicating greater satisfaction. The measure is probably ordinal in character but probably approximates interval level characteristics sufficiently to justify the application of Pearson correlation. We conclude this based on our successful use of the measure in our prior research and the Monte Carlo research of Havlicek and Peterson (1977), which indicates that Pearson correlation is relatively robust to violations of intervalness (see also Davison & Sharma, 1988).

We performed a correlational analysis between relationship satisfaction and the number of instances of reported sexual intercourse during the past 6 months for only those teens whose mothers believed that their child had never been sexually active. The first step in the analysis was to gain an appreciation for the response distributions of the two variables. Figure 14.2 presents a grouped frequency histogram for each variable, with a normal distribution superimposed in both cases. Both variables exhibit marked skewness, suggesting that the assumption of bivariate normality is violated. This is because normality of each variable is a necessary condition for bivariate normality. However, given our sample size ($N = 450$), the t test of the correlation coefficient will be robust to even this degree of bivariate nonnormality in the population (Van den Brink, 1988).

The few extreme scores on the sexual intercourse variable raise concerns about potential effects of outliers on the correlation coefficient. Because of this, we decided to supplement our analyses with correlational analyses that are resistant to outliers. One such method is rank-order correlation, which will be discussed in Chapter 16. Another method is based on *M estimators* (see Hamilton, 1992, for a discussion). Both of these analyses yielded conclusions similar to those we report here based on traditional Pearson correlation.

The assumption of homogeneous variances for Y scores at each value of X was examined using procedures described by Judge and colleagues (1985). The data indicated sufficient heterogeneity to suggest that the correlation analysis should be approached with caution but not abandoned. However, the rank-order and M estimation analytic

**FIGURE 14.2   Grouped Frequency Histograms of (a) Sexual Intercourse
and (b) Relationship Satisfaction**

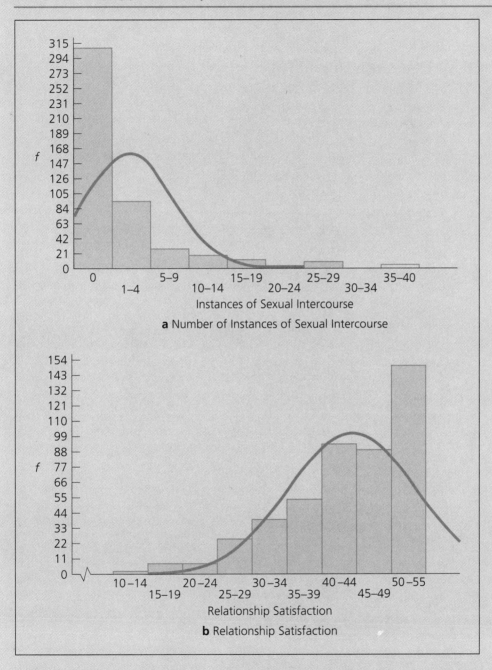

a Number of Instances of Sexual Intercourse

b Relationship Satisfaction

methods are not adversely affected by such heterogeneity, and all three analyses pointed to the same conclusion.

Table 14.3 presents the computer printout for the correlation analysis. The results are presented in the form of a correlation matrix (discussed in Chapter 5). We are interested in the *(continued)*

TABLE 14.3   Computer Output for Correlation and Regression for Sexual Intercourse and Relationship Satisfaction

```
- -  Correlation Coefficients  - -

                    NUMBER6              RSATIS

    NUMBER6         1.0000              -.1771
                    (   452)            (   450)
                    P = .               P = .000


    RSATIS          -.1771              1.0000
                    (   450)            (   467)
                    P = .000            P = .


    (Coefficient  /  (Cases)  /  2-tailed Significance)

    " . " is printed if a coefficient cannot be computed
```

entries where the row variable RSATIS (which is the relationship satisfaction variable) intersects with the column variable NUMBER6 (which is the sexual intercourse variable). The first entry is the correlation coefficient, which was −.18. The entry in parentheses below it is the number of cases on which the correlation was based.* The third entry is the *p* value for the correlation, assuming a nondirectional test. Because this is less than .05, the correlation is statistically significant. Even for teens whose mothers believe that their child has never engaged in sexual intercourse, there is a negative correlation between relationship satisfaction and sexual activity: The poorer the maternal relationship, the more sexually active the teen tends to be.

Figure 14.3 presents the scatterplot of these variables, with the least squares regression line indicated. This plot has considerable "noise" in it, and it is difficult to discern any linear trend. This is typical of scatterplots of real data. Neat, easily interpreted scatterplots such as those presented in Chapter 5 are typically found only for hypothetical data in statistics textbooks! Because of this, statisticians have developed *smoothing techniques* that remove much of the noise from the scatterplot and

thereby provide a greater sense of the fundamental trends in the data (Moore & McCabe, 1993). One smoothing technique is to present a scatterplot between the *X* variable and the mean *Y* score that occurs at each *X* value. Figure 14.4 presents such a graph. The linear trend is more apparent in the smoothed than in the original scatterplot. The smoothed scatterplot also suggests an "outlier" at a relationship satisfaction score of 20 (with a corresponding sexual intercourse score of 22). It turns out that this point represents a single individual whose impact on the correlation is not substantial, given the sample size.

In sum, the data are not consistent with the theory that teen sexual activity causes a deterioration in the parent–teen relationship as the parent learns of the teen's sexual behavior. Even for teens whose mothers believe that their child does not engage in sex, there is a statistically significant negative correlation between relationship satisfaction and sexual activity. This result does not rule out the possibility that sexual activity causes a deterioration in the parent–teen relationship. It only questions the particular mechanism that was suggested as being the basis of the causal direction (i.e., the

* Note that the correlation of NUMBER6 with itself is based on *N* = 452 and the correlation of RSATIS with itself is based on *N* = 467. Since the two variables had missing cases for different individuals, the sample size for the correlation of the two variables with each other was less than for either variable with itself.

FIGURE 14.3   Scatterplot of Sexual Intercourse and Relationship Satisfaction

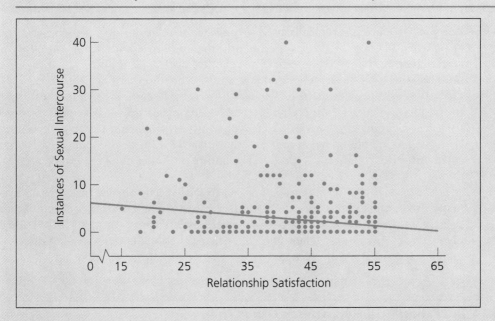

mother finds out about the teen's sexual activity and this causes the relationship to deteriorate). There are other possible explanations for the influence of sexual activity on relationship satisfaction. For example, adolescents who become highly committed to an intense, sexual relationship with another person may spend less time on homework (thereby causing their school performance to get worse) and less time with their parents, and this may cause the parent–teen relationship to deteriorate. These possibilities can be explored in more fine-grained analysis.

FIGURE 14.4   Smoothed Scatterplot of Sexual Intercourse and Relationship Satisfaction

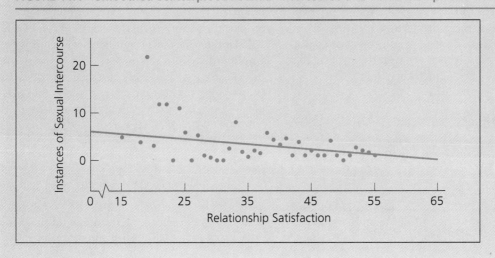

Summary

Pearson correlation is typically used to analyze the relationship between two variables when (1) both variables are quantitative in nature and are measured on a level that at least approximates interval characteristics, (2) the two variables have been measured on the same individuals, and (3) the observations on each variable are between-subjects in nature. The test from sample data of the hypothesis that the population correlation is equal to 0 is based on a sampling distribution of the correlation coefficient. This distribution permits an application of basic hypothesis testing principles. The strength of the linear relationship is indicated by the square of the correlation coefficient. The nature of the relationship is indicated by the sign of the correlation coefficient.

Regression equations can be applied to individuals outside of the sample to make predictions about their scores on variable Y (the criterion variable) from their scores on variable X (the predictor variable). The average error that will be made across individuals when making such predictions is estimated by the estimated standard error of estimate.

Appendix 14.1 Testing Null Hypotheses Other Than $\rho = 0$

Occasionally, an investigator will want to test a null hypothesis that ρ is equal to a value other than 0. As noted in Section 14.2, when ρ does not equal 0, the sampling distribution of the correlation coefficient is skewed. Fisher (1950) derived a logarithmic transformation of r, which we will symbolize r' (read "r prime") that has two desirable properties: (1) the sampling distribution of r' is approximately normally distributed irrespective of the value of ρ and (2) the standard error of r' is essentially independent of ρ.* Because of these properties, it is possible to convert a sample r into r' and then use the normal distribution to test the null hypothesis that ρ is equal to some value other than 0.

The formula for r' is

$$r' = .50[\log_e(1 + r) - \log_e(1 - r)] \qquad [14.7]$$

where $\log_e$ indicates the calculation of the natural logarithm of a number and r is the sample

correlation coefficient. All terms in Equation 14.7 are constants except for r. Thus, the value of r' is completely dependent on r, and r' is simply a rescaling of r. It is not necessary to actually calculate r' when testing a given null hypothesis because this has been done in Appendix I, which presents a table of values of r' for selected values of r.

Figure 14.5 presents sampling distributions of r' for $\rho = -.80$, $\rho = 0$, and $\rho = +.80$, for $N = 10$. Compare these distributions with the comparable sampling distributions of r in Figure 14.1. Unlike the sampling distributions of r, the sampling distributions of r' are similar in shape and variability. This reflects the fact that, as noted above, the standard error of r' is independent of ρ. Specifically, the standard error of r' (symbolized $\sigma_{r'}$) is equal to

$$\sigma_{r'} = \frac{1}{\sqrt{N - 3}} \qquad [14.8]$$

---

* Fisher referred to the r transformation using the symbol Z. Thus, you might see reference in the literature to "Fisher's r to Z transform." We use the symbol r' to avoid confusing the transformation with a z score.

FIGURE 14.5   **Sampling Distribution of *r'* for Three Values of ρ for *N* = 10**

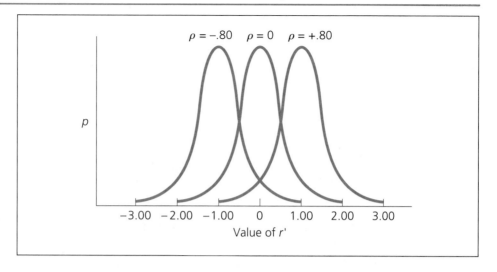

The test of a null hypothesis that ρ is equal to some value other than 0 is based on the following equation:

$$z = \frac{r' - \rho'}{\sigma_{r'}} \qquad [14.9]$$

where ρ′ is the log-transformed value corresponding to the hypothesized value of ρ, and *r'* and $\sigma_{r'}$ are as defined above. The observed value of *z* is then compared with the appropriate critical value(s) of *z* from the table of the normal distribution in Appendix B.

As an example of the above procedures, consider the following null and alternative hypotheses:

$$H_0: \quad \rho = .25$$

$$H_1: \quad \rho \neq .25$$

Suppose an *r* of .45 is observed for *N* = 100. Then the standard error of *r'* equals

$$\sigma_{r'} = \frac{1}{\sqrt{N - 3}}$$

$$= \frac{1}{\sqrt{100 - 3}} = .10$$

From Appendix I, the transformed value (*r'*) that corresponds to an *r* of .45 is .485, and the transformed value (ρ′) corresponds to a ρ of .25 is .255. The observed *z* score is thus

$$z = \frac{r' - \rho'}{\sigma_{r'}}$$

$$= \frac{.485 - .255}{.10} = 2.30$$

For an alpha level of .05, nondirectional test, the critical values of *z* are ±1.96. The observed *z* is greater than +1.96 and the null hypothesis is therefore rejected. Since the sample correlation of .45 is greater than the hypothesized population correlation of .25, we conclude that the population correlation between the two variables under study is greater than .25.

Exercises

Answers to asterisked (*) exercises appear at the back of the book. Answers to excercises with two asterisks are also worked out step by step in the Study Guide.

Exercises to Review Concepts

1. Under what conditions is Pearson correlation typically used to analyze a bivariate relationship?

*2. Explain the equivalence between the *t* test procedure for testing the significance of a correlation coefficient using Equation 14.2 and the procedure for testing the significance of a correlation coefficient by comparing the observed value of *r* with critical values of *r*.

*3. State the critical value(s) of *r* for Pearson correlation for an alpha level of .05 under each of the following conditions:
   a. $H_0: \rho = 0$, $H_1: \rho \neq 0$, $N = 32$
   b. $H_0: \rho = 0$, $H_1: \rho \neq 0$, $N = 22$
   c. $H_0: \rho = 0$, $H_1: \rho > 0$, $N = 22$
   d. $H_0: \rho = 0$, $H_1: \rho < 0$, $N = 15$
   e. $H_0: \rho = 0$, $H_1: \rho \neq 0$, $N = 67$

4. What are the assumptions underlying Pearson correlation?

5. What is the relationship between eta-squared and the correlation coefficient?

Use the following data to complete Exercises 6–8.

Individual	X	Y
1	4	9
2	7	11
3	9	14
4	9	10
5	2	8
6	12	14
7	4	8
8	5	9
9	13	10
10	3	7

*6. Compute the Pearson correlation coefficient and test for a relationship between X and Y using a nondirectional test.

*7. Compute the value of r^2. Does the observed value represent a weak, moderate, or strong effect?

*8. What is the nature of the relationship between the two variables?

Use the following data to complete Exercises 9–11.

Individual	X	Y
1	4	5
2	8	2
3	3	4
4	9	10
5	2	4
6	1	2
7	7	8
8	4	8
9	1	5
10	7	9
11	6	5
12	3	6
13	1	2

9. Compute the Pearson correlation coefficient and test for a relationship between X and Y using a nondirectional test.

10. Compute the value of r^2. Does the observed value represent a weak, moderate, or strong effect?

11. What is the nature of the relationship between the two variables?

For each of the studies described in Exercises 12–16, indicate the appropriate statistical test for analyzing the relationship between the variables. Assume that the underlying assumptions of the tests have been satisfied.

*12. An investigator examined the effects of two drugs on learning. Thirty participants were randomly assigned to take the first drug and a different 30 were randomly assigned to take the second drug. The mean scores on a learning task were compared for the two groups.

*13. A consumer psychologist tested preference for three brands of ice cream. One hundred individuals rated each of the three brands on a scale ranging from 1 to 15. The mean ratings for the three brands were compared.

*14. A sociologist studied the effect of the nuclear accident at Three Mile Island on attitudes toward nuclear energy. One hundred people had been interviewed prior to the accident and their attitudes measured on a ten-point scale. Five days after the accident they were reinterviewed and their attitudes measured again. The mean attitude scores were compared across interviews.

15. A researcher tested the relationship between age and blood pressure in a sample of 250 adults.

16. An investigator was interested in whether different driving conditions require different amounts of gas. Twenty midsize cars were driven 50 miles on smooth roads, 20 were driven 50 miles on hilly roads, and 20 were driven 50 miles on mountainous roads. The amounts of gas used were measured and the mean gas consumptions compared for the three conditions.

*17. If a researcher suspects that the strength of the relationship between two variables in the population is .15 as indexed by the correlation coefficient squared, what sample size should she use in a study involving an alpha level of .05, directional test, to achieve a power of .90?

18. Suppose an investigator conducted a correlational analysis with $N = 32$. If the value of the correlation coefficient squared in the population was .20, what would the power of his statistical test be

at an alpha level of .05, nondirectional test?

19. What is the meaning of prediction in the context of regression?

*20. What information is conveyed by the estimated standard error of estimate?

*21. Compute the regression equation for predicting Y from X for the data used in Exercises 6–8. What is the predicted Y score for an X score of 3? What is the predicted Y score for an X score of 7? What is the predicted Y score for an X score of 11?

*22. Compute the estimated standard error of estimate for the data used in Exercises 6–8.

23. Compute the regression equation for predicting Y from X for the data used in Exercises 9–11. What is the predicted Y score for an X score of 2? What is the predicted Y score for an X score of 8? What is the predicted Y score for an X score of 8.50?

24. Compute the estimated standard error of estimate for the data used in Exercises 9–11.

Multiple-Choice Questions

*25. When $\rho = 0$, the sampling distribution of the correlation coefficient will be skewed.
  a. true   b. false

26. The error score (ε) in the linear model reflects all factors that are uncorrelated with X that influence Y.
  a. true   b. false

*27. The estimated standard error of estimate reflects discrepancies between
  a. scores on variable X and scores on variable Y
  b. scores on variable X and predicted scores on variable Y
  c. scores on variable Y and predicted scores on variable X
  d. scores on variable Y and predicted scores on variable Y

28. Researchers pursue regression analysis only if a test of Pearson correlation yields statistically nonsignificant results.
    a. true   b. false

29. If a researcher suspects that the strength of the relationship between two variables in the population is .05 as indexed by the correlation coefficient squared, what sample size should she use in a study involving an alpha level of .05, nondirectional test, to achieve a power of .80?
    a. 121   c. 195
    b. 153   d. 227

*30. What is the symbol for the coefficient of determination?
    a. r c. b
    b. r^2 d. ρ

*31. Smoothing techniques provide a greater sense of the fundamental trends in the data by removing "noise" from a scatterplot.
    a. true   b. false

32. The slope of the regression line is an index of the strength of the relationship between two variables.
    a. true   b. false

*33. In the context of regression, the variable from which predictions are made (X) is known as the _____ variable and the variable being predicted (Y) is known as the _____ variable.
    a. independent; predictor
    b. criterion; dependent
    c. criterion; predictor
    d. predictor; criterion

Use the following information to complete Exercises 34–43.

An exercise physiologist studied the relationship between age and the number of sit-ups women are capable of doing. A sample of 21 women aged 18 and older were shown the proper form and asked to do as many sit-ups as possible during a 5-minute period. The mean age of the participants was 36.37 with a sum of squares of 1,446.85, the mean number of sit-ups was 23.21 with a sum of squares of 3,018.96, and the sum of cross-products was −1,247.01.

*34. If the value of the correlation coefficient squared in the population is .35, what is the power of the statistical test at an alpha level of .05, nondirectional test?
    a. .67   c. .85
    b. .75   d. .90

*35. What is the observed value of r?
    a. −.36   c. −.75
    b. −.60   d. −.77

36. Based on this study, we can conclude that age and the number of sit-ups women are capable of doing are related.
    a. true   b. false

*37. It is possible that the test results reflect a Type I error.
    a. true   b. false

38. It is possible that the test results reflect a Type II error.
    a true   b. false

39. What is the strength of the relationship, as indexed by r^2?
    a. −.36   c. .56
    b. .36   d. .59

*40. What can we conclude about the relationship between age and the number of sit-ups women are capable of doing?
    a. The two variables approximate a direct linear relationship.
    b. The two variables approximate an inverse linear relationship.
    c. There is no linear relationship between the two variables.
    d. There may or may not be some degree of a linear relationship between the two variables.

*41. What is the regression equation for predicting the number of sit-ups women are capable of doing from age?
    a. $\hat{Y} = 38.12 − .41X$
    b. $\hat{Y} = 45.89 − .41X$
    c. $\hat{Y} = 54.49 − .86X$
    d. $\hat{Y} = 56.33 − .86X$

42. What is the estimated standard error of estimate for predicting the number of sit-

ups women are capable of doing from age?
a. 5.24   c. 7.56
b. 6.98   d. 10.08

43. What is the predicted number of sit-ups for a woman aged 42?
a. 18.37
b. 20.21
c. 28.67
d. none of the above

Exercises to Apply Concepts

**44. Behavioral scientists have extensively studied factors that determine an effective group leader in small-group problem-solving situations. Much of this work has used a scale developed by Fiedler (1967) to measure different leadership styles. The scale involves having individuals think of all the people with whom they have ever worked and singling out their least preferred coworker (LPC). They then rate this coworker on a series of dimensions, such as the extent to which they believe the person was pleasant, friendly, and cooperative. A total LPC score is obtained by summing these ratings. According to Fiedler, individuals with high LPC scores tend to see even a poor coworker in a relatively favorable light. These leaders tend to behave in a manner described as compliant, nondirective, and generally relaxed. In contrast, individuals with low LPC scores tend to be demanding, controlling, and managing in their group interactions.

It is generally recognized that the effectiveness of any leader depends not only on his or her leadership style but also on the type of problem being addressed and the situation in which the group finds itself. In one investigation, Fiedler examined the relationship between a leader's LPC score and group problem-solving performance in the form of minutes until solution of the problem. In this investigation, the task was relatively unstructured, there were good leader–member relations, and the leader had relatively little control over the rewards given to each group member. The hypothetical data presented in the table are representative of the results of the study. Analyze the data using a nondirectional test, draw a conclusion, and write up your results using principles discussed in the relevant Method of Presentation section. If the correlation is statistically significant, compute the regression equation and the estimated standard error of estimate for predicting minutes until solution from LPC scores and include these in your report.

Group	X (leader's LPC score)	Y (minutes until solution)
1	63	11
2	68	12
3	71	15
4	65	10
5	61	6
6	75	19
7	64	9
8	63	7
9	70	13
10	73	7

45. Borden (1978) investigated factors related to individuals' concern for the environment. He hypothesized that one factor might be the extent to which individuals value technology. People who value technological innovation and who believe that technology can solve most world problems should be less likely to perform environment-conserving behaviors than those who question technology as a panacea to environmental problems. To test this hypothesis, Borden administered two test scales to a group of individuals. One scale measured belief in

technology, and the other scale measured the extent to which participants performed ecological-related behaviors (e.g., saving energy). Scores on each scale could range from 0 to 20, with higher values representing greater belief in technology and the performance of more ecological-related behaviors, respectively. The hypothetical data presented in the table are representative of the outcome of this investigation. Analyze the data using a nondirectional test, draw a conclusion, and write up your results using principles discussed in the relevant Method of Presentation section. If the correlation is statistically significant, compute the regression equation and the estimated standard error of estimate for predicting ecological-related behavior from belief in technology and include these in your report.

Individual	X (belief in technology)	Y (ecological-related behavior)
1	12	13
2	17	7
3	13	14
4	15	10
5	10	16
6	14	12
7	16	8
8	19	5
9	16	10
10	7	11
11	16	6
12	8	17

46. One perspective on the relationship between time spent on an examination and test performance is that performance should be better as test takers spend increasingly more time thinking through their answers and double-checking their responses. A second perspective holds that the most knowledgeable test takers will require less time to complete an examination than will those who are less well prepared. According to the first view, test-completion times and test performance should be positively related; according to the second, they should be negatively related. In one investigation, Becker and Suls (1982) determined the number of minutes that undergraduate students spent on a course test and their corresponding test scores. The hypothetical data in the table are representative of the results of the study. Analyze the data using a nondirectional test, draw a conclusion, and write up your results using principles discussed in the relevant Method of Presentation section. If the correlation is statistically significant, compute the regression equation and the estimated standard error of estimate for predicting test scores from test-completion times and include these in your report.

Student	X (test-completion time)	Y (test score)
1	43	88
2	37	91
3	41	86
4	46	76
5	50	94
6	48	83
7	34	77
8	47	76
9	50	82
10	49	86
11	44	93
12	38	84
13	46	80
14	49	95
15	41	81
16	43	74
17	47	88
18	50	75

**47. Greater numbers of students are attending graduate school than ever before. Very few graduate programs can admit all applicants and, accordingly, most have a screening committee to select those applicants with the most promise. What kinds of criteria distinguish good graduate students from poor ones? Willingham (1974) reviewed 43 studies from 1952 to 1972 that addressed this question. In his review, Willingham noted that the most commonly used selection criteria were (1) Graduate Record Examination quantitative ability scores (GRE-Q), verbal ability scores (GRE-V), and advanced scores (GRE-A) that measure mastery and comprehension of material basic to graduate study in a specific major field; (2) grade point average (GPA) during undergraduate education; and (3) letters of recommendation. Some of these criteria were found to be reasonable discriminators of the performance of graduate students, whereas others were quite poor in their prediction of graduate school success.

The hypothetical data in Set I represent GRE-A scores and graduate students' grade point averages after 2 years of graduate study in a particular program. Compute the Pearson correlation coefficient and test for a relationship between GRE-A scores and graduate GPA using a nondirectional test. If this is statistically significant, compute the regression equation for predicting graduate GPA from GRE-A scores. If only students who are predicted to maintain a 2-year grade point average of 3.00 (B) or better are to be admitted to the program, which of the applicants in Set II are viable candidates for admission?

Set I		
Student	X (GRE-A)	Y (GPA)
1	533	3.11
2	497	2.89
3	612	3.66
4	564	3.50
5	582	3.29
6	476	3.34
7	607	3.61
8	621	3.74
9	590	3.42
10	512	2.61

Set II	
Applicant	X (GRE-A)
1	532
2	478
3	589
4	483
5	527
6	493
7	546

48. A company that specializes in hand-assembly of decorative ornaments needs each assembler to put together a minimum of 40 ornaments a day if it is to make a profit. As part of the hiring process, an industrial-organizational psychologist gives job applicants a test of manual dexterity. Scores on this test can range from 0 to 50, with higher values indicating greater dexterity. The hypothetical data in Set I represent manual dexterity and job productivity (mean number of ornaments assembled per day) for assemblers who have worked for the company. Compute the Pearson correlation coefficient and test for a relationship between manual dexterity and job productivity using a nondirectional test. If this is statistically significant, compute the regression equation for predicting job productivity from manual dexterity scores. If only individuals who are predicted to assemble a minimum of 40 ornaments a day are to be hired, which of the applicants in Set II are viable candidates for employment?

	Set I			Set II	
Assembler	X (manual dexterity)	Y (job productivity)		Applicant	X (manual dexterity)
1	40	35.21		1	38
2	33	37.34		2	42
3	30	34.85		3	30
4	35	43.02		4	34
5	31	42.37		5	36
6	38	46.83		6	31
7	46	52.28		7	44
8	44	48.24		8	35
9	34	38.27			
10	47	50.56			
11	31	34.78			
12	42	51.19			
13	38	44.76			
14	35	40.85			

Chi-Square Test

15.1 Use of the Chi-Square Test

The **chi-square test** is typically used to analyze the relationship between two variables when these conditions are met:

1. Both variables are qualitative in nature (that is, measured on a nominal level).

2. The two variables have been measured on the same individuals.

3. The observations on each variable are *between-subjects* in nature.

As an example of an investigation that meets these criteria, suppose that a researcher interested in the relationship between gender and political party identification in a particular geographic area conducts a survey in which 170 residents indicate whether they identify themselves as Democrats, Republicans, or Independents. This investigation concerns the relationship between two qualitative variables: (1) gender (male or female) and (2) political party identification (Democrat, Republican, or Independent). The two variables are measured on the same individuals, and the observations on each dimension are between-subjects in nature. Given these conditions, the chi-square test is the statistical technique that would typically be used to analyze the relationship between the variables.

15.2 Two-Way Contingency Tables

Unlike the parametric statistical tests that we have considered in previous chapters, the chi-square test (which is a nonparametric test) analyzes relationships between variables using frequency information. This is because the chi-square test is designed for use with qualitative variables, and it is not appropriate to compute means or the measures of variability discussed previously for variables of this type.

The basis of analysis for the chi-square test is a **contingency table** (also called a *frequency* or *crosstabulation table*). A *two-way* contingency table is illustrated in Table 15.1, which examines the relationship between gender (male or female) and political party identification (Democrat, Republican, or Independent) for the sample of 170 individuals. This table is called a *two-way* table because it examines two variables. We can also call it a 2×3 (read "two by three") table, where the first number indicates the number of rows (excluding the "Totals" row) and the second number indicates the number of columns (excluding the "Totals" column). These are equivalent to the number of levels of the two variables. The assignment of one variable as the row variable and the other variable as the column variable is arbitrary. If gender had been represented in columns and political party identification in rows, we would have a 3×2 rather than a 2×3 table, but we would apply the identical analytical procedures described below.

Each unique combination of variables in a contingency table is referred to as a **cell.** The entries within the cells represent the number of individuals in the sample who are characterized by the corresponding levels of the variables and are referred to as **observed frequencies.** In this example, the observed frequency is 63 for male Democrats, 17 for male Republicans, 10 for male Independents, 35 for female Democrats, 20 for female Republicans, and 25 for female Independents. The numbers in the last column and the bottom row of Table 15.1 indicate how many individuals have each separate characteristic. For example, there are 90 men and 80 women. Also, there are 98 Democrats, 37 Republicans, and 35 Independents. These frequencies are referred to as **marginal frequencies** and are the sums of the frequencies in the corresponding rows (for example, $63 + 17 + 10 = 90$) or columns (for example, $63 + 35 = 98$). Finally, the number in the lower right-hand corner is the overall sample size (in this case, 170).

TABLE 15.1   Contingency Table for Gender and Political Party Identification Study

Gender	Party Identification			Totals
	Democrat	Republican	Independent	
Male	63	17	10	90
Female	35	20	25	80
Totals	98	37	35	170

15.3 Chi-Square Tests of Independence and Homogeneity

Since participants in the gender and political party identification study were selected without regard to their gender or political party identification, the marginal frequencies for both of these variables were free to vary, or **random.** As noted, the actual sample contained 90 men and 80 women, of whom 98 were Democrats, 37 were Republicans, and 35 were Independents. However, assuming that the researcher wanted to recruit exactly 170 participants, the sampling procedures could have hypothetically produced such other combinations of marginal frequencies as 72 men and 98 women for the gender variable, or 81 Democrats, 44 Republicans, and 45 Independents for the party identification variable. In contrast, suppose the researcher had explicitly selected the sample so that it would consist of a specified number of individuals of each gender (for instance, 100 men and 100 women) and had then classified each person into one of the three party identification categories. In this instance, the marginal frequencies for the party identification variable would still be random, but the marginal frequencies for the gender variable would be predetermined, or **fixed.** Alternatively, the researcher might have designed the study so that a predetermined number of people from each party identification category would be included (for instance, 50 Democrats, 50 Republicans, and 50 Independents) and then classified these individuals as being either male or female. In this case, the marginal frequencies for gender would be random but the marginal frequencies for party identification would be fixed.

When the marginal frequencies of both variables under study are random, the chi-square test described in this chapter is known as the **chi-square test of independence.** When the marginal frequencies are random for one variable and fixed for the other, the analytical procedures are identical but the test is referred to as the **chi-square test of homogeneity.** Although the two names are often used interchangeably in the literature, they refer to different situations and you are encouraged to apply them appropriately. Since the two tests are identical in all other respects, the general terminology *chi-square test* is used in the remainder of this book to encompass both situations.

15.4 Inference of a Relationship Using the Chi-Square Test

The logic underlying the chi-square test focuses on the concept of **expected frequencies.** This can be illustrated using the coin flipping example discussed in Chapter 8. Suppose you are given a coin and asked to determine whether it is fair. If you assume that the coin is fair, your best guess about the outcome of 100 flips would be 50 heads and 50 tails (since the probability of each is .50). These expected frequencies can then be compared with the frequencies observed when you actually flip the coin. Suppose that after 100 flips you had observed 53 heads and 47 tails. Although the observed frequencies are not exactly 50/50, you would be hesitant to conclude that the coin is biased because of your knowledge of sampling error. But suppose the observed frequencies had been 65/35. These are much more discrepant from the expected frequencies. At some point the observed frequencies will be so discrepant from the expected frequencies that you will conclude that the coin is biased. This is analogous to the logic of the chi-square test whereby we (1) assume a null hypothesis is true, (2) derive a set of expected frequencies based on this assumption, (3) compare the expected frequencies with the frequencies observed in the investigation, and (4) reject the null hypothesis if the overall difference between observed frequencies and expected frequencies is large (as defined by a given alpha level). We use the example on gender and political party identification to illustrate the chi-square test.

Null and Alternative Hypotheses

The null hypothesis for a chi-square test states that the variables of interest are unrelated in the population, whereas the alternative hypothesis states that there is a population relationship between the two variables.* Since this relationship can take a number of different forms, the alternative hypothesis for the chi-square test, as with the F test in analysis of variance, is nondirectional. The null and alternative hypotheses can be stated as follows for the study on gender and political party identification:

H_0: Gender and political party identification are unrelated in the population

H_1: Gender and political party identification are related in the population

Expected Frequencies and the Chi-Square Statistic

Application of the chi-square test requires computation of an expected frequency for each cell under the assumption of no relationship between the two variables. Consider the Democrat category for the party identification variable. If party identification and gender are unrelated, we would expect the proportion of men who are Democrats to be the same as the proportion of women who are Democrats. In our example, 98/170, or 57.65%, of the *total* sample are Democrats. This represents an estimate of the percentage of Democrats in the population. If gender and party identification are unrelated, we would expect 57.65% of the men to be Democrats and 57.65% of the women to be Democrats. In the example, there are 90 men, and 57.65% of 90 is 51.88. This is the expected frequency for Democratic men. There are 80 women in the example, and 57.65% of 80 is 46.12. This is the expected frequency for Democratic women.

  Now consider the Republican category for the party identification variable. In our example, 37/170, or 21.76%, of the total sample are Republicans. If gen-

---

* If we let A represent a given outcome for gender and B represent a given outcome for political party identification, then the null hypothesis can be stated formally using Equation 6.6: H_0: $p(A, B) = p(A)p(B)$ for all values of A and B. The corresponding alternative hypothesis is H_1: $p(A, B) \neq p(A)p(B)$ for all values of A and B.

der and party identification are unrelated, then 21.76% of the 90 men should be Republicans (yielding an expected frequency of 19.59) and 21.76% of the 80 women should be Republicans (yielding an expected frequency of 17.41). The same logic holds for the Independent category.

The foregoing steps involved in calculating expected frequencies based on the frequencies observed in a contingency table can be summarized as follows:

1. For the cell in question, divide the total of the column (the column marginal frequency) in which it appears by the total number of observations.

2. Multiply this value by the total of the row (the row marginal frequency) in which the cell appears.

This can be represented symbolically as

$$E_j = \left(\frac{CMF_j}{N}\right)(RMF_j) \qquad [15.1]$$

where E_j is the expected frequency for cell j, CMF_j is the column marginal frequency associated with the cell in question, N is the overall sample size, and RMF_j is the row marginal frequency associated with the cell in question. For instance, the expected frequency for Independent males is $(35/170)(90) = 18.53$.

The third column of Table 15.2 presents the expected frequency for each cell in our example. To determine how discrepant an observed frequency (column 2) is from the corresponding expected frequency, we compute the difference between the two. This has been done for each cell in column 4. We next combine the discrepancies into an index that reflects the overall difference between the observed and the expected frequencies. We do this using the **chi-square statistic**, which is defined as follows:

$$\chi^2 = \Sigma \frac{(O_j - E_j)^2}{E_j} \qquad [15.2]$$

TABLE 15.2   **Data and Computation of Chi-Square for Gender and Political Party Identification Study**

Cell	Observed frequency (O)	Expected frequency (E)	O − E	(O − E)²	(O − E)²/E
Democratic males	63	51.88	11.12	123.65	2.38
Democratic females	35	46.12	−11.12	123.65	2.68
Republican males	17	19.59	−2.59	6.71	.34
Republican females	20	17.41	2.59	6.71	.39
Independent males	10	18.53	−8.53	72.76	3.93
Independent females	25	16.47	8.53	72.76	4.42
					$\chi^2 = 14.14$

where χ^2 is the chi-square statistic, O_j is the observed frequency for cell j, E_j is the expected frequency for cell j derived using Equation 15.1, and the summation is across all cells.*

The calculation of chi-square for our example is worked through in columns 4–6 of Table 15.2. First, the difference between the observed and expected frequencies is determined for each cell (column 4). Second, each of these difference scores is squared (column 5). The squared difference for each cell is then divided by the corresponding expected frequency (column 6). Finally, column 6 is summed, yielding a chi-square value of 14.14.

Sampling Distribution of the Chi-Square Statistic

If two variables are unrelated in the population, the population value of chi-square will equal 0. However, because of sampling error, a chi-square computed from sample data might be greater than 0 even when the null hypothesis is true. Consider a population in which there is no relationship between the two variables. A sample of size N is randomly selected, and the chi-square statistic is computed. The value of chi-square might be .50. Now suppose we repeat this procedure with a second random sample of size N and we obtain a chi-square of 1.17. In principle, we could continue to do this a large number of times with the result being a large number of chi-square values. The sample chi-square values, if computed for all possible random samples of a given size, would form a **sampling distribution of the chi-square statistic.** This distribution closely approximates, under certain conditions to be discussed shortly, a theoretical distribution called the **chi-square distribution.**

The chi-square distribution differs depending on the degrees of freedom associated with it. For the chi-square test that we are considering, $df = (r-1)(c-1)$, where r is the number of levels of the row variable and c is the number of levels of the column variable. In the gender and party identification example, there are two rows and three columns, so there are $(2-1)(3-1) = 2$ degrees of freedom.

Figure 15.1 presents examples of chi-square distributions for 1, 2, and 8 degrees of freedom. Analogous to the previous sampling distributions that we have considered, probability statements can be made with respect to scores in the chi-square distribution. It is therefore possible to set an alpha level and define a critical value such that if the null hypothesis is true, the probability of obtaining a chi-square value larger than that critical value is less than alpha. Since all discrepancies from expected frequencies are reflected in the upper tail of the chi-square distribution (as defined by the critical value), the chi-square test is, by its nature, nondirectional.

Appendix J presents a table of critical values for the chi-square distribution. For the gender and political party identification investigation, the critical value of chi-square for an alpha level of .05 and 2 degrees of freedom is 5.991. The observed chi-square of 14.14 is greater than this value, and the null hypothesis is therefore rejected. We conclude that there is a relationship between gender and political party identification.

---

* For the sake of simplicity, the j subscript is excluded from the O and E notation throughout this chapter except where necessary for clarity.

FIGURE 15.1   **Chi-Square Distributions for Various Degrees of Freedom**

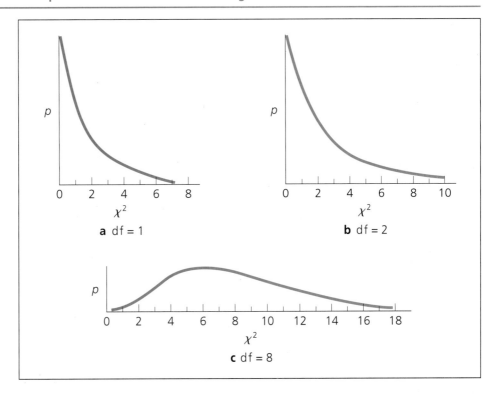

a df = 1

b df = 2

c df = 8

Study Exercise 15.1

A researcher was interested in the relationship between the social class of parents and the discipline style they use with their children. Groups of working-class, middle-class, and upper-class parents were studied. Each participant was classified as using primarily physical discipline (emphasizing physical punishment), primarily psychological discipline (emphasizing psychological punishment such as scolding and withdrawal of affection), or a mixture of physical and psychological discipline. The following contingency table was observed:

Social Class	Discipline Style			Totals
	Physical	Psychological	Mixed	
Working class	60	29	48	137
Middle class	24	49	25	98
Upper class	18	31	16	65
Totals	102	109	89	300

Use a chi-square test to test for a relationship between social class and discipline style.

Answer The null and alternative hypotheses can be stated as follows:

H_0: Social class and discipline style are unrelated in the population

H_1: Social class and discipline style are related in the population

The calculations for the chi-square statistic are summarized in the table below. We find that χ^2 = 25.71. For an alpha level of .05 and $(r - 1)(c - 1) = (3 - 1)(3 - 1) = 4$ degrees of freedom, the critical value of chi-square from Appendix J is 9.488. Since the observed chi-square value of 25.71 is greater than 9.488, we reject the null hypothesis and conclude that a relationship exists between social class and discipline style.

Cell	O	E	$O - E$	$(O - E)^2$	$(O - E)^2/E$
Physical—working class	60	46.58	13.42	180.10	3.87
Physical—middle class	24	33.32	−9.32	86.86	2.61
Physical—upper class	18	22.10	−4.10	16.81	.76
Psychological—working class	29	49.78	−20.78	431.81	8.67
Psychological—middle class	49	35.61	13.39	179.29	5.03
Psychological—upper class	31	23.62	7.38	54.46	2.31
Mixed—working class	48	40.64	7.36	54.17	1.33
Mixed—middle class	25	29.07	−4.07	16.56	.57
Mixed—upper class	16	19.28	−3.28	10.76	.56
					$\chi^2 = 25.71$

Assumptions of the Chi-Square Test

The chi-square test is based on two assumptions. These assumptions are important because they ensure that the sampling distribution of the chi-square statistic approximates a chi-square distribution. Specifically, the following are assumed:

1. The observations are independently and randomly sampled from the population of all possible observations.

2. The *expected* frequency for each cell is nonzero. In fact, although the issue is controversial, statisticians generally recommend that the lowest expected frequency one should have in order to use the chi-square test is somewhere around 5, with the required frequency decreasing as the dimensions of the contingency table increase. An exception is that expected frequencies for 2×2 tables can be relatively small.

Note that the second assumption is concerned with *expected* frequencies, not observed frequencies. Thus, as long as the expected frequency assumption is met, observed frequencies can be as low as 0. As you might anticipate, expected frequencies tend to increase as the size of the overall sample increases.

Alternatives have been proposed to the traditional chi-square test when expected frequencies are small. The most promising is a set of analytic procedures based on a technique called *Fisher's exact test*. These are discussed in Mehta and Patel (1992).

15.5   2 × 2 Tables

The analysis of 2×2 contingency tables raises several unique issues, which we now consider.

Yates' Correction for Continuity

When both of the variables under study have only two levels, the sampling distribution of the chi-square statistic corresponds less closely to a chi-square distribution than when one or both variables have more than two levels. In this case, a correction factor to be incorporated into the formula for computing the chi-square statistic has been suggested that supposedly yields a sampling distribution similar to the chi-square distribution. This correction factor is known as **Yates' correction for continuity** and involves subtracting .5 from the *absolute value* of $O_j - E_j$ before these quantities are squared, divided by E_j, and summed across cells. Symbolically, the formula for computing the chi-square statistic incorporating Yates' correction is

$$\chi^2 = \Sum \frac{(|\,O_j - E_j\,| - .5)^2}{E_j} \tag{15.3}$$

Whether one should apply Yates' correction has been the subject of considerable debate among statisticians (see Camilli & Hopkins, 1978; Conover, 1974a, 1974b; Mantel, 1974). In practice, the correction is applied more often than not. However, recent studies clearly indicate that the correction should *not* be used because it tends to reduce statistical power while adding little control over Type I errors. It is presented here merely for those who might encounter it in other sources.

Computational Formula for 2 × 2 Tables

In cases where both variables have only two levels, the following efficient computational formula can be used to derive the chi-square statistic:

$$\chi^2 = \frac{N(ad - bc)^2}{(a + b)(c + d)(a + c)(b + d)} \tag{15.4}$$

where the letters a, b, c, and d represent the observed frequencies for the four cells as follows:

	Variable 2	
	a	b
Variable 1		
	c	d

Consider the following contingency table:

	Variable 2		Totals
	25	16	41
Variable 1			
	15	23	38
Totals	40	39	79

According to Equation 15.4, the observed value of chi-square is

$$\chi^2 = \frac{79[(25)(23) - (16)(15)]^2}{(25 + 16)(15 + 23)(25 + 15)(16 + 23)}$$

$$= \frac{(79)(335)^2}{2,430,480} = 3.65$$

This is the same value of chi-square that we would obtain if we were to analyze the data using the more general Equation 15.2.

Alternatives to the Chi-Square Test

Methods other than the chi-square test have been suggested for testing the relationship between variables in 2×2 tables. One of the more popular alternatives is *Fisher's exact test*. Research has generally found the Fisher test to be preferable to the chi-square test, although they converge to the same result as N increases. Unfortunately, calculation of the test is sufficiently complex that it is best done with a computer.

15.6   Strength of the Relationship

A number of indexes have been proposed for measuring the strength of the relationship between two variables in a contingency table. Statisticians do not agree on which of these statistics is most appropriate. Among the more popular are Pearson's index of mean square contingency, the phi coefficient, gamma, the coefficient of contingency, and the Goodman–Kruskal index of predictive association. Probably the most common index of the strength of association is a measure known as the **fourfold point correlation coefficient** (as it is called when applied to the relationship between variables with two levels each) or **Cramer's statistic** (as it is called when one or both variables have more than two levels). The strength of association using this measure is defined as

$$V = \sqrt{\frac{\chi^2}{N(L - 1)}} \tag{15.5}$$

where V is the fourfold point correlation coefficient/Cramer's statistic, χ^2 is the observed chi-square, N is the overall sample size, and L is the number of levels of the variable that has the fewer values. For the study on gender and political party identification, $\chi^2 = 14.14$; $N = 170$; and $L = 2$ because the gender variable has two levels, the party identification variable has three levels, and 2 is less than 3. Thus,

$$V = \sqrt{\frac{14.14}{170(2-1)}} = .29$$

The fourfold point correlation coefficient/Cramer's statistic can range from 0 to 1.00, where a value of 0 indicates no relationship and a value of 1.00 indicates a perfect relationship. The magnitude of V is interpreted like that of the Pearson correlation coefficient. To illustrate the relationship between the Pearson correlation coefficient and the present index, consider a contingency table that has two levels of variable A and two levels of variable B. For each participant in the study, a score of 1 is assigned if the individual is characterized by level 1 of variable A and a score of 2 is assigned if the individual is characterized by level 2. The same is done for variable B. If the Pearson correlation coefficient between the variables were computed, the absolute value of the result would be equivalent to the fourfold point correlation coefficient. Conceptually, a large value of V indicates a tendency for particular categories of one variable to be associated with particular categories of the other variable.

15.7 Nature of the Relationship

If the null hypothesis of no relationship is rejected, additional steps are required to discern more fully the nature of the relationship. The test of the chi-square statistic applies to the data taken as a whole and provides no information about which cells are responsible for rejecting the null hypothesis. Just as the Tukey HSD test can be applied to break down the overall relationship following a statistically significant analysis of variance, comparable tests can be applied following a statistically significant chi-square test. Which test to use is controversial among statisticians. Procedures have been suggested by Cohen (1967), Cox and Key (1993), Goodman (1964), Levy (1977), Ryan (1959), Siegel and Castellan (1988), and Wike (1971). Goodman's procedure is popular among statisticians but quite conservative.

Appendix 15.1 discusses one approach to determining the nature of the relationship following a statistically significant chi-square test. This approach uses a *modified Bonferroni procedure* applied to tests of proportions. It appears in the appendix because the computational procedures are somewhat complicated.

From an intuitive perspective, insight into the nature of the relationship can be gained by examining the $(O - E)^2/E$ and $O - E$ values for each cell. For the gender and political party identification investigation, these correspond to columns 6 and 4 of Table 15.2. The numbers in the former column are summed to yield the overall chi-square statistic. The smaller numbers in the column contribute less to the rejection of the null hypothesis than the larger numbers. The largest values in our example are for Independent females (4.42) and Independent males (3.93). Examination of the observed minus expected frequencies ($O - E$) suggests that men are less likely to be Independents than expected (-8.53) and women are more likely to be Independents than expected (8.53), given the assumption of no relationship between gender and party identification. The $(O - E)^2/E$ values for Democratic males (2.38) and Democratic females (2.68) are also relatively large, and the corresponding values of $O - E$ suggest that men are more likely to be Democrats than expected (11.12) and women are less likely to be Democrats than expected (−11.12). The values of $(O - E)^2/E$ for Republican males (.34) and Republican females (.39) are relatively small and suggest that these groups contribute little to the rejection of the null hypothesis.

Study Exercise 15.2

Calculate Cramer's statistic and discern the nature of the relationship between social class and discipline style for the contingency table in Study Exercise 15.1.

Answer   Since both variables have three levels, L in this case is equal to 3. Applying Equation 15.5, we have

$$V = \sqrt{\frac{\chi^2}{N(L-1)}}$$

$$= \sqrt{\frac{25.71}{300(3-1)}} = .21$$

The nature of the relationship can be discerned by examining columns 6 and 4 of the calculations table in Study Exercise 15.1. The largest $(O - E)^2/E$ values are for the psychological—working class (8.67), psychological—middle class (5.03), and physical—working class (3.87) cells. Examination of the observed minus expected frequencies suggests that working class parents are less likely to use psychological discipline than expected (−20.78) and more likely to use physical discipline (13.42). In contrast, the frequency of psychological discipline used by middle-class parents was greater than expected (13.39).

15.8   Methodological Considerations

In Section 5.4, we noted three reasons two variables might be related: (1) the first variable might cause the second variable, (2) the second variable might cause the first variable, or (3) some additional variable(s) might cause both variables. A fourth reason two variables might be related is that one of them bears a relationship to some additional variable(s) that cause(s) the second variable. This is probably the case in the gender and political party identification investigation. Even though the chi-square test was statistically significant, this does not necessarily mean that gender *causes* individuals to identify with one political party or another. It is more likely that the relationship between the two variables is due to the different experiences that men and women have in their socialization, learning, and personal development. The fact that the strength of the relationship, as indexed by Cramer's statistic, was only .29 suggests that in addition to gender and associated factors, disturbance variables have an important influence on political party identification. We can speculate that these include such characteristics as race, social class, and parents' political party identification.

15.9 **Numerical Example**

Social psychologists have studied variables that influence altruistic behavior. In one experiment, a researcher investigated the effect of a role model on people's willingness to donate money to charity. Three hundred individuals were approached in a shopping center and asked to make a charitable donation. One hundred of these individuals were assigned to a positive-model condition such that they were approached just after they had seen an assistant of the experimenter make a donation. Another 100 individuals were assigned to a negative-model condition such that they were approached just after they had seen an assistant of the experimenter refuse to make a donation. Finally, the last 100 individuals were approached without having observed a model. The numbers of people who did and did not donate money in each condition are indicated in Table 15.3.

The null and alternative hypotheses can be stated as follows:

H_0: Model status and outcome are unrelated in the population

H_1: Model status and outcome are related in the population

The calculations for the chi-square statistic are summarized in Table 15.4. We find that $\chi^2 = 16.83$. The critical value of chi-square for an alpha level of .05 and $(r - 1)(c - 1) = (2 - 1)(3 - 1) = 2$ degrees of freedom is 5.991. Since the observed chi-square value of 16.83 is greater than 5.991, we reject the null hypothesis and conclude that there is a relationship between model status and outcome.

The strength of the relationship is indicated by Cramer's statistic. Since the model status variable has three levels and the outcome variable has two levels, L in this case is equal to 2. Using Equation 15.5, we find that

$$V = \sqrt{\frac{\chi^2}{N(L - 1)}}$$

$$= \sqrt{\frac{16.83}{300(2 - 1)}} = .24$$

The nature of the relationship can be discerned by examining columns 6 and 4 of Table 15.4. It can be seen that the no-model condition contributed very little to the overall chi-square [$(O-E)^2/E$ values of .002]. The other $(O - E)^2/E$ and $O - E$ values suggest that a positive model increases donating behavior whereas a

---

TABLE 15.3   **Contingency Table for Model Status and Outcome Experiment**

Model Status	Outcome		
	Donated	Did not donate	Totals
Positive	63	37	100
Negative	34	66	100
No model	49	51	100
Totals	146	154	300

TABLE 15.4   **Data and Computation of Chi-Square for Model Status and Outcome Experiment**

Cell	O	E	$O - E$	$(O - E)^2$	$(O - E)^2/E$
Positive—donated	63	48.67	14.33	205.35	4.22
Positive—did not donate	37	51.33	−14.33	205.35	4.00
Negative—donated	34	48.67	−14.67	215.21	4.42
Negative—did not donate	66	51.33	14.67	215.21	4.19
No model—donated	49	48.67	.33	.11	.002
No model—did not donate	51	51.33	−.33	.11	.002
					$\chi^2 = 16.83$

negative model decreases donating behavior. These results, of course, should be interpreted in the context of relevant methodological considerations. Do any of these come to mind?

15.10   Use of Quantitative Variables in the Chi-Square Test

Although the chi-square test is typically used to analyze the relationship between two qualitative variables, it can also be applied when one or both variables are quantitative. For instance, an investigator concerned with the relationship between the number of siblings one has (a quantitative variable) and preference for group versus individual activity (a qualitative variable) might use this approach. A common procedure involves collapsing scores on a quantitative variable into a small number of categories before applying the chi-square test. The following table for analyzing the relationship between men's ages at the time of marriage and the number of years that their marriages lasted illustrates this strategy:

	Duration of Marriage (Years)			
Age at Marriage	<5	5–9	10–14	≥15
<19	42	28	16	18
19–24	33	26	23	23
25–34	12	9	17	15
≥35	14	14	10	11

When possible, it is usually preferable to analyze quantitative variables with the parametric tests discussed in previous chapters rather than with the chi-square test. For instance, an alternative way to approach the relationship between age at marriage and marriage duration would be to perform a Pearson correlation between individuals' marriage-age and marriage-duration scores. Parametric tests are usually preferred over the chi-square test because they tend to be more powerful. The power of the chi-square test is further reduced when scores on quantitative variables are collapsed into categories because considerable information is likely to be lost by placing individuals with different scores (for instance, an age at marriage of 25 and an age at marriage of 34 in the example) into the same group. Although this approach implicitly assumes that all individuals assigned to a given category are equivalent on the underlying dimension, this may not in fact be the case. An advantage of the chi-square approach, however, is that a quantitative variable need be measured on only an ordinal level as opposed to the approximately interval level required for parametric tests.

15.11   Planning an Investigation Using the Chi-Square Test

Appendix E.5 contains tables of the sample sizes necessary to achieve various levels of power for the chi-square test. A set of tables is presented for each research design involving contingency tables of different dimensions. Table 15.5 reproduces the portion of this appendix for a 3×3 contingency table and an alpha level of .05. The first column presents various values of power, and the column headings are values of Cramer's statistic in the population. To illustrate the use of this table, suppose the desired level of power is .80 and the researcher suspects that the strength of the relationship in the population corresponds to a Cramer's statistic of .30. Then the number of participants that should be sampled is 66. Inspection of Table 15.5 and Appendix E.5 provides a general appreciation for the relationships among the dimensions of the table, the alpha level, sample size, the strength of the relationship one is trying to detect in the population, and power for the chi-square test.

TABLE 15.5   **Approximate Sample Sizes Necessary to Achieve Selected Levels of Power for 3 × 3 Table and Alpha = .05 As a Function of Population Values of Cramer's Statistic**

	Population Value of Cramer's Statistic								
Power	.10	.20	.30	.40	.50	.60	.70	.80	.90
.25	154	39	17	10	6	4	3	2	2
.50	321	80	36	20	13	9	7	5	4
.60	396	99	44	25	16	11	8	6	5
.70	484	121	54	30	19	13	10	8	6
.75	536	134	60	34	21	15	11	8	7
.80	597	149	66	37	24	17	12	9	7
.85	671	168	75	42	27	19	14	10	8
.90	770	193	86	48	31	21	16	12	10
.95	929	232	103	58	37	26	19	15	11
.99	1,262	316	140	79	50	35	26	20	16

15.12 Method of Presentation

When presenting the results of a chi-square analysis, the researcher should report the alpha level that was used, the degrees of freedom, the overall sample size, the number of observations for each cell, the observed value of chi-square, and the significance level. If the analysis is statistically significant, the strength and nature of the relationship should also be reported.

Textual presentation of a chi-square test is typically supplemented by a two-way contingency table of the observed frequencies. Such a table would have a format similar to that of Table 15.1. In manuscripts submitted for publication, tables are placed at the end of the research report and are referred to in the text, where appropriate, by number, starting with Table 1. Of course, there should be a clear tie-in between the content of a table and the text. Quoting from page 125 of the *Publication Manual of the American Psychological Association* (American Psychological Association, 1994): "An informative table supplements—not duplicates—the text. In the text, refer to every table and tell the reader what to look for. Discuss only the table's highlights; if you discuss every item of the table in text, the table is unnecessary."

The results for the study on gender and political party identification discussed earlier in this chapter might be reported as follows:

Results

A chi-square test was applied to the relationship between gender and political party identification and found to be statistically significant, χ^2(2, $\underline{N}$ = 170) = 14.14, $\underline{p}$ < .01, using an alpha level of .05. The observed frequencies for the six cells can be found in Table 1.

As indexed by Cramer's statistic, the strength of the relationship was .29. This reflects primarily the fact that men are less likely to be Independents and more likely to be Democrats than expected, and women are more likely to be Independents and less likely to be Democrats than expected.

The first sentence specifies the statistical test that was used to analyze the data. The chi-square statistic is not underlined because American Psychological Association format dictates that Greek letters not be italicized in published form. The first number in the parentheses is the degrees of freedom, and the number indicated by N is the overall sample size. This is followed by the observed chi-square value and the significance level. As noted above, the table referred to in the second sentence would take a form similar to Table 15.1.

Cramer's statistic indicates the strength of the relationship. If the variables had two levels each, this would have been called a fourfold point correlation coefficient. The last sentence addresses the nature of the relationship.

15.13 Examples from the Literature

Religion and College Attendance

Several sociologists have argued that American Catholics have not contributed to the scientific and intellectual development of the United States relative to their numbers in the population. This deficiency is most commonly thought to result from the cultural values of American Catholics, which are believed to impede intellectual achievement. Warkov and Greeley (1966), however, have argued that the problem may be due to the social conditions surrounding early American Catholic immigrants. According to Warkov and Greeley, Catholic immigrants were too poor to be concerned with much beyond economic survival. They argue that immigrants had little time for matters such as intellectualism and going to college because all of their effort had to be directed toward establishing an adequate economic base. This should not, however, be true for more recent generations who have had the opportunity to work from the economic base established by their ancestors.

Based on this logic, Warkov and Greeley hypothesized that there should be a relationship between religion (Catholic versus Protestant) and college attendance for older individuals but not for younger individuals because Catholics in the latter group have had the economic freedom to concern themselves with education whereas Catholics in the former group did not. To examine this issue, they collected data on religion and college attendance for each of the two age groups. The older group consisted of individuals between 50 and 59 years of age. The contingency table for this group follows:

	College Attendance	
Religion	Attended	Did not attend
Catholic	42	310
Protestant	24	73

The younger group consisted of individuals between 23 and 29 years of age. Here is the contingency table for this group:

	College Attendance	
Religion	Attended	Did not attend
Catholic	99	253
Protestant	265	642

If we were to conduct a chi-square test on the data for the older group, we would find a statistically significant relationship between religion and college attendance, $\chi^2(1, N = 449) = 9.95$, $p < .01$. The nature of the relationship was such that Protestants were more likely to have attended college than Catholics. The strength of the relationship, as indexed by the fourfold point correlation coefficient, was .15. In contrast, a chi-square test on the data for the younger group would be statistically nonsignificant, $\chi^2(1, N = 1{,}259) = .15, p > .70$. The fourfold point correlation coefficient in this instance was .01.

These findings are consistent with the reasoning of Warkov and Greeley. However, for the hypothesis to be tested without ambiguity, one would need to conduct a formal test comparing the two age groups, rather than separate tests for the two groups. This would involve an examination of the age, religion, and college attendance variables in the context of a three-way contingency table. The procedure for simultaneously analyzing three or more between-subjects variables of this type is *log-linear analysis*, which will be discussed in Chapter 18.

Gender Differences in the Importance of Romantic Love for Marriage

Do men and women have different feelings about the importance of romantic love as a prerequisite for establishing a marital relationship? In one attempt to address this issue, Simpson, Campbell, and Berscheid (1986) asked male and female college students to respond to the question "If a man (woman) had all the other qualities you desired, would you marry this person if you were not in love with him (her)?" Response categories were no, yes, and undecided. These data were obtained:

	Response		
Gender	No	Yes	Undecided
Male	148	3	22
Female	141	6	19

A chi-square test applied to this contingency table was not statistically significant, $\chi^2(2, N = 339) = 1.24, p > .50$. Thus, the importance of romantic love as a prerequisite for marriage appears to be independent of gender (although one must be careful not to accept the null hypothesis!). The strength of the relationship, as indexed by Cramer's statistic, was .06. Examination of the observed frequencies shows a clear consensus among both genders about the importance of romantic love; no matter how ideal a prospective partner might otherwise be, marriage is not believed to be a viable option unless one is in love.

Additional questions assessed the importance of romantic love as a prerequisite for *maintaining* a marital relationship. Using the response categories agree, disagree, and neutral, participants were asked to respond to the statements "If love has completely disappeared from a marriage, I think it is probably best for a couple to make a clean break and start new lives" and "In my opinion, the disappearance of love is not a sufficient reason for ending a marriage, and should not be viewed as such." As before, chi-square tests involving the two sets of responses were statistically nonsignificant, with a sizeable portion of both genders indicating that remaining in love is an important condition for the continuation of a marriage.

15.14 Chi-Square Goodness-of-Fit Test

In addition to its role in the chi-square test discussed in the preceding sections, the chi-square statistic is involved in the application of the **goodness-of-fit test.** The question addressed by this test is whether frequencies across categories for a variable in a population are distributed in a specified manner.* For example, suppose the relative frequencies of marital status for the population of American women under 40 years of age are as follows:

Marital status	*rf*
Married	.60
Single	.23
Separated	.04
Divorced	.12
Widowed	.01

Furthermore, suppose an investigator was interested in whether marital status is distributed the same way in the population of female executives under age 40 as it is in the general population of women of this age. The null and alternative hypotheses in this instance can be stated as follows:

H_0: Marital status is distributed the same way in the population of female executives under 40 as it is in the general population of women of this age

H_1: Marital status is not distributed the same way in the population of female executives under 40 as it is in the general population of women of this age

A sample of 200 female executives under age 40 are surveyed and the following frequencies observed for the marital status categories:

Marital status	O
Married	100
Single	44
Separated	16
Divorced	36
Widowed	4
Total	200

---

* Since the goodness-of-fit test is concerned with the distribution of frequencies for a single variable, unlike the other techniques discussed in Part 2 of this book, it is not a bivariate test.

Based on the relative frequencies in the general population noted above, we can compute the frequency that would be expected in the sample of executives for each marital status category if the null hypothesis is true and compare these with the frequencies that are actually observed. The expected frequencies are equal to the respective relative frequencies multiplied by the overall sample size. Symbolically,

$$E_j = (rf_j)(N) \qquad\qquad [15.6]$$

where E_j is the expected frequency for category j, rf_j is the relative frequency in the population for category j, and N is the overall sample size.

In our example, the expected frequency for the married category is $(.60)(200) = 120.00$. Similarly, the expected frequency for the single category is $(.23)(200) = 46.00$. The results of this procedure for all five categories are listed in column 3 of Table 15.6. The chi-square statistic is derived by applying Equation 15.2 to the observed and expected frequencies in the same manner as we did previously. The computations are summarized in Table 15.6, where it can be seen that $\chi^2 = 19.42$.

The degrees of freedom for the goodness-of-fit test are equal to $k - 1$, where k is the number of categories. From Appendix J, the critical value of chi-square for an alpha level of .05 and $5 - 1 = 4$ degrees of freedom is 9.488. The observed chi-square of 19.42 is greater than 9.488, so we reject the null hypothesis and conclude that the distribution of marital status in the population of female executives under age 40 differs from that of the general population of women of this age.

If we examine the last column in Table 15.6, we can gain an appreciation for which categories are most responsible for the rejection of the null hypothesis. Specifically, the categories with the largest values of $(O - E)^2/E$ contribute the most to the overall chi-square. In this case, the largest $(O - E)^2/E$ values are for the separated and divorced categories. Examination of the observed minus expected frequencies for these categories suggests that in the population, female executives under the age of 40 are more likely than expected to be separated or divorced.

A common use of the goodness-of-fit test is to test whether frequencies in the population are evenly distributed across the categories under study. For example, a consumer psychologist interested in whether people prefer oblong, round, or irregularly shaped potato chips might ask each of 60 individuals to indicate a preference for one of these three shapes. The null and alternative hypotheses in this instance can be stated as follows:

TABLE 15.6   **Data and Computation of Chi-Square for Marital Status Study**

Marital status	O	E	$O - E$	$(O - E)^2$	$(O - E)^2/E$
Married	100	120.00	−20.00	400.00	3.33
Single	44	46.00	−2.00	4.00	.09
Separated	16	8.00	8.00	64.00	8.00
Divorced	36	24.00	12.00	144.00	6.00
Widowed	4	2.00	2.00	4.00	2.00
					$\chi^2 = 19.42$

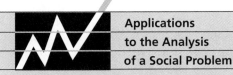

**Applications
to the Analysis
of a Social Problem**

Some conservative politicians have suggested that a major source of adolescent problem behavior is the breakdown of the family. Some have gone so far as to state that one cause of this breakdown is mothers who work. As a consequence of pursuing a career and working full-time, the argument goes, the mother has difficulty finding quality time to

(continued)

H_0: Oblong, round, and irregularly shaped potato chips are equally preferred in the population

H_1: Oblong, round, and irregularly shaped potato chips are not equally preferred in the population

Under the null hypothesis that no differentiation is made between the three shapes in the population, 20 participants should indicate a preference for oblong potato chips, 20 participants should indicate a preference for round potato chips, and 20 should prefer irregularly shaped potato chips. Thus, the expected frequency is 20.00 in each instance. If the observed frequencies were 17 for oblong chips, 24 for round chips, and 19 for irregularly shaped chips, the observed value of chi-square would be as computed in Table 15.7. For an alpha of .05 and $k - 1 = 3 - 1 = 2$ degrees of freedom, the critical value of chi-square is 5.991. Since the observed chi-square of 1.30 does not exceed the critical value, we fail to reject the null hypothesis that preference for the three shapes of potato chips is evenly distributed in the population.

The chi-square goodness-of-fit test is based on two assumptions. First, the observations must be independently and randomly sampled from the population of all possible observations. In addition, the expected frequency for each category must be a minimum size. Although the issue is controversial, statisticians generally recommend that the lowest *expected* frequency for any category be somewhere between 5 and 10 if only two categories are involved and about 5 if more than two categories are involved.

The method of presentation for the goodness-of-fit test is similar to that for the chi-square test except that there is no strength of relationship information to report.

TABLE 15.7   **Data and Computation of Chi-Square for Potato Chip Preference Study**

Preference	O	E	$O - E$	$(O - E)^2$	$(O - E)^2/E$
Oblong	17	20.00	−3.00	9.00	.45
Round	24	20.00	4.00	16.00	.80
Irregular	19	20.00	−1.00	1.00	.05
					$\chi^2 = 1.30$

spend with her children and this, in turn, has a negative impact on the development of the child. Is there a relationship between employment status of the mother and teen sexual behavior?

In the parent–teen communication study described in Chapter 1, we classified mothers into four groups based on their employment status: (1) employed full-time (35 hours or more a week), (2) employed part-time, (3) unemployed but looking for work, and (4) unemployed and not looking for work. We then performed a chi-square test to determine whether there is a relationship between employment status and whether or not the mother's teen has ever engaged in sexual in-

tercourse. Table 15.8 presents the computer output for the analysis. The contingency table appears at the top. In this table, the mother's employment status is called MEMPLOY, and whether or not the mother's teen has had sexual intercourse is called HADSEX. The first number in each cell is the observed frequency for that cell, and the number directly beneath it is the expected frequency. Column and row marginal frequencies are also provided, as is the overall sample size. Using the information contained in this table, we calculated some informative percentages: The percentage of full-time employed mothers whose teens have had sexual intercourse is 54.1%, the percentage of part-time

TABLE 15.8   Computer Output for Chi-Square Test for Mother's Employment Status and Teen's Intercourse Experience

```
MEMPLOY  by HADSEX


                                        HADSEX

                              Not                          Row
     MEMPLOY               Had Sex     Had Sex           Total

     Full time               150         177               327
                            138.4       188.6

     Part time                53          73               126
                             53.3        72.7

     Unemployed-looking       49          84               133
                             56.3        76.7

     Unemployed - not looking 54          83               137
                             58.0        79.0

           Column Total       306         417               723


     Chi-Square         Value         DF         Significance
     ------------       -------       -----      ----------------
     Pearson            3.80128        3            .28374


     Minimum Expected Frequency -  53.328


     Approximate
      Statistic         Value
     ------------       -------


     Cramer's V         .07251


     Number of Missing Observations:  28
```

employed mothers whose teens have had sexual intercourse is 57.9%, the percentage of unemployed mothers looking for work whose teens have had sexual intercourse is 63.2%, and the percentage of unemployed mothers not looking for work whose teens have had sexual intercourse is 60.6%.

Results for the chi-square test are presented beneath the contingency table. "Pearson" is the name of the person who developed the chi-square test. As indicated by the fact that the *p* value (labeled "significance") is not less than the alpha level of .05, the observed value of chi-square (3.80) is not statistically significant. These data are not consistent with the proposition that working mothers are more likely to have children who engage in sex.

The label "Minimum Expected Frequency" beneath the chi-square results identifies the smallest expected frequency that occurred in the contingency table. If this is less than around 5, then the chi-square test may be problematic because the chi-square test assumes that expected frequencies meet a minimum size requirement, as discussed in Section 15.4. The value of Cramer's statistic (.07) is provided at the bottom of the page. Finally, the output indicates that there were 28 cases with missing data for one or both of the variables.

Given that we failed to reject the null hypothesis, there is a possibility of a Type II error. To evaluate this possibility, we calculated the statistical power for a 4 × 2 chi-square test having an alpha level of .05 and an overall sample size of 723 for a population effect size that we considered to be small—namely, a Cramer's *V* of .20. We used the computational method described in Cohen (1988), although a rough estimate of power can be obtained by using the tables in Appendix E.5. The statistical power exceeded .99, which indicates that a Type II error is unlikely.*

---

* The table in Appendix E.5 for a 4 × 2 contingency table and an alpha level of .05 shows that the power for a population Cramer's *V* of .20 will exceed .99 if there are 588 or more participants in the study. Our overall sample size was 723, so, consistent with the result of the Cohen procedure, we conclude that our power exceeds .99.

Summary

The chi-square test is typically used to analyze the relationship between two variables when (1) both variables are qualitative in nature, (2) the two variables have been measured on the same individuals, and (3) the observations on each variable are between-subjects in nature. Although the analytical procedures are identical, the test is referred to as a test of independence when the marginal frequencies of both variables are random and as a test of homogeneity when the marginal frequencies are random for one variable and fixed for the other. In both instances, the test is based on the analysis of discrepancies between observed frequencies and frequencies expected under the null hypothesis. When certain conditions are met, the chi-square statistic has a sampling distribution that closely approximates a chi-square distribution.

The strength of the relationship is measured using an index referred to as the fourfold point correlation coefficient when both variables have two levels and as Cramer's statistic when one or both variables have more than two levels. The nature of the relationship is determined by examining the $(O - E)^2/E$ and $O - E$ values for the individual cells or, more formally, using a modified Bonferroni procedure.

In addition to its role in the chi-square test of independence/homogeneity, the chi-square statistic is involved in the application of the goodness-of-fit test. The question addressed by this test is whether frequencies across categories for a variable in a population are distributed in a specified manner. The analytical procedures parallel those for the two-variable case.

Appendix 15.1 Determining the Nature of the Relationship Using a Modified Bonferroni Procedure

---

Although presented in terms of frequencies, the chi-square test of independence/homogeneity can be thought of as a test of differences in proportions (Glass & Hopkins, 1984). For the gender and political party identification study discussed earlier in this chapter, one might ask the following questions: (1) Is the proportion of Democrats who are male the same as the proportion of Republicans who are male? (2) Is the proportion of Democrats who are male the same as the proportion of Independents who are male? and (3) Is the proportion of Republicans who are male the same as the proportion of Independents who are male? The overall chi-square test is an omnibus test of the null hypothesis that all of the above proportions are identical in the population. If the overall test is statistically significant, then we will conduct additional analyses (analogous to the way the Tukey HSD test is used to follow up a statistically significant analysis of variance) to determine which pairs of proportions are statistically significantly different. To do this, we first identify the proportions from the overall contingency table that are of conceptual interest. We then formulate null and alternative hypotheses regarding pairwise differences between these proportions.

In the gender and party identification example, there are three proportions of interest:

Proportion of male Democrats (p_{MD})
= 63/98 = .643

Proportion of male Republicans (p_{MR})
= 17/37 = .459

Proportion of male Independents (p_{MI})
= 10/35 = .286

We can formulate null and alternative hypotheses for the pairwise contrasts between these proportions as follows:

H_0: $\pi_{MD} = \pi_{MR}$
H_1: $\pi_{MD} \neq \pi_{MR}$

H_0: $\pi_{MD} = \pi_{MI}$
H_1: $\pi_{MD} \neq \pi_{MI}$

H_0: $\pi_{MR} = \pi_{MI}$
H_1: $\pi_{MR} \neq \pi_{MI}$

where π (lowercase Greek p, called "pi") represents the population proportion. Although these hypotheses have been stated in terms of men, they could have been stated with respect to women instead. Because gender has only two values, the proportion for women will always equal 1 minus the proportion for men. Therefore, the test of proportions for men will yield identical conclusions as the test of proportions for women in terms of statistical significance. Thus, it is unnecessary to perform both sets of tests, and the choice as to how to characterize the proportions is arbitrary.

For each null hypothesis, a z test is performed using the following formula (see Guilford, 1965):

$$z = \frac{p_1 - p_2}{\sqrt{p'(1 - p')\left[\dfrac{n_1 + n_2}{(n_1)(n_2)}\right]}} \qquad [15.7]$$

where p_1 is the first proportion of interest, p_2 is the second proportion of interest, n_1 is the total number of cases on which p_1 is based, n_2 is the total number of cases on which p_2 is based, and p' equals

$$p' = \frac{n_1 p_1 + n_2 p_2}{n_1 + n_2} \qquad [15.8]$$

For the comparison between Democrats and Republicans,

$$p' = \frac{(98)(.643) + (37)(.459)}{98 + 37} = .593$$

and

$$z = \frac{.643 - .459}{\sqrt{(.593)(1 - .593)\left[\dfrac{98 + 37}{(98)(37)}\right]}} = 1.94$$

For the comparison between Democrats and Independents, z is equal to 3.64, and for the comparison between Republicans and Independents, z is equal to 1.52. Each of these z scores has a probability associated with it, and these p values can be found in column 4 of Appendix B. For a z of 1.94, the corresponding p value is .052; for a z of 3.64, the corresponding p value is interpolated to be .0003; and for a z of 1.52, the corresponding p value is .129. The absolute z values are ordered in a column from largest to smallest, and the corresponding p values are each compared against a "critical alpha." If the p value is less than the "critical alpha," the null hypothesis in question is rejected.

The value of the "critical alpha" changes for each successive z value as one proceeds from the top to the bottom of the column. The largest z value must yield a p value less than the overall alpha level that we adopt for the set of comparisons divided by the total number of comparisons—in this case, three—for the null hypothesis to be rejected. The next largest z value must yield a p value less than the overall alpha level divided by the number of contrasts minus 1 for the null hypothesis to be rejected. The next largest z value must yield a p value less than the overall alpha level divided by the number of contrasts minus 2 for the null hypothesis to be rejected. And so on, subtracting an additional unit from the total number of comparisons with each successive contrast. If a contrast is statistically significant, the nature of the relationship is determined by comparing the magnitude of the two proportions. As soon as the first statistically nonsignificant contrast is encountered, as one proceeds from the largest to the smallest absolute z value, all remaining comparisons are declared statistically nonsignificant. The rationale for this modified Bonferroni procedure is described in Holland and Copenhaver (1988).

The results for the modified Bonferroni procedure for the gender and political party identification study can be summarized in a table as follows. In deriving this table, we used an overall alpha level of .05. Based on the information in the table and examination of the proportions, we conclude that the proportion of Democrats who are male (.643) is statistically significantly greater than the proportion of Independents who are male (.286). However, we cannot confidently conclude that the proportion of Republicans who are male (.459) differs from either of these proportions.

Null hypothesis tested	Absolute value of z	p value	Critical alpha	Null hypothesis rejected?
$\pi_{MD} = \pi_{MI}$	3.64	.0003	.05/3 = .017	Yes
$\pi_{MD} = \pi_{MR}$	1.94	.052	.05/2 = .025	No
$\pi_{MR} = \pi_{MI}$	1.52	.129	.05/1 = .050	No

Exercises

Answers to asterisked (*) exercises appear at the back of the book. Answers to exercises with two asterisks are also worked out step by step in the Study Guide.

Exercises to Review Concepts

1. Under what conditions is the chi-square test typically used to analyze a bivariate relationship?
2. How many rows are there in a 3×4 contingency table? How many columns? How many cells?
3. Differentiate between the chi-square test of independence and the chi-square test of homogeneity.
*4. Differentiate between observed frequencies and expected frequencies.

Use the following contingency table summarizing movie attendance and movie preference for a sample of individuals to complete Exercises 5–8.

Movie Attendance	Movie Preference		
	Romance	Comedy	Drama
Infrequent	10	20	30
Moderate	15	30	40
Frequent	20	40	50

*5. What is the marginal frequency for frequent moviegoers?
6. What is the marginal frequency for people who prefer comedies?
*7. What is the overall sample size?
*8. What is the expected frequency for each cell under the assumption of no relationship between movie attendance and movie preference?
*9. State the critical value of chi-square for a chi-square test for an alpha level of .05 under each of the following conditions:

a. $r = 2, c = 2$
b. $r = 3, c = 3$
c. $r = 2, c = 3$
d. $r = 4, c = 4$

10. What are the assumptions underlying the chi-square test?
*11. What is the disadvantage of using Yates' correction for continuity when analyzing 2×2 contingency tables?
12. What is the relationship between the fourfold point correlation coefficient and Cramer's statistic?

Use the following contingency table summarizing smoking behavior and cause of death for a sample of people who recently died to complete Exercises 13–16.

Cause of Death	Smoking Behavior	
	Smoker	Nonsmoker
Cancer	46	25
Other	34	45

*13. Test for a relationship between smoking behavior and cause of death using a chi-square test per Equation 15.2.
*14. Compute the value of the fourfold point correlation coefficient.
*15. Discern the nature of the relationship between smoking behavior and cause of death by examining the $(O - E)^2/E$ and $O - E$ values.
*16. Conduct a chi-square test using Equation 15.4. Compare the observed value of chi-square with that obtained in Exercise 13 using Equation 15.2.

Use the following contingency table summarizing political party identification and residential area for a sample of individuals to complete Exercises 17–19.

Party Identification	Residential Area		
	City	Suburbs	Country
Democrat	40	60	30
Republican	40	20	10
Independent	20	20	10

17. Test for a relationship between party identification and residential area using a chi-square test.

18. Compute the value of Cramer's statistic.

19. Discern the nature of the relationship between party identification and residential area by examining the $(O - E)^2/E$ and $O - E$ values.

*20. What is the advantage of analyzing quantitative variables with parametric tests rather than the chi-square test? What is an advantage of the chi-square approach?

For each of the studies described in Exercises 21–25, indicate the appropriate statistical test for analyzing the relationship between the variables. Assume that the underlying assumptions of the tests have been satisfied.

*21. A researcher tested the relationship between college students' need for achievement as assessed on a 20-item test and their grade point averages.

*22. A consumer psychologist studied the relationship between gender and preference for Ford, Chevrolet, and Chrysler cars. One hundred men and 100 women were interviewed and asked which make they preferred.

23. Thirty psychology majors, 30 business majors, and 30 biology majors responded to a social anxiety scale. The mean anxiety scores for the three groups were compared.

24. An educational psychologist examined the effect of a special summer school learning program on reading skills. A test of reading skills was administered to the same group of students before and after the program, and the mean performances at the two times were compared.

25. An economist tested the relationship between marital status and attitudes toward tax reform. Four hundred individuals were surveyed regarding their marital status and whether they were for or against such reform.

26. If a researcher suspects that the strength of the relationship between two variables in the population is .50 as indexed by Cramer's statistic, what sample size should he use in a study involving a 2 × 4 contingency table and an alpha level of .05 to achieve a power of .95?

*27. Suppose an investigator conducted a study involving a 2 × 2 contingency table with $N = 100$. If the value of the fourfold point correlation coefficient in the population was .30, what would the power of her statistical test be at an alpha level of .05?

**28. Suppose that the relative frequencies of preference for the ABC, NBC, and CBS evening news programs as determined by national ratings are as follows:

Preference	rf
ABC	.20
NBC	.23
CBS	.21
No preference	.36

Furthermore, suppose that a survey of the preferences of 1,000 college students yielded the following observed frequencies:

Preference	O
ABC	177
NBC	252
CBS	240
No preference	331

Test whether the distribution of evening news program preference for the population of college students is the same as national ratings using a goodness-of-fit test.

*29. Based on the data in Exercise 28, discern how the distribution of evening news program preference for the population of college students differs from the national ratings by examining the $(O - E)^2/E$ and $O - E$ values.

30. Consider the following eye colors for a sample of individuals:

Eye color	O
Blue	23
Brown	52
Green	45

Test whether blue, brown, and green eyes are equally common in the population of interest using a goodness-of-fit test.

31. Based on the data in Exercise 30, discern how the distribution of eye color differs in the population of interest from an even distribution by examining the $(O - E)^2/E$ and $O - E$ values.

32. What are the assumptions underlying the goodness-of-fit test?

Multiple-Choice Questions

33. The alternative hypothesis for a chi-square test states that
    a. the variables of interest are unrelated in the population
    b. the variables of interest are related in the population
    c. the population means are not all equal to one another
    d. the marginal frequencies for one variable differ from the marginal frequencies for the other variable

*34. If a researcher suspects that the strength of the relationship between two variables in the population is .40 as indexed by Cramer's statistic, what sample size should he use in a study involving a

2 × 3 contingency table and an alpha level of .05 to achieve a power of .85?
    a. 60   c. 79
    b. 68   d. 96

35. The magnitude of the fourfold point correlation coefficient/Cramer's statistic is interpreted like that of the Pearson correlation coefficient.
    a. true   b. false

*36. A common use of the goodness-of-fit test is to test whether frequencies in the population are evenly distributed across the categories under study.
    a. true   b. false

*37. The nature of the relationship following a statistically significant chi-square test can be determined using
    a. Yates' correction for continuity
    b. Cramer's statistic
    c. Fisher's exact test
    d. a modified Bonferroni procedure

38. An assumption of the chi-square test is that the observed frequency for each cell is nonzero.
    a. true   b. false

Use the following contingency table summarizing race and voting behavior during an election for a sample of individuals to complete Exercises 39–45.

	Voting Behavior	
Race	Voted	Did not vote
African-American	60	40
White	70	30
Other	75	25

39. If the value of Cramer's statistic in the population is .20, what is the power of the statistical test at an alpha level of .05?
    a. between .25 and .50
    b. between .70 and .75
    c. between .75 and .80
    d. between .85 and .90

*40. What is the observed value of chi-square?
   a. 2.32   c. 5.99
   b. 5.39   d. 29.05

*41. Based on this study, we can conclude that race and voting behavior are related.
   a. true   b. false

42. It is possible that the test results reflect a Type I error.
   a. true   b. false

*43. It is possible that the test results reflect a Type II error.
   a. true   b. false

*44. What is the strength of the relationship, as indexed by Cramer's statistic?
   a. .02   c. .13
   b. .09   d. .31

45. What can we conclude about the relationship between race and voting behavior?
   a. African Americans are more likely than expected not to vote.
   b. African Americans are more likely than expected not to vote, and members of "other" races are less likely than expected not to vote.
   c. Race and voting behavior are not related.
   d. Race and voting behavior may or may not be related.

Exercises to Apply Concepts

**46. Since the inception of television, behavioral scientists have been concerned with its influence on the social development of children. Violence in programming and its effects on aggressive behavior have been a major target of investigation. However, behavioral scientists have also studied how television affects gender stereotyping in children. For instance, McArthur and Eisen (1976) sought to document the amount of gender stereotyping that occurs on television. In one analysis, McArthur and Eisen classified commercials shown on Saturday morning children's television programs in terms of whether the central character was male or female and whether he or she was portrayed as an authority (that is, an expert about the product) or simply as a user of the product. This is important because Barcus (1971) has estimated that by age 17, the average viewer has seen approximately 350,000 commercials on television. The data for the investigation are presented in the table. Analyze these data, draw a conclusion, and write up your results using principles discussed in the Method of Presentation section.

Gender	Role	
	Authority	User
Male	138	114
Female	20	43

47. It is commonly believed that the "winner" of a televised presidential debate will benefit greatly at the polls. Research has indicated, however, that the assessment of who wins a debate is not entirely objective. In one study, a sample of individuals were asked whether they thought the Democratic or the Republican candidate won a presidential debate. These individuals had been interviewed prior to the debate to determine their political party identification. The data for the study are presented in the table. Analyze these data, draw a conclusion, and write up your results using principles discussed in the Method of Presentation section.

Party Identification	Judged Winner	
	Democratic candidate	Republican candidate
Democrat	71	10
Republican	17	58
Independent	35	32

48. Bushman (1988) studied the effect of attire on compliance by varying the clothing worn by a woman as she made a request of pedestrians walking near a shopping center. In one condition, the woman was dressed as a panhandler; in a second condition, she was dressed as a business executive; in a third condition, she wore an "ambiguous, but salient, uniform." In all cases, this individual was stationed about 30 feet from a car parked beside an expired parking meter. As pedestrians approached, the woman pointed at a man standing next to the car and said, "This fellow is overparked at the meter and doesn't have any change. Give him a nickel!" The man was dressed in casual clothing and was searching his pockets for change. The numbers of people who did and did not comply with the request in each condition are presented in the table. Analyze these data, draw a conclusion, and write up your results using principles discussed in the Method of Presentation section.

	Attire		
Outcome	Panhandler	Executive	Uniform
Complied	26	24	36
Did not comply	24	26	14

CHAPTER 16

Nonparametric Statistics

With the exception of the chi-square tests discussed in Chapter 15, all of the statistical tests that we have considered require assumptions about the distribution of scores in the populations from which the samples are selected. Tests of this type are called **parametric statistics**. As we have seen, the most common assumptions concern the normal distribution of scores and homogeneity of variance. Under certain conditions, parametric tests are robust to violations of these assumptions. However, there are instances where this is not the case, and then other analytic strategies must be pursued.

Nonparametric statistics are a class of statistical tests that, in general, make fewer distributional assumptions than parametric statistics. Because of this, they are often called *distribution-free statistics*. Nonparametric tests can be used to analyze quantitative variables that are measured on an ordinal level.* In contrast, parametric tests are usually applied to quantitative variables that are measured on a level that at least approximates interval characteristics.

Like the parametric tests considered in this book, many nonparametric tests are designed to examine the relationship between variables. However, rather than

---

* Nonparametric tests can also be used to analyze qualitative variables. The chi-square tests discussed in Chapter 15 are an example of such nonparametric statistics.

comparing groups in terms of means (as with *t* tests and analysis of variance), the nonparametric procedures compare groups in terms of differences between *distributions of scores*. Thus, one factor that is important in deciding whether to use a parametric or nonparametric test is what aspect of the distribution of the dependent variable one wishes to focus on. If one wants to compare means across conditions, a parametric test is the choice, unless distributional or measurement requirements preclude such analysis. If one wants to compare other features of the dependent variable distribution, then a nonparametric test might be used.

Most of the parametric tests that we have considered have a nonparametric counterpart in the sense that the structure of the independent variables is comparable for the two methods of analysis. For example, an independent groups *t* test analyzes the relationship between two variables in terms of means where the independent variable has two levels and is between-subjects in nature. A nonparametric counterpart is the Wilcoxon rank sum test. This nonparametric procedure analyzes the relationship between two variables in terms of ranks where the independent variable has two levels and is between-subjects in nature. Actually, there are two major types of nonparametric tests: *rank tests* and *sign tests*. We consider only tests based on ranks because they are the more popular of the two. The relevant sign tests are discussed in Marascuilo and McSweeney (1977).

Column 2 of Table 16.1 presents the nonparametric tests that will be considered in this chapter. Each of these is a counterpart to the parametric test listed in the first column. In contrast to the focus on differences between population means for the first four parametric tests, the corresponding nonparametric tests are designed to detect differences between *distributions of scores*. Analogous to Pearson correlation, Spearman correlation tests whether rank scores for the two variables under study covary in the population.

16.1 Rank Scores

All of the tests discussed in this chapter use rank scores. To illustrate ranking procedures, consider the following scores on an intelligence test given to five individuals:

Individual	Score	Rank
1	142	5
2	139	4
3	138	3
4	130	2
5	126	1

We can rank order the scores from least intelligence to most intelligence by assigning the number 1 to the lowest intelligence score and then assigning successive integers in increasing order of intelligence test scores, as has been done in the table.

TABLE16.1   **Parametric Tests and Their Nonparametric Counterparts**

Parametric test	Nonparametric Counterpart
Independent groups *t* test	Wilcoxon rank sum test/ Mann–Whitney *U* test
Correlated groups *t* test	Wilcoxon signed-rank test
One-way between-subjects analysis of variance	Kruskal–Wallis test
One-way repeated measures analysis of variance	Friedman analysis of variance by rank
Pearson correlation	Spearman rank-order correlation

Alternatively, we could rank the scores from most intelligence to least intelligence, if we wished. Then the intelligence test score of 142 would be assigned a rank of 1, the intelligence test score of 139 would be assigned a rank of 2, and so forth. Although the computational procedures described in this chapter will be the same regardless of whether scores are ranked from low to high or from high to low, the *interpretation* of results is facilitated if scores are ranked such that lower scores receive the lower ranks and higher scores receive the higher ranks. This is the strategy that we follow in the remainder of this chapter.

In some instances, we will encounter tied score values. Consider the following distribution:

Individual	Score	Rank
1	142	5
2	135	3.5
3	135	3.5
4	130	2
5	126	1

In cases where a tie occurs, the tied scores are assigned the average rank that the tied values occupy. In the above example, a tie occurred where ranks 3 and 4 were to be assigned. The average of these ranks is $(3 + 4)/2 = 3.5$. Thus, a rank of 3.5 is assigned to the two scores. The next highest score is assigned a rank of 5 because ranks 3 and 4 have already been "used."

Rank-order tests assume that quantitative variables are continuous in nature. Given this assumption, tied scores are viewed as being the result of the imprecision of measurement because, in principle, more precise measures would eliminate ties. Nevertheless, tied scores often occur in practice and must be dealt with when applying a nonparametric test. The most common approach is to rank the scores using the preceding procedure (assigning the average rank for ties) and then to use a *correction term* to adjust for ties. Unfortunately, statisticians do not agree on the optimal correction terms for the various tests. The statistical tests in this chapter are developed in pure form where no ties exist. As long as the number of ties is minimal, the sampling distributions for the test statistics are not affected greatly by tied ranks. However, this may be problematic when a large number of ties occur. Appendix 16.1 presents formulas for selected tests based on the correction terms most frequently used in this instance.

16.2 Nonparametric Statistics and Outliers

A benefit of many nonparametric tests is their relative lack of sensitivity to outliers. Because data are converted to ranks, outliers do not play the kind of havoc that they can on such statistics as means and correlations. Consider the following income scores and their rank order:

Income	Rank
$25,000	6
22,000	5
15,000	4
14,500	3
14,000	2
13,000	1

Now suppose that the $25,000 income was $250,000 instead. Here are the new data and their ranks:

Income	Rank
$250,000	6
22,000	5
15,000	4
14,500	3
14,000	2
13,000	1

Note that the very extreme value of $250,000 has no effect on the ranked data. A rank of 6 is assigned to the highest score regardless of whether it is $25,000 or $250,000. Such an extreme value might prove to be problematic for parametric analysis, but it poses no particular problem for many nonparametric analyses. Because of this property, nonparametric tests are often said to be *outlier resistant*.

16.3 Analysis of Ranked Data Using Parametric Formulas

In an important article, Conover and Iman (1981) noted interesting parallels between traditional parametric analyses and nonparametric analyses. They introduced the **rank transformation approach**, which involves converting a set of scores on a variable to ranks (using the guidelines described above) and then analyzing the rank scores using the traditional parametric formulas. For example, investigators who wished to apply a nonparametric test to the analysis of a dependent variable in a two-group between-subjects design could first convert the scores to ranks and then analyze these rank-transformed scores using the independent groups t test, exactly as described in Chapter 10. The resulting conclusions with respect to the null hypothesis would be similar to the conclusions made using the nonparametric Wilcoxon rank sum test/Mann–Whitney U test described below. However, the normality and homogeneity of variance assumptions are not applicable, and strictly ordinal data can be analyzed. A similar relationship exists for every nonparametric test discussed in this chapter. For example, a one-way between-subjects analysis of variance applied to rank-transformed data will tend to yield the same conclusion as the Kruskal–Wallis test. The similarity of the two approaches makes nonparametric analyses easy to conduct once one has a background in standard parametric analysis. We present nonparametric statistics in this chapter using the more traditional approaches. However, interested readers are referred to Conover (1980), Conover and Iman (1981), and Zimmerman and Zumbo (1993) for the analysis of rank-transformed data by means of parametric formulas.

16.4 Rank Tests for Two Independent Groups

The **Wilcoxon rank sum test** and the **Mann–Whitney U test** are the nonparametric counterparts of the independent groups t test. They are typically used to analyze the relationship between two variables when (1) scores on the dependent variable are in the form of ranks, (2) the independent variable is between-subjects in nature (it can be either qualitative or quantitative), and (3) the independent variable has two and only two levels. Although the computational procedures are somewhat different, the two tests yield the same result.

Analysis of the Relationship

Consider an investigation in which ten men and ten women were administered a questionnaire designed to measure prejudice against women. Scores on this questionnaire could range from 0 to 50, with higher values indicating greater prejudice. The scores for the 20 participants are listed in column 3 of Table 16.2. The first step of the analysis is to rank order the scores from lowest (rank of 1) to highest (rank of 20) across all individuals. This has been done in column 4 of

TABLE 16.2   **Data for Wilcoxon Rank Sum Test/Mann–Whitney *U* Test Example**

Individual	Gender	Prejudice score	Prejudice score rank
1	M	48	20
2	M	45	19
3	M	35	16
4	M	33	15
5	M	32	14
6	M	27	9
7	M	31	13
8	M	20	4
9	M	24	6
10	M	25	7
11	F	43	18
12	F	28	10
13	F	29	11
14	F	30	12
15	F	40	17
16	F	19	3
17	F	15	1
18	F	17	2
19	F	26	8
20	F	23	5

$$R_1 = 20 + 19 + 16 + 15 + 14 + 9 + 13 + 4 + 6 + 7 = 123$$
$$R_2 = 18 + 10 + 11 + 12 + 17 + 3 + 1 + 2 + 8 + 5 = 87$$

Table 16.2. Next, the sum of the ranks (R_j) is computed separately for each group. Let R_1 represent the sum of the ranks for group 1 (men) and R_2 represent the sum of the ranks for group 2 (women). From Table 16.2 we find that $R_1 = 123$ and $R_2 = 87$.

Given a null hypothesis of no relationship between gender and prejudice against women, any difference between R_1 and R_2 would reflect sampling error. There is no reason, under the assumption of the null hypothesis, to expect high scores to be concentrated in one group or the other, so we should find that high- and low-ranked scores are intermingled among the two groups. Thus, when $n_1 = n_2$, as in the present example, R_1 should equal R_2 except for sampling error. Specifically, we would expect to find that R_1 and R_2 are both approximately equal to the sum of ranks 1 through 20 divided by 2, or $(1 + 2 + \cdots + 19 + 20)/2 = 105.00$. In practice, we would not actually perform the above calculations because we can determine the expected sum of ranks for group j through a simple computational formula:

$$E_j = \frac{n_j(n_1 + n_2 + 1)}{2} \qquad\qquad [16.1]$$

where E_j is the expected rank sum for group j and n_j is the sample size for group j. Since both groups in our example have the same number of cases, E_j is the same for men and women:

$$E_j = E_1 = E_2 = \frac{10(10 + 10 + 1)}{2} = 105.00$$

Note that R_1 and R_2 are equidistant from E_j ($R_1 - E_j = 123 - 105.00 = 18.00$; $R_2 - E_j = 87 - 105.00 = -18.00$). This will occur whenever $n_1 = n_2$.

When $n_1 \neq n_2$, the values of E_j will differ for the two groups. Either value can be used in the *Wilcoxon rank sum test* presented in Equation 16.3 as long as the value of R_j is calculated on the same group. This value of R_j is known as the **R statistic**. Regardless of which group the R statistic is based on, its sampling distribution will have a mean equal to E_j and a standard deviation equal to

$$\sigma_R = \sqrt{\frac{n_1 n_2 (n_1 + n_2 + 1)}{12}} \qquad [16.2]$$

In our example,

$$\sigma_R = \sqrt{\frac{(10)(10)(10 + 10 + 1)}{12}} = 13.23$$

When the sample sizes of both groups are ten or more, the shape of the sampling distribution of R_j approximates a normal distribution, so we can convert the R statistic to a z score using the following formula (Ferguson, 1976):

$$z = \frac{\mid R_j - E_j \mid - 1}{\sigma_R} \qquad [16.3]$$

The same value of z will be obtained regardless of whether R_j and E_j are derived from group 1 or group 2. In terms of the women's scores,

$$z = \frac{\mid 87 - 105 \mid - 1}{13.23} = 1.28$$

in our example. For an alpha level of .05, nondirectional test, the critical values of z from Appendix B are ±1.96. Since 1.28 is neither less than −1.96 nor greater than +1.96, we fail to reject the null hypothesis of no relationship between gender and prejudice against women.

The Wilcoxon rank sum test is applicable when the sample sizes of both groups are ten or more. When one or both sample sizes are smaller than ten, the shape of the sampling distribution of the R statistic does not approximate a normal distribution, so application of the Wilcoxon rank sum test is not appropriate. Under this circumstance, the data can be analyzed using the *Mann–Whitney U test*. This involves the determination of the **U statistic,** which is the smaller of the two values of U_1 and U_2 as calculated by the following formulas:

$$U_1 = n_1 n_2 + \frac{n_1 (n_1 + 1)}{2} - R_1 \qquad [16.4]$$

$$U_2 = n_1 n_2 + \frac{n_2(n_2 + 1)}{2} - R_2 \qquad\qquad [16.5]$$

The hypothesis testing procedure consists of comparing the observed value of U with the relevant critical value of U from Appendix K. The observed U is statistically significant if it is *equal to or less than* the critical U.

As noted above, when one or both sample sizes are smaller than ten, application of the Wilcoxon rank sum test is not appropriate and the Mann–Whitney U test must be used instead. When the sample sizes of both groups are ten or more, either test can be applied as long as a critical value for U can be obtained from Appendix K. Application of the Mann–Whitney U test to our example would proceed as follows:

$$U_1 = (10)(10) + \frac{(10)(10 + 1)}{2} - 123 = 32$$

and

$$U_2 = (10)(10) + \frac{(10)(10 + 1)}{2} - 87 = 68$$

Since 32 is less than 68, U in this case is equal to 32. Referring to Appendix K, we find that the critical value of U for an alpha level of .05, nondirectional test, and $n_1 = n_2 = 10$ is 23. Since 32 is not equal to or less than 23, as with the Wilcoxon rank sum test, we fail to reject the null hypothesis of no relationship between gender and prejudice against women.

Regardless of whether the Wilcoxon rank sum test or the Mann–Whitney U test is used, the strength of the relationship between the two variables can be measured using the **Glass rank biserial correlation coefficient** (symbolized r_g). This coefficient is derived using the following formula:

$$r_g = \frac{2(\bar{R}_2 - \bar{R}_1)}{N} \qquad\qquad [16.6]$$

where $\bar{R}_1$ is the mean rank for group 1, $\bar{R}_2$ is the mean rank for group 2, and N is the total sample size. The value of r_g can range from -1.00 to $+1.00$, and its magnitude is interpreted like that of the Pearson correlation coefficient. In our example, $\bar{R}_1 = R_1/n_1 = 123/10 = 12.30$ and $\bar{R}_2 = R_2/n_2 = 87/10 = 8.70$. Thus,

$$r_g = \frac{2(8.70 - 12.30)}{20} = -.36$$

The nature of the relationship between the two variables is ascertained by inspecting $\bar{R}_1$ and $\bar{R}_2$. If we had rejected the null hypothesis in this example, the appropriate conclusion would be that men are more prejudiced against women than are women (since $\bar{R}_1 = 12.30$ and $\bar{R}_2 = 8.70$, and higher rank scores represent greater prejudice).

Method of Presentation

The results of a rank test for two independent groups are reported in terms of the Wilcoxon rank sum test when the R statistic is converted into a z score and in terms of the Mann–Whitney U test when the U statistic is calculated. Reports should include information about the alpha level that was used, the sample sizes, the mean ranks, the value of z or U (whichever is appropriate), the significance level, and, if the test is statistically significant, the strength and nature of the relationship. Unfortunately, the strength of the relationship is seldom reported. This can be done by including the value of the Glass rank biserial correlation coefficient. The results for the study on gender and prejudice against women might be presented as follows:

Results

A Wilcoxon rank sum test compared men's ($\underline{n}$ = 10) and women's ($\underline{n}$ = 10) prejudice against women. For an alpha level of .05, the mean ranks (12.30 for men and 8.70 for women) were found to not significantly differ, $\underline{z}$ = 1.28, $\underline{p}$ > .20.

If the null hypothesis had been rejected, additional sentences would have stated that the Glass rank biserial correlation coefficient was –.36 and that men are more prejudiced against women than are women.

16.5 Rank Test for Two Correlated Groups

Use of the Rank Test for Two Correlated Groups

The **Wilcoxon signed-rank test** is the nonparametric counterpart of the correlated groups t test. It is typically used to analyze the relationship between two variables when (1) scores on the dependent variable are in the form of ranked differences, (2) the independent variable is within-subjects in nature (it can be either qualitative or quantitative), and (3) the independent variable has two and only two levels.

Analysis of the Relationship

Consider an investigation in which the attitudes toward nuclear energy for each of ten individuals were measured before a nuclear accident and again just after it. Scores on the attitude scale could range from 0 to 30, with higher values indicating more favorable attitudes. The investigator was interested in comparing attitudes before and after the accident. The scores for the ten participants are presented in columns 2 and 3 of Table 16.3.

TABLE 16.3   Data for Wilcoxon Signed-Rank Test Example

Individual	Before accident	After accident	Difference	Ranked difference
1	27	15	12	9
2	26	16	10	8
3	24	16	8	7
4	20	26	−6	6
5	19	15	4	4
6	18	20	−2	2
7	15	14	1	1
8	13	16	−3	3
9	11	16	−5	5
10	9	9	0	—

$$R_p = 9 + 8 + 7 + 4 + 1 = 29$$
$$R_n = 6 + 5 + 3 + 2 = 16$$

The analysis begins with the computation of the difference between scores in the two conditions for each individual. This has been done in column 4 of Table 16.3. The differences are then rank ordered from smallest to largest, disregarding the sign of the difference and ignoring differences of 0, as in column 5. We then separately sum the rank scores for the positive differences (R_p) and the rank scores for the negative differences (R_n). This has been done in Table 16.3, and we find that $R_p = 29$ and $R_n = 16$. If the null hypothesis of no relationship between the time of assessment and attitudes toward nuclear energy is true, we would expect the scores for a given individual to be approximately the same in the two conditions. Since unfavorable and favorable attitude changes should be equally likely to occur and since the average size of the changes in each direction should be approximately equal, we would expect the sum of the ranks of the positive differences (R_p) to equal the sum of the ranks of the negative differences (R_n), except for sampling error. The symbol E is used to represent the expected value of each sum of ranks. In our example, E is equal to $(29 + 16)/2$, or 22.5. The question then becomes: Under the assumption that the null hypothesis is true, how likely is it that we would obtain rank sums of 29 and 16 when we expected to find rank sums of 22.5? The answer to this question depends on the value of the T statistic, which is equal to the smaller of R_p and R_n. In this instance, $T = R_n = 16$.

It is possible to construct sampling distributions of the T statistic under the assumption of the null hypothesis and to define critical values of T for the purpose of hypothesis testing. This has been done by Wilcoxon (1949), and a table of critical values is presented in Appendix L for sample sizes (N) of 50 or less.

In the present instance, we are interested in the critical value for $N = 9$ rather than $N = 10$ because we disregarded the individual whose difference score was 0. For an alpha level of .05, nondirectional test, the critical value of T is 5. In order for the null hypothesis to be rejected, the observed value of T must be *equal to or less than* the critical value. Since the observed T value of 16 is not equal to or less than 5, the appropriate decision is to fail to reject the null hypothesis of no relationship between the time of assessment and attitudes toward nuclear energy.

When the sample size is larger than 40, the sampling distribution of T is normally distributed with a standard deviation equal to

$$\sigma_T = \sqrt{\frac{N(N+1)(2N+1)}{24}} \qquad [16.7]$$

When the sample size requirement is met, we can use these properties to convert the T statistic to a z score and then test the null hypothesis using the appropriate critical values from Appendix B. The formula for the z score transformation is

$$z = \frac{T-E}{\sigma_T} \qquad [16.8]$$

The strength of the relationship for the Wilcoxon signed-rank test can be measured using the **matched-pairs rank biserial correlation coefficient** (symbolized r_C). This coefficient is computed using the following formula:

$$r_C = \frac{4\,|\,T-E\,|}{N\,|\,N+1\,|} \qquad [16.9]$$

where all terms are as previously defined. The value of r_C can range from 0 to 1.00, and its magnitude is interpreted in the same manner as the Pearson correlation coefficient. In our example,

$$r_C = \frac{4\,|\,16-22.5\,|}{9\,|\,9+1\,|} = .29$$

The nature of the relationship between the two variables is addressed by inspecting the rank sums. In our example, if we had rejected the null hypothesis, we would have concluded that attitudes toward nuclear energy are less favorable after a nuclear accident than before because the rank sum for the negative differences ($R_n = 16$) was smaller than the rank sum for the positive differences ($R_p = 29$).

Method of Presentation

When presenting the results of a Wilcoxon signed-rank test, one should include information about the alpha level that was used, the sample size, the rank sums, the value of the T statistic, and the significance level. If the test is statistically significant, the strength and nature of the relationship should also be addressed. The strength of the relationship can be discussed in terms of the matched-pairs rank biserial correlation coefficient. The results for the study on the time of assessment and attitudes toward nuclear energy might be reported as follows:

```
                        Results

    A Wilcoxon signed-rank test was applied to individu-
als' attitudes toward nuclear energy before versus after
the nuclear accident. For an alpha level of .05, the rank
sums (29 for the positive differences and 16 for the
```

negative differences) were found to not significantly differ, $\underline{N} = 9$, $\underline{T} = 16$, $\underline{p} > .10$.

16.6 Rank Test for Three or More Independent Groups

Use of the Rank Test for Three or More Independent Groups

The **Kruskal–Wallis test** is the nonparametric counterpart of one-way between-subjects analysis of variance. It is typically used to analyze the relationship between two variables when (1) scores on the dependent variable are in the form of ranks, (2) the independent variable is between-subjects in nature (it can be either qualitative or quantitative), and (3) the independent variable has three or more levels.

Analysis of the Relationship

Consider an experiment on the effect of the number of roommates one has on attitudes toward living in dormitories. Eighteen students were randomly assigned to dormitory rooms with one, two, or three roommates. After living with their roommates for one semester, the students were administered a questionnaire to measure their attitudes toward living in dormitories. Scores on this questionnaire could range from 0 to 30, with higher values indicating more favorable attitudes. The scores for the 18 students are presented in column 3 of Table 16.4, and they are rank ordered from lowest (rank of 1) to highest (rank of 18) in column 4.

If the null hypothesis of no relationship between the number of roommates one has and attitudes toward living in dormitories is true, we would expect the mean ranks for the three conditions to be the same, within the constraints of sampling error. To the extent that the mean ranks are different, the null hypothesis is questionable. The logic underlying the measurement of differences in mean ranks using the Kruskal–Wallis test can be illustrated by analogy (Friedman, 1972). Consider the following sets of two numbers that each sum to 10: 5, 5; 4, 6; 3, 7; 2, 8; 1, 9; 0, 10. Note what happens when each number is squared and the squares are summed for each set:

$$5^2 + 5^2 = 25 + 25 = 50 \qquad 1^2 + 9^2 = 1 + 81 = 82$$

$$4^2 + 6^2 = 16 + 36 = 52 \qquad 0^2 + 10^2 = 0 + 100 = 100$$

$$3^2 + 7^2 = 9 + 49 = 58$$

$$2^2 + 8^2 = 4 + 64 = 68$$

Note that as the difference between the two numbers increases, the sum of the squared values also increases.

The foregoing property was used by Kruskal and Wallis (1952) in proposing the following test statistic (the *H* **statistic**) to reflect differences between groups:

TABLE 16.4   **Data for Kruskal–Wallis Test Example**

Student	Number of roommates	Attitude score	Attitude score rank
1	1	29	17
2	1	24	12
3	1	27	15
4	1	22	10
5	1	30	18
6	1	25	13
7	2	23	11
8	2	26	14
9	2	21	9
10	2	17	5
11	2	28	16
12	2	19	7
13	3	18	6
14	3	15	3
15	3	16	4
16	3	20	8
17	3	14	2
18	3	13	1

$$R_1 = 17 + 12 + 15 + 10 + 18 + 13 = 85$$
$$R_2 = 11 + 14 + 9 + 5 + 16 + 7 = 62$$
$$R_3 = 6 + 3 + 4 + 8 + 2 + 1 = 24$$

$$H = \left[\frac{12}{N(N+1)} \right] \left(\sum \frac{R_j^2}{n_j} \right) - 3(N+1) \qquad [16.10]$$

where N is the total sample size, R_j is the sum of the rank scores for group j, n_j is the sample size for group j, and the summation of R_j^2/n_j is across all groups. According to this formula, the sum of the ranks in each group is to be squared and then divided by the number of individuals in that group. The sum of these values combines with the other terms in the equation to yield a statistic, H, that has a sampling distribution that approximates a chi-square distribution with $k - 1$ degrees of freedom, where k is the number of groups.

For the data in Table 16.4, the sum of the rank scores is 85 for the one-roommate group, 62 for the two-roommate group, and 24 for the three-roommate group. The value of H is thus

$$H = \left[\frac{12}{18(18+1)} \right] \left(\frac{85^2}{6} + \frac{62^2}{6} + \frac{24^2}{6} \right) - 3(18+1)$$

$$= (.035)(1{,}940.83) - 57 = 10.93$$

For an alpha level of .05 and $k - 1 = 3 - 1 = 2$ degrees of freedom, the critical value of H from the chi-square table in Appendix J is 5.991. Since 10.93 is greater than 5.991, we reject the null hypothesis and conclude that there is a relationship between the number of roommates one has and attitudes toward living in dormitories.

The strength of the relationship for the Kruskal–Wallis test can be measured using an index known as **epsilon-squared.** Epsilon-squared can range from 0 to 1.00, where a value of 0 indicates no relationship between the independent and dependent variables and a value of 1.00 indicates a perfect relationship. The formula for epsilon-squared (symbolized E_R^2) is

$$E_R^2 = \frac{H}{(N^2 - 1)/(N + 1)} \qquad [16.11]$$

where all terms are as previously defined. In our example,

$$E_R^2 = \frac{10.93}{(18^2 - 1)/(18 + 1)} = .64$$

Analogous to one-way analysis of variance, rejection of the null hypothesis when a nonparametric test is applied to more than two groups tells us only that at least two of the k groups differ in ranks; it does not indicate the nature of these differences. Several procedures have been proposed for analyzing the nature of the relationship following a statistically significant Kruskal–Wallis test (for instance, see Dunn, 1964; Miller, 1966; Ryan, 1959; Steel, 1960; Wike, 1971). Which one to use, however, is controversial among statisticians. We suggest the procedure recommended by Dunn (1964). The **Dunn procedure** involves using the Wilcoxon rank sum test or the Mann–Whitney U test to compare the mean ranks for all possible pairs of conditions, just as the Tukey HSD test was applied to all possible pairs of means following a statistically significant analysis of variance. In order to maintain the overall Type I error rate at the desired level of alpha, however, Dunn recommends that the critical value for each comparison be determined based on a revised alpha level equal to alpha/C, where C is the number of pairs of mean ranks to be tested.*

Since the desired alpha level in our example is .05 and there are three pairs of mean ranks (those for the one-roommate versus the two-roommate condition; those for the one-roommate versus the three-roommate condition; those for the two-roommate versus the three-roommate condition), the revised alpha level to be used with the Dunn procedure in this instance is .05/3 = .017. If we were to apply this procedure, we would find that the mean rank for the three-roommate condition (4.00) is significantly lower than the mean rank for the one-roommate condition (14.17), which suggests that the attitudes toward living in dormitories of students who live with three roommates are less favorable than the attitudes of students who live with one roommate. However, we cannot confidently conclude

---

* Critical values for the Wilcoxon rank sum test or the Mann–Whitney U test corresponding to values of alpha/C can be interpolated based on the critical values reported in Appendix B or Appendix K, respectively, for tabled levels of alpha.

that the attitudes of dormitory students who live with two roommates (mean rank of 10.33) differ from those who live with one or three roommates.

Method of Presentation

Reports of a Kruskal–Wallis test should include information about the alpha level that was used, the degrees of freedom, the sample sizes, the mean ranks, the value of the H statistic, and the significance level. Following the format for the chi-square test, the degrees of freedom and the total sample size are given in parentheses following the symbol for the test statistic. If the test is statistically significant, the strength of the relationship can be addressed by epsilon-squared and the nature of the relationship can be addressed in terms of the Dunn procedure. The results for the experiment on the number of roommates one has and attitudes toward living in dormitories might be presented as follows:

```
                          Results

    A Kruskal-Wallis test was applied to the ranked data
(n = 6 per condition) relating the number of roommates one
has to attitudes toward living in dormitories. For an
alpha level of .05, the resulting value of H was found to
be statistically significant, H(2, N = 18) = 10.93,
p < .01. The strength of the relationship, as indexed by
epsilon-squared, was .64. A follow-up procedure suggested
by Dunn (1964) indicated that the attitudes of dormitory
students who live with three roommates (mean rank of 4.00)
are less favorable than the attitudes of dormitory stu-
dents who live with one roommate (mean rank of 14.17). The
mean rank for dormitory students who live with two room-
mates (10.33) did not significantly differ from either of
these groups.
```

Note that a reference has been provided for the Dunn procedure. This should be done anytime a procedure is not widely known because doing so will allow interested readers to locate information about the technique.

16.7 Rank Test for Three or More Correlated Groups

Use of the Rank Test for Three or More Correlated Groups

Friedman analysis of variance by ranks is the nonparametric counterpart of one-way repeated measures analysis of variance. It is typically used to analyze the relationship between two variables when (1) scores on the dependent variable are in the form of ranks across conditions for each participant, (2) the independent variable is within-subjects in nature (it can be either qualitative or quantitative), and (3) the independent variable has three or more levels.

Analysis of the Relationship

Consider an experiment in which each of ten individuals attempted to solve a different complex problem under quiet, slightly noisy, and noisy circumstances. Performance in each condition was scored 0 to 20, with higher values indicating better performance. The three problems had been pretested to ensure that they were of equal difficulty, and the order of the conditions was counterbalanced. Table 16.5 presents the experimental data.

Friedman analysis of variance by ranks involves first rank ordering the scores *for each participant* across the research conditions, as has been done in Table 16.6. For example, the first individual had a score of 10 in the quiet condition, a score of 6 in the slightly noisy condition, and a score of 5 in the noisy condition. These scores correspond to ranks of 3, 2, and 1, respectively.

If the null hypothesis that the level of noise is unrelated to problem solving performance is true, then differences in the rank scores are merely a function of sampling error, and we would expect the sums of the ranks in the three conditions to be approximately equal. Friedman (1972) has suggested the following statistic for ascertaining the magnitude of differences in rank sums:

$$\chi_r^2 = \frac{12 \sum R_j^2}{Nk(k+1)} - 3N(k+1) \qquad [16.12]$$

where R_j is the sum of the rank scores in condition j, N is the sample size, k is the number of conditions, and the summation of R_j^2 is across all conditions. The χ_r^2 statistic has a sampling distribution that approximates a chi-square distribution with $k-1$ degrees of freedom.

In our example, the sum of the rank scores is 25 for the quiet condition, 20 for the slightly noisy, and 15 for the noisy. The value of χ_r^2 is thus

$$\chi_r^2 = \frac{12(25^2 + 20^2 + 15^2)}{(10)(3)(3+1)} - (3)(10)(3+1)$$

$$= 125.00 - 120 = 5.00$$

TABLE 16.5   Data for Friedman Analysis of Variance by Ranks Example

Participant	Quiet	Slightly noisy	Noisy
1	10	6	5
2	15	10	9
3	15	14	8
4	15	13	17
5	8	5	3
6	6	5	2
7	8	7	3
8	9	8	6
9	9	14	15
10	8	14	9

TABLE 16.6   **Ranked Data for Friedman Analysis of Variance by Ranks Example**

Participant	Quiet	Slightly noisy	Noisy
1	3	2	1
2	3	2	1
3	3	2	1
4	2	1	3
5	3	2	1
6	3	2	1
7	3	2	1
8	3	2	1
9	1	2	3
10	1	3	2
	$R_1 = 25$	$R_2 = 20$	$R_3 = 15$

For an alpha level of .05 and $k - 1 = 3 - 1 = 2$ degrees of freedom, the critical value of χ_r^2 from the chi-square table in Appendix J is 5.991. Since 5.00 does not exceed 5.991, we fail to reject the null hypothesis of no relationship between noise level and problem solving performance.

The strength of the relationship for Friedman analysis of variance by ranks can be measured using an index known as the **concordance coefficient** (symbolized W):

$$W = \frac{\chi_r^2}{N(k - 1)} \qquad [16.13]$$

where all terms are as previously defined. Like the epsilon-squared measure presented in Section 16.6, this index can range from 0 to 1.00, with higher values indicating a stronger relationship between the independent and dependent variables. In our example,

$$W = \frac{5.00}{10(3 - 1)} = .25$$

If we had rejected the null hypothesis, it would have been necessary to conduct additional analyses to discern the nature of the relationship. The procedure recommended by Dunn (1964) could be used in such an instance. This would involve comparing the rank sums for all possible pairs of conditions using the Wilcoxon signed-rank test, but setting the alpha level for defining the critical value for each comparison at alpha/C (where C is the number of pairs of rank sums to be tested) as a means of maintaining the overall Type I error rate at the desired level of alpha.*

---

* Critical values for the Wilcoxon signed-rank test corresponding to values of alpha/C can be interpolated based on the critical values reported in Appendix L for tabled levels of alpha.

Method of Presentation

The method of presentation for Friedman analysis of variance by ranks parallels that for the Kruskal–Wallis test as discussed in Section 16.6, except that the values of χ_r^2 and W are reported instead of H and E_R^2, and the nature of the relationship is discussed in terms of the application of the Dunn procedure using the Wilcoxon signed-rank test rather than the Wilcoxon rank sum test/Mann–Whitney U test.

16.8 Rank Test for Correlation

Use of the Rank Test for Correlation

Spearman rank-order correlation, more simply known as *Spearman correlation*, is a nonparametric counterpart of Pearson correlation. It is typically used to analyze the relationship between two variables when (1) scores on both variables are in the form of ranks, (2) the two variables have been measured on the same individuals, and (3) the observations on each variable are between-subjects in nature.

Analysis of the Relationship

Consider a study in which an investigator examined the relationship between pollution and cancer mortality. A pollution index was developed and applied to 20 cities in the United States. Scores on this index could range from 0 to 4, with higher values indicating more pollution. For each city, the cancer mortality per 100,000 people was also obtained. The data for the 20 cities are presented in columns 2 and 3 of Table 16.7.

The first step of the analysis is to rank order the scores separately for each variable such that the lower scores receive the lower ranks and the higher scores receive the higher ranks. The rank scores for the first variable (pollution) in our example are presented in column 4 of Table 16.7 and the rank scores for the second variable (cancer mortality) are presented in column 5. The Spearman rank-order correlation procedure involves computing a correlation coefficient with respect to the ranked data. This coefficient can range from −1.00 through 0 to +1.00 and is interpreted like the Pearson correlation coefficient. In fact, consistent with the discussion in Section 16.3, the Spearman correlation coefficient can be derived by applying the formula for the Pearson correlation coefficient to the ranked data using the procedures discussed in Chapter 5. In practice, however, the Spearman correlation between two variables is most commonly calculated using a formula available for this purpose.

The first step in calculating the Spearman correlation coefficient using this formula is to compute the difference between the rank scores (D) for each individual. This has been done in column 6 of Table 16.7. The differences are then squared, as in column 7. Lastly, the Spearman correlation coefficient (symbolized r_s) is calculated as follows:

$$r_s = 1 - \frac{6 \sum D^2}{N(N^2 - 1)}$$

[16.14]

TABLE 16.7   **Data for Spearman Rank-Order Correlation Example**

City	Pollution	Cancer mortality	Pollution rank	Cancer mortality rank	Rank difference (D)	D^2
1	1.0	124	1	1	0	0
2	1.5	131	4	7	−3	9
3	2.0	135	8	10	−2	4
4	2.5	139	11	14	−3	9
5	3.0	145	15	18	−3	9
6	1.6	125	5	2	3	9
7	2.8	130	14	6	8	64
8	1.2	136	2	11	−9	81
9	1.7	140	6	15	−9	81
10	2.2	146	9	19	−10	100
11	2.6	126	12	3	9	81
12	3.2	129	16	5	11	121
13	3.7	137	18	12	6	36
14	4.0	141	20	16	4	16
15	1.3	133	3	9	−6	36
16	1.9	127	7	4	3	9
17	2.4	132	10	8	2	4
18	2.7	138	13	13	0	0
19	3.4	142	17	17	0	0
20	3.8	148	19	20	−1	1
						$\Sigma D^2 = 670$

where N is the sample size and D is as defined above. For our data, the Spearman correlation between pollution and cancer mortality is

$$r_s = 1 - \frac{(6)(670)}{20(20^2 - 1)}$$

$$= 1 - \frac{4,020}{7,980} = .50$$

The null hypothesis for Spearman correlation states that there is no linear relationship between the two rank variables—that is, that the population correlation, ρ_s, between the two rank variables is 0. Under the assumption that this hypothesis is true, when the sample size is greater than ten, the sample correlation coefficient can be converted to a statistic that has a sampling distribution that closely approximates the t distribution with $N - 2$ degrees of freedom as follows:

$$t = \frac{r_s}{\sqrt{(1 - r_s^2)/(N - 2)}}$$ [16.15]

Note that this formula parallels the formula for converting r to t presented in Equation 14.2.

In our example,

$$t = \frac{.50}{\sqrt{(1 - .50^2)/(20 - 2)}} = 2.50$$

The critical values of t for an alpha level of .05, nondirectional test, and $N - 2 = 20 - 2 = 18$ degrees of freedom are ± 2.101. Since the observed t value of 2.50 is greater than 2.101, we reject the null hypothesis and conclude that there is some degree of a linear relationship between pollution and cancer mortality.

Appendix M presents critical values of r_s that can be used to test for linear relationships between rank variables for selected sample sizes when $N \leq 30$. This approach is easier to use than the t test approach and *must* be used when $N \leq 10$ because application of the t test is not appropriate under this circumstance.

The strength of the relationship for Spearman correlation is indicated by the magnitude of the correlation coefficient, and the nature of the relationship is indicated by the sign of the correlation coefficient. In our example, $r_s = .50$, so the two variables approximate a direct linear relationship; as rank scores on one variable increase, rank scores on the other tend to also. This interpretation was facilitated by the original scores for both variables being ranked from low to high. If one variable had been ranked from low to high and the other had been ranked from high to low, the magnitude of r_s would be the same but the sign would be negative, indicating that as rank scores on the first variable increase (high ranks representing high levels of the variable), rank scores on the second variable tend to decrease (low ranks representing high levels of the variable). To avoid the possible misinterpretation that might arise from situations of this type, it is important that lower scores consistently receive the lower ranks and higher scores consistently receive the higher ranks when Spearman correlation is applied.

An alternative index of correlation for rank scores is a statistic proposed by Kendall called *tau*. The computation of tau is considerably more complex than that of r_s, but it has certain advantageous statistical properties. For a comparison of the two techniques, see Glass and Stanley (1970).

Method of Presentation

Reports of an analysis using Spearman correlation should include statements of the alpha level that was used, the degrees of freedom, the observed value of r_s, and the significance level. It is not necessary to explicitly state the strength of the relationship between the variables because this is indicated by the magnitude of the correlation coefficient. Although the nature of the relationship is indicated by the sign of the correlation coefficient, researchers often choose to make this explicit by stating it verbally. The results for the study on pollution and cancer mortality might be reported as follows:

Results

A Spearman rank-order correlation addressed the relationship between rank scores for the pollution and cancer mortality measures for the 20 cities. For an alpha level of .05, the observed correlation was found to be statistically significant, $\underline{r}_s$(18) = .50, $\underline{p}$ < .05, suggesting that as pollution in cities increases, so does mortality due to cancer.

16.9 Examples from the Literature

Severity of the Final Verdict As a Function of the Order of Consideration of Lenient Versus Harsh Verdicts

Consider the following scenario (Greenberg, Williams, & O'Brien, 1986):

> A defendant in a criminal case is charged with first-degree murder in the stabbing death of a woman. Before deliberation, a judge instructs the jury first to consider whether or not the defendant is guilty of first degree murder. If they cannot agree to this verdict, they should next consider whether the defendant is guilty of second degree murder. Then, if the jurors cannot agree to this verdict, they are instructed to consider the verdict of voluntary manslaughter. If they still do not agree they are to consider involuntary manslaughter. Finally, if the jurors still have not reached agreement, they are instructed to find the defendant not guilty. (p. 41)

This procedure of having juries consider a series of increasingly less severe verdicts against a defendant during their deliberations is common in criminal trials. However, it is possible that certain cognitive biases associated with this process lead to harsher verdicts than if the jury were to reverse the order of deliberation and consider the verdicts from least severe to most severe instead.

To study this issue, Greenberg, Williams, and O'Brien (1986) had 16 college students read a condensed version of evidence presented at an actual murder trial. Students in the harsh-to-lenient condition were instructed to consider the verdict of "guilty of murder in the first degree" first and then to proceed progressively down the list of less severe verdicts noted above until a final verdict was reached. In contrast, students in the lenient-to-harsh condition were instructed to reach a final verdict by considering the verdict of "not guilty" first and then proceeding progressively up the list of more severe verdicts as necessary. This was done on an individual basis by each experiment participant.

A Mann–Whitney *U* test showed that students' verdicts were significantly harsher in the harsh-to-lenient than in the lenient-to-harsh condition, $U = 7$, $p < .01$. To the extent that this finding is generalizable to actual jury trials, it raises serious questions about the fairness of the judicial process. Greenberg and colleagues did not report any information about the strength of the relationship.

Effect of Exposure to an Aggressive Model on Observers' Aggressive Behavior

What is the effect of exposure to an aggressive model on observers' aggressive behavior? Bandura, Ross, and Ross (1963) studied this issue by assigning 96 nursery school children either to a no-model control condition or to conditions where they observed an aggressive adult model on film, an aggressive adult model live, or an aggressive cartoon character. All models performed identical aggressive acts of repeatedly punching, kicking, hitting, and throwing a Bobo doll (an inflatable doll that bounces back when it is knocked down). Children in the four conditions were matched on the basis of their teacher's ratings of their characteristic levels of aggressive behavior. Following the model manipulation, children in all conditions spent 20 minutes alone in a room that contained a variety of toys, including a Bobo doll. The number of aggressive acts directed toward the Bobo doll and the other toys during this time span served as the measure of aggressive behavior.

In Chapter 11, it was noted that matched-subjects designs are analyzed as if the independent variable is within-subjects in nature. Given this and the fact that the researchers were concerned with violations of distributional assumptions, the aggressiveness scores were analyzed using Friedman analysis of variance by ranks. This test was found to be statistically significant, $\chi_r^2(3, N = 24) = 9.06$, $p < .05$.* The strength of the relationship, as indexed by epsilon-squared, was .04. Follow-up tests showed that the level of aggressive behavior exhibited by children in the three model conditions did not significantly differ, but that children in all three model conditions behaved significantly more aggressively than children in the no-model condition. This basic finding has been replicated in other studies.

Closeness of Relationships and Intimacy of Touching Behavior

One factor that has been hypothesized to affect nonverbal intimacy is the nature of the relationship between the persons involved in the interaction. In a study of this issue, Heslin and Boss (1980) observed 103 travelers at an airport and a randomly selected member from the party who had come to greet or send off each one. It was proposed that the closeness of the relationship between the traveler and the other person would be positively related to the intimacy of their touching.

The researchers determined the nature of each relationship after observation of the interaction by asking a nonobserved member of the greeting or sendoff party. Each relationship was subsequently classified into one of seven ranked categories reflecting familiarity, involvement, and having influence over one another. Nonverbal intimacy was assessed by classifying each interaction into one of six ranked touch categories ranging from "none" to "extended kissing and embracing." Consistent with the research hypothesis, a Spearman rank-order correlation indicated a statistically significant direct relationship between closeness of the relationship and intimacy of touching behavior, $r_s(101) = .54$, $p < .01$.

---

* The total sample size for the purpose of the statistical test is 24 rather than 96 because the 96 children were matched and assigned to conditions such that 24 scores were provided in each one.

Applications
to the Analysis
of a Social Problem

In Chapter 14, we used Pearson correlation to analyze the relationship between satisfaction with the maternal relationship and teen sexual activity for the parent–teen data set described in Chapter 1. Our analyses revealed that outliers and violation of the underlying statistical assumptions might be problematic, so the decision was made to pursue alternative analyses that are outlier resistant and free of distributional assumptions. Spearman rank-order correlation is one such method. We calculated the Spearman rank-order correlation between the two variables using the data discussed in Chapter 14. The resulting correlation of −.19 was statistically significant, yielding a conclusion comparable to that made using Pearson correlation.

In Chapter 12, we analyzed the relationship between the teen's age and maternal relationship satisfaction using a one-way between-subjects analysis of variance. We can also analyze these data using the Kruskal–Wallis test.* Table 16.8 presents the computer printout for this analysis. The top portion of the printout provides the mean rank and the sample size for each age group. The bottom left portion provides the value of the H statistic (labeled "Chi-Square"), the degrees of freedom, and the p value (labeled "Significance"). The bottom right portion contains the same information, but for a Kruskal–Wallis test that is corrected for ties. Neither version of the test was statistically significant, so the conclusion is comparable to that for the analysis of variance in Chapter 12. Given that we failed to reject the null hypothesis, we also calculated the statistical power for a Kruskal–Wallis test for an alpha level of .05, four groups, and our sample sizes for a small population effect size. The power exceeded .95, which indicates a negligible likelihood of a Type II error.

---

* Maxwell and Delaney (1990, p. 707) have shown that if we assume that homogeneity of variance of the underlying continuous variable holds across age groups, then the Kruskal–Wallis test is a test of differences in population medians.

**TABLE 16.8   Computer Output for Kruskal–Wallis Test
for Satisfaction with Maternal Relationship**

```
- - - - - Kruskal-Wallis 1-Way Anova
   SMOTHER  by AGE

   Mean Rank      Cases
       401.71        114        AGE = 14
       356.22        270        AGE = 15
       364.57        220        AGE = 16
       374.24        133        AGE = 17
                     ----
                      737
```

| | | | | Corrected for ties | | |
Chi-Square	D.F.	Significance		Chi-Square	D.F.	Significance
3.8406	3	.2792		3.8500	3	.2781

Summary

Nonparametric statistics are a class of statistical tests that, in general, make fewer assumptions about population distributions than parametric statistics. These tests can be used to analyze quantitative variables that are measured on an ordinal level. A benefit of many of these tests is their relative lack of sensitivity to outliers. The major rank-order nonparametric tests for analyzing bivariate relationships are the Wilcoxon rank sum test/Mann–Whitney U test, the Wilcoxon signed-rank test, the Kruskal–Wallis test, Friedman analysis of variance by ranks, and Spearman rank-order correlation. Each of these is a counterpart to a parametric test, as indicated in Table 16.1. When scores are ranked for a given variable, the interpretation of results will be facilitated if ranks are assigned such that lower scores receive the lower ranks and higher scores receive the higher ranks.

Appendix 16.1   Corrections for Ties for Nonparametric Rank Tests

All of the tests that we have considered in this chapter deal with ties by assigning tied scores the average rank that the tied values occupy. However, some of the procedures require that correction terms be used in the derivation of the test statistics when a large number of ties occur, in order to closely approximate the relevant sampling distribution. Two tests where this is the case are the Wilcoxon rank sum test and the Kruskal–Wallis test.

For the Wilcoxon rank sum test, the formula for z when numerous ties occur is

$$z = \frac{|R_j - E_j| - 1}{\hat{s}_R} \qquad [16.16]$$

where

$$\hat{s}_R = \sqrt{\left[\frac{n_1 n_2}{N(N-1)}\right]\left[\frac{N^3 - N}{12} - \Sigma\left(\frac{t^3 - t}{12}\right)\right]}$$

$$[16.17]$$

In this formula, t is the number of scores tied at a particular rank, and the summation of $(t^3 - t)/12$ extends across all groups of ties. All other terms in Equations 16.16 and 16.17 are as defined in Section 16.4.

For the Kruskal–Wallis test, when numerous ties occur, the value of H is computed by the following formula:

$$H = \frac{\left[\dfrac{12}{N(N+1)}\right]\left(\Sigma \dfrac{R_j^2}{n_j}\right) - 3(N+1)}{1 - \dfrac{\Sigma\left(\dfrac{t^3 - t}{12}\right)}{N^3 - N}} \qquad [16.18]$$

where, again, t is the number of scores tied at a particular rank, and the summation of $(t^3 - t)/12$ extends across all groups of ties. All other terms in Equation 16.18 are as defined in Section 16.6.

Exercises

Exercises to Review Concepts

1. How do parametric and nonparametric statistical tests differ?

2. Match each parametric test in the first column with its nonparametric counterpart in the second column.

Parametric test	Nonparametric test
1. correlated groups t test	a. Wilcoxon rank sum test/ Mann–Whitney U test
2. Pearson correlation	
3. one-way between-subjects ANOVA	b. Friedman analysis of variance by ranks
4. independent groups t test	c. Spearman rank-order correlation
5. one-way repeated measures ANOVA	d. Wilcoxon signed-rank test
	e. Kruskal–Wallis test

*3. Rank order each of the following sets of scores such that the lowest score in each set receives a rank of 1:

Set I	Set II	Set III	Set IV
135	131	130	104
132	131	136	102
136	136	135	102
131	135	130	102
134	132	135	105
134	134	134	106

4. Rank order each of the following sets of scores such that the lowest score in each set receives a rank of 1:

Set I	Set II	Set III	Set IV
12	6	4	10
16	3	1	6
9	4	1	13
16	7	2	12
9	11	8	6
13	10	3	10
8	2	7	5

*5. Explain the rank transformation approach to nonparametric analysis.

6. What is the relationship between the Wilcoxon rank sum test and the Mann–Whitney U test?

**7. A high school counselor studied the relationship between car ownership and performance in school by comparing the grade point averages of 15 students who own cars and 15 students who do not own cars. The GPAs for the two groups are presented in the table. Conduct a nondirectional Wilcoxon rank sum test and specify the nature of the relationship between the two variables.

Owns car	Does not own car
3.10	2.20
2.50	2.40
2.75	2.60
2.96	2.80
3.50	3.20
2.10	3.00
2.30	2.19
2.15	2.98
3.30	2.77
3.70	2.66
2.90	2.00
3.90	3.40
2.70	3.60
2.64	4.00
1.90	3.80

*8. Compute the value of the Glass rank biserial correlation coefficient for the data in Exercise 7.

**9. A researcher compared the attitudes of men and women toward an experimental male birth control pill. Eight members of each gender read a description of the characteristics of the pill and rated how they would feel about its use (either by themselves or by their partners) on a 0 to 100 scale, where higher values indicated more favorable attitudes. The data for the study are presented in the table. Conduct a nondirectional Mann–Whitney U test and specify the nature of the relationship between the two variables.

Males	Females
51	52
48	50
74	76
56	63
59	61
65	66
67	86
82	84

*10. Compute the value of the Glass rank biserial correlation coefficient for the data in Exercise 9.

11. A researcher examined the effect of marijuana on recognition of auditory stimuli. Eight participants were presented an auditory stimulus that was increased in intensity until it could be heard (as signaled by the participant). The amount of time, in seconds, it took each person to recognize the stimulus was measured both while under the influence of marijuana and while not under the influence of marijuana. The times for the study are presented in the table. Conduct a nondirectional Wilcoxon signed-rank test and specify the nature of the relationship between the two variables.

Participant	Under influence of marijuana	Not under influence of marijuana
1	2.73	2.63
2	2.83	2.65
3	2.69	2.64
4	2.67	2.41
5	2.72	2.71
6	2.43	2.44
7	2.85	2.83
8	3.02	2.86

12. Compute the value of the matched-pairs rank biserial correlation coefficient for the data in Exercise 11.

13. A sociologist compared the amount of violence on the three major television networks during prime time. During a week, eight randomly selected television shows for each network were watched and rated in terms of their violence on a scale of 0 ("no violence") to 50 ("considerable violence"). The ratings for the three networks are presented in the table. Conduct a Kruskal–Wallis test.

Network A	Network B	Network C
20	21	17
28	29	27
32	33	31
35	34	30
38	39	43
40	41	26
42	45	50
15	13	16

14. Compute the value of epsilon-squared for the data in Exercise 13.

**15. A consumer psychologist compared the quality of three brands of picture tubes. Ten television repairers rated each brand on a scale of 0 to 50, where higher scores indicated higher quality. Conduct a Friedman analysis of variance by ranks.

Repairer	Brand A	Brand B	Brand C
1	15	17	19
2	27	28	29
3	31	33	32
4	18	20	22
5	41	39	36
6	44	35	37
7	13	14	16
8	43	44	45
9	24	26	30
10	20	23	25

*16. Compute the value of the concordance coefficient for the data in Exercise 15.

**17. A researcher studied the relationship between crime rates in cities and the size of a city's police force. For 18 large cities, the rank order of the crime rate (1 = lowest and 18 = highest) and the rank order of the size of the police force (1 = smallest and 18 = largest) were determined. The ranks for the study are presented in the table. Compute the Spearman rank-order correlation coefficient, test for a relationship between the two variables using a nondirectional test, and specify the strength and nature of the relationship.

City	Crime rate	Size of police force
1	3	4
2	15	16
3	18	5
4	1	2
5	8	12
6	14	15
7	13	6
8	2	1
9	7	11
10	12	7
11	16	18
12	6	10
13	11	9
14	17	14
15	10	8
16	4	17
17	9	13
18	5	3

Multiple-Choice Questions

*18. When would we be most likely to introduce a correction term into the formula for a nonparametric test?
  a. Sample sizes are unequal.
  b. Data are measured on an ordinal level.
  c. There are more than two groups of scores.

  d. There are tied ranks.
19. Many nonparametric tests are relatively insensitive to outliers.
  a. true   b. false
20. The nature of the relationship following a statistically significant Kruskal–Wallis test can be analyzed using
  a. epsilon-squared
  b. the H statistic
  c. the Wilcoxon signed-rank test
  d. the Dunn procedure
*21. Such nonparametric procedures as the Mann–Whitney U test and the Kruskal–Wallis test are designed to detect differences between distributions of scores.
  a. true   b. false
*22. Which of the following is a nonparametric alternative to the correlated groups t test?
  a. Spearman rank-order correlation
  b. Wilcoxon signed-rank test
  c. Kruskal–Wallis test
  d. Mann–Whitney U test
23. Which of the following is a nonparametric alternative to one-way between-subjects analysis of variance?
  a. Wilcoxon rank sum test
  b. Friedman analysis of variance by ranks
  c. Mann–Whitney U test
  d. Kruskal–Wallis test
*24. Which of the following is a nonparametric alternative to Pearson correlation?
  a. Spearman rank-order correlation
  b. Wilcoxon signed-rank test
  c. Wilcoxon rank sum test
  d. Friedman analysis of variance by ranks
*25. Which of the following is a nonparametric alternative to one-way repeated measures analysis of variance?
  a. Friedman analysis of variance by ranks
  b. Wilcoxon signed-rank test
  c. Mann–Whitney U test
  d. Kruskal–Wallis test

26. Which of the following is a nonparametric alternative to the independent groups *t* test?
    a. Spearman rank-order correlation
    b. Friedman analysis of variance by ranks
    c. Kruskal–Wallis test
    d. Wilcoxon rank sum test

Exercises 27–29 involve the application of a nondirectional Mann–Whitney U test to the following set of unranked data:

Condition 1	Condition 2
97	48
82	87
68	51
73	66
42	49
100	

27. What is the observed value of the *U* statistic?
    a. 8
    b. 22
    c. 23
    d. 43
28. Based on this study, we can conclude that the independent variable is related to the dependent variable.
    a. true
    b. false
29. What is the strength of the relationship, as indexed by the Glass rank biserial correlation coefficient?
    a. −.36
    b. −.47
    c. −.86
    d. none of the above

Exercises 30–32 involve the application of a nondirectional Wilcoxon signed-rank test to the following set of unranked data:

Participant	Condition 1	Condition 2
1	32	34
2	36	47
3	23	28
4	18	22
5	27	27
6	40	46
7	30	27
8	20	30

*30. What is the observed value of the *T* statistic?
    a. 2   c. 24
    b. 14   d. 26
*31. Based on this study, we can conclude that the independent variable is related to the dependent variable.
    a. true   b. false
*32. What is the strength of the relationship, as indexed by the matched-pairs rank biserial correlation coefficient?
    a. .23   c. .86
    b. .67   d. none of the above

Exercises 33–35 involve the application of the Kruskal–Wallis test to the following set of unranked data:

Condition 1	Condition 2	Condition 3
6	16	28
13	12	24
8	19	31

33. What is the observed value of the *H* statistic?
    a. 4.28   c. 36.49
    b. 6.49   d. 261.67
34. Based on this study, we can conclude that the independent variable is related to the dependent variable.
    a. true   b. false
35. What is the strength of the relationship, as indexed by epsilon-squared?
    a. .20   c. .90
    b. .66   d. none of the above

Exercises 36–38 involve the application of Friedman analysis of variance by ranks to the following set of unranked *data:*

Participant	Condition 1	Condition 2	Condition 3	Condition 4
1	7	6	11	9
2	6	3	8	7
3	4	12	10	8

*36. What is the observed value of χ_r^2?
   a. 1.68
   b. 2.95
   c. 3.81
   d. 4.20

*37. Based on this study, we can conclude that the independent variable is related to the dependent variable.
   a. true   b. false

*38. What is the strength of the relationship, as indexed by the concordance coefficient?
   a. .12
   b. .47
   c. .69
   d. none of the above

Exercises 39 and 40 involve the application of a nondirectional Spearman rank-order correlation to the following set of unranked *data:*

Individual	Variable 1	Variable 2
1	6	27
2	3	60
3	9	31
4	4	44
5	2	68
6	12	24

39. What is the observed value of the Spearman rank-order correlation coefficient?
   a. −.28
   b. −.64
   c. −.75
   d. −.94

40. Based on this study, we can conclude that the two variables are related.
   a. true   b. false

Exercises to Apply Concepts

41. A researcher interested in the intelligence of criminals who had committed violent versus nonviolent crimes administered an intelligence test to eight criminals of each type. The test scores for the study are presented in the table. Analyze these data using a nondirectional Wilcoxon rank sum test, draw a conclusion, and write up your results using the principles given in the relevant Method of Presentation section.

Violent crime	Nonviolent Crime
88	98
96	95
91	106
104	103
97	90
100	99
84	85
93	92
89	102
101	105
	94
	108

**42. A psychologist tested the effect of a group encounter session on individuals' self-esteem. A scale measuring self-esteem was administered to ten individuals just before they participated in the encounter session and again just after. Scores on this scale could range from 10 to 80, with higher values indicating higher self-esteem. The scores for the study are presented in the table. Analyze these data using a nondirectional Wilcoxon signed-rank test, draw a conclusion, and write up your results using the principles given in the relevant Method of Presentation section.

Individual	Before session	After session
1	60	67
2	38	42
3	43	37
4	39	42
5	34	26
6	30	39
7	54	43
8	48	50
9	72	71
10	49	62

**43. A psychologist studied the effect of the affective content of words on memory. Fifteen participants were read a list of 25 words and asked to recall as many as they could. For five participants the words were all positive (e.g., "intelligent"), for another five the words were all negative (e.g., "conceited"), and for the remaining five the words were all neutral (e.g., "historical"). The words in the three lists were equated on potentially relevant dimensions such as length and the frequency with which they occur in literature. The numbers of words correctly recalled in each condition are presented in the table. Analyze these data as completely as possible using the Kruskal–Wallis test, draw a conclusion, and write up your results using the principles given in the relevant Method of Presentation section.

Positive words	Negative words	Neutral words
20	15	16
8	10	21
2	9	14
17	7	13
19	11	12

44. An educator examined the effect of viewing a film about drugs on students' knowledge of drug-related health issues. A health knowledge test was administered to ten students before viewing the film, immediately after viewing the film, and again 1 week later. Scores on this test could range from 0 to 100, with higher values indicating greater knowledge. The scores for the study are presented in the table. Analyze these data as completely as possible using Friedman analysis of variance by ranks, draw a conclusion, and write up your results using the principles given in the relevant Method of Presentation section.

Individual	Before film	Immediately after film	1 week after film
1	28	31	29
2	32	39	37
3	14	23	21
4	20	17	15
5	36	35	33
6	40	43	41
7	24	27	25
8	16	19	18
9	10	13	11
10	44	47	45

45. A researcher tested the relationship between political conservatism and support for gun control. Twenty individuals were administered a scale measuring political conservatism and a scale measuring attitudes toward gun control. Scores on the first scale could range from 0 to 100, and scores on the second scale could range from 0 to 50, with higher values indicating greater conservatism and more positive attitudes, respectively. The scores for the study are presented in the table. Analyze these data using Spearman correlation, nondirectional test, draw a conclusion, and write up your results using the principles given in the relevant Method of Presentation section.

Individual	Conservatism	Attitude toward gun control
1	74	10
2	67	48
3	75	28
4	56	14
5	64	24
6	66	26
7	85	47
8	79	35
9	87	46
10	60	23
11	68	32
12	52	11
13	54	15
14	58	16
15	83	44
16	81	45
17	77	34
18	73	30
19	62	22
20	70	31

Additional Topics

CHAPTER 17

Two-Way Between-Subjects Analysis of Variance

Our focus to this point has been on the application of statistical techniques to analyze the relationship between two variables. For instance, we have seen how one-way between-subjects analysis of variance can be used to examine the relationship between a qualitative or quantitative independent variable and a quantitative dependent variable. In real life, however, it is rare that a given dependent variable is influenced by only one independent variable. For instance, a person's attitude toward abortion might be affected by the religion she was raised in, how religious she is, the type of upbringing she had, and so on. Thus, statistical techniques have been developed to analyze the relationship between a dependent variable and two or more independent variables. This chapter considers one such technique, called **two-way between-subjects analysis of variance** or, more simply, *two-way analysis of variance* (abbreviated as "two-way ANOVA").*

---

* Our discussion of two-way analysis of variance focuses on the situation where the sample sizes in the individual groups under study are all equal. The case of unequal sample sizes is considered in Section 17.9. We use the cell mean model of ANOVA rather than the treatment effect model of ANOVA (Kirk, 1995).

17.1 Factorial Designs

Suppose an investigator interested in factors that influence the number of children people want to have in their completed families posits that one important factor is religious affiliation. To examine this issue, he might conduct a study in which 50 Catholics and 50 Protestants are asked what they consider to be the ideal number of children in a family. Suppose that the data are analyzed using the independent groups *t* test discussed in Chapter 10 (the independent variable being religion and the dependent variable being ideal family size) and the results indicate that Catholics want more children than Protestants.

Suppose the same investigator is also interested in studying the effect of a second independent variable—how religious an individual is—on ideal family size. Suppose further that the investigator has a measure of religiosity that reliably categorizes individuals into one of two categories: religious versus nonreligious. Again, a study could be conducted in which 50 religious and 50 nonreligious individuals are asked what they consider to be the ideal number of children in a family. Suppose this is done, the independent groups *t* test is applied, and the results indicate that religious individuals want more children than nonreligious individuals.

From the two studies, we would conclude that both religion and religiosity are related to ideal family size. However, these studies do not tell us anything about how the two variables *act in conjunction with each other* to exert a joint influence. Fortunately, though, the joint effects of two independent variables on a dependent variable can be studied using **factorial designs.** Table 17.1 illustrates a factorial design for the present example. There are four groups of individuals in the study: (1) religious Catholics, (2) nonreligious Catholics, (3) religious Protestants, and (4) nonreligious Protestants. The four groups are defined by combining the two levels of each of the independent variables, or **factors,** as they are commonly called in two-way analysis of variance. That is, the first factor, religion, has two levels, Catholic versus Protestant. The second factor, religiosity, also has two levels, religious versus nonreligious.* If we combine the two levels of religion with

TABLE 17.1   **Example of a Two-Way Factorial Design**

Religion	Religiosity	
	Religious	Nonreligious
Catholic	Religious Catholics	Nonreligious Catholics
Protestant	Religious Protestants	Nonreligious Protestants

---

* The designation of religion as the first factor and religiosity as the second factor is arbitrary. If we had wished, we could have specified religiosity as the first factor and religion as the second factor instead.

the two levels of religiosity, we have a total of four groups, each of which is represented by a unique combination of variables, or **cells,** in Table 17.1.

Because it represents two independent variables, the design in Table 17.1 is called a *two-way* factorial design. We can also refer to this as a 2 × 2 (read "two by two") factorial design. More generally, a factorial design that has two factors can be represented by the notation $a \times b$, where a refers to the number of levels of the first factor and b refers to the number of levels of the second factor. If the present study had involved three levels of religion instead of two (for instance, Catholic, Protestant, and Jewish), the design would have been a 3 × 2 factorial involving six groups of individuals instead of four: (1) religious Catholics, (2) nonreligious Catholics, (3) religious Protestants, (4) nonreligious Protestants, (5) religious Jews, and (6) nonreligious Jews. The number of groups included in a between-subjects factorial design is simply the product of the number of levels of each factor. In a 2 × 2 factorial design, the number of groups is 2 multiplied by 2, or four. In a 4 × 2 factorial design, the number of groups is 4 multiplied by 2, or eight.

This chapter is restricted to the study of the relationship between two independent variables and one dependent variable. However, factorial designs can also be used to examine the relationship between a dependent variable and three or more independent variables. For example, in addition to religion and religiosity, an investigator might study the relationship of gender to ideal family size. This would involve examining three factors in a 2 × 2 × 2 factorial design. The first factor, consisting of two levels, would be religion; the second factor, also consisting of two levels, would be religiosity; and the third factor, again consisting of two levels, would be gender. This 2 × 2 × 2 factorial design would include eight [(2)(2)(2) = 8] groups of individuals.

17.2 Use of Two-Way Between-Subjects Analysis of Variance

The conditions for using two-way between-subjects analysis of variance are similar to those for the independent groups t test and one-way between-subjects analysis of variance, except that two independent variables rather than one independent variable are studied. Thus, two-way analysis of variance is typically applied when these conditions are met:

1. The dependent variable is quantitative in nature and is measured on a level that at least approximates interval characteristics.

2. The independent variables are both *between-subjects* in nature (they can be either qualitative or quantitative).

3. The independent variables both have two or more levels.

4. The independent variables are combined to form a factorial design.

Consider the example involving the effects of religion and religiosity on ideal family size. Ideal family size is the dependent variable, and it is quantitative in nature. Religion is one of the independent variables; it is between-subjects in nature and has two levels (Catholic versus Protestant). Religiosity is the other independent variable; it is also between-subjects in nature and has two levels (religious versus nonreligious). Finally, the two factors are combined to yield a 2×2 factorial design. Given these conditions, two-way between-subjects analysis of variance is the statistical technique that would typically be used to analyze the relationship between the variables.

17.3   The Concepts of Main Effects and Interactions

As noted above, two-way factorial designs allow us to study the relationship between a dependent variable and two independent variables. Specifically, they allow us to address three issues, phrased here in terms of the example relating religion and religiosity to ideal family size:

1. Is there a relationship between religion, considered alone, and ideal family size?

2. Is there a relationship between religiosity, considered alone, and ideal family size?

3. Is there a relationship between religion and religiosity, considered *in combination,* and ideal family size, independent of the effects of religion alone and religiosity alone?

   In practice, we address these issues by examining sample means and using this information to make inferences about the state of affairs in the population. However, so that we can develop the relevant theoretical concepts without regard to sampling error, the examples in this section are in terms of *population* means. The appropriate inferential procedures are then developed in Section 17.4.
   The first two questions are addressed in terms of **main effects.** The main effect for religion is the comparison of the mean ideal family size for Catholics with the mean ideal family size for Protestants, *collapsed across,* or disregarding, religiosity. The main effect for religiosity is the comparison of the mean ideal family size for religious individuals with the mean ideal family size for nonreligious individuals, *collapsed across,* or disregarding, religion.
   Each main effect in a two-way analysis of variance has a null hypothesis and an alternative hypothesis associated with it. In each instance, the null hypothesis states that the population means for the groups that constitute that effect are equal to each other, and the alternative hypothesis states that these means are not equal to each other. These hypotheses take the same form as they would for an independent groups *t* test (if there are two levels of an independent variable) or for a one-way between-subjects analysis of variance (if there are three or more levels). In our example, the null and alternative hypotheses for the main effect of religion are

H_0: $\mu_C = \mu_P$

H_1: $\mu_C \neq \mu_P$

where the subscript "C" denotes the Catholic group and the subscript "P" denotes the Protestant group. The null and alternative hypotheses for the main effect of religiosity are

$$H_0: \quad \mu_R = \mu_N$$

$$H_1: \quad \mu_R \neq \mu_N$$

where the subscripts "R" and "N" represent the religious and the nonreligious groups, respectively.

Table 17.2 presents hypothetical population means for the four groups in question as well as the corresponding main effect means. If there are equal numbers of individuals in each group, then the main effect means can be calculated by taking the average of the two cells in a given row or column. In this instance, there is a relationship between religion and ideal family size because $\mu_C \neq \mu_P$. The nature of this relationship is such that Catholics ($\mu = 3.00$) want more children than Protestants ($\mu = 2.00$). There is also a relationship between religiosity and ideal family size as indicated by $\mu_R \neq \mu_N$. On the average, religious individuals ($\mu = 3.00$) want more children than nonreligious individuals ($\mu = 2.00$).

The third question addressed in factorial designs concerns **interaction effects.** An interaction effect refers to the comparison of cell means in terms of whether the nature of the relationship between one of the independent variables and the dependent variable differs as a function of the other independent variable. Stated informally, the null hypothesis for an interaction effect is that the mean differences on the dependent variable as a function of one independent variable are the same at each level of the other independent variable. The alternative hypothesis is that the mean differences on the dependent variable as a function of one independent variable differ, depending on the levels of the other independent variable. In our example, one way to phrase the null hypothesis is that the difference in mean ideal family size as a function of religion is the same for religious and nonreligious individuals. The corresponding alternative hypothesis is that the difference in mean ideal family size as a function of religion is not the same for religious and nonreligious individuals. We discuss the specification of null and alternative hypotheses for interaction effects in formal statistical terms in Section 17.4.

TABLE 17.2   **Population Means for Ideal Family Size As a Function of Religion and Religiosity**

Religion	Religiosity		Main effect of religion
	Religious	Nonreligious	
Catholic	4.00	2.00	3.00
Protestant	2.00	2.00	2.00
Main Effect of Religiosity	3.00	2.00	

Examine the relationship between religion and ideal family size for just religious individuals in Table 17.2. *When we consider only religious individuals,* there is a relationship between religion and ideal family size: Catholics ($\mu = 4.00$) want more children than Protestants ($\mu = 2.00$). Now examine the same relationship for just nonreligious individuals. For these individuals, religion is unrelated to ideal family size. On the average, Catholics want the same number of children as Protestants (2.00 in each case). This illustrates an **interaction.** An interaction refers to the case where the nature of the relationship between one of the independent variables and the dependent variable differs as a function of the other independent variable. In our example, the relationship between religion and ideal family size depends on religiosity. For religious individuals, Catholics want more children than Protestants. For nonreligious individuals, however, Catholics want the same number of children as Protestants. An important feature of factorial designs is that they allow us to test for such effects.

Although we have stated that the nature of the relationship between religion and ideal family size depends on religiosity, we could have stated the reverse instead—that the nature of the relationship between religiosity and ideal family size differs as a function of religion. Consider just Catholics. Inspection of Table 17.2 reveals that religious Catholics ($\mu = 4.00$) want more children than nonreligious Catholics ($\mu = 2.00$). Now consider just Protestants. Table 17.2 shows that religious Protestants want the same number of children as nonreligious Protestants (2.00 in each case). The nature of the relationship between religiosity and ideal family size depends on religion. For Catholics, religious individuals want more children than nonreligious individuals. For Protestants, religious individuals want the same number of children as nonreligious individuals.

The perspective from which we state an interaction effect (and, if statistically significant, an interaction) is guided by theory or the research questions being investigated. The important point is that religion and religiosity are interacting in our example such that their joint influence uniquely affects the dependent variable.

Identifying Main Effects and Interactions

In statistical jargon, a main effect is said to be present if the null hypothesis concerning that effect is rejected. If an investigator concludes that there is "a main effect of religiosity" but "no main effect of religion" for a set of data, this would mean that the null hypothesis was rejected for the religiosity factor but not for the religion factor. Similarly, the "presence" of an interaction means that the null hypothesis of no interaction effect was rejected.

Figure 17.1 presents examples of various population means for the four groups defined by the two levels of religion and the two levels of religiosity. The examples are also depicted on graphs. On the abscissas of the graphs are demarcations for the two levels of religion, Catholic and Protestant. On the ordinates of the graphs are demarcations for values of the dependent variable, ideal family size. Directly above each abscissa demarcation for Catholics is a dot corresponding to the mean ideal family size for religious Catholics and directly above each abscissa demarcation for Protestants is a dot corresponding to the mean ideal family size

FIGURE 17.1   Examples of Main Effects and Interactions for a 2 x 2 Factorial Design (Adapted from Johnson & Liebert, 1977)

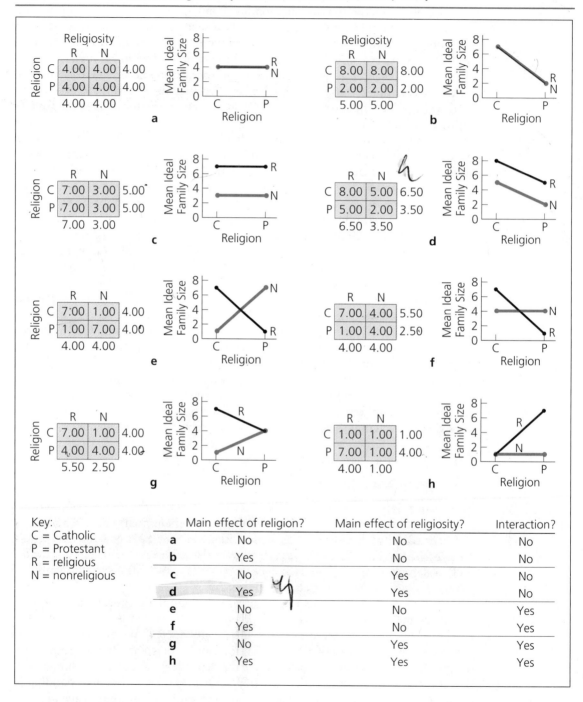

Key:
C = Catholic
P = Protestant
R = religious
N = nonreligious

	Main effect of religion?	Main effect of religiosity?	Interaction?
a	No	No	No
b	Yes	No	No
c	No	Yes	No
d	Yes	Yes	No
e	No	No	Yes
f	Yes	No	Yes
g	No	Yes	Yes
h	Yes	Yes	Yes

for religious Protestants.* For each graph, these two points are connected by a line, as are the corresponding points for nonreligious individuals.

Since we are dealing with population values, the most straightforward way to determine whether there is a main effect of religion for a given example is to compare the means of the two columns. If they are the same, no main effect is present; otherwise, one is present. The presence or absence of a main effect of religiosity is similarly determined by comparing the means of the two rows. When dealing with population means, we can readily determine whether an interaction is present by examining the slopes of the lines in a given graph. One of these lines represents the relationship between religion and ideal family size for religious individuals, and the other line represents this relationship for nonreligious individuals. If the lines are parallel (or, as in Figures 17.1a and 17.1b, overlapping), the nature of the relationship is the same for both religious and nonreligious individuals. Thus, there is no interaction. If the lines are not parallel, the nature of the relationship between religion and ideal family size depends on religiosity, which means that an interaction is present.

Figure 17.1a represents a case where there are neither main effects nor an interaction. In Figure 17.1b, there is a main effect of religion (Protestants, $\mu = 2.00$, want fewer children than Catholics, $\mu = 8.00$) but no main effect of religiosity and no interaction. As indicated by the parallel lines, there are also no interactions in Figures 17.1c and 17.1d. However, there is a main effect of religiosity in Figure 17.1c (religious individuals, $\mu = 7.00$, want more children than nonreligious individuals, $\mu = 3.00$) and main effects of both religion (Protestants, $\mu = 3.50$, want fewer children than Catholics, $\mu = 6.50$) and religiosity (religious individuals, $\mu = 6.50$, want more children than nonreligious individuals, $\mu = 3.50$) in Figure 17.1d.

Figures 17.1e–17.1h represent cases where interactions occur. Note that in each graph, the lines are nonparallel. This signifies that the nature of the relationship between religion and ideal family size differs as a function of religiosity. In Figure 17.1e, religious Protestants ($\mu = 1.00$) want fewer children than religious Catholics ($\mu = 7.00$), but nonreligious Protestants ($\mu = 7.00$) want more children than nonreligious Catholics ($\mu = 1.00$). In Figure 17.1f, religious Protestants ($\mu = 1.00$) again want fewer children than religious Catholics ($\mu = 7.00$); however, both nonreligious Protestants and nonreligious Catholics have an ideal family size of 4.00. There is also a main effect of religion in Figure 17.1f such that, collapsing across religiosity, Protestants ($\mu = 2.50$) want fewer children than Catholics ($\mu = 5.50$). See if you can interpret the nature of the main effects and interactions for Figures 17.1g and 17.1h.

Main Effects and Interactions in Designs with More Than Two Levels for One or Both Factors

The principles for detecting main effects and interactions are readily generalizable from 2×2 designs to designs with more than two levels for one or both factors.

---

* The representation of religion on the abscissa and religiosity in the body of the graphs is arbitrary. If we had wished, we could have represented religiosity on the abscissa and religion in the body of the graphs instead.

Figure 17.2 presents examples of a 3×2 factorial design that has three levels of religion rather than two: Catholic, Protestant, and Jewish. Whenever the lines are parallel, a lack of interaction is indicated. This is the case in Figures 17.2b and 17.2d. For these examples, the nature of the relationship between one independent variable and the dependent variable is the same irrespective of the level of the other independent variable. When the lines are not parallel, the presence of an interaction is indicated. This is the case in Figures 17.2a and 17.2c. For these examples, the nature of the relationship between religion and ideal family size differs as a function of religiosity. In all four examples, there are main effects of both religion and religiosity.

Main Effects and Interactions with Sample Data

In the following section, we examine inferential techniques for determining the existence of main effects and interactions in populations, based on sample data. At this point, we wish to emphasize that when dealing with sample data, we *cannot* make these determinations by mere visual inspection of sample means or slopes of lines in a graph. Rather, the existence of a given population main effect or interaction can be affirmed only by a significant statistical test. Thus, as with all of the previous designs that we have considered, because of the role of sampling error, nonequivalent sample means do not necessarily indicate nonequivalent population means. For the same reason, we cannot assume that an interaction exists in the population just because the lines representing the sample means are nonparallel. If an interaction effect *is* statistically significant, however, these lines will indeed have different slopes, and an examination of them can help to clarify the nature of the relationship between the variables.

17.4   Inference of Relationships Using Two-Way Between-Subjects Analysis of Variance

Null and Alternative Hypotheses

Table 17.3 presents data for the 2×2 factorial design in Table 17.1, which we use as a numerical example. Unlike the situation in Section 17.3, we are now working with sample data and will continue to do so for the remainder of the chapter. The first factor, which we will refer to as factor A, is religion; the second factor, referred to as factor B, is religiosity; and the dependent variable is ideal family size. The main effect means for the Catholic and Protestant levels of factor A are represented as $\overline{X}_C$ and $\overline{X}_P$, respectively. The main effect means for the religious and nonreligious levels of factor B are similarly represented as $\overline{X}_R$ and $\overline{X}_N$.

The notation for the mean scores for the four groups in the 2×2 design requires two subscripts. The first subscript identifies the level of factor A, and the second subscript identifies the level of factor B. For instance, the cell mean for nonreligious Catholics (or, consistent with the order of the subscripts, Catholic–nonreligious individuals) is represented as $\overline{X}_{C,N}$ and the cell mean for religious Protestants (that is, Protestant–religious individuals) is represented as $\overline{X}_{P,R}$.

FIGURE 17.2   Examples of Main Effects and Interactions for a 3 x 2 Factorial Design

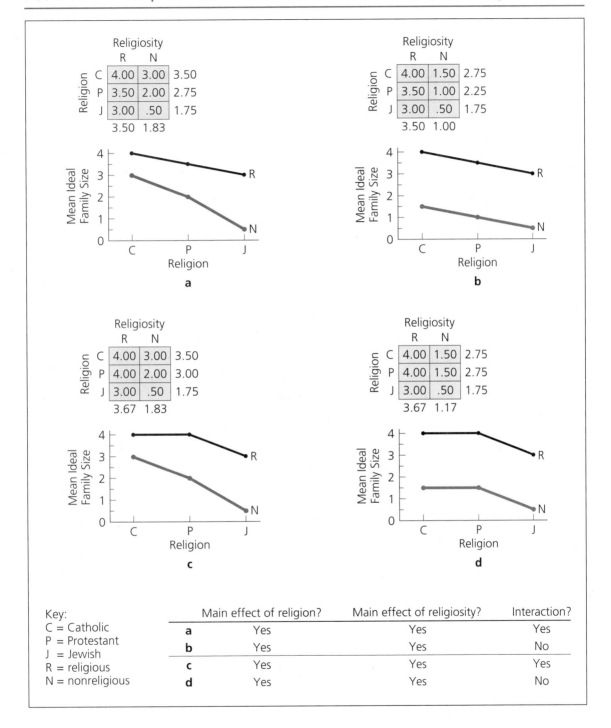

Key:
C = Catholic
P = Protestant
J = Jewish
R = religious
N = nonreligious

	Main effect of religion?	Main effect of religiosity?	Interaction?
a	Yes	Yes	Yes
b	Yes	Yes	No
c	Yes	Yes	Yes
d	Yes	Yes	No

TABLE 17.3 Data and Computation of Sums of Squares for Religion, Religiosity, and Ideal Family Size Study

RELIGION (A)	RELIGIOSITY (B) Religious		Nonreligious		MAIN EFFECT OF RELIGION
	X	X^2	X	X^2	
Catholic	7	49	5	25	$T_C = 40$
	4	16	2	4	$\overline{X}_C = 4.00 \rightarrow$ avg.
	3	9	1	1	$T_C^2 = 1{,}600$
	5	25	3	9	x^2
	6	36	4	16	
	$T_{CR} = 25$		$T_{CN} = 15$		
	$\overline{X}_{CR} = 5.00$		$\overline{X}_{CN} = 3.00$		
	$T_{CR}^2 = 625$		$T_{CN}^2 = 225$		
Protestant	0	0	2	4	$T_P = 20$
	4	16	4	16	$\overline{X}_P = 2.00$
	2	4	0	0	$T_P^2 = 400$
	2	4	2	4	
	2	4	2	4	
	$T_{PR} = 10$		$T_{PN} = 10$		
	$\overline{X}_{PR} = 2.00$		$\overline{X}_{PN} = 2.00$		
	$T_{PR}^2 = 100$		$T_{PN}^2 = 100$		
MAIN EFFECT OF RELIGIOSITY	$T_R = 35$		$T_N = 25$		
	$\overline{X}_R = 3.50$		$\overline{X}_N = 2.50$		
	$T_R^2 = 1{,}225$		$T_N^2 = 625$		

$$\sum X^2 = 49 + 16 + \cdots + 4 + 4 = 246$$

$$\sum X = 7 + 4 + \cdots + 2 + 2 = 60$$

$$\frac{(\sum X)^2}{N} = \frac{60^2}{20} = 180$$

$$\frac{\sum T_{A_i}^2}{nb} = \frac{1{,}600 + 400}{(5)(2)} = 200$$

$$\frac{\sum T_{B_j}^2}{na} = \frac{1{,}225 + 625}{(5)(2)} = 185$$

$$\frac{\sum T_{A_i B_j}^2}{n} = \frac{625 + 225 + 100 + 100}{5} = 210$$

As stated in Section 17.3, the null and alternative hypotheses for the main effect of religion are

$$H_0: \quad \mu_C = \mu_P$$

$$H_1: \quad \mu_C \neq \mu_P$$

and the null and alternative hypotheses for the main effect of religiosity are

$$H_0: \quad \mu_R = \mu_N$$

$$H_1: \quad \mu_R \neq \mu_N$$

When we have a 2×2 factorial design, the specification of the null and alternative hypotheses for the interaction effect in formal statistical terms is straightforward. Let us develop these by considering only religious individuals. The effect that religious affiliation has on ideal family size for these individuals is reflected in the mean difference $\mu_{C,R} - \mu_{P,R}$. If the difference between these two means is 0, religious affiliation has no effect on ideal family size in the sense that the population mean for Catholics is the same as the population mean for Protestants. If this difference is nonzero, religious affiliation does have an effect on ideal family size. The same can be said for nonreligious individuals: The effect that religious affiliation has on ideal family size for these individuals is reflected in the mean difference $\mu_{C,N} - \mu_{P,N}$. Now suppose that the effect of religious affiliation is exactly the same for religious and nonreligious individuals. This would mean that no interaction is present, as reflected in the fact that $\mu_{C,R} - \mu_{P,R}$ would equal $\mu_{C,N} - \mu_{P,N}$. This leads to our null hypothesis for the interaction effect:

$$H_0: \quad \mu_{C,R} - \mu_{P,R} = \mu_{C,N} - \mu_{P,N}$$

The alternative hypothesis turns the null hypothesis into an inequality—namely,

$$H_1: \quad \mu_{C,R} - \mu_{P,R} \neq \mu_{C,N} - \mu_{P,N}$$

If the alternative hypothesis is true, then there is an interaction because the effect of religious affiliation on ideal family size is not the same for religious and nonreligious individuals.

In the above example, the null and alternative hypotheses for the interaction effect are stated in terms of the relationship between religious affiliation and ideal family size as a function of religiosity. They could have instead been framed in terms of the relationship between religiosity and ideal family size as a function of religion. If we specified the null and alternative hypotheses in this way, they would be

$$H_0: \quad \mu_{C,R} - \mu_{C,N} = \mu_{P,R} - \mu_{P,N}$$

$$H_1: \quad \mu_{C,R} - \mu_{C,N} \neq \mu_{P,R} - \mu_{P,N}$$

The formal specification of the hypotheses for interaction effects for larger designs is cumbersome, but these can be stated informally, as discussed in Section 17.3.

Partitioning of Variability

In Chapter 12, the relationship between a between-subjects independent variable and a dependent variable was analyzed by defining the total variability in the dependent variable in terms of two components: between-group variability and within-group variability. Specifically, it will be recalled that the sum of squares total was partitioned into a sum of squares between and a sum of squares within. This was represented symbolically as $SS_{TOTAL} = SS_{BETWEEN} + SS_{WITHIN}$. The quantities $SS_{BETWEEN}$ and SS_{WITHIN} were divided by their respective degrees of freedom to obtain a mean square between and a mean square within. An F ratio was then formed by dividing $MS_{BETWEEN}$ by MS_{WITHIN}.

The sum of squares total in two-way analysis of variance can be similarly partitioned into a sum of squares between and a sum of squares within:

$$SS_{TOTAL} = SS_{BETWEEN} + SS_{WITHIN} \qquad [17.1]$$

However, the overall sum of squares between is further partitioned into three components: (1) variability due to the first independent variable (designated as "factor A"), (2) variability due to the second independent variable (designated as "factor B"), and (3) variability due to the interaction of factors A and B (symbolized "$A \times B$"). Thus,

$$SS_{TOTAL} = SS_{BETWEEN} + SS_{WITHIN}$$
$$\swarrow \quad \downarrow \quad \searrow$$
$$SS_A \quad SS_B \quad SS_{A \times B}$$

where SS_A is the sum of squares defining between-group variability for factor A, SS_B is the sum of squares defining between-group variability for factor B, and $SS_{A \times B}$ is the sum of squares defining between-group variability for the $A \times B$ interaction. Stated more formally,

$$SS_{BETWEEN} = SS_A + SS_B + SS_{A \times B} \qquad [17.2]$$

The sum of squares total can thus be represented as

$$SS_{TOTAL} = SS_A + SS_B + SS_{A \times B} + SS_{WITHIN} \qquad [17.3]$$

In two-way analysis of variance, each of the components of the sum of squares between is divided by its respective degrees of freedom, and the resulting mean squares, MS_A, MS_B, and $MS_{A \times B}$, are then divided by the mean square within to yield F ratios. The F ratio formed by MS_A/MS_{WITHIN} is used to test the null hypothesis with regard to the main effect of factor A. The F ratio formed by MS_B/MS_{WITHIN} is used to test the null hypothesis with respect to the main effect of factor B. Finally, the F ratio formed by $MS_{A \times B}/MS_{WITHIN}$ is used to test the null hypothesis with respect to the $A \times B$ interaction.

Computation of the Sums of Squares

The sum of squares total is calculated using the procedures discussed in Chapter 12 for applying the standard computational formula for a sum of squares to an entire data set. Specifically,

$$SS_{TOTAL} = \Sigma X^2 - \frac{(\Sigma X)^2}{N} \qquad [17.4]$$

where ΣX^2 is the sum of the squared scores in the data set, $(\Sigma X)^2$ is the square of the summed scores, and N is the total sample size. Looking at Table 17.3, we find that the sum of the 20 X^2 scores in our example is

$$\Sigma X^2 = 49 + 16 + \cdots + 4 + 4 = 246$$

and the sum of the 20 X scores is

$$\Sigma X = 7 + 4 + \cdots + 2 + 2 = 60$$

Squaring the latter value and dividing by N, we obtain

$$\frac{(\sum X)^2}{N} = \frac{60^2}{20} = 180$$

Thus, the sum of squares total in this instance is

$$SS_{TOTAL} = \sum X^2 - \frac{(\sum X)^2}{N}$$
$$= 246 - 180 = 66$$

The formula for calculating the sum of squares for between-group variability due to factor A (religion) is

$$SS_A = \frac{\sum T_{A_i}^2}{nb} - \frac{(\sum X)^2}{N} \qquad\qquad [17.5]$$

where $T_{A_i}^2$ is the square of the sum of the scores at level i of factor A, n is the per-cell sample size (remember that we are focusing on the situation where sample sizes in the individual groups being considered are all equal), b is the number of levels of factor B, and $(\sum X)^2$ and N are as defined previously. In our example, $T_{A_i}^2$ equals 1,600 for Catholics and 400 for Protestants. Thus,

$$\frac{\sum T_{A_i}^2}{nb} = \frac{1{,}600 + 400}{(5)(2)} = 200$$

so

$$SS_A = \frac{\sum T_{A_i}^2}{nb} - \frac{(\sum X)^2}{N}$$
$$= 200 - 180 = 20$$

The formula for calculating the sum of squares for between-group variability due to factor B (religiosity) is

$$SS_B = \frac{\sum T_{B_j}^2}{na} - \frac{(\sum X)^2}{N} \qquad\qquad [17.6]$$

where $T_{B_j}^2$ is the square of the sum of the scores at level j of factor B, a is the number of levels of factor A, and all other terms are as defined above. In our example, $T_{B_j}^2$ equals 1,225 for religious individuals and 625 for nonreligious individuals. Thus,

$$\frac{\sum T_{B_j}^2}{na} = \frac{1{,}225 + 625}{(5)(2)} = 185$$

so

$$SS_B = \frac{\sum T_{B_j}^2}{na} - \frac{(\sum X)^2}{N}$$
$$= 185 - 180 = 5$$

The formula for calculating the sum of squares for between-group variability due to the interaction of factor A and factor B is

$$SS_{A \times B} = \frac{(\sum X)^2}{N} + \frac{\sum T_{A_i B_j}^2}{n} - \frac{\sum T_{A_i}^2}{nb} - \frac{\sum T_{B_j}^2}{na}$$ [17.7]

where $T_{A_i B_j}^2$ is the square of the sum of the scores in cell $A_i B_j$ and all other terms are as defined above. In our example, $T_{A_i B_j}^2$ equals 625 for religious Catholics, 225 for nonreligious Catholics, 100 for religious Protestants, and 100 for nonreligious Protestants. Thus,

$$\frac{\sum T_{A_i B_j}^2}{n} = \frac{625 + 225 + 100 + 100}{5} = 210$$

so

$$SS_{A \times B} = \frac{(\sum X)^2}{N} + \frac{\sum T_{A_i B_j}^2}{n} - \frac{\sum T_{A_i}^2}{nb} - \frac{\sum T_{B_j}^2}{na}$$

$$= 180 + 210 - 200 - 185 = 5$$

The formula for calculating the sum of squares within is

$$SS_{\text{WITHIN}} = \sum X^2 - \frac{\sum T_{A_i B_j}^2}{n}$$ [17.8]

where all terms are as defined above. In our example,

$$SS_{\text{WITHIN}} = 246 - 210 = 36$$

Of course, given that the sum of squares total is equal to the sum of squares for factor A plus the sum of squares for factor B plus the sum of squares for the $A \times B$ interaction plus the sum of squares within, if four of these quantities have already been calculated, the fifth can be determined through simple algebraic manipulation. For instance, we could have obtained the sum of squares for the $A \times B$ interaction as follows:

$$SS_{A \times B} = SS_{\text{TOTAL}} - SS_A - SS_B - SS_{\text{WITHIN}}$$

$$= 66 - 20 - 5 - 36 = 5$$

However, it is usually a good idea to calculate all five sums of squares using the above formulas. This allows us to check for computational errors by determining whether the calculated values of the other four sums of squares add up to the calculated value of the sum of squares total.

Derivation of the Summary Table

Each of the above sums of squares has a certain number of degrees of freedom associated with it. These are calculated as follows:

$$df_A = a - 1$$ [17.9]

$$df_B = b - 1 \qquad\qquad [17.10]$$

$$df_{A \times B} = (a - 1)(b - 1) \qquad\qquad [17.11]$$

$$df_{\text{WITHIN}} = (a)(b)(n - 1) \qquad\qquad [17.12]$$

As with the sums of squares, the degrees of freedom are additive:

$$df_{\text{TOTAL}} = df_A + df_B + df_{A \times B} + df_{\text{WITHIN}} \qquad\qquad [17.13]$$

Since $df_A = a - 1$, $df_B = b - 1$, $df_{A \times B} = (a - 1)(b - 1)$, and $df_{\text{WITHIN}} = (a)(b)(n - 1)$, and $(a - 1) + (b - 1) + [(a - 1)(b - 1)] + [(a)(b)(n - 1)] = a - 1 + b - 1 + (ab - a - b + 1) + (abn - ab) = abn - 1 = N - 1$, the degrees of freedom total can be calculated directly as

$$df_{\text{TOTAL}} = N - 1 \qquad\qquad [17.14]$$

In our example,

$$df_A = 2 - 1 = 1$$

$$df_B = 2 - 1 = 1$$

$$df_{A \times B} = (2 - 1)(2 - 1) = 1$$

$$df_{\text{WITHIN}} = (2)(2)(5 - 1) = 16$$

$$df_{\text{TOTAL}} = 20 - 1 = 19$$

The relevant mean squares are obtained by dividing the sum of squares for each source of between-group variability by its degrees of freedom:

$$MS_A = \frac{SS_A}{df_A} \qquad\qquad [17.15]$$

$$= \frac{20}{1} = 20.00$$

$$MS_B = \frac{SS_B}{df_B} \qquad\qquad [17.16]$$

$$= \frac{5}{1} = 5.00$$

$$MS_{A \times B} = \frac{SS_{A \times B}}{df_{A \times B}} \qquad\qquad [17.17]$$

$$= \frac{5}{1} = 5.00$$

$$MS_{\text{WITHIN}} = \frac{SS_{\text{WITHIN}}}{df_{\text{WITHIN}}} \qquad\qquad [17.18]$$

$$= \frac{36}{16} = 2.25$$

• The F ratios used to test the null hypotheses for the two main effects and the $A \times B$ interaction are computed as follows:

$$F_A = \frac{MS_A}{MS_{WITHIN}}$$ [17.19]

$$= \frac{20.00}{2.25} = 8.89$$

$$F_B = \frac{MS_B}{MS_{WITHIN}}$$ [17.20]

$$= \frac{5.00}{2.25} = 2.22$$

$$F_{A \times B} = \frac{MS_{A \times B}}{MS_{WITHIN}}$$ [17.21]

$$= \frac{5.00}{2.25} = 2.22$$

All of the preceding calculations can be summarized in a summary table:

Source	SS	df	MS	F
A (religion)	20.00	1	20.00	8.89
B (religiosity)	5.00	1	5.00	2.22
A × B	5.00	1	5.00	2.22
Within	36.00	16	2.25	
Total	66.00	19		

This table is similar in format to the summary tables in Chapter 12, but the sum of squares between is represented by its three components: the main effect for factor A, the main effect for factor B, and the $A \times B$ interaction. Let us now examine how we can use the information in this table to address the three questions about a relationship for each of these sources of variability.

Inference of Relationships

The test of the null hypothesis for the main effect of religion is made with reference to the F value derived from MS_A/MS_{WITHIN}. This F equals 8.89 with 1 and 16 degrees of freedom (when an F value is reported, it is conventional to state the degrees of freedom for the effect followed by the degrees of freedom within). Appendix F indicates that the critical value of F is 4.49 for these degrees of freedom and an alpha level of .05. Since the observed F value is greater than 4.49, we reject the null hypothesis and conclude that there is a relationship between religion and ideal family size.

The test of the null hypothesis for the main effect of religiosity is made with reference to the F value derived from MS_B/MS_{WITHIN}. This F equals 2.22 with 1 and 16 degrees of freedom. This does not exceed the critical value of 4.49 and, hence, we fail to reject the null hypothesis. We cannot confidently conclude that there is a relationship between religiosity and ideal family size.

The test of the null hypothesis for the interaction effect is made with reference to the F value derived from $MS_{A \times B}/MS_{WITHIN}$. This F equals 2.22 with 1 and 16 degrees of freedom. This does not exceed the critical value of 4.49, so we fail to reject the null hypothesis. We cannot confidently conclude that there is an interaction between religion and religiosity.

Assumptions of the *F* Tests

The preceding F tests are based on the same assumptions that underlie one-way between-subjects analysis of variance as discussed in Chapter 12:

1. The samples are independently and randomly selected from their respective populations.

2. The scores in each population are normally distributed.

3. The scores in each population have equal variances.

In addition, the dependent variable should be quantitative in nature and measured on a level that at least approximates interval characteristics.

For the F tests to be valid, it is important that the assumption of independent and random selection be met. As with one-way analysis of variance, under certain conditions the F tests are robust to violations of the normality and homogeneity of variance assumptions. In fact, the statements made with respect to one-way analysis of variance are also applicable to two-way analysis of variance: The F tests are particularly robust when the sample sizes in each cell are moderate (greater than 20) to large and the same for all groups. Under these conditions, the F tests are quite robust to even marked departures from normality. As the number of groups in the factorial design increases, heterogeneity of population variances becomes increasingly problematic. Even with a relatively large number of groups, the F tests are relatively robust when the population variance for one cell is as much as 2 to 3 times larger than the population variances for the other cells. However, the Type I error rate begins to become unacceptable when the population variance for one cell is 4 or more times larger than the population variances for the other cells. In the case of highly different sample sizes, the robustness of the F tests to nonnormality and variance heterogeneity often diminishes, sometimes considerably (see Harwell, Rubinstein & Hayes, 1992; Milligan, Wong & Thompson, 1987).

17.5 Strength of the Relationships

The strength of the relationship between the dependent variable and each of the three sources of between-group variability is computed using the following formulas for eta-squared:

$$eta_A^2 = \frac{SS_A}{SS_{TOTAL}} \qquad [17.22]$$

$$eta_B^2 = \frac{SS_B}{SS_{TOTAL}} \qquad [17.23]$$

$$eta_{A \times B}^2 = \frac{SS_{A \times B}}{SS_{TOTAL}} \qquad [17.24]$$

where eta_A^2 is eta-squared for the main effect of factor A, eta_B^2 is eta-squared for the main effect of factor B, and $eta_{A \times B}^2$ is eta-squared for the interaction effect. For our example,

$$eta_A^2 = \frac{20}{66} = .30$$

$$eta_B^2 = \frac{5}{66} = .08$$

$$eta_{A \times B}^2 = \frac{5}{66} = .08$$

These values represent the proportion of variability in the dependent variable that is associated with the particular source of between-group variability. The proportion of variability in ideal family size that is associated with the main effect of religion is .30. This represents a strong effect. The proportion of variability in ideal family size that is associated with the main effect of religiosity and with the interaction between religion and religiosity is .08 in both instances. These represent weak to moderate effects.

17.6 Nature of the Relationships

Analysis of Main Effects

When a statistically significant main effect has only two levels, the nature of the relationship is determined in the same fashion as for the independent groups t test. This involves making inferences about the two population means by examining the two sample means. In the present example, the nature of the relationship between religion and ideal family size is such that Catholics ($\overline{X}_C = 4.00$) want more children than Protestants ($\overline{X}_P = 2.00$).

When a statistically significant main effect has three or more levels, the nature of the relationship is determined using a Tukey HSD test conceptually identical to the one discussed in Chapter 12. This involves computing the absolute difference between all possible pairs of sample means that make up the main effect and then comparing each of these against a critical difference.

For the main effect of factor A, the critical difference is defined as

$$CD = q\sqrt{\frac{MS_{WITHIN}}{nb}} \qquad [17.25]$$

where q is a Studentized range value obtained from Appendix G, MS_{WITHIN} is the mean square within from the summary table, n is the per-cell sample size, and b is the number of levels of factor B. The value of q is determined with reference to the overall alpha level for the effect, the degrees of freedom within from the summary table, and the number of levels of *factor A* (symbolized k in Appendix G). We follow the practice of adopting an overall alpha level of .05 for each main effect that we analyze.

For the main effect of factor B, the critical difference is defined as

$$CD = q\sqrt{\frac{MS_{WITHIN}}{na}} \qquad [17.26]$$

where a is the number of levels of factor A and all other terms are as defined previously. In this case, however, q is determined with reference to the overall alpha level for the effect, the degrees of freedom within, and the number of levels of *factor B* (again symbolized k in Appendix G).

If the absolute difference between a given pair of sample means exceeds the critical difference, we conclude that the corresponding population means differ from each other. If the absolute difference between the sample means does not exceed the critical difference, we are unable to draw this conclusion. We consider a numerical example involving a factor with three levels in Section 17.8.

Analysis of Interactions

When an interaction effect is statistically significant, the nature of the interaction can be determined using a number of statistical procedures. One popular procedure is an analytic method called *simple main effects analysis*. A second is an approach called **interaction comparisons**. In general, statisticians recommend the latter strategy and, hence, we consider it here.

Interaction comparisons focus on 2×2 subtables within a factorial design. When the overall design is 2×2 in nature, the nature of the interaction is determined by examining the difference between the cell means for one independent variable at each level of the other independent variable. In this example, the interaction effect was not statistically significant. If the interaction effect had been statistically significant, the table of cell means would have been of interest:

	Religious	Nonreligious
Catholic	5.00	3.00
Protestant	2.00	2.00

In this table, the effect of religion on ideal family size is more pronounced for religious individuals (5.00 − 2.00 = 3.00) than for nonreligious individuals (3.00 − 2.00 = 1.00). However, because the interaction effect was not statistically significant, the difference between 3.00 and 1.00 can be attributed to sampling error rather than a true interaction.

For more complex designs, interaction comparisons involve breaking an overall factorial design into a series of 2 × 2 subtables that are interpreted as above. For example, for a 3 × 2 factorial design involving three levels of religion (Catholic, Protestant, Jewish) and two levels of religiosity (religious, nonreligious), there are three possible 2 × 2 subtables:

	Religious	Nonreligious
Catholic		
Protestant		

	Religious	Nonreligious
Catholic		
Jewish		

	Religious	Nonreligious
Protestant		
Jewish		

Interaction comparisons involve performing a separate 2 × 2 analysis of variance for each of these subtables using the procedures to be described. The entire series of 2 × 2 tests is then used to discern the nature of the interaction, as illustrated in Section 17.8.

Let the four cells in a 2 × 2 subtable, referred to as subtable K, be identified with letters as follows:

	Variable 2	
	a	b
Variable 1		
	c	d

The sum of squares for the interaction effect for the 2 × 2 subtable is equal to

$$SS_{A \times B(K)} = \frac{n(\overline{X}_a + \overline{X}_d - \overline{X}_b - \overline{X}_c)^2}{4} \qquad [17.27]$$

The mean square for between-group variability due to the interaction will always equal the sum of squares for the interaction because an interaction effect for a 2×2 subtable always has a single degree of freedom associated with it. This mean square is divided by the mean square within *from the overall summary table* to yield an F ratio for the 2×2 interaction. Symbolically,

$$F_{A \times B(K)} = \frac{MS_{A \times B(K)}}{MS_{WITHIN}}$$

[17.28]

The F ratio has degrees of freedom equal to 1 and df_{WITHIN} from the overall summary table. If the F value is statistically significant, the null hypothesis associated with the 2×2 table is rejected.

The interaction-comparisons approach involves conducting separate analyses of this type for all possible 2×2 subtables. Each statistically significant subtable is then interpreted to reveal the nature of the overall interaction. This strategy is illustrated in Section 17.8 for a 2×3 analysis of variance, with a **modified Bonferroni procedure** used to control the Type I error rate across the set of comparisons.

17.7 Methodological Considerations

Several methodological considerations are worth noting in the context of the study relating religion and religiosity to ideal family size. First are the usual issues of uncontrolled disturbance variables and generalizability of results. There is also the problem of confounding associated with nonrandom assignment to groups when observational independent variables such as religion and religiosity are studied. Beyond this, however, we can use the investigation to illustrate an important methodological strategy. In Chapter 12, one of the examples used to demonstrate one-way between-subjects analysis of variance was concerned with the relationship between religion and ideal family size. In that chapter, it was pointed out that religiosity was acting as a disturbance variable and thus creating within-group variability (that is, it was increasing the size of the sum of squares within). In the present investigation, religiosity was combined with religion to form four groups, and the within-group variability was based on the variability of scores within each of the four groups separately. As such, the disturbance effects of religiosity did not enter into the computation of within-group variability because, like religion, it was held constant within groups. This highlights an important advantage of factorial designs: Not only do they allow us to assess the interaction between independent variables, but they also "remove" the individual and joint effects of these variables from the within-group variability. This makes the tests of the main effects more sensitive than if one of the variables were left uncontrolled and took on the role of a disturbance variable. Thus, an addition to the strategies for dealing with disturbance variables discussed in Chapters 9 and 11 is to bring a variable into the research design by including it as a factor.

17.8 Numerical Example

Social psychologists have studied extensively the variables that influence the ability of a speaker to persuade an audience to take the speaker's position on an issue. One important factor that influences the amount of attitude change a speaker can generate is the discrepancy between the position advocated by the speaker and the position of the audience. Up to a point, the more discrepant the speaker's position, the greater the attitude change that will result. However, if the speaker's position becomes too discrepant, the speaker loses credibility and the message is less persuasive.

It has been hypothesized that the nature of the relationship between message discrepancy and attitude change differs, depending on the expertise of the speaker, formally referred to as the *source*. According to this perspective, speakers with high expertise can take much more discrepant positions than speakers with low expertise and still obtain large amounts of attitude change. As an example of how this proposition could be tested, consider the following hypothetical experiment.

College students evaluated the quality of a passage of poetry on a 21-point scale and then listened to a taped message concerning this passage that was presented as representing the opinion of either an expert (a famous poetry critic) or a nonexpert (an undergraduate student enrolled in a creative writing class). The messages were identical except for which source they were attributed to. In addition, the messages were constructed to be either slightly discrepant, moderately discrepant, or highly discrepant from students' initial ratings of quality. For example, in the large-discrepancy condition, if a student rated the passage as being relatively high in quality, the message argued that the passage was low in quality. After listening to the message, students rerated the poetry. The resulting design was a 3×2 factorial with three levels of message discrepancy (small, medium, or large) and two levels of source expertise (high versus low). The dependent variable was the amount of change in the quality ratings after listening to the message. Scores could range from -20 to $+20$, with higher values indicating greater attitude change in the direction advocated by the source. The data for the experiment are presented in Table 17.4 along with intermediate statistics necessary to calculate the sums of squares.

Null and Alternative Hypotheses

The null and alternative hypotheses for the main effect of message discrepancy are

H_0: $\mu_S = \mu_M = \mu_L$

H_1: The three population means are not all equal

where the subscript "S" represents the small-discrepancy condition, the subscript "M" represents the medium-discrepancy condition, and the subscript "L" represents the large-discrepancy condition.

The null and alternative hypotheses for the main effect of source expertise are

H_0: $\mu_H = \mu_L$

H_1: $\mu_H \neq \mu_L$

TABLE 17.4   Data and Computation of Sums of Squares for Message Discrepancy, Source Expertise, and Attitude Change Experiment

MESSAGE DISCREPANCY (A)	SOURCE EXPERTISE (B)				MAIN EFFECT OF MESSAGE DISCREPANCY
	High		Low		
	X	X^2	X	X^2	
Small	3	9	1	1	$T_S = 20$
	4	16	0	0	$\overline{X}_S = 2.00$
	2	4	2	4	$T_S^2 = 400$
	3	9	1	1	
	3	9	1	1	
	$T_{SH} = 15$		$T_{SL} = 5$		
	$\overline{X}_{SH} = 3.00$		$\overline{X}_{SL} = 1.00$		
	$T_{SH}^2 = 225$		$T_{SL}^2 = 25$		
Medium	8	64	3	9	$T_M = 50$
	7	49	2	4	$\overline{X}_M = 5.00$
	7	49	3	9	$T_M^2 = 2,500$
	7	49	4	16	
	6	36	3	9	
	$T_{MH} = 35$		$T_{ML} = 15$		
	$\overline{X}_{MH} = 7.00$		$\overline{X}_{ML} = 3.00$		
	$T_{MH}^2 = 1,225$		$T_{ML}^2 = 225$		
Large	9	81	0	0	$T_L = 50$
	8	64	1	1	$\overline{X}_L = 5.00$
	10	100	1	1	$T_L^2 = 2,500$
	9	81	2	4	
	9	81	1	1	
	$T_{LH} = 45$		$T_{LL} = 5$		
	$\overline{X}_{LH} = 9.00$		$\overline{X}_{LL} = 1.00$		
	$T_{LH}^2 = 2,025$		$T_{LL}^2 = 25$		
MAIN EFFECT OF SOURCE EXPERTISE	$T_H = 95$		$T_L = 25$		
	$\overline{X}_H = 6.33$		$\overline{X}_L = 1.67$		
	$T_H^2 = 9,025$		$T_L^2 = 625$		

$$\sum X^2 = 9 + 16 + \cdots + 4 + 1 = 762$$

$$\sum X = 3 + 4 + \cdots + 2 + 1 = 120$$

$$\frac{(\sum X)^2}{N} = \frac{120^2}{30} = 480.00$$

$$\frac{\sum T_{A_i}^2}{nb} = \frac{400 + 2,500 + 2,500}{(5)(2)} = 540.00$$

$$\frac{\sum T_{B_j}^2}{na} = \frac{9,025 + 625}{(5)(3)} = 643.33$$

$$\frac{\sum T_{A_i B_j}^2}{n} = \frac{225 + 25 + 1,225 + 225 + 2,025 + 25}{5}$$

$$= 750.00$$

where the subscripts "H" and "L" respectively denote the high-expertise and the low-expertise conditions.

Stated informally, the null hypothesis for the interaction effect is that the mean differences in attitude change as a function of message discrepancy are the same for the high-expertise and the low-expertise sources. The alternative hypothesis is that the mean differences in attitude change as a function of message discrepancy are not the same for the high-expertise and the low-expertise sources.

Computation of the Sums of Squares

The sum of squares for variability due to factor A (message discrepancy) is

$$SS_A = \frac{\sum T^2_{A_i}}{nb} - \frac{(\sum X)^2}{N}$$

$$= 540.00 - 480.00 = 60.00$$

The sum of squares for variability due to factor B (source expertise) is

$$SS_B = \frac{\sum T^2_{B_j}}{na} - \frac{(\sum X)^2}{N}$$

$$= 643.33 - 480.00 = 163.33$$

The sum of squares for variability due to the interaction of factor A and factor B is

$$SS_{A \times B} = \frac{(\sum X)^2}{N} + \frac{\sum T^2_{A_i B_j}}{n} - \frac{\sum T^2_{A_i}}{nb} - \frac{\sum T^2_{B_j}}{na}$$

$$= 480.00 + 750.00 - 540.00 - 643.33 = 46.67$$

The sum of squares within is

$$SS_{WITHIN} = \sum X^2 - \frac{\sum T^2_{A_i B_j}}{n}$$

$$= 762 - 750.00 = 12.00$$

Finally, the sum of squares total is

$$SS_{TOTAL} = \sum X^2 - \frac{(\sum X)^2}{N}$$

$$= 762 - 480.00 = 282.00$$

Derivation of the Summary Table

The degrees of freedom are

$$df_A = a - 1 = 3 - 1 = 2$$

$$df_B = b - 1 = 2 - 1 = 1$$

$$df_{A \times B} = (a - 1)(b - 1) = (3 - 1)(2 - 1) = 2$$

$$df_{\text{WITHIN}} = (a)(b)(n - 1) = (3)(2)(5 - 1) = 24$$

$$df_{\text{TOTAL}} = N - 1 = 30 - 1 = 29$$

The relevant mean squares are

$$MS_A = \frac{SS_A}{df_A} = \frac{60.00}{2} = 30.00$$

$$MS_B = \frac{SS_B}{df_B} = \frac{163.33}{1} = 163.33$$

$$MS_{A \times B} = \frac{SS_{A \times B}}{df_{A \times B}} = \frac{46.67}{2} = 23.34$$

$$MS_{\text{WITHIN}} = \frac{SS_{\text{WITHIN}}}{df_{\text{WITHIN}}} = \frac{12.00}{24} = .50$$

The F ratios are thus

$$F_A = \frac{MS_A}{MS_{\text{WITHIN}}} = \frac{30.00}{.50} = 60.00$$

$$F_B = \frac{MS_B}{MS_{\text{WITHIN}}} = \frac{163.33}{.50} = 326.66$$

$$F_{A \times B} = \frac{MS_{A \times B}}{MS_{\text{WITHIN}}} = \frac{23.34}{.50} = 46.68$$

These calculations yield the following summary table:

Source	SS	df	MS	F
A (message discrepancy)	60.00	2	30.00	60.00
B (source expertise)	163.33	1	163.33	326.66
A × B	46.67	2	23.34	46.68
Within	12.00	24	.50	
Total	282.00	29		

Inference of Relationships

The observed value of F for the main effect of message discrepancy is 60.00. The critical value of F from Appendix F for an alpha level of .05 and 2 and 24 degrees of freedom is 3.40. Since the observed F value is greater than 3.40, we reject the null hypothesis and conclude that a relationship exists between message discrepancy and attitude change.

The observed value of F for the main effect of source expertise is 326.66. The critical value of F from Appendix F for an alpha level of .05 and 1 and 24 degrees of freedom is 4.26. Since the observed F value is greater than 4.26, we reject the null hypothesis and conclude that a relationship exists between source expertise and attitude change.

The observed F value for the interaction effect is 46.68. Since this is greater than the critical value of 3.40 for an alpha level of .05 and 2 and 24 degrees of freedom, we reject the null hypothesis and conclude that there is an interaction between message discrepancy and source expertise.

Strength of the Relationships

The strength of the relationship between the dependent variable and each of the three sources of between-group variability is computed using Equations 17.22–17.24:

$$eta_A^2 = \frac{SS_A}{SS_{TOTAL}} = \frac{60.00}{282.00} = .21$$

$$eta_B^2 = \frac{SS_B}{SS_{TOTAL}} = \frac{163.33}{282.00} = .58$$

$$eta_{A \times B}^2 = \frac{SS_{A \times B}}{SS_{TOTAL}} = \frac{46.67}{282.00} = .17$$

The strongest effect is for the main effect of source expertise. The proportion of variability in attitude change that is associated with this factor is .58. This represents a strong effect. The proportion of variability in attitude change that is associated with message discrepancy is .21. This also represents a strong effect. The interaction between message discrepancy and source expertise accounts for 17% of the variability in attitude change. This again represents a strong effect.

Nature of the Relationships

Since it has only two levels, the nature of the relationship for the main effect of source expertise is determined by examining the sample means. This examination shows that messages will produce more attitude change when they are attributed to a high-expertise source ($\overline{X}_H = 6.33$) than when they are attributed to a low-expertise source ($\overline{X}_L = 1.67$).

The main effect of message discrepancy has three levels and, hence, it is necessary to apply the Tukey HSD test to determine the nature of the relationship between message discrepancy and attitude change. Since message discrepancy is factor A, the value of the critical difference is established using Equation 17.25. From Appendix G, the value of q for an overall alpha level of .05, $df_{WITHIN} = 24$, and $k = a = 3$ is 3.53. Thus,

$$CD = q\sqrt{\frac{MS_{WITHIN}}{nb}} = 3.53\sqrt{\frac{.50}{(5)(2)}} = .79$$

Mean differences are identified using the procedures reviewed in Section 17.6. As summarized in the following table, this entails comparing the absolute difference between each pair of main effect means with the critical difference:

Null hypothesis tested	Absolute difference between sample means	Value of CD	Null hypothesis rejected?
$\mu_S = \mu_M$	$|2.00 - 5.00| = 3.00$	.79	Yes
$\mu_S = \mu_L$	$|2.00 - 5.00| = 3.00$	.79	Yes
$\mu_M = \mu_L$	$|5.00 - 5.00| = .00$	.79	No

The nature of the relationship is such that medium- ($\overline{X}_M = 5.00$) and large-discrepancy ($\overline{X}_L = 5.00$) messages will produce more attitude change than small-discrepancy messages ($\overline{X}_S = 2.00$). However, we cannot confidently conclude that the amount of attitude change produced by medium- versus large-discrepancy messages will differ.

Because the interaction effect was statistically significant, we perform a set of interaction comparisons to explore the nature of the interaction. We will control the Type I error rate using a modified Bonferroni procedure. There are three 2×2 subtables. We indicate them here, along with the cell means:

	High expertise	Low expertise
Small discrepancy	3.00	1.00
Medium discrepancy	7.00	3.00

	High expertise	Low expertise
Small discrepancy	3.00	1.00
Large discrepancy	9.00	1.00

	High expertise	Low expertise
Medium discrepancy	7.00	3.00
Large discrepancy	9.00	1.00

Using Equation 17.27, we calculate the sum of squares for the interaction effect for each 2×2 subtable. For the first subtable, the sum of squares is

$$SS_{A \times B(1)} = \frac{n(\overline{X}_a + \overline{X}_d - \overline{X}_b - \overline{X}_c)^2}{4}$$

$$= \frac{5(3.00 + 3.00 - 1.00 - 7.00)^2}{4} = 5.00$$

For the second subtable, the sum of squares is

$$SS_{A \times B(2)} = \frac{5(3.00 + 1.00 - 1.00 - 9.00)^2}{4} = 45.00$$

and for the third subtable, the sum of squares is

$$SS_{A \times B(3)} = \frac{5(7.00 + 1.00 - 3.00 - 9.00)^2}{4} = 20.00$$

The mean square for each of these sum of squares is simply the sum of squares divided by 1. We form an F ratio for each subtable, using these mean squares as the numerator and the mean square within from the overall summary table as the denominator, per Equation 17.28:

$$F_{A \times B(1)} = \frac{5.00}{.50} = 10.00$$

$$F_{A \times B(2)} = \frac{45.00}{.50} = 90.00$$

$$F_{A \times B(3)} = \frac{20.00}{.50} = 40.00$$

To apply the modified Bonferroni procedure, we determine a p value for each of the above F ratios (based on 1 and 24 degrees of freedom). This usually must be done with the aid of a computer. The F ratios are ordered in a column from largest to smallest, and the corresponding p values are each compared against a "critical alpha." If the p value is less than the "critical alpha," the interaction comparison is statistically significant.

The value of the "critical alpha" changes for each successive F ratio as one proceeds from the top to the bottom column. The largest F ratio must yield a p value less than the overall alpha level divided by the total number of interaction comparisons (in this case, 3) for the comparison to be statistically significant. The next largest F ratio must yield a p value less than the overall alpha level divided by the number of interaction comparisons minus 1 for the comparison to be statistically significant. The next largest F ratio must yield a p value less than the overall alpha level divided by the number of interaction comparisons minus 2 for the comparison to be statistically significant. And so on, subtracting an additional unit from the total number of contrasts with each successive comparison. If a comparison is statistically significant, the nature of the interaction is determined by examining the differences between the cell means for the relevant subtable. As soon as the first statistically nonsignificant interaction comparison is encountered as one proceeds from the largest to the smallest F ratio, all remaining comparisons are declared statistically nonsignificant. The rationale for this modified Bonferroni procedure is described in Holland and Copenhaver (1988).

The results for the modified Bonferroni procedure can be summarized in a table as follows:

Interaction comparison	F ratio	p value	Critical alpha	Null hypothesis rejected?
$A \times B_{(2)}$	90.00	.000000001	.05/3 = .017	Yes
$A \times B_{(3)}$	40.00	.000002	.05/2 = .025	Yes
$A \times B_{(1)}$	10.00	.004	.05/1 = .050	Yes

In deriving this table, we used an overall alpha level of .05. All three subtables yielded statistically significant interaction effects. Examination of the differences between the cell means for each subtable leads to the following statements about the interaction: Medium-discrepancy messages will produce more attitude change than small-discrepancy messages when source expertise is high (7.00 − 3.00 = 4.00) than when it is low (3.00 − 1.00 = 2.00). Similarly, large-discrepancy messages will produce more attitude change than small-discrepancy messages when source expertise is high (9.00 − 3.00 = 6.00) as compared to low (1.00 − 1.00 = .00). Finally, large-discrepancy messages will produce more attitude change in the advocated direction than medium-discrepancy messages when source expertise is high (9.00 − 7.00 = 2.00) as opposed to low (1.00 − 3.00 = −2.00). Figure 17.3 presents the relevant means graphically.

FIGURE 17.3   **Mean Attitude Change As a Function of Message Discrepancy and Source Expertise**

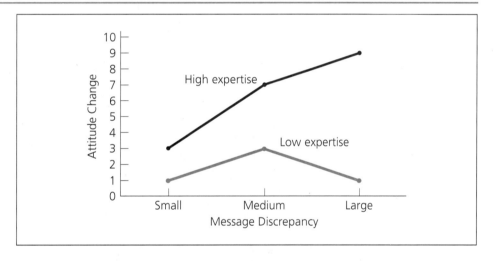

17.9 Unequal Sample Sizes

In the examples thus far, the cell sizes (n) have been equal. However, this is not always the case in behavioral science research. For example, in animal experimentation, a subject might be lost because of disease or sickness. Or a person scheduled to participate in a particular research condition might not show up. Unequal cell sizes necessitate modifications in the analysis of variance techniques discussed in this chapter.

When the sample sizes are the same in all cells, the two independent variables are unrelated to each other in the sample. Consider the following sample sizes for a 2 × 2 factorial design:

	Religion	
Gender	Catholic	Protestant
Male	10	10
Female	10	10

In this study, there are equal sample sizes in all cells, and the two independent variables, *in the context of this data set,* are therefore unrelated to each other. This can be thought of in terms of conditional probabilities: The probability of being a given religion is the same for males and females. The variable of religion is independent of the variable of gender.

Now consider the case of unequal sample sizes:

	Religion	
Gender	Catholic	Protestant
Male	6	13
Female	13	8

In this instance, a relationship exists in the sample between gender and religion. For example, if you know that an individual is a man, you know it is more likely that he is a Protestant than a Catholic. If you know that an individual is a woman, you know it is more likely that she is a Catholic than a Protestant. Thus, there is a relationship between the two independent variables.*

The introduction of a relationship between the independent variables creates a number of statistical and conceptual issues in testing the two main effects and the $A \times B$ interaction. Consider factor A in a design with unequal sample sizes. On the basis of an analysis of variance, we might conclude that factor A is related to the dependent variable in the population and accounts for some of the variability in it. However, because factor A is related to factor B due to the unequal sample sizes, some of the variability in the dependent variable that we attribute to factor A may actually be due to factor B. The problem faced by statisticians is what to do about the between-group variability that is common to both factor A and fac-

---

* Another way of thinking about this is that if we were to perform a chi-square analysis on the cell frequencies (as discussed in Chapter 15), χ^2 would equal 0 in the case of equal n's (indicating the absence of a relationship in the sample) but would be greater than 0 in the case of unequal n's (indicating the presence of a relationship in the sample). Unequal sample sizes do *not* introduce a relationship between the two independent variables when the sample sizes for the groups that make up one factor are proportional across levels of the other factor (for example, when the ratio of Catholics to Protestants is 3 to 1 for both men and women).

tor *B*. A technique called *least squares analysis of variance* is usually applied in this case, which focuses on explained variability that is unique to a given factor (i.e., common explained variability is excluded when calculating the sum of squares associated with a given factor). Details of this approach can be found in Kirk (1995).

17.10   Planning an Investigation Using Two-Way Between-Subjects Analysis of Variance

Appendix E.2 contains tables used in the calculation of the per-cell sample sizes necessary to achieve various values of power for a given effect for two-way between-subjects analysis of variance. In two-way analysis of variance, one must refer to three sample size tables: that for the main effect of factor *A*, that for the main effect of factor *B*, and that for the interaction effect. Appendix E.2 presents sets of tables for various numbers of degrees of freedom for the effect of interest. Table 17.5 reproduces a portion of this appendix for $df_{EFFECT} = 1$ and an alpha level of .05. The first column presents various values of power, and the column headings are values of eta-squared in the population.

The sample sizes given in Appendix E.2 must be adjusted to take into account the nature of the factorial design (for example, 2×3, 3×3, or 2×4). The tabled values are adjusted as follows:

$$n' = \frac{(n_T - 1)(df_{EFFECT} + 1)}{ab} \qquad [17.29]$$

where n' is the adjusted per-cell sample size, n_T is the tabled per-cell sample size, *a* is the number of levels of factor *A*, *b* is the number of levels of factor *B*, and df_{EFFECT} is the degrees of freedom for the effect in question [*a* − 1, *b* − 1, or

TABLE 17.5   Approximate Sample Sizes Necessary to Achieve Selected Levels of Power for $df_{EFFECT} = 1$ and Alpha = .05 As a Function of Population Values of Eta-Squared

Power	\multicolumn{11}{c}{Population Eta-Squared}
	.01	.03	.05	.07	.10	.15	.20	.25	.30	.35	.40
.10	22	8	5	4	3	2	2	2	—	—	—
.50	193	63	38	27	18	12	9	7	5	5	4
.70	310	101	60	42	29	18	13	10	8	7	6
.80	393	128	76	53	36	23	17	13	10	8	7
.90	526	171	101	71	48	31	22	17	13	11	9
.95	651	211	125	87	60	38	27	21	16	13	11
.99	920	298	176	123	84	53	38	29	22	18	15

$(a - 1)(b - 1)]$. The value of n' is rounded up to the nearest integer. The determination of the per-cell sample size necessary to ensure that the power requirement is met for each effect requires that we calculate the necessary adjusted cell size for each effect individually and then use the *largest* sample size of the three. Possible compromises to this strategy are discussed in Cohen (1977).

As an example of these procedures, consider a 2×2 factorial design for which the investigator suspects that the population value of eta-squared is .15 for factor A, .25 for factor B, and .10 for the $A \times B$ interaction, and desires a .80 level of power for each effect for an alpha level of .05. Since $df_A = a - 1 = 2 - 1 = 1$, the value of n_T for the main effect of factor A can be obtained from Table 17.5. This is equal to 23. Thus, the adjusted per-cell sample size using Equation 17.29 is

$$n' = \frac{(23 - 1)(1 + 1)}{(2)(2)} = 11.00$$

Since $df_B = b - 1 = 2 - 1 = 1$ and $df_{A \times B} = (a - 1)(b - 1) = (2 - 1)(2 - 1) = 1$, Table 17.5 can also be used to obtain the values of n_T for the main effect of factor B and for the interaction effect. For the main effect of factor B, $n_T = 13$, so

$$n' = \frac{(13 - 1)(1 + 1)}{(2)(2)} = 6.00$$

For the interaction effect, $n_T = 36$, so

$$n' = \frac{(36 - 1)(1 + 1)}{(2)(2)} = 17.50$$

or, rounded up to the nearest integer, 18. The largest adjusted cell size dictates the number of participants required per group. In this instance, the investigator should include 18 participants in each of the four groups. Note that this will increase the power of the F tests beyond the .80 level for the two main effects.

17.11 Method of Presentation

The results for a two-way analysis of variance are reported in much the same way as those for a one-way analysis of variance. As with the one-way case, it is not necessary to explicitly state that the design was between-subjects in nature because, unless stated otherwise, it is assumed that a between-subjects analysis was performed. For each effect, the degrees of freedom, the observed value of F, the significance level, the sample means, and the standard deviation estimates should be reported. In addition, a statement should be made of the alpha level that was used for the three F tests. If an effect is statistically significant, the strength and nature of the relationship for the effect should also be addressed. Furthermore, the overall alpha level for the modified Bonferroni procedure should be reported if the interaction effect is statistically significant.

It is often more efficient to present the means and standard deviation estimates for the interaction effect in a table rather than in the text. As a way of avoiding "statistics laden text that is difficult to read" (American Psychological Association, 1994, p. 130), it is sometimes also preferable to present the analysis of variance statistics in a table. This might take a similar form as the summary tables presented earlier. When information is presented in tables, the guidelines described in Section 15.12 should be followed. If the interaction is statistically significant, it is sometimes beneficial to also depict the cell means in a graph like Figure 17.3. In American Psychological Association format, graphs are consecutively numbered starting with Figure 1 and are so referenced in the text.

The results for the experiment relating message discrepancy and source expertise to attitude change discussed earlier in this chapter might be reported as follows:

Results

Attitude change scores were subjected to a two-way analysis of variance having three levels of message discrepancy (small, medium, and large) and two levels of source expertise (high versus low). For an alpha level of .05, all effects were found to be statistically significant. The main effect of message discrepancy yielded an F ratio of $F(2, 24) = 60.00$, $p < .01$. The strength of the relationship, as indexed by eta^2, was .21. A Tukey HSD test revealed that the mean for the small-discrepancy messages ($M = 2.00$, $SD = 1.25$) was significantly lower than the means for both the medium- ($M = 5.00$, $SD = 2.21$) and the large- ($M = 5.00$, $SD = 4.27$) discrepancy messages. The means for the latter two conditions did not significantly differ. The main effect of source expertise was such that the messages from the high-expertise source ($M = 6.33$, $SD = 2.66$) produced significantly more attitude change than the messages from the low-expertise source ($M = 1.67$, $SD = 1.18$), $F(1, 24) = 326.66$, $p < .01$. The strength of the relationship, as indexed by eta^2, was .58.

The interaction effect, $F(2, 24) = 46.68$, $p < .01$, was analyzed using interaction comparisons in conjunction with a modified Bonferroni procedure (Holland & Copenhaver, 1988) based on an overall alpha level of .05. The relevant means and standard deviations can be found in Table 1. All 2 × 2 subtables yielded statistically significant interaction effects. The medium-discrepancy messages produced more attitude change than the small-discrepancy messages when source expertise was high (7.00 - 3.00 = 4.00) as compared to low (3.00 - 1.00 = 2.00). Similarly, the large-discrepancy messages produced more attitude change than the small-discrepancy messages when source expertise was high (9.00 - 3.00 = 6.00) than when it was low (1.00 - 1.00 = .00). Finally, the large-discrepancy messages produced more attitude change in the advocated direction than the medium-discrepancy messages when source expertise was high (9.00 - 7.00 = 2.00) as opposed to low (1.00 - 3.00 = -2.00). The strength of the overall interaction effect, as indexed by eta^2, was .17.

The table of means and standard deviation estimates for the interaction might appear as follows:

Table 1

Mean Attitude Change As a Function of Message Discrepancy and Source Expertise

	Message discrepancy		
Source expertise	Small	Medium	Large
High	3.00 (.71)	7.00 (.71)	9.00 (.71)
Low	1.00 (.71)	3.00 (.71)	1.00 (.71)

Note. Standard deviations are in parentheses. For each cell, n = 5.

The note at the bottom of the table informs the reader that the values in parentheses are the standard deviation estimates and that the sample size for all of the cells of the factorial design is 5. According to American Psychological Association format, general information that relates to the table as a whole is placed below the table and designated by the word "Note" underlined and followed by a period.

17.12 Examples from the Literature

Attributions for Success As a Function of the Gender of the Performer and the Type of Task

When someone is successful at a task, we sometimes attribute that success to the person's ability. Alternatively, we may simply think that the person "got lucky" and that the success had little to do with ability.

Johnson (1976) was interested in the extent to which these two attributions are used to explain male and female task success. Participants listened to tape recordings indicating that either a male or a female target person had succeeded at either a traditionally masculine task (identifying mechanical objects such as wrenches and screwdrivers) or a traditionally feminine task (identifying household objects such as mops and pots). Participants were then asked to make attributions about the target person's task performance on a 13-point scale. On this scale, a score of 1 indicated that participants thought task performance was due entirely to luck, and a score of 13 indicated that participants thought task performance was due entirely to ability. The design was thus a 2×2 factorial, with the gender of the performer and the type of task (masculine versus feminine) as the independent variables and ratings of luck/ability as the dependent variable.

It was hypothesized that for the masculine task, male success would be attributed more to ability than would female success, but the reverse would be true for the feminine task. Thus, Johnson predicted an interaction, with the relationship between the gender of the performer and ability/luck attributions depending on the type of task.

The main effect for the gender of the performer was statistically significant, $F(1, 96) = 4.38$, $p < .05$, and indicated that successful performance by men ($\overline{X} = 9.24$) will be attributed more to ability than will successful performance by women ($\overline{X} = 8.40$). The strength of the relationship, as indexed by eta-squared, was only .04, however. This represents a weak effect. The main effect for the type of task was not statistically significant, $F(1, 96) = 2.46$, $p > .05$, with the mean for the masculine task being 8.50 and the mean for the feminine task being 9.14. The strength of the relationship, as indexed by eta-squared, was .02; this again represents a weak effect.

Although a statistically significant interaction effect was observed, $F(1, 96) = 9.08$, $p < .01$, the nature of the interaction was somewhat different from what was expected. Follow-up analyses indicated that for the masculine task, the man's performance ($\overline{X} = 9.53$) was attributed more to ability than was the woman's performance on the same task ($\overline{X} = 7.48$). However, for the feminine task, the mean ratings for the male ($\overline{X} = 8.95$) and the female ($\overline{X} = 9.32$) performers were not significantly different. The strength of the overall interaction effect, as indexed by eta-squared, was .08. This represents a weak to moderate effect. What type of interpretation might you give these findings?

Restaurant Tipping as a Function of Interpersonal Touch and Diners' Gender

Crusco and Wetzel (1984) examined the effects of two types of touch on restaurant tipping. Restaurant diners were randomly assigned to conditions where, while returning their change, their waitress twice touched their palms with her fingers for .5 second (hand-touch condition), placed her hands on their shoulders for 1 to 1.5 seconds (shoulder-touch condition), or did not touch them (no-touch condition). It was predicted that the hand touch would produce positive feelings toward the waitress and thus increase the amount of her tip relative to the no-touch condition. Since a touch on the shoulder can be construed as a sign of dominance, it was felt that this might not be viewed as positively and that tipping might therefore also be greater in the hand-touch condition than in the shoulder-touch condition. So that possible gender differences in these effects could be determined, separate observations were made for male and female diners. The design was thus

In Chapter 10, we noted that not all parents are adamantly opposed to premarital sexual intercourse on the part of their teenage children. For example, some parents believe that if it is with someone who is special and whom the teen has known for a long time, then sexual intercourse is permissible. Interestingly, results reported in Chapter 10 indicate that teens perceive mothers to be more disapproving of premarital sexual intercourse for adolescent girls than for boys. In this earlier analysis, the age of the teenager was ignored. However, it is possible that the perceived difference in disapproval for boys versus girls may dissipate or, alternatively, be magnified over the adolescent years. For example, it may be the case that mothers become more accepting of the inevitability of sexual intercourse as the teen ages. This may be especially true in regard to adolescent girls, who, as reported in Chapter 10, tend to experience stronger maternal sanctions against sexual intercourse to begin with.

To examine the potential interaction of age and gender in relation to this important predictor of sexual behavior, we conducted a 2 × 4 analysis of variance on perceived maternal disapproval of sexual intercourse using data from the parent–teen communication study described in Chapter 1. The two independent variables were the teens' gender and age (14, 15, 16, or 17 years old). The dependent variable was the maternal attitude measure described in Chapter 10. Recall that the teens were asked to indicate their agreement with the following five statements:

1. My mother would disapprove of me having sex at this time in my life.
2. My mother has specifically told me *not* to have sex.
3. My mother thinks I definitely should *not* be sexually active (having sexual intercourse) at this time in my life.
4. If it were with someone who was special to me and whom I knew well, like a steady boyfriend (girlfriend), my mother would not mind if I had sexual intercourse at this time in my life.
5. My mother thinks it is fine for me to be sexually active (having sexual intercourse) at this time in my life.

Each statement was rated on a five-point scale having the response options strongly disagree, moderately disagree, neither agree nor disagree, moderately agree, and strongly agree. The responses to

a 3 × 2 factorial, with the type of touch (hand, shoulder, or none) and the gender of the diner as the independent variables. The dependent variable was the percentage of the bill left as a tip.

A two-way analysis of variance yielded a main effect for the gender of the diner, $F(1, 108) = 3.93$, $p < .05$, such that men ($\overline{X} = 15.3\%$) tipped more than women ($\overline{X} = 12.6\%$). The main effect for the type of touch was also statistically significant, $F(2, 108) = 3.45$, $p < .05$. Analysis of the three touch means showed that tipping was higher in the hand-touch ($\overline{X} = 16.7\%$) and shoulder-touch ($\overline{X} = 14.4\%$) conditions than in the no-touch condition ($\overline{X} = 12.2\%$). The hand-touch and shoulder-touch means did not significantly differ. The interaction effect also failed to attain statistical significance, $F(2, 108) < 1$. It is conventional to use the < 1 notation when the observed value of F is less than 1.00. It is not necessary to indicate that the F value is statistically nonsignificant because an F of less than 1.00 can never lead to rejection of the null hypothesis. (Can you think of why this is the case?) The values of eta-squared were not provided in the research report.

the first three statements were scored from 1 ("strongly disagree") to 5 ("strongly agree"), whereas the responses to the last two items were reverse scored (1 = "strongly agree" to 5 = "strongly disagree"). An overall index of the extent to which the teen perceived the mother as disapproving of the teen engaging in sexual intercourse was obtained by summing the responses to the five statements. This index of perceived maternal disapproval could range from 5 to 25, with higher scores indicating greater perceived maternal disapproval of engaging in sex.

The first step in the analysis was to examine the frequency histograms for each of the eight groups defined by the 2 × 4 design. Figure 17.4 *(continued)*

FIGURE 17.4 Frequency Histograms of Perceived Maternal Attitude Toward (a) 14-Year-Old Boys and (b) 17-Year-Old Girls Engaging in Sex

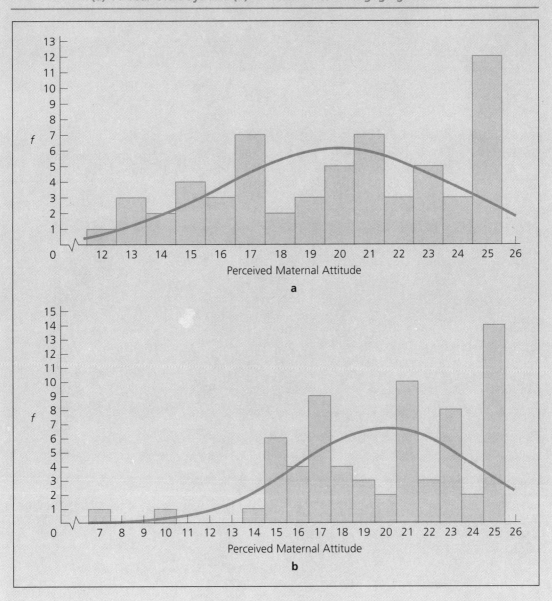

presents the histograms for the 14-year-old boys and the 17-year-old girls, with a normal distribution superimposed on each. These histograms typify the shapes of the other groups, which we omit in the interest of space. The distributions are non-normal, with a preponderance of teens perceiving strong disapproval (a score of 25) by their mothers. Given the relatively large sample sizes in each of our cells, the population nonnormality suggested by the nonnormality in the samples should not prove problematic because of the robustness of the F test (see Milligan, Wong, & Thompson, 1987). However, the histogram for the 17-year-old girls suggests two outliers. The interviews for these teens were examined more closely, but we could find no special considerations that set them apart from the other teens. Nevertheless, we decided to complement the 2 × 4 analysis reported here with additional outlier-resistant analyses. These pro-

duced comparable conclusions to the results reported below.

Table 17.6 presents the computer output for the analysis of variance. The first section of the printout reports the results of two tests of homogeneity of variance, one by Cochran and the other by Bartlett-Box. Although these tests are generally not as useful as the Levene test discussed in Chapters 10 and 12 due to their being sensitive to nonnormality, the computer program we used does not perform a Levene test. However, we conducted a Levene test using a different program and observed a statistically nonsignificant result. The Cochran test was also statistically nonsignificant, but the Bartlett-Box test did yield a statistically significant finding.

As discussed in Chapter 12, tests of variance homogeneity are somewhat problematic. A statistically nonsignificant result cannot be used to ac-

---

TABLE 17.6   **Computer Output for Two-Way Between-Subjects Analysis of Variance for Perceived Maternal Attitude Toward Teen Engaging in Sex**

```
Homogeneity of Variance Tests

Variable  . .  MDISAPPROVE

     Cochrans C(89,8) =                 .15596, P =    .359 (approx.)
     Bartlett-Box F(7,380724) =        2.07298, P =    .043

Combined Observed Means for GENDER

GENDER

     Female            19.11142
     Male              15.81440

AGE

     14                17.74528
     15                17.79623
     16                17.24424
     17                16.90152
```

Source of Variation	SS	DF	MS	F	Sig of F
GENDER	1830.77	1	1830.77	84.91	.000
AGE	104.09	3	34.70	1.61	.186
GENDER BY AGE	44.13	3	14.71	.68	.563
WITHIN	15351.85	712	21.56		
(Total)	17452.75	719	24.27		

cept the null hypothesis, and a statistically significant result may be an artifact of nonnormality in the data or may represent a real but trivial difference in population variances. However, we determined the variance estimates for the eight cells of the design and found that the largest one (27.04) was less than twice as large as the smallest (13.84). Computer simulation studies suggest that the F tests will be robust to such differences, even with our unequal sample sizes (Milligan, Wong, & Thompson, 1987; Moore & McCabe, 1993).

The next section of the printout in Table 17.6 presents the mean scores for the two main effects, followed by the summary table. Because we had unequal sample sizes, we conducted a least squares analysis of variance, a technique commonly used with unequal n's. Only the main effect of gender was statistically significant, with the F value of 84.91 having an associated p value of .000. The strength of the relationship, as indexed by eta-squared, was .10, which represents a moderate effect. The means indicate that girls perceive greater maternal disapproval than boys. The lack of a significant interaction suggests that this gender difference is uniform across the different age groups, although we can't strictly draw such a conclusion because it would be tantamount to accepting the null hypothesis. We *can* state that there is no compelling evidence to suggest that the gender effect is different across age groups and that whatever differences are observed as a function of age can reasonably be attributed to sampling error.

Given the statistically nonsignificant effects for age and the age × gender interaction, there is a possibility of Type II errors occurring. To evaluate this, we calculated the statistical power for both the main effect of age and the interaction effect for an alpha level of .05 using the sample sizes from our study, assuming a small effect size in the population (a population eta-squared of .03). We used the procedures described in Cohen (1988) for our power calculations. In both cases, the statistical power exceeded .95, which suggests that Type II errors were unlikely.

These results affirm the bias in maternal attitudes toward premarital sexual intercourse for boys versus girls and suggest that this bias occurs independent of the age of the adolescent (at least for the age range we studied).

Summary

Two-way between-subjects analysis of variance allows us to investigate the separate and joint effects of two independent variables on a dependent variable. This is accomplished through the analysis of main effects and interaction effects in the context of factorial designs. A two-way factorial design is one in which the a levels of one independent variable are combined with the b levels of a second independent variable to yield $a \times b$ groups. A test of a main effect refers to the test of the relationship between one of the independent variables and the dependent variable. A test of an interaction effect refers to the test of whether the nature of the relationship between one of the independent variables and the dependent variable differs as a function of the other independent variable. Two-way analysis of variance is typically applied when (1) the dependent variable is quantitative in nature and is measured on a level that at least approximates interval characteristics, (2) the independent variables are both between-subjects in nature, (3) the independent variables both have two or more levels, and (4) the independent variables are combined to form a factorial design.

The statistical procedures for two-way analysis of variance are similar to those for one-way analysis of variance except that the overall between-group variability is broken down into three components: between-group variability due to factor A, between-group variability due to factor B, and between-group variability due to the $A \times B$ interaction. Hypothesis testing procedures are then applied to each of these three sources of between-group variability. The strength of the relationships is measured using eta-squared. When a statistically significant main effect is obtained and there are two levels of the independent variable, the nature of the effect is determined in the same fashion as for the independent groups t test; when there are three or more levels of the independent variable, the Tukey HSD test is used. When a statistically significant interaction effect is obtained, the nature of the interaction is determined using interaction comparisons coupled with a modified Bonferroni procedure to control the Type I error rate.

The computational procedures presented in this chapter are applicable only when cells have equal numbers of participants. When sample sizes are unequal, other approaches must be used.

Exercises

Answers to asterisked (*) exercises appear at the back of the book. Exercises with two asterisks are also worked out by step in the Study Guide.

Exercises to Review Concepts

1. How many groups are required in a 3×3 factorial design? In a 2×5 factorial design? In a 4×3 factorial design?

*2. How many independent variables are in a 3×3 factorial design? In a 2×5 factorial design? In a $2 \times 2 \times 4$ factorial design?

3. Give an example of an investigation that would use a 2×3 factorial design.

4. Under what conditions is two-way between-subjects analysis of variance typically used to analyze the relationship between a dependent variable and two independent variables?

*5. What is a main effect? An interaction effect? An interaction?

*6. For each of the following sets of *population* means, indicate whether there is a main effect of factor A, a main effect of factor B, and/or an $A \times B$ interaction:

a.

	B_1	B_2
A_1	4.00	5.00
A_2	4.00	5.00

b.

	B_1	B_2
A_1	6.00	6.00
A_2	4.00	4.00
A_3	7.00	7.00

c.

	B_1	B_2	B_3
A_1	1.00	2.00	3.00
A_2	5.00	6.00	7.00
A_3	8.00	9.00	10.00

7. For each of the following sets of *population* means, indicate whether there is a main effect of factor A, a main effect of factor B, and/or an $A \times B$ interaction:

a.

	B_1	B_2
A_1	5.00	10.00
A_2	10.00	5.00

b.

	B_1	B_2	B_3
A_1	4.00	7.00	9.00
A_2	5.00	6.00	7.00
A_3	6.00	8.00	10.00

c.

	B_1	B_2	B_3	B_4
A_1	5.00	6.00	6.00	7.00
A_2	7.00	8.00	4.00	5.00
A_3	9.00	10.00	2.00	3.00

*8. Generate a set of population means for a 2×3 factorial design that reflects a main effect of factor A, no main effect of factor B, and an $A \times B$ interaction.

9. Generate a set of population means for a 2×2 factorial design that reflects no main effect of factor A, no main effect of factor B, and an $A \times B$ interaction.

*10. What do nonparallel lines indicate in a graph of population means? What do nonparallel lines indicate in a graph of sample means? What accounts for the difference in the two situations?

11. What are the three components of the sum of squares between in two-way between-subjects analysis of variance?

12. Distinguish among the sum of squares total, the sum of squares for factor A, the sum of squares for factor B, the sum of squares for the $A \times B$ interaction, and the sum of squares within. How are they interrelated?

*13. Insert the missing entries in the summary table for a 3×4 factorial design:

Source	SS	df	MS	F
A	—	—	—	—
B	45.00	—	—	—
$A \times B$	60.00	—	—	—
Within	216.00	—	—
Total	341.00	119		

14. Insert the missing entries in the summary table for a 3×3 factorial design:

Source	SS	df	MS	F
A	—	—	—	3.50
B	12.00	—	—	—
$A \times B$	40.00	—	—	—
Within	—	27	2.00	
Total	120.00	—		

*15. State the critical values of F for the main effect of factor A, the main effect of factor B, and the interaction effect for a two-way analysis of variance for an alpha level of .05 under each of the following conditions:
   a. $a = 2, b = 3, n = 11$
   b. $a = 2, b = 4, n = 7$
   c. $a = 3, b = 3, n = 11$
   d. $a = 4, b = 3, n = 6$

16. What are the assumptions underlying two-way between-subjects analysis of variance?

Refer to the following summary table for a two-way factorial design to do Exercises 17–22.

Source	SS	df	MS	F
A	50.00	1	50.00	10.00
B	40.00	2	20.00	4.00
$A \times B$	40.00	2	20.00	4.00
Within	240.00	48	5.00	
Total	370.00	53		

*17. How many levels of factor A are included in the design? How many levels of factor B? How many groups?

*18. What is the total sample size? Given that there are equal numbers of participants in each group, what is the per-group sample size?

*19. What is the total amount of between-group variability (that is, what is the value of the sum of squares between)?

*20. State the null and alternative hypotheses for the main effect of factor A, the main effect of factor B, and the interaction effect. (The hypotheses for the interaction effect may be stated from either perspective.)

*21. Test the viability of the null hypotheses with respect to the main effect of factor A, the main effect of factor B, and the interaction effect.

*22. Compute the values of eta-squared for the main effect of factor A, the main effect of factor B, and the interaction effect. Indicate whether each value represents a weak, moderate, or strong effect.

*23. How many 2×2 subtables would be included in the set of interaction comparisons for a 2×4 factorial design? For a 2×5 factorial design? For a 3×3 factorial design?

Refer to the following means and summary table for a 2×3 factorial design with 11 participants per cell to do Exercises 24–26.

	B_1	B_2	B_3
A_1	10.00	15.00	20.00
A_2	14.00	20.00	24.00

Source	SS	df	MS	F
A	240.00	1	240.00	24.00
B	910.00	2	455.00	45.50
$A \times B$	10.00	2	5.00	.50
Within	600.00	60	10.00	
Total	1,760.00	65		

*24. Test the viability of the null hypothesis with respect to the main effect of factor A. If the null hypothesis is rejected, discern the nature of the relationship between factor A and the dependent variable.

*25. Test the viability of the null hypothesis with respect to the main effect of factor B. If the null hypothesis is rejected, analyze the nature of the relationship between factor B and the dependent variable using the Tukey HSD test.

*26. Test the viability of the null hypothesis with respect to the interaction effect. If the null hypothesis is rejected, analyze the nature of the interaction between factor A and factor B using interaction comparisons coupled with a modified Bonferroni procedure.

Refer to the following means and summary table for a 2×3 factorial design with five participants per cell to do Exercises 27–29.

	B_1	B_2	B_3
A_1	10.00	12.00	14.00
A_2	12.00	8.00	4.00

Source	SS	df	MS	F
A	120.00	1	120.00	10.00
B	20.00	2	10.00	.83
$A \times B$	380.00	2	190.00	15.83
Within	288.00	24	12.00	
Total	808.00	29		

27. Test the viability of the null hypothesis with respect to the main effect of factor A. If the null hypothesis is rejected, discern the nature of the relationship between factor A and the dependent variable.

28. Test the viability of the null hypothesis with respect to the main effect of factor B. If the null hypothesis is rejected, analyze the nature of the relationship between factor B and the dependent variable using the Tukey HSD test.

29. Test the viability of the null hypothesis with respect to the interaction effect. If the null hypothesis is rejected, analyze the nature of the interaction between factor A and factor B using interaction comparisons coupled with a modified Bonferroni procedure.

Use the following numbers of job offers received by male and female college seniors in four academic majors to do Exercises 30–34.

	Academic Major			
Gender	Computer science	Business	Liberal arts	Behavioral sciences
Male	3	4	2	2
	4	4	2	3
	6	5	1	3
	3	2	3	1
	3	4	2	2
Female	2	1	3	2
	3	2	2	3
	3	2	1	5
	2	3	2	3
	1	1	3	4

30. Test the viability of the null hypotheses with respect to the main effect of gender, the main effect of academic major, and the interaction effect using a two-way analysis of variance.
31. Compute the values of eta-squared for the main effect of gender, the main effect of academic major, and the interaction effect. Indicate whether each value represents a weak, moderate, or strong effect.
32. Discern the nature of the relationship between gender and number of job offers.
33. Analyze the nature of the relationship between academic major and number of job offers using the Tukey HSD test.
34. Analyze the nature of the interaction between gender and academic major using interaction comparisons coupled with a modified Bonferroni procedure.

*35. Why are unequal cell sizes problematic for two-way between-subjects analysis of variance?
*36. If a researcher suspects that the strength of the relationship in the population is .15 for the main effect of factor A, .03 for the main effect of factor B, and .10 for the interaction effect as indexed by eta-squared, what sample size should he use per cell in a study involving three levels of factor A, four levels of factor B, and an alpha level of .05 in order to achieve a power of at least .90 for each effect?
37. If a researcher suspects that the strength of the relationship in the population is .20 for the main effect of factor A, .07 for the main effect of factor B, and .07 for the interaction effect as indexed by eta-squared, what sample size should she use per cell in a study involving two levels of factor A, five levels of factor B, and an alpha level of .05 in order to achieve a power of at least .85 for each effect?

Multiple-Choice Questions

38. As soon as the first statistically nonsignificant comparison is encountered as one proceeds from the largest to the smallest F ratio for a set of interaction comparisons, all remaining comparisons are declared statistically nonsignificant.
    a. true   b. false
*39. Suppose we have a 2×2 factorial design in which the two levels of the first factor, gender, are male (M) and female (F) and the two levels of the second factor, assertiveness, are low (L) and high (H). One way we can state the null and alternative hypotheses for the interaction effect is
    a. H_0: $\mu_{M,L} - \mu_{F,H} = \mu_{M,H} - \mu_{F,L}$
       H_1: $\mu_{M,L} - \mu_{F,H} \neq \mu_{M,H} - \mu_{F,L}$
    b. H_0: $\mu_{M,L} - \mu_{F,L} = \mu_{F,H} - \mu_{M,H}$
       H_1: $\mu_{M,L} - \mu_{F,L} \neq \mu_{F,H} - \mu_{M,H}$
    c. H_0: $\mu_{M,L} - \mu_{M,H} = \mu_{F,L} - \mu_{F,H}$
       H_1: $\mu_{M,L} - \mu_{M,H} \neq \mu_{F,L} - \mu_{F,H}$
    d. none of the above

40. Nonparallel lines in a graph of sample means always indicate the existence of an interaction.
    a. true   b. false
*41. The nature of a statistically significant interaction effect for a 2×2 factorial design is determined using interaction comparisons coupled with a modified Bonferroni procedure.
    a. true   b. false
42. In a 3×2 factorial design, we can test for _____ main effect(s) and _____ interaction(s).
    a. 1; 2   c. 3; 2
    b. 2; 1   d. 6; 1
*43. General information that relates to a table as a whole is placed above the table and underlined according to American Psychological Association format.
    a. true   b. false
*44. Least squares analysis of variance deals with the relationship introduced between the independent variables by unequal cell sizes by focusing on
    a. explained variability that is common to both factors
    b. explained variability that is unique to a given factor
    c. both common and unique explained variability
    d. variability that is unexplained by a given factor
45. A modified Bonferroni procedure is used with interaction comparisons to
    a. simplify the required calculations
    b. control the Type I error rate
    c. control the Type II error rate
    d. determine the nature of statistically significant main effects
46. $SS_{A \times B} = SS_A \times SS_B$
    a. true   b. false
*47. The values of the "critical alpha" for the modified Bonferroni procedure
    a. get smaller as the F ratios for the set of interaction comparisons get smaller
    b. get larger as the F ratios for the set of interaction comparisons get smaller

c. are equal to the overall alpha level
d. are equal to the overall alpha level divided by the total number of interaction comparisons in the set
48. A possible strategy for dealing with disturbance variables is to bring a variable into the research design by including it as a factor.
    a. true   b. false

Exercises to Apply Concepts

**49. Psychologists have studied extensively the factors that contribute to weight gain in people. It is currently believed that individuals use at least two sources of "cues" to decide that they are hungry and should eat. One source is internal cues in the form of changes in one's physiology that create or suggest hunger. A second source is external cues that occur in the environment and suggest to the individual that he or she should be hungry. For example, the approach of 12:00 noon might serve as such a cue by signifying that it is lunch time. Several researchers have suggested that one difference between overweight and normal-weight individuals is that normal-weight individuals attend mostly to internal rather than external cues as guidelines for eating, whereas the reverse is true for overweight individuals.

Suppose that in an investigation designed to study this issue, 15 overweight and 15 normal-weight individuals were instructed not to eat breakfast on the morning before participating in an experiment. When they arrived at the laboratory, the participants were asked to remove their watches under the pretense that part of the study would involve estimating time. After performing a number of tasks, the participants were asked to rate how hungry they felt on a 0 to 10 scale, with higher scores indicating greater hunger. Just prior to this, one-third of the participants were informed that it was 11:00, another third were informed that it was 12:00, the actual time, and the final third were informed that it was 1:00. The design was thus a 3×2 factorial with three levels of supposed time and two levels of weight.

If overweight individuals attend mostly to external cues, we would expect them to report greater hunger the later the supposed time because this implies that they have gone longer without food. If normal-weight individuals attend mostly to internal cues, the time manipulation should not affect their ratings of hunger. Hypothetical data for this experiment are presented in the table. Analyze these data as completely as possible, draw a conclusion for each effect, and report your results using the principles discussed in the Method of Presentation section.

	Overweight	Normal weight
11:00	5	6
	6	6
	7	7
	6	5
	6	6
12:00	9	6
	7	6
	8	5
	9	7
	10	6
1:00	10	7
	10	6
	9	7
	8	8
	9	5

50. All of us tend to classify people into social categories (e.g., "conservative businessman," "typical housewife," and so on), and frequently these categorizations influence how we behave toward one another. Psychologists have studied factors that influence social categorization and the effects of such categorizations on behavior. In one experiment (Rubovits & Maehr, 1973), white female undergraduates enrolled in a teacher-training course were asked to prepare a lesson for four seventh-grade students. The teachers were then given information about each of the students. For half of the teachers, a particular student was described as being "gifted" (i.e., extremely intelligent), whereas for the other half of the teachers, the same student was described as being "nongifted" (i.e., of average intelligence). This student was either African American or white, yielding a 2×2 factorial design with two levels of race and two levels of attributed intelligence.

Each teacher was observed interacting with the target student and the three other students during a 40-minute period. The number of interactions directed toward the target student was recorded, and this served as an index of how much attention the teacher gave him or her. Hypothetical data representative of the results of this experiment are presented in the table. Analyze these data as completely as possible, draw a conclusion for each effect, and report your results using the principles discussed in the Method of Presentation section.

	Gifted	Nongifted
African American	30	30
	29	29
	30	28
	30	30
	31	28
White	36	29
	36	32
	36	31
	35	30
	37	33

51. In order to study the effect of video game aggression on affect, Anderson and Ford (1986) conducted an experiment in which male and female college students either did not play a video game or played a mildly or a highly aggressive video game for 20 minutes. Game selection was based on ratings of content violence and graphics violence of 11 video games provided by a different sample of students during a preliminary study. The design was thus a 3×2 factorial with three levels of game type and two levels of gender.

After playing the assigned game, students completed the Multiple Affect Adjective Checklist (MAAC). This instrument measures hostility, anxiety, and depression as experienced by respondents at the time of assessment. Students in the no-game condition completed the MAAC with the understanding that they would be playing a video game later. We consider only the hostility ratings. Hypothetical data representative of the results of this experiment are presented in the table. Analyze these data as completely as possible, draw a conclusion for each effect, and report your results using the principles discussed in the Method of Presentation section.

	Male	Female
No game	6	6
	7	7
	7	6
	6	8
	5	6
	5	6
Mildly aggressive game	9	9
	9	9
	8	9
	10	8
	9	10
	10	9
Highly aggressive game	10	10
	10	10
	11	9
	11	11
	9	10
	10	10

Overview and Extension: Selecting the Appropriate Statistical Test for Analyzing Bivariate Relationships and Procedures for More Complex Designs

The statistical techniques that we have considered in this book have focused on the analysis of bivariate relationships. We maintain this focus in the first part of this chapter by considering issues involved in selecting the appropriate test for analyzing data in the two-variable case. The second half of the chapter is a brief introduction to several commonly used statistical tests when three or more variables are analyzed simultaneously.

18.1 Selecting the Appropriate Statistical Test for Analyzing Bivariate Relationships

We have considered a total of 12 statistical procedures for studying bivariate relationships.* Except for regression, three questions have guided our analysis in each instance: (1) Given sample data, can we infer that a relationship exists between two variables in the population? (2) If so, what is the strength of the relationship? (3) If so, what is the nature of the relationship? The specific statistics

---

* In reaching this total, we counted the Wilcoxon rank sum test and the Mann–Whitney U test as one technique.

used to address each question are summarized for each of the relevant inferential tests in Table 18.1.

Although a large number of tests are available for analyzing bivariate relationships, the task of selecting a test can be simplified if one adheres to the guidelines presented below. It must be emphasized that these guidelines are only rules of thumb, however; there are potential exceptions to each one. Nevertheless, they should be of heuristic value in the vast majority of cases. In each instance, the recommendation of a test assumes that the relevant requirements for application of that test have been met.

18.2 Case I: The Relationship Between Two Qualitative Variables

When both of the variables under study are qualitative and between-subjects in nature, the appropriate method of data analysis is the chi square test. This situation would occur, for example, when examining the relationship between gender and political party identification, as discussed in Chapter 15. If both variables are qualitative but one or both are within-subjects in nature, then none of the statistical tests that we have considered can be used. The appropriate statistical procedures in this instance are discussed in McNemar (1962) and Marascuilo and McSweeney (1977).

18.3 Case II: The Relationship Between a Qualitative Independent Variable and a Quantitative Dependent Variable

Table 18.2 presents a decision tree that can be used to select a statistical test in situations where the independent variable is qualitative and the dependent variable is quantitative. The three decision points represented in Table 18.2 are (1) use of a parametric versus a nonparametric test, (2) whether the independent variable is between-subjects or within-subjects in nature, and (3) the number of levels that characterize the independent variable.

Use of a Parametric Versus a Nonparametric Test

An important factor to consider when deciding whether to use a parametric or a nonparametric test is what aspect of the distribution one wishes to compare groups on (e.g., means, ranks). If means are the focus, then a parametric test is the analysis of choice. However, even with this focus, further consideration must be given to the number of values of the dependent variable, the level of measurement of the dependent variable, and the extent of violation of distributional assumptions before embarking on parametric analysis. If ranks are the focus, then nonparametric methods should be pursued.

TABLE 18.1   **Statistics Used to Address the Three Questions for the Analysis of Bivariate Relationships for Inferential Tests**

Independent Groups t Test

Inference of a relationship:	t statistic
Strength of the relationship:	Eta-squared
Nature of the relationship:	Inspection of group means

Correlated Groups t Test

Inference of a relationship:	t statistic
Strength of the relationship:	Eta-squared
Nature of the relationship:	Inspection of group means

One-Way Between-Subjects Analysis of Variance

Inference of a relationship:	F ratio
Strength of the relationship:	Eta-squared
Nature of the relationship:	Tukey HSD test

One-Way Repeated Measures Analysis of Variance

Inference of a relationship:	F ratio
Strength of the relationship:	Eta-squared
Nature of the relationship:	Tukey HSD test*

Pearson Correlation

Inference of a relationship:	r statistic
Strength of the relationship:	r^2
Nature of the relationship:	Sign of r

Chi-square Test

Inference of a relationship:	χ^2 statistic
Strength of the relationship:	Fourfold point correlation coefficient/Cramer's statistic
Nature of the relationship:	Modified Bonferroni procedure

Wilcoxon Rank Sum Test/ Mann–Whitney U Test

Inference of a relationship:	z statistic/ U statistic
Strength of the relationship:	Glass rank biserial correlation coefficient
Nature of the relationship:	Inspection of mean ranks

Wilcoxon Signed-Rank Test

Inference of a relationship:	T statistic/ z statistic
Strength of the relationship:	Matched-pairs rank biserial correlation coefficient
Nature of the relationship:	Inspection of rank sums

Kruskal–Wallis Test

Inference of a relationship:	H statistic
Strength of the relationship:	Epsilon-squared
Nature of the relationship:	Dunn procedure

Friedman Analysis of Variance by Ranks

Inference of a relationship:	χ_r^2 statistic
Strength of the relationship:	Concordance coefficient
Nature of the relationship:	Dunn procedure

Spearman Rank-Order Correlation

Inference of a relationship:	r_s statistic
Strength of the relationship:	Magnitude of r_s
Nature of the relationship:	Sign of r_s

* When the sphericity assumption is violated, a modified Bonferroni procedure is used instead of the HSD test.

The choice of which aspect of the distribution to compare across groups is often dictated by theory or the substantive question of interest. Alternatively, the observed distribution may influence one's decision about what feature to focus on. For example, if a dependent variable is highly skewed, then the decision might be

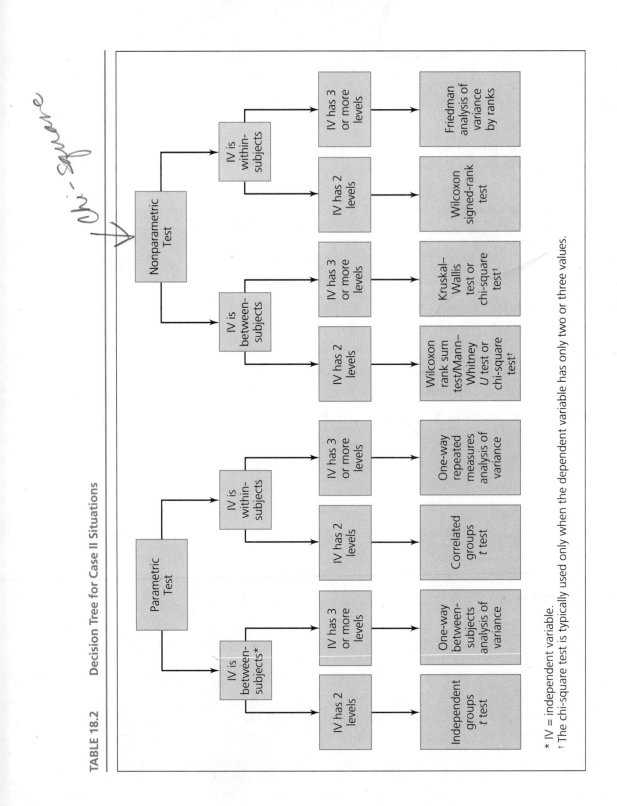

TABLE 18.2   Decision Tree for Case II Situations

Chi-square

* IV = independent variable.
† The chi-square test is typically used only when the dependent variable has only two or three values.

to focus on medians rather than means to analyze the relationship between two variables. Sometimes, an investigator will want to focus on multiple indexes (e.g., means, medians, ranks), in which case *both* parametric and nonparametric statistical tests would be pursued.

Let us explore in more depth the issue of the number of values of the dependent variable and how this can influence the choice of analytic strategy. Consider the case where the independent variable is geographic location (Northeast, South, Midwest, or West) and the dependent variable is attitude toward nuclear energy. The relevant attitude scale could conceivably have anywhere from two to a large number of response points. In situations where the dependent measure has only two or three values, most researchers will *not* use a parametric test. As noted in previous chapters, an assumption of parametric tests is that population scores are normally distributed. A dependent variable cannot be normally distributed if it has only two or three values and, hence, this assumption of parametric tests will be violated. Under this circumstance, nonparametric tests are typically used.*

When the dependent measure has only two or three values, the rank tests discussed in Chapter 16 might also be of questionable value. This is because with only two or three values on the dependent variable, there will be a large number of ties in ranks and, as noted in Chapter 16, numerous ties can create problems for rank-order approaches. The alternative in this case is to use correction terms for tied ranks or, if the independent variable is between-subjects in nature, to use the chi-square test. When applicable, most researchers choose the chi-square approach (or log-linear analysis, described below).

As an example of a situation where we might apply the chi-square test to a study that has a qualitative independent variable and a quantitative dependent variable with only two values, consider the case where the independent variable is occupation (blue collar, white collar, or clerical) and the dependent variable is the number of hands (one or two) with which one is dexterous. A 2×3 contingency table in which manual dexterity is the row variable and occupation is the column variable can be formulated for this problem, and the chi-square procedures developed in Chapter 15 and summarized in Table 18.1 applied to the observed frequencies.

Given a quantitative dependent variable with a sufficient number of values, additional characteristics of the data must be considered when determining whether a parametric test is appropriate. For example, if the dependent variable is measured on an ordinal level that seriously departs from interval level characteristics, then a nonparametric test might be pursued instead of a parametric test. A nonparametric test should also be pursued if there is a reason to believe that the distributional assumptions of the corresponding parametric test have been violated to the extent that the parametric test is not robust.

---

* There are parametric procedures that can be applied in this situation, but a discussion of these is beyond the scope of this book. Interested readers are referred to Hsu and Feldt (1969), Joreskog and Sorbom (1993), and Lunney (1970).

Using Table 18.2 to Select the Appropriate Statistical Test

To illustrate the use of Table 18.2, let us consider an investigation where the researcher asked 30 single, 30 married, 30 widowed, and 30 divorced individuals to indicate their life satisfaction on a seven-point scale. The independent variable (marital status) is qualitative and the dependent variable (life satisfaction) is quantitative; hence, this is a Case II situation.

The first step in using Table 18.2 is to decide whether to use a parametric or a nonparametric test. The life satisfaction measure has seven values, which is not restrictive. Also, suppose we can safely assume that it approximates interval characteristics. Given the equivalence and magnitude of the sample sizes, a parametric test would almost certainly be robust to any violations of distributional assumptions. The researcher concludes that the statistical conditions are appropriate for a parametric test and that it is theoretically meaningful to compare means.

The next decision in Table 18.2 is to classify the independent variable as between-subjects or within-subjects in nature. Since each group includes 30 different individuals, the independent variable in this instance is between-subjects.

Finally, the number of levels that characterize the independent variable has to be considered. In this instance, there are four. This dictates a one-way between-subjects analysis of variance as the analytic technique. Referring to Table 18.1, we would use an F ratio to infer the existence of a relationship between the independent and dependent variables, eta-squared to measure the strength of the relationship, and a Tukey HSD test to determine the nature of the relationship.

18.4   Case III: The Relationship Between a Quantitative Independent Variable and a Qualitative Dependent Variable

Generally speaking, the methods we have presented in this book are not conducive to analyzing quantitative independent variables and qualitative dependent variables. The exception is the chi-square test, and it is typically used when the independent variable is also qualitative. Depending on how many there are, the different values of the quantitative variable will either constitute separate levels of the contingency table or be collapsed into a small number of categories, as discussed in Section 15.10. Other statistical techniques for Case III situations are *logistic regression* and *polychotomous logistic regression*. These techniques are discussed in Koch and Edwards (1988).

18.5   Case IV: The Relationship Between Two Quantitative Variables

Given two between-subjects variables that are measured on a level that at least approximates interval characteristics, the most common method of analysis is Pearson correlation. As discussed in Chapters 5 and 14, Pearson correlation evaluates a *linear* relationship between variables. If the expected relationship is nonlinear, procedures for nonlinear relationships discussed in Pedhazur (1982) can be applied. When one or both variables are measured on an ordinal level that seriously departs from interval level characteristics, Spearman rank-order correlation or some other nonparametric correlational index is the test of choice.

When the independent variable is within-subjects in nature or has only a few values associated with it, the Case II statistics elaborated in Table 18.2 might be applicable. As an example, a researcher might investigate the relationship between age and intelligence by administering an intelligence test to 100 5-year-old children, 100 7-year-old children, and 100 9-year-old children. In this case, both variables are quantitative in nature, but age has only three values (5, 7, and 9). In fact, the study was explicitly structured to include these three particular values of the age variable. Under these circumstances, most investigators would use Table 18.2 to choose the appropriate statistical test. Since the independent variable, age, is between-subjects and has three levels, the method of analysis would be one-way between-subjects analysis of variance if a parametric test is appropriate or the Kruskal–Wallis test if a nonparametric test is required.

The use of Case II statistics is appropriate when a quantitative independent variable has fewer than five or so values and each value (level) has a sufficient number of cases to justify the analysis of means, medians, or the like. If the dependent variable has only two or three values, the chi-square test can be used.

18.6   Procedures for More Complex Designs

The focus of the statistical procedures that we have considered to this point has been the analysis of bivariate relationships. However, many research problems require that more than two variables be studied simultaneously, and statistical techniques have been developed to analyze the relationship between three or more variables. Because these techniques consider the variation among multiple variables, they are referred to as **multivariate statistics.** One example of a multivariate test is two-way between-subjects analysis of variance considered in Chapter 17. As we have seen, this technique is a direct extension of one-way between-subjects analysis of variance for the situation where there is one dependent variable and two independent variables. In the sections that follow, we briefly discuss several additional multivariate tests that are commonly used in the behavioral sciences.

Two-Way Repeated Measures and Two-Way Between-Within Analyses of Variance

Just as the joint influence of two between-subjects independent variables on a dependent variable can be studied using two-way between-subjects analysis of variance, it is also possible to study the joint influence of two within-subjects independent variables. For instance, in the wine label and perceived quality experiment discussed in Chapter 13, the researcher might decide to vary information regarding the year the wine was supposedly produced as well as information about the supposed country of origin. In the simplest case, she might present the wine as being either an old or a new vintage. Crossing the year of production information with the country of origin information (French, Italian, or American), we get a total of six conditions (for instance, old-vintage French wine, new-vintage French wine, and so forth). Now, instead of having each participant taste a total of three "different" wines, each participant will taste a total of six "different" wines. Since both independent variables are within-subjects in nature, the obtained data can be analyzed using a **two-way repeated measures analysis of variance.**

Sometimes research designs involve one between-subjects independent variable and one within-subjects independent variable. For instance, another way to conduct the study outlined above would be to use two groups of participants; one group tastes the supposed French, Italian, and American wines under circumstances where the labels indicate that they are all old vintage, and the other group tastes the wines under circumstances where the labels indicate that they are all new vintage. In this case, the country of origin variable is within-subjects and the vintage variable is between-subjects, so the appropriate means of analysis is a **two-way between-within analysis of variance.**

The procedures for analysis of variance with two independent variables can be extended to analyze the relationship between a dependent variable and three or more independent variables. Given the flexibility of analysis of variance procedures, it is not surprising that this is one of the most commonly used statistical approaches in the behavioral sciences.

Multivariate Analysis of Variance

A situation that frequently arises in behavioral science research is the analysis of two or more *dependent* variables. For instance, a researcher interested in the achievement concerns of high school, junior college, and college graduates might have members of each group complete measures assessing concern with social achievement, occupational achievement, and financial achievement. In this design, there is one independent variable (education) and three dependent variables (concern with social achievement, concern with occupational achievement, and concern with financial achievement).

One approach to analyzing the data would be to conduct three one-way between-subjects analyses of variance, one for each dependent variable. The problem with this approach is that when multiple analyses of variance are performed, the probability of making a Type I error for at least one of the tests increases beyond the probability specified by the alpha level for any single analysis. This is similar to the problem discussed in Chapter 12 with conducting a series of independent groups *t* tests to determine the nature of the relationship between the independent and dependent variables following a statistically significant one-way analysis of variance. To circumvent this problem, analysts often perform a **multivariate analysis of variance.**

Multivariate analysis of variance (abbreviated as "MANOVA") tests whether the research groups have different population means *on the dependent variables considered jointly.* It does this by calculating a *multivariate F ratio* that can then be compared with a critical F value. In the present example, the multivariate analysis of variance would enable us to infer whether the three education groups differ in their mean achievement concerns for all three dependent measures considered simultaneously. If the multivariate F test is statistically significant, we proceed with follow-up analyses for each dependent variable separately, using procedures that maintain the overall error rate at a specified level. If the multivariate F test is not statistically significant, the null hypothesis of equivalent population means on the dependent variables considered jointly cannot be rejected, so no additional analyses are warranted.

Multivariate analysis of variance can be used with any number of dependent variables and any number and combination of between-subjects and within-subjects independent variables. When applied to a single independent variable that has only two levels, it is called the **Hotelling T^2 test.**

Multiple Regression

In Chapter 14, we saw how a regression equation can be used to predict individuals' scores on one variable (the criterion variable, symbolized Y) from knowledge of their scores on a second variable (the predictor variable, symbolized X). **Multiple regression** extends these procedures to the prediction of a criterion variable from two or more predictor variables. For instance, a common use of multiple regression is to predict educational achievement from previous academic performance. Admissions officers at many colleges have established regression equations for predicting grade point averages at their institution from students' high school grade point average, SAT scores, class rank, teacher ratings, and the like. The rationale is that if several of these measures are considered simultaneously, more accurate predictions can be made than if only one predictor variable were studied. When this approach is used, predicted achievement scores are among the criteria considered when making admission decisions.

Multiple regression is a direct extension of regression with one predictor variable. Per Section 14.1, the bivariate linear model for a population is

$$Y = \alpha + \beta X + \varepsilon$$

In the case of multiple regression, the right-hand side of the equation is expanded to include more than one predictor variable. For example, if we had two predictors, X and Z, then the linear model would appear as follows:

$$Y = \alpha + \beta_1 X + \beta_2 Z + \varepsilon$$

In this equation, α represents the overall intercept. Each predictor has a "slope" (β) known as a **regression coefficient**. These coefficients represent the number of units the criterion variable is predicted to change for each unit change in a given predictor variable *when the effects of the other predictor variables are held constant*. For instance, a regression coefficient of .21 for high school grade point average would mean that, when all other predictor variables in the equation are held constant, college grade point average is predicted to increase by .21 unit for each unit increase in high school grade point average.

An index of the strength of the relationship between the criterion variable and the set of predictor variables is provided by the **squared multiple correlation coefficient,** symbolized R^2. The quantity R^2 is analogous to r^2 with one predictor variable and indicates the proportion of variability in the criterion variable that is associated with the predictor variables considered simultaneously.

Factor Analysis

The goal of **factor analysis** is to determine whether the correlations among a set of variables can be accounted for by one or more underlying dimensions, or **factors.** As an example of the use of factor analysis, suppose a study is conducted in which nine different beliefs about abortion are measured in a sample of participants. The investigator might hypothesize that individuals' responses to the belief statements reflect three underlying dimensions: (1) a concern with the physical effects of abortion, (2) a concern with the emotional effects of abortion, and (3) a concern with moral issues. Factor analysis tests hypotheses of this nature, albeit somewhat indirectly, by analyzing the patterning of correlations (or covariances) between all pairs of variables in the data set. If the investigator's hypothesis in the present study is correct, the correlations between the nine beliefs can be accounted for by the three hypothesized factors. If these three factors underlie the data, then the correlations between variables should have a predictable pattern. Factor analysis formally examines the patterning of correlations (or covariances) among variables and provides information on the type of *factor structure* (that is, the number and makeup of factors) that might underlie the data.

Log-Linear Analysis

Sometimes a research question requires that three or more qualitative variables be examined simultaneously. For instance, we might wish to extend the altruism experiment discussed in Chapter 15 by studying how willingness to donate money is influenced by one's gender in addition to the type of model one is exposed to. Since the three variables (gender, model status, and outcome) are all qualitative and between-subjects in nature, we could form a three-way contingency table of the observed frequencies. Although the chi-square test can be extended to multidimensional tables of this type, a statistical technique known as **log-linear analysis** is usually used instead. Though conceptually similar to chi-square analysis, log-linear analysis possesses statistical properties that make it more suitable for the simultaneous analysis of multiple between-subjects qualitative variables.

18.7   Alternative Approaches to Null Hypothesis Testing

This book has emphasized traditional statistical tests as applied in the behavioral sciences. The tests that we describe are firmly entrenched in the literature, and you will encounter them frequently in research journals. The approach we have used is frequently called "null hypothesis testing" or, more simply, "hypothesis testing."

Despite its widespread use, there are critics of the hypothesis testing approach. In an insightful article, Cohen (1994) articulated many of these criticisms (see also the set of rejoinders in the December 1995 issue of *American Psychologist*). He argues that what behavioral scientists typically want to know is "Given these data, what is the probability that the null hypothesis is true?", whereas hypothesis testing tells us "Given that the null hypothesis is true, what is the probability of these (or more extreme) data?" The two issues are quite different, and Cohen develops the implications of approaching science from the two perspectives. He essentially argues that the first of the three questions for the analysis of bivariate relationships (Is there a relationship between two variables?) is meaningless because (1) the null hypothesis of the means of interest being *exactly* the same or a correlation coefficient being *exactly* 0 is almost always false and (2) formal hypothesis testing procedures provide perspective on the probability of observing an empiricial outcome given that the null hypothesis is true rather than the more relevant question of whether the null hypothesis is true given the data. Cohen argues that the primary focus should be on the second and third questions (What are the strength and nature of a relationship?) and methods for adequately answering them, and he emphasizes the importance of effect sizes and developing adequate strategies for estimating effect sizes. We highly recommend Cohen's article and the rejoinders to it.

An approach that is quite distinct from null hypothesis testing is *Bayesian statistics*. This is a branch of statistics that emphasizes probability and *odds ratios*. The cornerstone of Bayesian statistics is Bayes' theorem, which we introduced in Section 6.5. The approach offers many advantages over traditional hypothesis testing frameworks, but it also has some drawbacks. Interested readers are referred to Howson and Urbach (1989). An additional approach, *interval estimation*, is more closely tied to null hypothesis testing but differs in important respects. This was discussed in Section 8.10 in the context of one-sample tests.

Summary

When one selects a test for analyzing a bivariate relationship, it is useful to distinguish among four situations: both variables qualitative (Case I), a qualitative independent variable and a quantitative dependent variable (Case II), a quantitative independent variable and a qualitative dependent variable (Case III), and both variables quantitative (Case IV). Given between-subjects variables, the appropriate method of analysis for Case I situations is the chi-square test. The selection of a statistical test in Case II situations involves deciding on the use of a parametric versus a nonparametric procedure, whether the independent variable is between-subjects or within-subjects in nature, and the number of levels that characterize the independent variable. Case III situations are analyzed using the chi-square test, logistic regression, or polychotomous logistic regression. Case IV situations most commonly utilize correlational procedures.

Statistical techniques that analyze the relationship between three or more variables are known as multivariate statistics. Among the most commonly used multivariate tests are variants of analysis of variance, including between-subjects, repeated measures, between–within, and multiple dependent variable procedures. Other multivariate tests include multiple regression, factor analysis, and log-linear analysis.

Multiple regression is an extension of regression with one predictor variable and allows for the prediction of a criterion variable from two or more predictor variables. Factor analysis tries to determine whether the correlations among a set of variables can be accounted for by one or more underlying dimensions (factors). Log-linear analysis is used to simultaneously analyze multiple between-subjects qualitative variables.

Exercises

Answers to asterisked (*) exercises appear at the back of the book.

*1. What is the appropriate statistical technique to use for analyzing the relationship between two between-subjects qualitative variables?

2. What are the three decision points for selecting a statistical test in situations where the independent variable is qualitative and the dependent variable is quantitative?

3. What factors should be considered in deciding whether to use a parametric or a nonparametric test in situations where the independent variable is qualitative and the dependent variable is quantitative?

*4. What methods are available for analyzing the relationship between two quantitative variables? Identify the conditions under which each of these is used.

Indicate whether each of the studies described in Exercises 5–22 represents a Case I, Case II, Case III, or Case IV situation, and why. For each study, indicate the appropriate statistical test for analyzing the relationship between the variables and state the reasons for your selection. If a parametric technique might be applicable, indicate which test you would use under conditions where (a) the underlying as-

sumptions of the parametric technique have been satisfied and (b) the underlying assumptions of the parametric technique have been violated to the extent that a nonparametric test is required.

*5. A health psychologist interested in whether changes in mood are associated with certain times of the year studied whether people tend to be more depressed in the winter or in the spring. One hundred individuals were administered a depression scale in December (winter) and again in May (spring). Scores on this scale can range from 0 to 50.

*6. A consumer psychologist examined the effect of the color of ice cream on taste ratings. Two hundred people tasted each of three "brands" of vanilla ice cream. The ice creams were actually identical to one another except for the shade of yellow used for coloring. The order in which individuals tasted each of the three "brands" was randomized. After tasting a given ice cream, participants rated the quality of its taste on a 1 to 10 scale.

*7. A researcher tested whether the noise level of music affects the growth of houseplants. Forty seeds were randomly assigned to one of two conditions. In one condition, plants were grown with a steady background of music playing at a low volume. In the other condition, plants were grown under identical conditions but with the music playing at a high volume. After 6 months, each plant's growth in inches was measured.

*8. A psychologist examined the relationship between parents' marital status and childhood imagination. For each of 100 first-graders, it was determined whether the parents were married, divorced, separated, widowed, or single and whether or not the child had an imaginary friend.

*9. An investigator tested the relationship between social class and how dogmatic individuals are. Dogmatism refers to closed-mindedness and the tendency to be inflexible in thought and intolerant of other viewpoints. Three hundred individuals were administered a dogmatism scale on which scores could range from 1 to 70. The social class of each individual was measured using an occupational index on which scores could range from 1 to 100.

*10. A social psychologist examined the relationship between the number of friends people have and how much social support they perceive. Forty individuals indicated the number of people they consider close friends and whether or not they feel these friends would, in general, "be there" for them in an emergency.

*11. A researcher studied the effect of social influence on drinking behavior. Sixty people were randomly assigned to one of three conditions where they were given a soft drink to consume while waiting for an experiment to supposedly begin. An experimental assistant, posing as another participant, was also given a drink. In one condition, the assistant sipped the drink at a faster rate than the research participant; in another condition, the assistant sipped the drink at the same rate as the participant; and in the last condition, the assistant sipped the drink at a

slower rate than the participant. The time it took each participant to consume his or her drink was measured.

*12. A psychologist tested whether hypnosis can influence responses on a biofeedback task. Fifty participants were randomly assigned to one of two conditions. In one condition, participants were hypnotized and then told to try to make one of their hands warmer by just thinking about it. The other group was given the same task but not placed under hypnosis. The change in hand temperature was measured in centigrade units to the nearest hundredth of a degree using a special temperature gauge.

*13. An investigator studied the relationship between the time between taking an exam and physiological arousal. Thirty individuals scheduled to take an exam on a Thursday were instructed to determine their scores on the Palmer Sweat Index (a measure of physiological arousal on which scores can range from 0 to 100) just before going to bed on Tuesday (2 days before the exam), Wednesday (the day before the exam), Thursday (the day of the exam), and Friday (the day after the exam).

14. A researcher examined the relationship between religious affiliation and belief in an afterlife. Fifteen Jews, 15 Catholics, and 15 Protestants were each asked whether or not they believe there is life after death.

15. In general, we feel weak in the morning when we awaken. A researcher studied whether people are actually weaker when they wake up compared with later in the day. Fifty people were instructed to squeeze a dynamometer (a device that measures grip strength) when they woke up in the morning and again 3 hours later. Dynamometer scores could range from 0 to 30.

16. An investigator examined the relationship between gender and marijuana use

in college students. Thirty male and 30 female college students were each asked whether or not they had ever used marijuana.

17. An investigator tested for race discrimination by loan officers at banks. Each of 120 loan officers was given background information on an applicant and asked how much money he or she would be willing to lend the individual. Forty of the loan officers were told that the applicant was African American, 40 were told she was white, and 40 were told she was Hispanic. Aside from this, the descriptions of the applicant were identical.

18. A sociologist studied the relationship between individuals' performance in college and their income 5 years later. Data were obtained on the grade point averages while in college of 250 people and their annual salaries after being out of college for 5 years.

19. A researcher tested the relationship between the number of children in a household and pet ownership. Fifty individuals indicated the number of children aged 18 or younger living at home and whether or not they have any pets.

20. A professor examined the relationship between how quickly students finish an exam and how well they perform. As students turned in their tests for a scheduled course examination, the professor kept track of the order in which they were received. The class was then divided into three groups: the first third of the class to turn in the exam, the middle third of the class to turn in the exam, and the last third of the class to turn in the exam. Scores on the exam could range from 0 to 100.

21. A consumer psychologist studied people's impressions of individuals who buy generic foods. Ninety participants were given a grocery list of a hypothetical shopper. For half of them, the list contained some generic brands, whereas for the other half, the list contained only national brands. All products on the two lists were otherwise identical. After reading the list, each participant rated the hypothetical shopper on a 1 to 100 scale in terms of how discriminating he was perceived to be in his food preferences.

22. An educational psychologist examined the relationship between psychology departments' national reputations and the number of scientific articles published by their faculties. A national ranking of the top 100 American psychology departments was obtained (where 1 represented the lowest-rated department and 100 represented the highest-rated department), and for each department the number of publications generated by its faculty was tabulated.

*23. What is the defining characteristic of multivariate statistics?

24. Differentiate among two-way between-subjects analysis of variance, two-way repeated measures analysis of variance, two-way between-within analysis of variance, multivariate analysis of variance, and the Hotelling T^2 test.

*25. What is the problem with conducting multiple analyses of variance when analyzing two or more dependent variables?

26. What is the rationale behind multiple regression?

*27. What is a regression coefficient? What is a squared multiple correlation coefficient?

*28. What is the goal of factor analysis? How is this accomplished?

29. Under what conditions is log-linear analysis appropriate?

30. What is a criticism of the hypothesis testing approach to data analysis?

Multiple-Choice Questions

31. Multiple regression is used to predict _____ variable(s) from _____ variable(s).
    a. a criterion; two or more predictor
    b. a predictor; two or more criterion

c. two or more criterion; a predictor

d. two or more predictor; a criterion

*32. Multivariate analysis of variance is *not* used when there

a. is only one independent variable

b. are two or more independent variables

c. is only one dependent variable

d. are two or more dependent variables

*33. The relationship between a quantitative independent variable and a qualitative dependent variable can be analyzed using logistic regression or polychotomous logistic regression.

a. true   b. false

34. The three questions regarding the existence of a relationship between two variables, the strength of the relationship, and the nature of the relationship are applicable to all of the bivariate procedures discussed in this book except for

a. the chi-square test

b. Friedman analysis of variance by ranks

c. regression

d. the independent groups *t* test

35. Given a qualitative independent variable and a quantitative dependent variable that has only two or three values, a parametric test is usually preferred over a nonparametric test.

a. true   b. false

*36. Which of the following is *not* a multivariate statistical test?

a. one-way repeated measures analysis of variance

b. two-way between-subjects analysis of variance

c. log-linear analysis

d. Hotelling T^2 test

*37. The approach to statistical analysis emphasized in this book is called

a. interval estimation

b. hypothesis testing

c. Bayesian statistics

d. none of the above

38. The relationship between a within-subjects qualitative variable and a between-subjects qualitative variable can be analyzed using the chi-square test.

a. true   b. false

*39. The statistics for analyzing the relationship between a qualitative independent variable and a quantitative dependent variable might be applicable to the analysis of two quantitative variables when the independent variable is within-subjects in nature or has fewer than five or so values associated with it.

a. true   b. false

40. In the linear model for multiple regression, α represents

a. the squared multiple correlation coefficient

b. a regression coefficient

c. the overall slope

d. the overall intercept

41. If a multivariate *F* test is statistically nonsignificant, follow-up analyses can be applied to each dependent variable separately.

a. true   b. false

Table of Random Numbers

The following table contains a listing of random numbers. Random number tables have many different uses. For example, suppose we had a list of 350 people in a population and wanted to select a random sample of 50 individuals. This could be accomplished using the following table by proceeding as follows: Arbitrarily number the 350 individuals from 1 to 350. Then enter the random number table at any point. Suppose we start in the upper left-hand corner with the digits 19612. Since we are concerned, at most, with a three-digit number, we examine only the leftmost three digits. (We could examine only the rightmost three digits or only the middle three digits instead if we wished.) This yields the number 196. Therefore, the individual who is number 196 is included in the sample. Next, move down one row. (We could move across one column if we wished.) The three-digit number is 391. Since there is no individual with this number, ignore it and continue moving. The next "valid" number is 035, meaning the individual who is numbered 35 is included in the sample. Continue this process until 50 individuals have been selected.

Reprinted from pages 3 and 5 of *A Million Random Digits with 100,000 Normal Deviates,* by The Rand Corporation. New York: The Free Press, 1955. Copyright ©1955 The Rand Corporation. Used by permission.

19612	78430	11661	94770	77603	65669	86868	12665	30012	75989
39141	77400	28000	64238	73258	71794	31340	26256	66453	37016
64756	80457	08747	12836	03469	50678	03274	43423	66677	82556
92901	51878	56441	22998	29718	38447	06453	25311	07565	53771
03551	90070	09483	94050	45938	18135	36908	43321	11073	51803
98884	66209	06830	53656	14663	56346	71430	04909	19818	05707
27369	86882	53473	07541	53633	70863	03748	12822	19360	49088
59066	75974	63335	20483	43514	37481	58278	26967	49325	43951
91647	93783	64169	49022	98588	09495	49829	59068	38831	04838
83605	92419	39542	07772	71568	75673	35185	89759	44901	74291
24895	88530	70774	35439	46758	70472	70207	92675	91623	61275
35720	26556	95596	20094	73750	85788	34264	01703	46833	65248
14141	53410	38649	06343	57256	61342	72709	75318	90379	37562
27416	75670	92176	72535	93119	56077	06886	18244	92344	31374
82071	07429	81007	47749	40744	56974	23336	88821	53841	10536
21445	82793	24831	93241	14199	76268	70883	68002	03829	17443
72513	76400	52225	92348	62308	98481	29744	33165	33141	61020
71479	45027	76160	57411	13780	13632	52308	77762	88874	33697
83210	51466	09088	50395	26743	05306	21706	70001	99439	80767
68749	95148	94897	78636	96750	09024	94538	91143	96693	61866
05184	75763	47075	88158	05313	53439	14908	08830	60096	21551
13651	62546	96892	25240	47511	58483	87342	78818	07855	39269
00566	21220	00292	24069	25072	29519	52548	54091	21282	21296
50958	17695	58072	68990	60329	95955	71586	63417	35947	67807
57621	64547	46850	37981	38527	09037	64756	03324	04986	83666
09282	25844	79139	78435	35428	43561	69799	63314	12991	93516
23394	94206	93432	37836	94919	26846	02555	74410	94915	48199
05280	37470	93622	04345	15092	19510	18094	16613	78234	50001
95491	97976	38306	32192	82639	54624	72434	92606	23191	74693
78521	00104	18248	75583	90326	50785	54034	66251	35774	14692
96345	44579	85932	44053	75704	20840	86583	83944	52456	73766
77963	31151	32364	91691	47357	40338	23435	24065	08458	95366
07520	11294	23238	01748	41690	67328	54814	37777	10057	42332
38423	02309	70703	85736	46148	14258	29236	12152	05088	65825
02463	65533	21199	60555	33928	01817	07396	89215	30722	22102
15880	92261	17292	88190	61781	48898	92525	21283	88581	60098
71926	00819	59144	00224	30570	90194	18329	06999	26857	19238
64425	28108	16554	16016	00042	83229	10333	36168	65617	94834
79782	23924	49440	30432	81077	31543	95216	64865	13658	51081
35337	74538	44553	64672	90960	41849	93865	44608	93176	34851
05249	29329	19715	94082	14738	86667	43708	66354	93692	25527
56463	99380	38793	85774	19056	13939	46062	27647	66146	63210
96296	33121	54196	34108	75814	85986	71171	15102	28992	63165
98380	36269	60014	07201	62448	46385	42175	88350	46182	49126
52567	64350	16315	53969	80395	81114	54358	64578	47269	15747
78498	90830	25955	99236	43286	91064	99969	95144	64424	77377
49553	24241	08150	89535	08703	91041	77323	81079	45127	93686
32151	07075	83155	10252	73100	88618	23891	87418	45417	20268
11314	50363	26860	27799	49416	83534	19187	08059	76677	02110
12364	71210	87052	50241	90785	97889	81399	58130	64439	05614

03991	10461	93716	16894	66083	24653	84609	58232	88618	19161
38555	95554	32886	59780	08355	60860	29735	47762	71299	23853
17546	73704	92052	46215	55121	29281	59076	07936	27954	58909
32643	52861	95819	06831	00911	98936	76355	93779	80863	00514
69572	68777	39510	35905	14060	40619	29549	69616	33564	60780
24122	66591	27699	06494	14845	46672	61958	77100	90899	75754
61196	30231	92962	61773	41839	55382	17267	70943	78038	70267
30532	21704	10274	12202	39685	23309	10061	68829	55986	66485
03788	97599	75867	20717	74416	53166	35208	33374	87539	08823
48228	63379	85783	47619	53152	67433	35663	52972	16818	60311
60365	94653	35075	33949	42614	29297	01918	28316	98953	73231
83799	42402	56623	34442	34994	41374	70071	14736	09958	18065
32960	07405	36409	83232	99385	41600	11133	07586	15917	06253
19322	53845	57620	52606	66497	68646	78138	66559	19640	99413
11220	94747	07399	37408	48509	23929	27482	45476	85244	35159
31751	57260	68980	05339	15470	48355	88651	22596	03152	19121
88492	99382	14454	04504	20094	98977	74843	93413	22109	78508
30934	47744	07481	83828	73788	06533	28597	20405	94205	20380
22888	48893	27499	98748	60530	45128	74022	84617	82037	10268
78212	16993	35902	91386	44372	15486	65741	14014	87481	37220
41849	84547	46850	52326	34677	58300	74910	64345	19325	81549
46352	33049	69248	93460	45305	07521	61318	31855	14413	70951
11087	96294	14013	31792	59747	67277	76503	34513	39663	77544
52701	08337	56303	87315	16520	69676	11654	99893	02181	68161
57275	36898	81304	48585	68652	27376	92852	55866	88448	03584
20857	73156	70284	24326	79375	95220	01159	63267	10622	48391
15633	84924	90415	93614	33521	26665	55823	47641	86225	31704
92694	48297	39904	02115	59589	49067	66821	41575	49767	04037
77613	19019	88152	00080	20554	91409	96277	48257	50816	97616
38688	32486	45134	63545	59404	72059	43947	51680	43852	59693
25163	01889	70014	15021	41290	67312	71857	15957	68971	11403
65251	07629	37239	33295	05870	01119	92784	26340	18477	65622
36815	43625	18637	37509	82444	99005	04921	73701	14707	93997
64397	11692	05327	82162	20247	81759	45197	25332	83745	22567
04515	25624	95096	67946	48460	85558	15191	18782	16930	33361
83761	60873	43253	84145	60833	25983	01291	41349	20368	07126
14387	06345	80854	09279	43529	06318	38384	74761	41196	37480
51321	92246	80088	77074	88722	56736	66164	49431	66919	31678
72472	00008	80890	18002	94813	31900	54155	83436	35352	54131
05466	55306	93128	18464	74457	90561	72848	11834	79982	68416
39528	72484	82474	25593	48545	35247	18619	13674	18611	19241
81616	18711	53342	44276	75122	11724	74627	73707	58319	15997
07586	16120	82641	22820	92904	13141	32392	19763	61199	67940
90767	04235	13574	17200	69902	63742	78464	22501	18627	90872
40188	28193	29593	88627	94972	11598	62095	36787	00441	58997
34414	82157	86887	55087	19152	00023	12302	80783	32624	68691
63439	75363	44989	16822	36024	00867	76378	41605	65961	73488
67049	09070	93399	45547	94458	74284	05041	49807	20288	34060
79495	04146	52162	90286	54158	34243	46978	35482	59362	95938
91704	30552	04737	21031	75051	93029	47665	64382	99782	93478

Proportions of Scores in a Normal Distribution

The following table reports proportions of scores in a normal distribution that occur within selected ranges of z. A given proportion represents the probability of obtaining scores within the range of interest. Column 1 lists values of z. Column 2 indicates the probability of observing z scores greater than or equal to $-z$ and less than or equal to $+z$. Column 3 indicates the probability of obtaining z scores greater than or equal to $+z$. This column can also be used to determine the probability of obtaining z scores less than or equal to $-z$. Column 4 indicates the probability of obtaining z scores less than or equal to $-z$ or greater than or equal to $+z$. Column 5 indicates the probability of obtaining z scores between 0 and the z score of interest.

As an example, consider the z score 1.96. In column 2, we see that the probability of obtaining z scores between -1.96 and $+1.96$ is .9500. In column 3, we see that the probability of obtaining z scores greater than or equal to $+1.96$ is .0250. Since the normal distribution is symmetrical, this column also indicates the probability of obtaining z scores less than or equal to -1.96. This probability is also .0250. In column 4, we see that the probability of obtaining z scores less than or equal to -1.96 or greater than or equal to $+1.96$ is .0500. In column 5, we see that the probability of obtaining z scores between 0 and $+1.96$ is .4750. Again, because the normal distribution is symmetrical, this also represents the probability of obtaining z scores between 0 and -1.96.

Adapted from *Fundamental Statistics for Psychology*, by R. B. McCall. Copyright ©1970 Harcourt Brace Jovanovich. Adapted with permission.

Column 1	Column 2	Column 3	Column 4	Column 5
		+z		0 +z
	−z +z	−z	−z +z	−z 0
				≥ 0 and $\leq +z$ (also use to find $\geq -z$ and ≤ 0)
z	$\geq -z$ and $\leq +z$	$\geq +z$ (also use to find $\leq -z$)	$\leq -z$ or $\geq +z$	
.00	.0000	.5000	1.0000	.0000
.01	.0080	.4960	.9920	.0040
.02	.0160	.4920	.9840	.0080
.03	.0240	.4880	.9760	.0120
.04	.0320	.4840	.9680	.0160
.05	.0398	.4801	.9602	.0199
.06	.0478	.4761	.9522	.0239
.07	.0558	.4721	.9442	.0279
.08	.0638	.4681	.9362	.0319
.09	.0718	.4641	.9282	.0359
.10	.0796	.4602	.9204	.0398
.11	.0876	.4562	.9124	.0438
.12	.0956	.4522	.9044	.0478
.13	.1034	.4483	.8966	.0517
.14	.1114	.4443	.8886	.0557
.15	.1192	.4404	.8808	.0596
.16	.1272	.4364	.8728	.0636
.17	.1350	.4325	.8650	.0675
.18	.1428	.4286	.8572	.0714
.19	.1506	.4247	.8494	.0753
.20	.1586	.4207	.8414	.0793
.21	.1664	.4168	.8336	.0832
.22	.1742	.4129	.8258	.0871
.23	.1820	.4090	.8180	.0910
.24	.1896	.4052	.8104	.0948
.25	.1974	.4013	.8026	.0987
.26	.2052	.3974	.7948	.1026
.27	.2128	.3936	.7872	.1064
.28	.2206	.3897	.7794	.1103
.29	.2282	.3859	.7718	.1141
.30	.2358	.3821	.7642	.1179
.31	.2434	.3783	.7566	.1217
.32	.2510	.3745	.7490	.1255
.33	.2586	.3707	.7414	.1293
.34	.2662	.3669	.7338	.1331
.35	.2736	.3632	.7264	.1368
.36	.2812	.3594	.7188	.1406
.37	.2886	.3557	.7114	.1443
.38	.2960	.3520	.7040	.1480

Column 1	Column 2	Column 3	Column 4	Column 5
				≥ 0 and $\leq +z$ (also use to find $\geq -z$ and ≤ 0)
z	$\geq -z$ and $\leq +z$	$\geq +z$ (also use to find $\leq -z$)	$\leq -z$ or $\geq +z$	
.39	.3034	.3483	.6966	.1517
.40	.3108	.3446	.6892	.1554
.41	.3182	.3409	.6818	.1591
.42	.3256	.3372	.6744	.1628
.43	.3328	.3336	.6672	.1664
.44	.3400	.3300	.6600	.1700
.45	.3472	.3264	.6528	.1736
.46	.3544	.3228	.6456	.1772
.47	.3616	.3192	.6384	.1808
.48	.3688	.3156	.6312	.1844
.49	.3758	.3121	.6242	.1879
.50	.3830	.3085	.6170	.1915
.51	.3900	.3050	.6100	.1950
.52	.3970	.3015	.6030	.1985
.53	.4038	.2981	.5962	.2019
.54	.4108	.2946	.5892	.2054
.55	.4176	.2912	.5824	.2088
.56	.4246	.2877	.5754	.2123
.57	.4314	.2843	.5686	.2157
.58	.4380	.2810	.5620	.2190
.59	.4448	.2776	.5552	.2224
.60	.4514	.2743	.5486	.2257
.61	.4582	.2709	.5418	.2291
.62	.4648	.2676	.5352	.2324
.63	.4714	.2643	.5286	.2357
.64	.4778	.2611	.5222	.2389
.65	.4844	.2578	.5156	.2422
.66	.4908	.2546	.5092	.2454
.67	.4972	.2514	.5028	.2486
.68	.5034	.2483	.4966	.2517
.69	.5098	.2451	.4902	.2549
.70	.5160	.2420	.4840	.2580
.71	.5222	.2389	.4778	.2611
.72	.5284	.2358	.4716	.2642
.73	.5346	.2327	.4654	.2673
.74	.5408	.2296	.4592	.2704
.75	.5468	.2266	.4532	.2734
.76	.5528	.2236	.4472	.2764
.77	.5588	.2206	.4412	.2794

Column 1	Column 2	Column 3	Column 4	Column 5
		+z		0 +z
	−z +z	−z	−z +z	−z 0
z	$\geq -z$ and $\leq +z$	$\geq +z$ (also use to find $\leq -z$)	$\leq -z$ or $\geq +z$	≥ 0 and $\leq +z$ (also use to find $\geq -z$ and ≤ 0)
.78	.5646	.2177	.4354	.2823
.79	.5704	.2148	.4296	.2852
.80	.5762	.2119	.4238	.2881
.81	.5820	.2090	.4180	.2910
.82	.5878	.2061	.4132	.2939
.83	.5934	.2033	.4066	.2967
.84	.5990	.2005	.4010	.2995
.85	.6046	.1977	.3954	.3023
.86	.6102	.1949	.3898	.3051
.87	.6156	.1922	.3844	.3078
.88	.6212	.1894	.3788	.3106
.89	.6266	.1867	.3734	.3133
.90	.6318	.1841	.3682	.3159
.91	.6372	.1814	.3628	.3186
.92	.6424	.1788	.3576	.3212
.93	.6476	.1762	.3524	.3238
.94	.6528	.1736	.3472	.3264
.95	.6578	.1711	.3422	.3289
.96	.6630	.1685	.3370	.3315
.97	.6680	.1660	.3320	.3340
.98	.6730	.1635	.3270	.3365
.99	.6778	.1611	.3222	.3389
1.00	.6826	.1587	.3174	.3413
1.01	.6876	.1562	.3124	.3438
1.02	.6922	.1539	.3078	.3461
1.03	.6970	.1515	.3030	.3485
1.04	.7016	.1492	.2984	.3508
1.05	.7062	.1469	.2938	.3531
1.06	.7108	.1446	.2892	.3554
1.07	.7154	.1423	.2846	.3577
1.08	.7198	.1401	.2802	.3599
1.09	.7242	.1379	.2758	.3621
1.10	.7286	.1357	.2714	.3643
1.11	.7330	.1335	.2670	.3665
1.12	.7372	.1314	.2628	.3686
1.13	.7416	.1292	.2584	.3708
1.14	.7458	.1271	.2542	.3729
1.15	.7498	.1251	.2502	.3749
1.16	.7540	.1230	.2460	.3770

Column 1	Column 2	Column 3	Column 4	Column 5
		+z		0 +z
	−z +z	−z	−z +z	−z 0
z	≥ −z and ≤ +z	≥ +z (also use to find ≤ −z)	≤ −z or ≥ +z	≥ 0 and ≤ +z (also use to find ≥ −z and ≤ 0)
1.17	.7580	.1210	.2420	.3790
1.18	.7620	.1190	.2380	.3810
1.19	.7660	.1170	.2340	.3830
1.20	.7698	.1151	.2302	.3849
1.21	.7738	.1131	.2262	.3869
1.22	.7776	.1112	.2224	.3888
1.23	.7814	.1093	.2186	.3907
1.24	.7850	.1075	.2150	.3925
1.25	.7888	.1056	.2112	.3944
1.26	.7924	.1038	.2076	.3962
1.27	.7960	.1020	.2040	.3980
1.28	.7994	.1003	.2006	.3997
1.29	.8030	.0985	.1970	.4015
1.30	.8064	.0968	.1936	.4032
1.31	.8098	.0951	.1902	.4049
1.32	.8132	.0934	.1868	.4066
1.33	.8164	.0918	.1836	.4082
1.34	.8198	.0901	.1802	.4099
1.35	.8230	.0885	.1770	.4115
1.36	.8262	.0869	.1738	.4131
1.37	.8294	.0853	.1706	.4147
1.38	.8324	.0838	.1676	.4162
1.39	.8354	.0823	.1646	.4177
1.40	.8384	.0808	.1616	.4192
1.41	.8414	.0793	.1586	.4207
1.42	.8444	.0778	.1556	.4222
1.43	.8472	.0764	.1528	.4236
1.44	.8502	.0749	.1498	.4251
1.45	.8530	.0735	.1470	.4265
1.46	.8558	.0721	.1442	.4279
1.47	.8584	.0708	.1416	.4292
1.48	.8612	.0694	.1388	.4306
1.49	.8638	.0681	.1362	.4319
1.50	.8664	.0668	.1336	.4332
1.51	.8690	.0655	.1310	.4345
1.52	.8714	.0643	.1286	.4357
1.53	.8740	.0630	.1260	.4370
1.54	.8764	.0618	.1236	.4382
1.55	.8788	.0606	.1212	.4394

Column 1	Column 2	Column 3	Column 4	Column 5
		+z		0 +z
	−z +z	−z	−z +z	−z 0
z	≥ −z and ≤ +z	≥ +z (also use to find ≤ −z)	≤ −z or ≥ +z	≥ 0 and ≤ +z (also use to find ≥ −z and ≤ 0)
1.56	.8812	.0594	.1188	.4406
1.57	.8836	.0582	.1164	.4418
1.58	.8858	.0571	.1142	.4429
1.59	.8882	.0559	.1118	.4441
1.60	.8904	.0548	.1096	.4452
1.61	.8926	.0537	.1074	.4463
1.62	.8948	.0526	.1052	.4474
1.63	.8968	.0516	.1032	.4484
1.64	.8990	.0505	.1010	.4495
1.65	.9010	.0495	.0990	.4505
1.66	.9030	.0485	.0970	.4515
1.67	.9050	.0475	.0950	.4525
1.68	.9070	.0465	.0930	.4535
1.69	.9090	.0455	.0910	.4545
1.70	.9108	.0446	.0892	.4554
1.71	.9128	.0436	.0872	.4564
1.72	.9146	.0427	.0854	.4573
1.73	.9164	.0418	.0836	.4582
1.74	.9182	.0409	.0818	.4591
1.75	.9198	.0401	.0802	.4599
1.76	.9216	.0392	.0784	.4608
1.77	.9232	.0384	.0764	.4616
1.78	.9250	.0375	.0750	.4625
1.79	.9266	.0367	.0734	.4633
1.80	.9282	.0359	.0718	.4641
1.81	.9298	.0351	.0702	.4649
1.82	.9312	.0344	.0688	.4656
1.83	.9328	.0336	.0672	.4664
1.84	.9342	.0329	.0658	.4671
1.85	.9356	.0322	.0644	.4678
1.86	.9372	.0314	.0628	.4686
1.87	.9386	.0307	.0614	.4693
1.88	.9398	.0301	.0602	.4699
1.89	.9412	.0294	.0588	.4706
1.90	.9426	.0287	.0574	.4713
1.91	.9438	.0281	.0562	.4719
1.92	.9452	.0274	.0548	.4726
1.93	.9464	.0268	.0536	.4732
1.94	.9476	.0262	.0524	.4738

z	Column 2 ≥ −z and ≤ +z	Column 3 ≥ +z (also use to find ≤ −z)	Column 4 ≤ −z or ≥ +z	Column 5 ≥ 0 and ≤ +z (also use to find ≥ −z and ≤ 0)
1.95	.9488	.0256	.0512	.4744
1.96	.9500	.0250	.0500	.4750
1.97	.9512	.0244	.0488	.4756
1.98	.9522	.0239	.0478	.4761
1.99	.9534	.0233	.0466	.4767
2.00	.9544	.0228	.0456	.4772
2.01	.9556	.0222	.0444	.4778
2.02	.9566	.0217	.0434	.4783
2.03	.9576	.0212	.0424	.4788
2.04	.9586	.0207	.0414	.4793
2.05	.9596	.0202	.0404	.4798
2.06	.9606	.0197	.0394	.4803
2.07	.9616	.0192	.0384	.4808
2.08	.9624	.0188	.0376	.4812
2.09	.9634	.0183	.0366	.4817
2.10	.9642	.0179	.0358	.4821
2.11	.9652	.0174	.0348	.4826
2.12	.9660	.0170	.0340	.4830
2.13	.9668	.0166	.0332	.4834
2.14	.9676	.0162	.0324	.4838
2.15	.9684	.0158	.0316	.4842
2.16	.9692	.0154	.0308	.4846
2.17	.9700	.0150	.0300	.4850
2.18	.9708	.0146	.0292	.4854
2.19	.9714	.0143	.0286	.4857
2.20	.9722	.0139	.0278	.4861
2.21	.9728	.0136	.0272	.4864
2.22	.9736	.0132	.0264	.4868
2.23	.9742	.0129	.0258	.4871
2.24	.9750	.0125	.0250	.4875
2.25	.9756	.0122	.0244	.4878
2.26	.9762	.0119	.0238	.4881
2.27	.9768	.0116	.0232	.4884
2.28	.9774	.0113	.0226	.4887
2.29	.9780	.0110	.0220	.4890
2.30	.9786	.0107	.0214	.4893
2.31	.9792	.0104	.0208	.4896
2.32	.9796	.0102	.0204	.4898
2.33	.9802	.0099	.0198	.4901

Column 1	Column 2	Column 3	Column 4	Column 5
		+z		0 +z
	−z +z	−z	−z +z	−z 0
				≥ 0 and ≤ +z (also use to find
		≥ +z (also use		≥ −z and ≤ 0)
z	≥ −z and ≤ +z	to find ≤ −z)	≤ −z or ≥ +z	
2.34	.9808	.0096	.0192	.4904
2.35	.9812	.0094	.0188	.4906
2.36	.9818	.0091	.0182	.4909
2.37	.9822	.0089	.0178	.4911
2.38	.9826	.0087	.0174	.4913
2.39	.9832	.0084	.0168	.4916
2.40	.9836	.0082	.0164	.4918
2.41	.9840	.0080	.0160	.4920
2.42	.9844	.0078	.0156	.4922
2.43	.9850	.0075	.0150	.4925
2.44	.9854	.0073	.0146	.4927
2.45	.9858	.0071	.0142	.4929
2.46	.9862	.0069	.0138	.4931
2.47	.9864	.0068	.0136	.4932
2.48	.9868	.0066	.0132	.4934
2.49	.9872	.0064	.0128	.4936
2.50	.9876	.0062	.0124	.4938
2.51	.9880	.0060	.0120	.4940
2.52	.9882	.0059	.0118	.4941
2.53	.9886	.0057	.0114	.4943
2.54	.9890	.0055	.0110	.4945
2.55	.9892	.0054	.0108	.4946
2.56	.9896	.0052	.0104	.4948
2.57	.9898	.0051	.0102	.4949
2.58	.9902	.0049	.0098	.4951
2.59	.9904	.0048	.0096	.4952
2.60	.9906	.0047	.0094	.4953
2.61	.9910	.0045	.0090	.4955
2.62	.9912	.0044	.0088	.4956
2.63	.9914	.0043	.0086	.4957
2.64	.9918	.0041	.0082	.4959
2.65	.9920	.0040	.0080	.4960
2.66	.9922	.0039	.0078	.4961
2.67	.9924	.0038	.0076	.4962
2.68	.9926	.0037	.0074	.4963
2.69	.9928	.0036	.0072	.4964
2.70	.9930	.0035	.0070	.4965
2.71	.9932	.0034	.0068	.4966
2.72	.9934	.0033	.0066	.4967

Column 1	Column 2	Column 3	Column 4	Column 5
		+z		0 +z
	−z +z	−z	−z +z	−z 0
z	$\geq -z$ and $\leq +z$	$\geq +z$ (also use to find $\leq -z$)	$\leq -z$ or $\geq +z$	≥ 0 and $\leq +z$ (also use to find $\geq -z$ and ≤ 0)
2.73	.9936	.0032	.0064	.4968
2.74	.9938	.0031	.0062	.4969
2.75	.9940	.0030	.0060	.4970
2.76	.9942	.0029	.0058	.4971
2.77	.9944	.0028	.0056	.4972
2.78	.9946	.0027	.0054	.4973
2.79	.9948	.0026	.0052	.4974
2.80	.9948	.0026	.0052	.4974
2.81	.9950	.0025	.0050	.4975
2.82	.9952	.0024	.0048	.4976
2.83	.9954	.0023	.0046	.4977
2.84	.9954	.0023	.0046	.4977
2.85	.9956	.0022	.0044	.4978
2.86	.9958	.0021	.0042	.4979
2.87	.9958	.0021	.0042	.4979
2.88	.9960	.0020	.0040	.4980
2.89	.9962	.0019	.0038	.4981
2.90	.9962	.0019	.0038	.4981
2.91	.9964	.0018	.0036	.4982
2.92	.9964	.0018	.0036	.4982
2.93	.9966	.0017	.0034	.4983
2.94	.9968	.0016	.0032	.4984
2.95	.9968	.0016	.0032	.4984
2.96	.9970	.0015	.0030	.4985
2.97	.9970	.0015	.0030	.4985
2.98	.9972	.0014	.0028	.4986
2.99	.9972	.0014	.0028	.4986
3.00	.9974	.0013	.0026	.4987
3.01	.9974	.0013	.0026	.4987
3.02	.9974	.0013	.0026	.4987
3.03	.9976	.0012	.0024	.4988
3.04	.9976	.0012	.0024	.4988
3.05	.9978	.0011	.0022	.4989
3.06	.9978	.0011	.0022	.4989
3.07	.9978	.0011	.0022	.4989
3.08	.9980	.0010	.0020	.4990
3.09	.9980	.0010	.0020	.4990
3.10	.9980	.0010	.0020	.4990
3.11	.9982	.0009	.0018	.4991

Column 1	Column 2	Column 3	Column 4	Column 5
z	$\geq -z$ and $\leq +z$	$\geq +z$ (also use to find $\leq -z$)	$\leq -z$ or $\geq +z$	≥ 0 and $\leq +z$ (also use to find $\geq -z$ and ≤ 0)
3.12	.9982	.0009	.0018	.4991
3.13	.9982	.0009	.0018	.4991
3.14	.9984	.0008	.0016	.4992
3.15	.9984	.0008	.0016	.4992
3.16	.9984	.0008	.0016	.4992
3.17	.9984	.0008	.0016	.4992
3.18	.9986	.0007	.0014	.4993
3.19	.9986	.0007	.0014	.4993
3.20	.9986	.0007	.0014	.4993
3.21	.9986	.0007	.0014	.4993
3.22	.9988	.0006	.0012	.4994
3.23	.9988	.0006	.0012	.4994
3.24	.9988	.0006	.0012	.4994
3.25	.9988	.0006	.0012	.4994
3.30	.9990	.0005	.0010	.4995
3.35	.9992	.0004	.0008	.4996
3.40	.9994	.0003	.0006	.4997
3.45	.9994	.0003	.0006	.4997
3.50	.9996	.0002	.0004	.4998
3.60	.9996	.0002	.0004	.4998
3.70	.9998	.0001	.0002	.4999
3.80	.9998	.0001	.0002	.4999
3.90	.9999	.00005	.00010	.49995
4.00	.99994	.00003	.00006	.49997

Factorials

The following table presents factorials for the numbers 0 through 20. For example, the value of 7! is 5,040.

Number	Factorial
0	1
1	1
2	2
3	6
4	24
5	120
6	720
7	5,040
8	40,320
9	362,880
10	3,628,800
11	39,916,800
12	479,001,600
13	6,227,020,800
14	87,178,291,200
15	1,307,674,368,000
16	20,922,789,888,000
17	355,687,428,096,000
18	6,402,373,705,728,000
19	121,645,100,408,832,000
20	2,432,902,008,176,640,000

Critical Values for the *t* Distribution

The following table presents critical values of *t* for directional (one-tailed) and nondirectional (two-tailed) tests. The column headed "df" lists the degrees of freedom associated with the *t* distribution of interest. The entries in the table are the *t* values that define the rejection regions. The values .10, .05, .025, .01, .005, and .0005 at the top of the table under the heading "Level of Significance for Directional Test" represent alpha levels for directional tests. For instance, if we were conducting a directional test having 15 degrees of freedom at an alpha level of .05, we would locate 15 in the df column and follow across that row until it intersected the .05 column for a directional test. The entry of 1.753 is the critical value of *t* for a directional test using the upper tail of the *t* distribution. Since the *t* distribution is symmetrical, the critical value of *t* for a directional test using the lower tail of the *t* distribution is −1.753.

The values .20, .10, .05, .02, .01, and .001 under the heading "Level of Significance for Nondirectional Test" represent alpha levels for nondirectional tests. For instance, if we were conducting a nondirectional test having 15 degrees of freedom at an alpha level of .05, we would locate 15 in the df column and follow across that row until it intersected the .05 column for a nondirectional test. The entry of 2.131 signifies that the positive critical value of *t* is +2.131 and the negative critical value of *t* is −2.131. The rejection region thus consists of all values of *t* less than −2.131 or greater than +2.131.

From Table III in R. A. Fisher and F. Yates, *Statistical Tables for Biological, Agricultural, and Medical Research*, Sixth Edition, published by Addison Wesley Longman Ltd. (1974). Reprinted with permission.

	Level of Significance for Directional Test					
	.10	.05	.025	.01	.005	.0005
	Level of Significance for Nondirectional Test					
df	.20	.10	.05	.02	.01	.001
1	3.078	6.314	12.706	31.821	63.657	636.619
2	1.886	2.920	4.303	6.965	9.925	31.598
3	1.638	2.353	3.182	4.541	5.841	12.941
4	1.533	2.132	2.776	3.747	4.604	8.610
5	1.476	2.015	2.571	3.365	4.032	6.859
6	1.440	1.943	2.447	3.143	3.707	5.959
7	1.415	1.895	2.365	2.998	3.499	5.405
8	1.397	1.860	2.306	2.896	3.355	5.041
9	1.383	1.833	2.262	2.821	3.250	4.781
10	1.372	1.812	2.228	2.764	3.169	4.587
11	1.363	1.796	2.201	2.718	3.106	4.437
12	1.356	1.782	2.179	2.681	3.055	4.318
13	1.350	1.771	2.160	2.650	3.012	4.221
14	1.345	1.761	2.145	2.624	2.977	4.140
15	1.341	1.753	2.131	2.602	2.947	4.073
16	1.337	1.746	2.120	2.583	2.921	4.015
17	1.333	1.740	2.110	2.567	2.898	3.965
18	1.330	1.734	2.101	2.552	2.878	3.922
19	1.328	1.729	2.093	2.539	2.861	3.883
20	1.325	1.725	2.086	2.528	2.845	3.850
21	1.323	1.721	2.080	2.518	2.831	3.819
22	1.321	1.717	2.074	2.508	2.819	3.792
23	1.319	1.714	2.069	2.500	2.807	3.767
24	1.318	1.711	2.064	2.492	2.797	3.745
25	1.316	1.708	2.060	2.485	2.787	3.725
26	1.315	1.706	2.056	2.479	2.779	3.707
27	1.314	1.703	2.052	2.473	2.771	3.690
28	1.313	1.701	2.048	2.467	2.763	3.674
29	1.311	1.699	2.045	2.462	2.756	3.659
30	1.310	1.697	2.042	2.457	2.750	3.646
40	1.303	1.684	2.021	2.423	2.704	3.551
60	1.296	1.671	2.000	2.390	2.660	3.460
120	1.289	1.658	1.980	2.358	2.617	3.373
∞	1.282	1.645	1.960	2.326	2.576	3.291

Power and Sample Size

The following sets of tables indicate the sample sizes necessary to achieve desired levels of power for the statistical techniques listed below. The procedure for using the tables for a given technique is described in the relevant chapter in the text.

Technique	Page
Independent or Correlated Groups t Test	564
One- or Two-Way Between-Subjects Analysis of Variance	566
One-Way Repeated Measures Analysis of Variance	576
Pearson Correlation	581
Chi-Square Test	583

APPENDIX E.1: Independent or Correlated Groups t Test

Directional Test, Alpha = .05

Power	.01	.03	.05	.07	.10	.15	.20	.25	.30	.35	.40	.45	.50	.55	.60	.65	.70	.75	.80
								POPULATION ETA-SQUARED											
.25	48	16	10	7	5	3	3	2	2	2	—	—	—	—	—	—	—	—	—
.50	136	45	26	19	13	8	6	5	4	3	3	2	2	2	2	—	—	—	—
.60	181	59	35	25	17	11	8	6	5	4	3	3	3	2	2	2	—	—	—
.67	216	70	42	29	20	13	9	7	6	5	4	3	3	3	2	2	2	—	—
.70	236	77	45	32	22	14	10	8	6	5	4	4	3	3	2	2	2	2	—
.75	270	88	52	36	25	16	11	9	7	6	5	4	3	3	3	2	2	2	—
.80	310	101	59	42	29	18	13	10	8	6	5	4	4	4	3	2	2	2	2
.85	360	117	69	48	33	21	15	11	9	7	6	5	4	4	3	3	2	2	2
.90	429	139	82	58	39	25	18	14	11	9	7	6	5	5	4	4	3	2	2
.95	542	176	104	73	49	31	22	17	13	11	9	7	6	5	4	4	3	3	2
.99	789	256	151	105	72	45	32	24	19	15	13	10	9	7	6	5	4	3	3

Directional Test, Alpha = .01

Power	.01	.03	.05	.07	.10	.15	.20	.25	.30	.35	.40	.45	.50	.55	.60	.65	.70	.75	.80
								POPULATION ETA-SQUARED											
.25	138	46	27	20	14	9	7	6	5	4	4	3	3	3	2	2	2	2	2
.50	272	89	53	37	26	17	12	10	8	7	6	5	4	4	3	3	3	2	2
.60	334	109	65	46	31	20	15	11	9	8	6	6	5	4	4	3	3	3	2
.67	382	127	75	53	36	23	17	13	11	9	7	6	5	5	4	4	3	3	2
.70	408	133	79	56	38	25	18	14	11	9	8	6	6	5	4	4	3	3	3
.75	452	147	87	61	42	27	20	15	12	10	8	7	6	5	5	4	3	3	3
.80	503	164	97	68	47	30	22	17	13	11	9	8	7	6	5	4	4	3	3
.85	567	184	109	77	52	34	24	18	15	12	10	8	7	6	5	5	4	4	3
.90	652	212	125	88	60	38	27	21	17	14	11	9	8	7	6	5	4	4	3
.95	790	257	151	106	72	46	33	25	20	16	13	11	9	8	7	6	5	4	3
.99	1,084	352	207	145	99	63	45	34	27	22	18	15	12	10	9	7	6	5	4

Nondirectional Test, Alpha = .05

Power							POPULATION ETA-SQUARED												
	.01	.03	.05	.07	.10	.15	.20	.25	.30	.35	.40	.45	.50	.55	.60	.65	.70	.75	.80
.25	84	28	17	12	8	6	5	3	3	3	2	2	2	2	2	—	—	—	—
.50	193	63	38	27	18	12	9	7	5	5	4	3	3	3	2	2	2	—	—
.60	246	80	48	34	23	15	11	8	7	6	5	4	3	3	3	2	2	2	2
.67	287	93	55	39	27	17	12	10	8	6	5	4	4	3	3	3	2	2	2
.70	310	101	60	42	29	18	13	10	8	7	6	5	4	4	3	3	2	2	2
.75	348	113	67	47	32	21	15	11	9	7	6	5	4	4	3	3	2	2	2
.80	393	128	76	53	36	23	17	13	10	8	7	6	5	4	4	3	3	2	2
.85	450	146	86	61	41	26	19	14	11	9	8	6	5	5	4	3	3	2	2
.90	526	171	101	71	48	31	22	17	13	11	9	7	6	5	5	4	3	3	2
.95	651	211	125	87	60	38	27	21	16	13	11	9	8	6	5	5	4	3	3
.99	920	298	176	123	84	53	38	29	22	18	15	12	10	9	7	6	5	4	3

Nondirectional Test, Alpha = .01

Power							POPULATION ETA-SQUARED												
	.01	.03	.05	.07	.10	.15	.20	.25	.30	.35	.40	.45	.50	.55	.60	.65	.70	.75	.80
.25	183	60	36	26	18	12	9	7	6	5	4	4	3	3	3	2	2	2	2
.50	333	109	65	46	31	20	15	11	9	8	6	6	5	4	4	3	3	3	2
.60	402	131	78	55	38	24	18	14	11	9	8	6	6	5	4	4	3	3	3
.67	454	148	87	62	42	27	20	15	12	10	8	7	6	5	5	4	3	3	3
.70	482	157	93	65	45	29	21	16	13	10	9	7	6	5	5	4	4	3	3
.75	528	172	102	72	49	31	23	17	14	11	9	8	7	6	5	4	4	3	3
.80	586	190	113	79	54	35	25	19	15	12	10	9	7	6	5	5	4	3	3
.85	654	213	126	88	60	38	28	21	17	14	11	9	8	7	6	5	4	4	3
.90	746	242	143	100	69	44	31	24	19	15	13	11	9	8	6	6	5	4	3
.95	892	290	171	120	82	52	37	28	22	18	15	12	10	9	7	6	5	4	4
.99	1,203	390	230	161	110	70	50	38	30	24	20	16	14	11	10	8	7	6	5

APPENDIX E.2: One- or Two-Way Between-Subjects Analysis of Variance

Degrees of Freedom Between or Effect = 1, Alpha = .05

POPULATION ETA-SQUARED

Power	.01	.03	.05	.07	.10	.15	.20	.25	.30	.35	.40	.45	.50	.55	.60	.65	.70	.75	.80
.10	22	8	5	4	3	2	2	2	—	—	—	—	—	—	—	—	—	—	—
.50	193	63	38	27	18	12	9	7	5	5	4	3	3	3	2	2	2	2	—
.70	310	101	60	42	29	18	13	10	8	7	6	5	4	4	3	3	2	2	2
.80	393	128	76	53	36	23	17	13	10	8	7	6	5	4	4	3	3	2	2
.90	526	171	101	71	48	31	22	17	13	11	9	7	6	5	5	4	3	3	2
.95	651	211	125	87	60	38	27	21	16	13	11	9	8	6	5	5	4	3	3
.99	920	298	176	123	84	53	38	29	22	18	15	12	10	9	7	6	5	4	3

Degrees of Freedom Between or Effect = 2, Alpha = .05

POPULATION ETA-SQUARED

Power	.01	.03	.05	.07	.10	.15	.20	.25	.30	.35	.40	.45	.50	.55	.60	.65	.70	.75	.80
.10	22	8	5	4	3	2	2	2	—	—	—	—	—	—	—	—	—	—	—
.50	165	55	32	23	16	10	8	6	5	4	3	3	3	2	2	2	2	2	—
.70	255	84	50	35	24	16	11	9	7	6	5	4	4	3	3	2	2	2	2
.80	319	105	62	44	30	19	14	11	9	7	6	5	4	4	3	3	2	2	2
.90	417	137	81	57	39	25	18	14	11	9	7	6	5	4	4	3	3	2	2
.95	511	168	99	69	47	30	22	16	13	11	9	7	6	5	4	4	3	3	2
.99	708	232	137	96	65	41	29	22	18	14	12	10	8	7	6	5	4	3	3

Degrees of Freedom Between or Effect = 3, Alpha = .05

POPULATION ETA-SQUARED

Power	.01	.03	.05	.07	.10	.15	.20	.25	.30	.35	.40	.45	.50	.55	.60	.65	.70	.75	.80
.10	21	7	5	4	3	2	2	2	—	—	—	—	—	—	—	—	—	—	—
.50	144	48	28	20	14	9	7	5	4	4	3	3	2	2	2	2	2	—	—
.70	219	72	43	30	21	13	10	8	6	5	4	4	3	3	2	2	2	2	2
.80	272	90	53	37	26	17	12	9	7	6	5	4	4	3	3	2	2	2	2
.90	351	115	68	48	33	21	15	12	9	8	6	5	5	4	3	3	3	2	2
.95	426	140	83	58	40	25	18	14	11	9	7	6	5	5	4	3	3	2	2
.99	583	191	113	79	54	34	24	19	15	12	10	8	7	6	5	4	4	3	2

Degrees of Freedom Between or Effect = 4, Alpha = .05

POPULATION ETA-SQUARED

Power	.01	.03	.05	.07	.10	.15	.20	.25	.30	.35	.40	.45	.50	.55	.60	.65	.70	.75	.80
.10	19	7	5	3	3	2	—	2	—	—	—	—	—	—	—	—	—	—	—
.50	128	43	25	18	13	8	6	5	4	3	3	3	2	2	2	2	2	—	—
.70	193	64	38	27	18	12	9	7	6	5	4	3	3	3	2	2	2	2	—
.80	238	78	46	33	23	15	10	8	7	5	5	4	3	3	3	2	2	2	2
.90	306	101	59	42	29	18	13	10	8	7	6	5	4	4	3	3	2	2	2
.95	369	121	72	50	34	22	16	12	10	8	7	6	5	4	3	3	3	2	2
.99	501	164	97	68	46	30	21	16	13	10	9	7	6	5	4	4	3	3	2

Degrees of Freedom Between or Effect = 5, Alpha = .05

POPULATION ETA-SQUARED

Power	.01	.03	.05	.07	.10	.15	.20	.25	.30	.35	.40	.45	.50	.55	.60	.65	.70	.75	.80
.10	18	7	4	3	3	2	—	—	—	—	—	—	—	—	—	—	—	—	—
.50	117	39	23	17	12	8	6	5	4	3	3	2	2	2	2	2	2	—	—
.70	174	57	34	24	17	11	8	6	5	4	4	3	3	2	2	2	2	2	—
.80	213	70	42	29	20	13	9	7	6	5	4	4	3	3	2	2	2	2	2
.90	273	90	53	37	26	17	12	9	7	6	5	4	4	3	3	2	2	2	2
.95	328	108	64	45	31	20	14	11	9	7	6	5	4	4	3	3	2	2	2
.99	442	145	86	60	41	26	19	14	11	9	8	6	5	5	4	3	3	2	2

Degrees of Freedom Between or Effect = 6, Alpha = .05

POPULATION ETA-SQUARED

Power	.01	.03	.05	.07	.10	.15	.20	.25	.30	.35	.40	.45	.50	.55	.60	.65	.70	.75	.80
.10	17	6	4	3	2	2	—	—	—	—	—	—	—	—	—	—	—	—	—
.50	107	36	21	15	11	7	5	4	4	3	3	2	2	2	2	2	—	—	—
.70	159	53	31	22	15	10	7	6	5	4	3	3	3	2	2	2	2	2	—
.80	194	64	38	27	19	12	9	7	6	5	4	3	3	3	2	2	2	2	2
.90	247	81	48	34	23	15	11	8	7	6	5	4	3	3	3	2	2	2	2
.95	296	97	58	41	28	18	13	10	8	7	5	5	4	3	3	3	2	2	2
.99	398	131	77	54	37	24	17	13	10	8	7	6	5	4	4	3	3	2	2

Degrees of Freedom Between or Effect = 8, Alpha = .05

POPULATION ETA-SQUARED

Power	.01	.03	.05	.07	.10	.15	.20	.25	.30	.35	.40	.45	.50	.55	.60	.65	.70	.75	.80
.10	16	6	4	3	2	2	—	—	—	—	—	—	—	—	—	—	—	—	—
.50	94	31	19	13	9	6	5	4	3	3	2	2	2	2	2	2	—	—	—
.70	137	45	27	19	13	9	6	5	4	4	3	3	2	2	2	2	2	—	—
.80	167	55	33	23	16	10	8	6	5	4	4	3	3	2	2	2	2	2	—
.90	211	70	41	29	20	13	9	7	6	5	4	4	3	3	2	2	2	2	2
.95	251	83	49	35	24	15	11	9	7	6	5	4	4	3	3	2	2	2	2
.99	335	110	65	46	31	20	14	11	9	7	6	5	4	4	3	3	2	2	2

Degrees of Freedom Between or Effect = 10, Alpha = .05

POPULATION ETA-SQUARED

Power	.01	.03	.05	.07	.10	.15	.20	.25	.30	.35	.40	.45	.50	.55	.60	.65	.70	.75	.80
.10	15	5	4	3	2	2	—	—	—	—	—	—	—	—	—	—	—	—	—
.50	84	28	17	12	9	6	4	4	3	3	2	2	2	2	2	—	—	—	—
.70	122	40	24	17	12	8	6	5	4	3	3	2	2	2	2	2	2	—	—
.80	147	49	29	21	14	9	7	5	4	4	3	3	2	2	2	2	2	—	—
.90	186	61	36	26	18	12	8	7	5	4	4	3	3	3	2	2	2	2	—
.95	221	73	43	30	21	14	10	8	6	5	5	4	3	3	3	2	2	2	2
.99	292	96	57	40	27	18	13	10	8	6	5	5	4	3	3	3	2	2	2

Degrees of Freedom Between or Effect = 12, Alpha = .05

POPULATION ETA-SQUARED

Power	.01	.03	.05	.07	.10	.15	.20	.25	.30	.35	.40	.45	.50	.55	.60	.65	.70	.75	.80
.10	14	5	3	3	2	2	—	—	—	—	—	—	—	—	—	—	—	—	—
.50	77	26	16	11	8	5	4	3	3	2	2	2	2	2	2	—	—	—	—
.70	111	37	22	16	11	7	5	4	4	3	3	2	2	2	2	2	—	—	—
.80	133	44	26	19	13	9	6	5	4	3	3	3	2	2	2	2	2	—	—
.90	168	55	33	23	16	11	8	6	5	4	4	3	3	2	2	2	2	2	—
.95	198	65	39	27	19	12	9	7	6	5	4	3	3	3	2	2	2	2	—
.99	261	86	51	36	25	16	11	9	7	6	5	4	4	3	3	2	2	2	2

Degrees of Freedom Between or Effect = 15, Alpha = .05

POPULATION ETA-SQUARED

Power	.01	.03	.05	.07	.10	.15	.20	.25	.30	.35	.40	.45	.50	.55	.60	.65	.70	.75	.80
.10	13	5	3	3	2	2	—	—	—	—	—	—	—	—	—	—	—	—	—
.50	68	23	14	10	7	5	4	3	3	2	2	2	2	2	—	—	—	—	—
.70	98	33	20	14	10	7	5	4	3	3	2	2	2	2	2	2	—	—	—
.80	118	39	23	17	12	8	6	5	4	3	3	2	2	2	2	2	2	—	—
.90	147	49	29	21	14	9	7	5	4	4	3	3	2	2	2	2	2	—	—
.95	174	57	34	24	17	11	8	6	5	4	4	3	3	2	2	2	2	2	—
.99	227	75	44	31	22	14	10	8	6	5	4	4	3	3	3	2	2	2	2

Degrees of Freedom Between or Effect = 1, Alpha = .01

POPULATION ETA-SQUARED

Power	.01	.03	.05	.07	.10	.15	.20	.25	.30	.35	.40	.45	.50	.55	.60	.65	.70	.75	.80
.10	84	28	17	12	9	6	5	4	3	3	2	2	2	2	2	—	—	—	—
.50	333	109	65	46	31	20	15	11	9	8	6	6	5	4	4	3	3	3	2
.70	482	157	93	65	45	29	21	16	13	10	9	7	6	5	5	4	4	3	3
.80	586	190	113	79	54	35	25	19	15	12	10	9	7	6	5	5	4	3	3
.90	746	242	143	100	69	44	31	24	19	15	13	11	9	8	6	6	5	4	3
.95	892	289	171	120	82	52	37	28	22	18	15	12	10	9	7	6	5	4	4
.99	1,203	390	230	161	110	70	50	38	30	24	20	16	14	11	10	8	7	6	5

Degrees of Freedom Between or Effect = 2, Alpha = .01

POPULATION ETA-SQUARED

Power	.01	.03	.05	.07	.10	.15	.20	.25	.30	.35	.40	.45	.50	.55	.60	.65	.70	.75	.80
.10	77	26	16	11	8	5	5	3	3	2	2	2	2	2	2	—	—	—	—
.50	272	89	53	37	26	16	13	9	7	6	5	4	4	3	3	2	2	2	2
.70	383	126	74	52	36	23	17	13	10	8	7	6	5	4	4	3	3	2	2
.80	459	151	89	62	43	27	20	15	12	10	8	7	6	5	4	3	3	3	2
.90	576	189	111	78	53	34	25	18	15	12	10	8	7	6	5	4	3	3	2
.95	683	224	132	93	63	40	29	22	17	14	11	9	8	7	6	5	4	3	3
.99	906	297	175	122	83	53	38	28	22	18	15	12	10	8	7	6	5	4	3

Degrees of Freedom Between or Effect = 3, Alpha = .01

POPULATION ETA-SQUARED

Power	.01	.03	.05	.07	.10	.15	.20	.25	.30	.35	.40	.45	.50	.55	.60	.65	.70	.75	.80
.10	70	23	14	10	7	5	4	3	3	2	2	2	2	2	—	—	—	—	—
.50	232	76	45	32	22	14	11	8	6	5	4	4	3	3	3	2	2	2	2
.70	323	106	63	44	30	19	14	11	9	7	6	5	4	4	3	3	2	2	2
.80	384	126	75	52	36	23	17	13	10	8	7	6	5	4	4	3	3	2	2
.90	478	157	93	65	44	28	21	15	12	10	8	7	6	5	4	4	3	3	2
.95	563	184	109	76	52	33	24	18	14	12	10	8	7	6	5	4	3	3	2
.99	740	242	143	100	68	43	31	23	18	15	12	10	8	7	6	5	4	3	3

Degrees of Freedom Between or Effect = 4, Alpha = .01

POPULATION ETA-SQUARED

Power	.01	.03	.05	.07	.10	.15	.20	.25	.30	.35	.40	.45	.50	.55	.60	.65	.70	.75	.80
.10	64	21	13	9	7	5	4	3	2	2	2	2	2	2	—	—	—	—	—
.50	204	67	40	28	19	13	10	7	6	5	4	4	3	3	2	2	2	2	2
.70	280	92	55	38	26	17	13	9	8	6	5	4	4	3	3	3	2	2	2
.80	333	109	65	46	31	20	15	11	9	7	6	5	4	4	3	3	2	2	2
.90	412	135	80	56	38	25	18	13	11	9	7	6	5	4	4	3	3	2	2
.95	483	158	94	66	45	29	21	16	12	10	8	7	6	5	4	4	3	3	2
.99	631	207	122	86	58	37	27	20	16	13	11	9	7	6	5	4	4	3	3

Degrees of Freedom Between or Effect = 5, Alpha = .01

POPULATION ETA-SQUARED

Power	.01	.03	.05	.07	.10	.15	.20	.25	.30	.35	.40	.45	.50	.55	.60	.65	.70	.75	.80
.10	59	20	12	9	6	4	4	3	2	2	2	2	2	—	—	—	—	—	—
.50	183	61	36	25	18	11	9	7	5	4	4	3	3	3	2	2	2	2	—
.70	251	83	49	35	24	15	11	9	7	6	5	4	4	3	3	2	2	2	2
.80	296	97	58	41	28	18	13	10	8	7	5	5	4	3	3	3	2	2	2
.90	365	120	71	50	34	22	16	12	10	8	7	5	5	4	3	3	3	2	2
.95	426	140	83	58	40	25	18	14	11	9	7	6	5	5	4	3	3	2	2
.99	554	182	107	75	51	33	24	18	14	11	9	8	7	6	5	4	3	3	2

Degrees of Freedom Between or Effect = 6, Alpha = .01

POPULATION ETA-SQUARED

Power	.01	.03	.05	.07	.10	.15	.20	.25	.30	.35	.40	.45	.50	.55	.60	.65	.70	.75	.80
.10	55	19	11	8	6	4	3	3	2	2	2	2	2	—	—	—	—	—	—
.50	168	55	33	23	16	11	8	6	5	4	4	3	3	2	2	2	2	2	—
.70	228	75	45	31	22	14	10	8	6	5	4	4	3	3	3	2	2	2	2
.80	268	88	52	37	25	16	12	9	7	6	5	4	4	3	3	2	2	2	2
.90	329	108	64	45	31	20	14	11	9	7	6	5	4	4	3	3	2	2	2
.95	384	126	74	52	36	23	17	13	10	8	7	6	5	4	4	3	3	2	2
.99	497	163	96	68	46	29	21	16	13	10	9	7	6	5	4	4	3	3	2

Degrees of Freedom Between or Effect = 8, Alpha = .01

POPULATION ETA-SQUARED

Power	.01	.03	.05	.07	.10	.15	.20	.25	.30	.35	.40	.45	.50	.55	.60	.65	.70	.75	.80
.10	49	17	10	7	5	4	3	2	2	2	2	2	—	—	—	—	—	—	—
.50	145	48	29	20	14	9	7	5	4	4	3	3	2	2	2	2	2	—	—
.70	195	64	38	27	19	12	9	7	6	5	4	3	3	3	2	2	2	2	—
.80	228	75	45	31	22	14	10	8	6	5	4	4	3	3	3	2	2	2	2
.90	279	92	54	38	26	17	12	9	8	6	5	4	4	3	3	3	2	2	2
.95	323	106	63	44	30	19	14	11	9	7	6	5	4	4	3	3	2	2	2
.99	416	136	81	57	39	25	18	14	11	9	7	6	5	4	4	3	3	2	2

Degrees of Freedom Between or Effect = 10, Alpha = .01

POPULATION ETA-SQUARED

Power	.01	.03	.05	.07	.10	.15	.20	.25	.30	.35	.40	.45	.50	.55	.60	.65	.70	.75	.80
.10	45	15	9	7	5	3	3	2	2	2	2	2	—	—	—	—	—	—	—
.50	128	43	25	18	13	8	6	5	4	3	3	3	2	2	2	2	2	—	—
.70	172	57	34	24	17	11	8	6	5	4	4	3	3	2	2	2	2	2	—
.80	201	66	39	28	19	12	9	7	6	5	4	3	3	3	2	2	2	2	2
.90	244	80	48	34	23	15	11	8	7	6	5	4	3	3	3	2	2	2	2
.95	283	93	55	39	27	17	12	10	8	6	5	4	4	3	3	3	2	2	2
.99	361	119	70	49	34	22	16	12	9	8	6	5	5	4	3	3	3	2	2

Degrees of Freedom Between or Effect = 12, Alpha = .01

POPULATION ETA-SQUARED

Power	.01	.03	.05	.07	.10	.15	.20	.25	.30	.35	.40	.45	.50	.55	.60	.65	.70	.75	.80
.10	41	14	9	6	5	3	3	2	2	2	2	—	—	—	—	—	—	—	—
.50	117	39	23	17	12	8	6	5	4	3	3	2	2	2	2	2	2	—	—
.70	155	51	31	22	15	10	7	6	5	4	3	3	3	2	2	2	2	2	—
.80	181	60	35	25	17	11	8	6	5	4	4	3	3	2	2	2	2	2	—
.90	219	72	43	30	21	13	10	8	6	5	4	4	3	3	2	2	2	2	2
.95	253	83	49	35	24	15	11	9	7	6	5	4	4	3	3	2	2	2	2
.99	322	106	63	44	30	19	14	11	9	7	6	5	4	4	3	3	2	2	2

Degrees of Freedom Between or Effect = 15, Alpha = .01

POPULATION ETA-SQUARED

Power	.01	.03	.05	.07	.10	.15	.20	.25	.30	.35	.40	.45	.50	.55	.60	.65	.70	.75	.80
.10	37	13	8	6	4	3	2	2	2	2	2	—	—	—	—	—	—	—	—
.50	103	34	21	15	10	7	5	4	3	3	3	2	2	2	2	2	—	—	—
.70	137	45	27	19	13	9	6	5	4	4	3	3	2	2	2	2	2	—	—
.80	157	52	31	22	15	10	7	6	5	4	3	3	3	2	2	2	2	2	—
.90	191	63	38	27	18	12	9	7	5	5	4	3	3	3	2	2	2	2	—
.95	220	73	43	30	21	14	10	8	6	5	4	4	3	3	2	2	2	2	2
.99	279	92	54	38	26	17	12	9	8	6	5	4	4	3	3	3	2	2	2

APPENDIX E.3: One-Way Repeated Measures Analysis of Variance

Degrees of Freedom IV = 2, Alpha = .05

POPULATION ETA-SQUARED

Power	.01	.03	.05	.07	.10	.15	.20	.25	.30	.35	.40	.45	.50	.55	.60	.65	.70	.75	.80
.10	32	11	7	5	4	3	2	2	2	2	—	—	—	—	—	—	—	—	—
.50	247	81	48	34	23	15	11	8	7	6	5	4	3	3	3	2	2	2	2
.70	382	125	74	52	36	23	16	13	10	8	7	6	5	4	4	3	3	2	2
.80	478	157	93	65	44	28	20	15	12	10	8	7	6	5	4	4	3	3	2
.90	627	206	121	85	58	37	26	20	16	13	10	9	7	6	5	4	4	3	3
.95	765	251	148	104	70	45	32	24	19	15	13	10	9	7	6	5	4	4	3
.99	1,060	347	204	143	97	62	44	33	26	21	17	14	12	10	8	7	6	5	4

Degrees of Freedom IV = 3, Alpha = .05

POPULATION ETA-SQUARED

Power	.01	.03	.05	.07	.10	.15	.20	.25	.30	.35	.40	.45	.50	.55	.60	.65	.70	.75	.80
.10	27	9	6	4	3	2	2	2	2	—	—	—	—	—	—	—	—	—	—
.50	191	63	37	27	18	12	9	7	5	5	4	3	3	3	2	2	2	2	—
.70	291	96	57	40	27	18	13	10	8	6	5	5	4	3	3	3	2	2	2
.80	361	118	70	49	34	22	16	12	9	8	6	5	5	4	3	3	3	2	2
.90	469	154	91	64	44	28	20	15	12	10	8	7	6	5	4	4	3	3	2
.95	568	186	110	77	53	33	24	18	14	12	10	8	7	6	5	4	3	3	2
.99	777	254	150	105	72	45	32	25	19	16	13	11	9	7	6	5	4	4	3

Degrees of Freedom IV = 4, Alpha = .05

POPULATION ETA-SQUARED

Power	.01	.03	.05	.07	.10	.15	.20	.25	.30	.35	.40	.45	.50	.55	.60	.65	.70	.75	.80
.10	24	8	5	4	3	2	2	2	2	—	—	—	—	—	—	—	—	—	—
.50	160	53	31	22	15	10	7	6	5	4	3	3	3	2	2	2	2	2	—
.70	241	79	47	33	23	15	11	8	7	5	5	4	3	3	3	2	2	2	2
.80	297	98	58	41	28	18	13	10	8	7	5	5	4	3	3	3	2	2	2
.90	382	126	74	52	36	23	16	13	10	8	7	6	5	4	4	3	3	2	2
.95	461	151	89	63	43	27	20	15	12	10	8	7	6	5	4	3	3	3	2
.99	626	205	121	85	58	37	26	20	16	13	10	9	7	6	5	4	4	3	3

Degrees of Freedom IV = 5, Alpha = .05

POPULATION ETA-SQUARED

Power	.01	.03	.05	.07	.10	.15	.20	.25	.30	.35	.40	.45	.50	.55	.60	.65	.70	.75	.80
.10	21	8	5	4	3	2	2	2	—	—	—	—	—	—	—	—	—	—	—
.50	139	46	28	20	14	9	7	5	4	4	3	3	2	2	2	2	2	—	—
.70	208	69	41	29	20	13	9	7	6	5	4	4	3	3	2	2	2	2	2
.80	225	84	50	35	24	16	11	9	7	6	5	4	4	3	3	3	2	2	2
.90	327	108	64	45	31	20	14	11	9	7	6	5	4	4	3	3	3	2	2
.95	393	129	76	54	37	23	17	13	10	8	7	6	5	4	4	3	3	2	2
.99	530	174	103	72	49	31	22	17	13	11	9	8	6	5	5	4	3	3	2

Degrees of Freedom IV = 6, Alpha = .05

POPULATION ETA-SQUARED

Power	.01	.03	.05	.07	.10	.15	.20	.25	.30	.35	.40	.45	.50	.55	.60	.65	.70	.75	.80
.10	20	7	5	4	3	2	2	2	—	—	—	—	—	—	—	—	—	—	—
.50	125	41	25	18	12	8	6	5	4	3	3	3	2	2	2	2	2	—	—
.70	185	61	36	26	18	12	8	7	5	4	4	3	3	3	2	2	2	2	—
.80	226	74	44	31	21	14	10	8	6	5	4	4	3	3	3	2	2	2	2
.90	288	95	56	40	27	17	13	10	8	6	5	5	4	3	3	3	2	2	2
.95	345	113	67	47	32	21	15	11	9	7	6	5	4	4	3	3	2	2	2
.99	464	152	90	63	43	27	20	15	12	10	8	7	6	5	4	4	3	3	2

Degrees of Freedom IV = 2, Alpha = .01

POPULATION ETA-SQUARED

Power	.01	.03	.05	.07	.10	.15	.20	.25	.30	.35	.40	.45	.50	.55	.60	.65	.70	.75	.80
.10	115	38	23	16	11	8	6	4	4	3	3	2	2	2	2	2	—	—	—
.50	406	133	79	55	38	24	17	13	11	9	7	6	5	4	4	3	3	2	2
.70	574	188	111	78	53	34	24	18	14	12	10	8	7	6	5	4	3	3	2
.80	688	225	133	93	63	40	29	22	17	14	11	9	8	7	6	5	4	3	3
.90	864	283	167	117	79	50	36	27	21	17	14	12	10	8	7	6	5	4	3
.95	1,023	335	197	138	94	60	42	32	25	20	16	14	11	9	8	7	5	4	4
.99	1,358	444	261	183	124	79	56	42	33	26	22	18	15	12	10	8	7	6	4

Degrees of Freedom IV = 3, Alpha = .01

POPULATION ETA-SQUARED

Power	.01	.03	.05	.07	.10	.15	.20	.25	.30	.35	.40	.45	.50	.55	.60	.65	.70	.75	.80
.10	92	31	18	13	9	6	5	4	3	3	2	2	2	2	2	—	—	—	—
.50	308	101	60	42	29	19	13	10	8	7	6	5	4	4	3	3	2	2	2
.70	429	141	83	58	40	25	18	14	11	9	7	6	5	5	4	3	3	2	2
.80	511	168	99	69	47	30	22	16	13	11	9	7	6	5	4	4	3	3	2
.90	636	208	123	86	59	37	27	20	16	13	11	9	7	6	5	4	4	3	3
.95	749	245	145	101	69	44	31	24	19	15	12	10	9	7	6	5	4	4	3
.99	985	323	190	133	90	57	41	31	24	19	16	13	11	9	8	6	5	4	3

Degrees of Freedom IV = 4, Alpha = .01

POPULATION ETA-SQUARED

Power	.01	.03	.05	.07	.10	.15	.20	.25	.30	.35	.40	.45	.50	.55	.60	.65	.70	.75	.80
.10	79	26	16	11	8	5	4	3	3	2	2	2	2	2	2	—	—	—	—
.50	254	84	50	35	24	15	11	9	7	6	5	4	4	3	3	2	2	2	2
.70	350	115	68	48	33	21	15	12	9	8	6	5	5	4	3	3	3	2	2
.80	416	136	81	57	39	25	18	14	11	9	7	6	5	4	4	3	3	2	2
.90	514	169	100	70	48	30	22	17	13	11	9	7	6	5	4	4	3	3	2
.95	603	198	117	82	56	35	25	19	15	12	10	8	7	6	5	4	4	3	3
.99	788	258	152	107	73	46	33	25	20	16	13	11	9	8	6	5	4	4	3

Degrees of Freedom IV = 5, Alpha = .01

POPULATION ETA-SQUARED

Power	.01	.03	.05	.07	.10	.15	.20	.25	.30	.35	.40	.45	.50	.55	.60	.65	.70	.75	.80
.10	70	24	14	10	7	5	4	3	3	2	2	2	2	2	—	—	—	—	—
.50	219	72	43	30	21	14	10	8	6	5	4	4	3	3	2	2	2	2	2
.70	300	99	58	41	28	18	13	10	8	7	6	5	4	3	3	3	2	2	2
.80	355	117	69	48	33	21	15	12	9	8	6	5	5	4	3	3	3	2	2
.90	437	143	85	60	41	26	19	14	11	9	8	6	5	5	4	3	3	2	2
.95	511	168	99	69	47	30	22	16	13	11	9	7	6	5	4	4	3	3	2
.99	664	218	128	90	61	39	28	21	17	13	11	9	8	6	5	5	4	3	3

Degrees of Freedom IV = 6, Alpha = .01

POPULATION ETA-SQUARED

Power	.01	.03	.05	.07	.10	.15	.20	.25	.30	.35	.40	.45	.50	.55	.60	.65	.70	.75	.80
.10	64	21	13	9	7	5	4	3	2	2	2	2	2	2	—	—	—	—	—
.50	195	64	38	27	19	12	9	7	6	5	4	3	3	3	2	2	2	2	—
.70	265	87	52	36	25	16	12	9	7	6	5	4	4	3	3	2	2	2	2
.80	312	103	61	43	29	19	14	10	8	7	6	5	4	4	3	3	2	2	2
.90	383	126	74	52	36	23	16	13	10	8	7	6	5	4	4	3	3	2	2
.95	447	147	87	61	42	27	19	15	12	9	8	7	6	5	4	3	3	3	2
.99	579	190	112	79	54	34	24	19	15	12	10	8	7	6	5	4	4	3	2

APPENDIX E.4:   **Pearson Correlation**

Directional Test, Alpha = .05

POPULATION CORRELATION COEFFICIENT SQUARED

Power	.01	.03	.05	.07	.10	.15	.20	.25	.30	.35	.40	.45	.50	.55	.60	.65	.70	.75	.80
.25	99	34	21	15	11	8	6	6	5	4	4	4	3	3	3	3	3	3	2
.50	277	92	56	40	28	19	14	11	9	8	7	6	6	5	5	4	4	4	3
.60	368	122	73	52	36	24	18	14	12	10	9	8	7	6	5	5	5	4	4
.67	430	142	85	61	42	28	21	16	13	11	10	9	8	7	6	5	5	4	4
.70	470	156	93	66	46	30	22	18	14	12	10	9	8	7	6	6	5	5	4
.75	537	177	106	75	52	34	25	20	16	14	12	10	9	8	7	6	6	5	5
.80	618	204	121	86	60	39	29	22	18	15	13	11	10	9	8	7	6	6	5
.85	727	239	143	101	70	46	33	26	21	18	15	13	11	10	9	8	7	6	5
.90	864	284	169	120	83	54	39	31	25	21	18	15	13	11	10	9	8	7	6
.95	1,105	363	216	152	105	68	50	39	31	26	22	19	16	14	12	11	10	8	7
.99	1,585	520	308	218	150	97	70	55	44	36	31	26	22	19	17	15	13	11	10

Directional Test, Alpha = .01

POPULATION CORRELATION COEFFICIENT SQUARED

Power	.01	.03	.05	.07	.10	.15	.20	.25	.30	.35	.40	.45	.50	.55	.60	.65	.70	.75	.80
.25	273	91	55	39	27	18	14	12	9	8	7	6	6	5	5	4	4	4	3
.50	540	178	106	76	52	34	25	20	16	14	12	10	9	8	7	6	6	5	5
.60	663	219	130	92	64	42	31	24	20	16	14	12	11	9	8	7	7	6	5
.67	757	249	148	105	73	47	35	28	22	18	16	13	12	10	9	8	7	6	6
.70	809	266	158	112	77	50	37	29	23	19	17	14	12	11	10	8	8	7	6
.75	897	295	175	124	86	56	41	32	26	21	18	16	14	12	10	9	8	7	6
.80	998	328	195	138	95	62	45	36	28	24	20	17	15	13	11	10	9	8	7
.85	1,126	370	220	155	107	69	51	40	32	26	22	19	16	14	13	11	10	8	7
.90	1,296	425	253	178	123	80	58	45	36	30	25	22	19	16	14	12	11	9	8
.95	1,585	520	308	218	150	97	70	55	44	36	31	26	22	19	17	15	13	11	10
.99	2,154	706	418	295	203	131	95	74	59	49	41	35	30	26	22	19	17	14	12

Nondirectional Test, Alpha = .05

POPULATION CORRELATION COEFFICIENT SQUARED

Power	.01	.03	.05	.07	.10	.15	.20	.25	.30	.35	.40	.45	.50	.55	.60	.65	.70	.75	.80
.25	166	56	34	25	17	12	9	8	6	6	5	5	4	4	4	3	3	3	3
.50	384	127	76	54	38	25	19	15	12	10	9	8	7	6	6	5	5	4	4
.60	489	162	97	69	48	31	23	18	15	13	11	9	8	7	7	6	5	5	4
.67	570	188	112	80	55	36	27	21	17	14	12	11	9	8	7	7	6	5	5
.70	616	203	121	86	59	39	29	23	18	15	13	11	10	9	8	7	6	6	5
.75	692	228	136	96	67	43	32	25	20	17	14	12	11	10	9	8	7	6	5
.80	783	258	153	109	75	49	36	28	23	19	16	14	12	11	9	8	7	7	6
.85	895	294	175	124	85	56	41	32	26	21	18	16	14	12	10	9	8	7	6
.90	1,046	344	204	144	100	65	47	37	30	25	21	18	15	13	12	10	9	8	7
.95	1,308	429	255	180	124	80	58	46	37	30	26	22	19	16	14	12	11	10	8
.99	1,828	599	355	251	172	111	81	63	50	42	35	30	26	22	19	17	14	13	11

Nondirectional Test, Alpha = .01

POPULATION CORRELATION COEFFICIENT SQUARED

Power	.01	.03	.05	.07	.10	.15	.20	.25	.30	.35	.40	.45	.50	.55	.60	.65	.70	.75	.80
.25	362	120	72	51	36	24	18	15	12	10	9	7	7	6	5	5	4	4	4
.50	662	218	130	92	64	42	31	24	19	16	14	12	11	9	8	7	7	6	5
.60	797	262	156	111	76	50	36	29	23	19	16	14	12	11	9	8	7	7	6
.67	901	296	176	125	86	56	41	32	26	21	18	16	14	12	10	9	8	7	6
.70	957	315	187	132	91	59	43	34	27	23	19	17	14	13	11	10	9	8	7
.75	1,052	346	205	145	100	65	47	37	30	25	21	18	16	14	12	10	9	8	7
.80	1,163	382	227	160	110	72	52	41	33	27	23	20	17	15	13	11	10	8	8
.85	1,299	426	253	179	123	80	58	45	36	30	25	22	19	16	14	12	11	9	8
.90	1,480	485	288	203	140	91	66	51	41	34	29	24	21	18	16	14	12	11	9
.95	1,790	587	348	246	169	109	79	62	49	41	34	29	25	22	19	16	14	12	11
.99	2,390	783	464	327	225	145	105	82	65	54	45	38	33	28	24	21	18	16	13

APPENDIX E.5:   Chi-Square Test

Type of Table: 2 × 2, Alpha = .05

POPULATION FOURFOLD POINT CORRELATION COEFFICIENT

Power	.10	.20	.30	.40	.50	.60	.70	.80	.90
.25	165	41	18	10	7	5	3	3	2
.50	384	96	43	24	15	11	8	6	5
.60	490	122	54	31	20	14	10	8	6
.70	617	154	69	39	25	17	13	10	8
.75	694	175	77	43	28	19	14	11	9
.80	785	196	87	49	31	22	16	12	10
.85	898	224	100	56	36	25	18	14	11
.90	1,051	263	117	66	42	29	21	16	13
.95	1,300	325	144	81	52	36	27	20	16
.99	1,837	459	204	115	73	51	37	29	23

Type of Table: 2 × 3, Alpha = .05

POPULATION VALUE OF CRAMER'S STATISTIC

Power	.10	.20	.30	.40	.50	.60	.70	.80	.90
.25	226	56	25	14	9	6	5	4	3
.50	496	124	55	31	20	14	10	8	6
.60	621	155	69	39	25	17	13	10	8
.70	770	193	86	48	31	21	16	12	10
.75	859	215	95	54	34	24	18	13	11
.80	964	241	107	60	39	27	20	15	12
.85	1,092	273	121	68	44	30	22	17	13
.90	1,265	316	141	79	51	35	26	20	16
.95	1,544	386	172	97	62	43	32	24	19
.99	2,140	535	238	134	86	59	44	33	26

Type of Table: 2 × 4, Alpha = .05

POPULATION VALUE OF CRAMER'S STATISTIC

Power	.10	.20	.30	.40	.50	.60	.70	.80	.90
.25	258	65	29	16	10	7	5	4	3
.50	576	144	64	36	23	16	12	9	7
.60	715	179	79	45	29	20	15	11	9
.70	879	220	98	55	35	24	18	14	11
.75	976	244	108	61	39	27	20	15	12
.80	1,090	273	121	68	44	30	22	17	13
.85	1,230	308	137	77	49	34	25	19	15
.90	1,417	354	157	89	57	39	29	22	17
.95	1,717	429	191	107	69	48	35	27	21
.99	2,352	588	261	147	94	65	48	37	29

Type of Table: 3 × 3, Alpha = .05

POPULATION VALUE OF CRAMER'S STATISTIC

Power	.10	.20	.30	.40	.50	.60	.70	.80	.90
.25	154	39	17	10	6	4	3	2	2
.50	321	80	36	20	13	9	7	5	4
.60	396	99	44	25	16	11	8	6	5
.70	484	121	54	30	19	13	10	8	6
.75	536	134	60	34	21	15	11	8	7
.80	597	149	66	37	24	17	12	9	7
.85	671	168	75	42	27	19	14	10	8
.90	770	193	86	48	31	21	16	12	10
.95	929	232	103	58	37	26	19	15	11
.99	1,262	316	140	79	50	35	26	20	16

Type of Table: 3 × 4, Alpha = .05

POPULATION VALUE OF CRAMER'S STATISTIC

Power	.10	.20	.30	.40	.50	.60	.70	.80	.90
.25	185	46	21	12	7	5	4	3	2
.50	375	94	42	23	15	10	8	6	5
.60	460	115	51	29	18	13	9	7	6
.70	557	139	62	35	22	15	11	9	7
.75	615	154	68	38	25	17	13	10	8
.80	681	170	76	43	27	19	14	11	8
.85	763	191	85	48	31	21	16	12	9
.90	871	218	97	54	35	24	18	14	11
.95	1,043	261	116	65	42	29	21	16	13
.99	1,403	351	156	88	56	39	29	22	17

Type of Table: 4 × 4, Alpha = .05

POPULATION VALUE OF CRAMER'S STATISTIC

Power	.10	.20	.30	.40	.50	.60	.70	.80	.90
.25	148	37	16	9	6	4	3	2	2
.50	294	73	33	18	12	8	6	5	4
.60	357	89	40	22	14	10	7	6	4
.70	430	107	48	27	17	12	9	7	5
.75	472	118	52	30	19	13	10	7	6
.80	522	130	58	33	21	14	11	8	6
.85	582	145	65	36	23	16	12	9	7
.90	661	165	73	41	26	18	13	10	8
.95	786	197	87	49	31	22	16	12	10
.99	1,046	262	116	65	42	29	21	16	13

Type of Table: 2 × 2, Alpha = .01

POPULATION FOURFOLD POINT CORRELATION COEFFICIENT

Power	.10	.20	.30	.40	.50	.60	.70	.80	.90
.25	362	90	40	23	14	10	7	6	4
.50	664	166	74	41	27	18	14	10	8
.60	800	200	89	50	32	22	16	13	10
.70	961	240	107	60	38	27	20	15	12
.75	1,056	264	117	66	42	29	22	17	13
.80	1,168	292	130	73	47	32	24	18	14
.85	1,305	326	145	82	52	36	27	20	16
.90	1,488	372	165	93	60	41	30	23	18
.95	1,781	445	198	111	71	49	36	28	22
.99	2,403	601	267	150	96	67	49	38	30

Type of Table: 2 × 3, Alpha = .01

POPULATION VALUE OF CRAMER'S STATISTIC

Power	.10	.20	.30	.40	.50	.60	.70	.80	.90
.25	467	117	52	29	19	13	10	7	6
.50	819	205	91	51	33	23	17	13	10
.60	975	244	108	61	39	27	20	15	12
.70	1,157	289	129	72	46	32	24	18	14
.75	1,264	316	140	79	51	35	26	20	16
.80	1,388	347	154	87	56	39	28	22	17
.85	1,540	385	171	96	62	43	31	24	19
.90	1,743	436	194	109	70	48	36	27	22
.95	2,065	516	229	129	83	57	42	32	25
.99	2,742	685	305	171	110	76	56	43	34

Type of Table: 2 × 4, Alpha = .01

POPULATION VALUE OF CRAMER'S STATISTIC

Power	.10	.20	.30	.40	.50	.60	.70	.80	.90
.25	544	136	60	34	22	15	11	8	7
.50	931	233	103	58	37	26	19	15	11
.60	1,101	275	122	69	44	31	22	17	14
.70	1,297	324	144	81	52	36	26	20	16
.75	1,412	353	157	88	56	39	29	22	17
.80	1,546	386	172	97	62	43	32	24	19
.85	1,709	427	190	107	68	47	35	27	21
.90	1,925	481	214	120	77	53	39	30	24
.95	2,267	567	252	142	91	63	46	35	28
.99	2,983	746	331	186	119	83	61	47	37

Type of Table: 3 × 3, Alpha = .01

POPULATION VALUE OF CRAMER'S STATISTIC

Power	.10	.20	.30	.40	.50	.60	.70	.80	.90
.25	304	76	34	19	12	8	6	5	4
.50	512	128	57	32	20	14	10	8	6
.60	602	151	67	38	24	17	12	9	7
.70	706	177	78	44	28	20	14	11	9
.75	767	192	85	48	31	21	16	12	9
.80	824	206	92	52	33	23	17	13	10
.85	924	231	103	58	37	26	19	14	11
.90	1,037	259	115	65	41	29	21	16	13
.95	1,217	304	135	76	49	34	25	19	15
.99	1,590	398	177	99	64	44	32	25	20

Type of Table: 3 × 4, Alpha = .01

POPULATION VALUE OF CRAMER'S STATISTIC

Power	.10	.20	.30	.40	.50	.60	.70	.80	.90
.25	357	89	40	22	14	10	7	6	4
.50	588	147	65	37	24	16	12	9	7
.60	687	172	76	43	27	19	14	11	8
.70	801	200	89	50	32	22	16	13	10
.75	867	217	96	54	35	24	18	14	11
.80	944	236	105	59	38	26	19	15	12
.85	1,037	259	115	65	41	29	21	16	13
.90	1,159	290	129	72	46	32	24	18	14
.95	1,352	338	150	85	54	38	28	21	17
.99	1,751	438	195	109	70	49	36	27	22

Type of Table: 4 × 4, Alpha = .01

POPULATION VALUE OF CRAMER'S STATISTIC

Power	.10	.20	.30	.40	.50	.60	.70	.80	.90
.25	280	70	31	18	11	8	6	4	3
.50	453	113	50	28	18	13	9	7	6
.60	526	132	58	33	21	15	11	8	6
.70	610	153	68	38	24	17	12	10	8
.75	658	165	73	41	26	18	13	10	8
.80	714	179	79	45	29	20	15	11	9
.85	782	196	87	49	31	22	16	12	10
.90	871	218	97	54	35	24	18	14	11
.95	1,010	253	112	63	40	28	21	16	12
.99	1,296	324	144	81	52	36	26	20	16

Critical Values for the *F* Distribution

The following table presents critical values of *F* for alpha levels of .05 and .01. The .05 critical values are in roman type, and the .01 critical values are in **boldface.** The values at the top of the table under the heading "Degrees of Freedom for Numerator" represent the degrees of freedom associated with the numerator of the *F* ratio for the analysis of interest. This will be $df_{BETWEEN}$ for one-way between-subjects analysis of variance; df_{IV} for one-way repeated measures analysis of variance; and df_A, df_B, or $df_{A \times B}$ for two-way between-subjects analysis of variance. The values at the left of the table under the heading "Degrees of Freedom for Denominator" represent the degrees of freedom associated with the denominator of the *F* ratio for the analysis of interest. This will be df_{WITHIN} for one- and two-way between-subjects analysis of variance, and df_{ERROR} for one-way repeated measures analysis of variance. For example, the critical value of *F* for a one-way between-subjects analysis of variance for $df_{BETWEEN} = 2$, $df_{WITHIN} = 11$, and an alpha level of .05 is 3.98.

Reprinted by permission from *Statistical Methods*, by George W. Snedecor and William G. Cochran, Seventh Edition, copyright ©1980 The Iowa State University Press, Ames, Iowa 50010.

Degrees of Freedom for Numerator

df	1	2	3	4	5	6	7	8	9	10	11	12	14	16	20	24	30	40	50	75	100	200	500	∞
1	161 / 4,052	200 / 4,999	216 / 5,403	225 / 5,625	230 / 5,764	234 / 5,859	237 / 5,928	239 / 5,981	241 / 6,022	242 / 6,056	243 / 6,082	244 / 6,106	245 / 6,142	246 / 6,169	248 / 6,208	249 / 6,234	250 / 6,258	251 / 6,286	252 / 6,302	253 / 6,323	253 / 6,334	254 / 6,352	254 / 6,361	254 / 6,366
2	18.51 / 98.49	19.00 / 99.01	19.16 / 99.17	19.25 / 99.25	19.30 / 99.30	19.33 / 99.33	19.36 / 99.34	19.37 / 99.36	19.38 / 99.38	19.39 / 99.40	19.40 / 99.41	19.41 / 99.42	19.42 / 99.43	19.43 / 99.44	19.44 / 99.45	19.45 / 99.46	19.46 / 99.47	19.47 / 99.48	19.47 / 99.48	19.48 / 99.49	19.49 / 99.49	19.49 / 99.49	19.50 / 99.50	19.50 / 99.50
3	10.13 / 34.12	9.55 / 30.81	9.28 / 29.46	9.12 / 28.71	9.01 / 28.24	8.94 / 27.91	8.88 / 27.67	8.84 / 27.49	8.81 / 27.34	8.78 / 27.23	8.76 / 27.13	8.74 / 27.05	8.71 / 26.92	8.69 / 26.83	8.66 / 26.69	8.64 / 26.60	8.62 / 26.50	8.60 / 26.41	8.58 / 26.30	8.57 / 26.27	8.56 / 26.23	8.54 / 26.18	8.54 / 26.14	8.53 / 26.12
4	7.71 / 21.20	6.94 / 18.00	6.59 / 16.69	6.39 / 15.98	6.26 / 15.52	6.16 / 15.21	6.09 / 14.98	6.04 / 14.80	6.00 / 14.66	5.96 / 14.54	5.93 / 14.45	5.91 / 14.37	5.87 / 14.24	5.84 / 14.15	5.80 / 14.02	5.77 / 13.93	5.74 / 13.83	5.71 / 13.74	5.70 / 13.69	5.68 / 13.61	5.66 / 13.57	5.65 / 13.52	5.64 / 13.48	5.63 / 13.46
5	6.61 / 16.26	5.79 / 13.27	5.41 / 12.06	5.19 / 11.39	5.05 / 10.97	4.95 / 10.67	4.88 / 10.45	4.82 / 10.27	4.78 / 10.15	4.74 / 10.05	4.70 / 9.96	4.68 / 9.89	4.64 / 9.77	4.60 / 9.68	4.56 / 9.55	4.53 / 9.47	4.50 / 9.38	4.46 / 9.29	4.44 / 9.24	4.42 / 9.17	4.40 / 9.13	4.38 / 9.07	4.37 / 9.04	4.36 / 9.02
6	5.99 / 13.74	5.14 / 10.92	4.76 / 9.78	4.53 / 9.15	4.39 / 8.75	4.28 / 8.47	4.21 / 8.26	4.15 / 8.10	4.10 / 7.98	4.06 / 7.87	4.03 / 7.79	4.00 / 7.72	3.96 / 7.60	3.92 / 7.52	3.87 / 7.39	3.84 / 7.31	3.81 / 7.23	3.77 / 7.14	3.75 / 7.09	3.72 / 7.02	3.71 / 6.99	3.69 / 6.94	3.68 / 6.90	3.67 / 6.88
7	5.59 / 12.25	4.74 / 9.55	4.35 / 8.45	4.12 / 7.85	3.97 / 7.46	3.87 / 7.19	3.79 / 7.00	3.73 / 6.84	3.68 / 6.71	3.63 / 6.62	3.60 / 6.54	3.57 / 6.47	3.52 / 6.35	3.49 / 6.27	3.44 / 6.15	3.41 / 6.07	3.38 / 5.98	3.34 / 5.90	3.32 / 5.85	3.29 / 5.78	3.28 / 5.75	3.25 / 5.70	3.24 / 5.67	3.23 / 5.65
8	5.32 / 11.26	4.46 / 8.65	4.07 / 7.59	3.84 / 7.01	3.69 / 6.63	3.58 / 6.37	3.50 / 6.19	3.44 / 6.03	3.39 / 5.91	3.34 / 5.82	3.31 / 5.74	3.28 / 5.67	3.23 / 5.56	3.20 / 5.48	3.15 / 5.36	3.12 / 5.28	3.08 / 5.20	3.05 / 5.11	3.03 / 5.06	3.00 / 5.00	2.98 / 4.96	2.96 / 4.91	2.94 / 4.88	2.93 / 4.86
9	5.12 / 10.56	4.26 / 8.02	3.86 / 6.99	3.63 / 6.42	3.48 / 6.06	3.37 / 5.80	3.29 / 5.62	3.23 / 5.47	3.18 / 5.35	3.13 / 5.26	3.10 / 5.18	3.07 / 5.11	3.02 / 5.00	2.98 / 4.92	2.93 / 4.80	2.90 / 4.73	2.86 / 4.64	2.82 / 4.56	2.80 / 4.51	2.77 / 4.45	2.76 / 4.41	2.73 / 4.36	2.72 / 4.33	2.71 / 4.31
10	4.96 / 10.04	4.10 / 7.56	3.71 / 6.55	3.48 / 5.99	3.33 / 5.64	3.22 / 5.39	3.14 / 5.21	3.07 / 5.06	3.02 / 4.95	2.97 / 4.85	2.94 / 4.78	2.91 / 4.71	2.86 / 4.60	2.82 / 4.52	2.77 / 4.41	2.74 / 4.33	2.70 / 4.25	2.67 / 4.17	2.64 / 4.12	2.61 / 4.05	2.59 / 4.01	2.56 / 3.96	2.55 / 3.93	2.54 / 3.91
11	4.84 / 9.65	3.98 / 7.20	3.59 / 6.22	3.36 / 5.67	3.20 / 5.32	3.09 / 5.07	3.01 / 4.88	2.95 / 4.74	2.90 / 4.63	2.86 / 4.54	2.82 / 4.46	2.79 / 4.40	2.74 / 4.29	2.70 / 4.21	2.65 / 4.10	2.61 / 4.02	2.57 / 3.94	2.53 / 3.86	2.50 / 3.80	2.47 / 3.74	2.45 / 3.70	2.42 / 3.66	2.41 / 3.62	2.40 / 3.60
12	4.75 / 9.33	3.88 / 6.93	3.49 / 5.95	3.26 / 5.41	3.11 / 5.06	3.00 / 4.82	2.92 / 4.65	2.85 / 4.50	2.80 / 4.39	2.76 / 4.30	2.72 / 4.22	2.69 / 4.16	2.64 / 4.05	2.60 / 3.98	2.54 / 3.86	2.50 / 3.78	2.46 / 3.70	2.42 / 3.61	2.40 / 3.56	2.36 / 3.49	2.35 / 3.46	2.32 / 3.41	2.31 / 3.38	2.30 / 3.36
13	4.67 / 9.07	3.80 / 6.70	3.41 / 5.74	3.18 / 5.20	3.02 / 4.86	2.92 / 4.62	2.84 / 4.44	2.77 / 4.30	2.72 / 4.19	2.67 / 4.10	2.63 / 4.02	2.60 / 3.96	2.55 / 3.85	2.51 / 3.78	2.46 / 3.67	2.42 / 3.59	2.38 / 3.51	2.34 / 3.42	2.32 / 3.37	2.28 / 3.30	2.26 / 3.27	2.24 / 3.21	2.22 / 3.18	2.21 / 3.16
14	4.60 / 8.86	3.74 / 6.51	3.34 / 5.56	3.11 / 5.03	2.96 / 4.69	2.85 / 4.46	2.77 / 4.28	2.70 / 4.14	2.65 / 4.03	2.60 / 3.94	2.56 / 3.86	2.53 / 3.80	2.48 / 3.70	2.44 / 3.62	2.39 / 3.51	2.35 / 3.43	2.31 / 3.34	2.27 / 3.26	2.24 / 3.21	2.21 / 3.14	2.19 / 3.11	2.16 / 3.06	2.14 / 3.02	2.13 / 3.00
15	4.54 / 8.68	3.68 / 6.36	3.29 / 5.42	3.06 / 4.89	2.90 / 4.56	2.79 / 4.32	2.70 / 4.14	2.64 / 4.00	2.59 / 3.89	2.55 / 3.80	2.51 / 3.73	2.48 / 3.67	2.43 / 3.56	2.39 / 3.48	2.33 / 3.36	2.29 / 3.29	2.25 / 3.20	2.21 / 3.12	2.18 / 3.07	2.15 / 3.00	2.12 / 2.97	2.10 / 2.92	2.08 / 2.89	2.07 / 2.87
16	4.49 / 8.53	3.63 / 6.23	3.24 / 5.29	3.01 / 4.77	2.85 / 4.44	2.74 / 4.20	2.66 / 4.03	2.59 / 3.89	2.54 / 3.78	2.49 / 3.69	2.45 / 3.61	2.42 / 3.55	2.37 / 3.45	2.33 / 3.37	2.28 / 3.25	2.24 / 3.18	2.20 / 3.10	2.16 / 3.01	2.13 / 2.96	2.09 / 2.89	2.07 / 2.86	2.04 / 2.80	2.02 / 2.77	2.01 / 2.75
17	4.45 / 8.40	3.59 / 6.11	3.20 / 5.18	2.96 / 4.67	2.81 / 4.34	2.70 / 4.10	2.62 / 3.93	2.55 / 3.79	2.50 / 3.68	2.45 / 3.59	2.41 / 3.52	2.38 / 3.45	2.33 / 3.35	2.29 / 3.27	2.23 / 3.16	2.19 / 3.08	2.15 / 3.00	2.11 / 2.92	2.08 / 2.86	2.04 / 2.79	2.02 / 2.76	1.99 / 2.70	1.97 / 2.67	1.96 / 2.65
18	4.41 / 8.28	3.55 / 6.01	3.16 / 5.09	2.93 / 4.58	2.77 / 4.25	2.66 / 4.01	2.58 / 3.85	2.51 / 3.71	2.46 / 3.60	2.41 / 3.51	2.37 / 3.44	2.34 / 3.37	2.29 / 3.27	2.25 / 3.19	2.19 / 3.07	2.15 / 3.00	2.11 / 2.91	2.07 / 2.83	2.04 / 2.78	2.00 / 2.71	1.98 / 2.68	1.95 / 2.62	1.93 / 2.59	1.92 / 2.57
19	4.38 / 8.18	3.52 / 5.93	3.13 / 5.01	2.90 / 4.50	2.74 / 4.17	2.63 / 3.94	2.55 / 3.77	2.48 / 3.63	2.43 / 3.52	2.38 / 3.43	2.34 / 3.36	2.31 / 3.30	2.26 / 3.19	2.21 / 3.12	2.15 / 3.00	2.11 / 2.92	2.07 / 2.84	2.02 / 2.76	2.00 / 2.70	1.96 / 2.63	1.94 / 2.60	1.91 / 2.54	1.90 / 2.51	1.88 / 2.49
20	4.35 / 8.10	3.49 / 5.85	3.10 / 4.94	2.87 / 4.43	2.71 / 4.10	2.60 / 3.87	2.52 / 3.71	2.45 / 3.56	2.40 / 3.45	2.35 / 3.37	2.31 / 3.30	2.28 / 3.23	2.23 / 3.13	2.18 / 3.05	2.12 / 2.94	2.08 / 2.86	2.04 / 2.77	1.99 / 2.69	1.96 / 2.63	1.92 / 2.56	1.90 / 2.53	1.87 / 2.47	1.85 / 2.44	1.84 / 2.42

DEGREES OF FREEDOM FOR DENOMINATOR

Degrees of Freedom for Numerator

Each cell shows the upper value (α = .05) and lower value (α = .01).

df (Denom.)	1	2	3	4	5	6	7	8	9	10	11	12	14	16	20	24	30	40	50	75	100	200	500	∞
21	4.32 / 8.02	3.47 / 5.78	3.07 / 4.87	2.84 / 4.37	2.68 / 4.04	2.57 / 3.81	2.49 / 3.65	2.42 / 3.51	2.37 / 3.40	2.32 / 3.31	2.28 / 3.24	2.25 / 3.17	2.20 / 3.07	2.15 / 2.99	2.09 / 2.88	2.05 / 2.80	2.00 / 2.72	1.96 / 2.63	1.93 / 2.58	1.80 / 2.51	1.87 / 2.47	1.84 / 2.42	1.82 / 2.38	1.81 / 2.36
22	4.30 / 7.94	3.44 / 5.72	3.05 / 4.82	2.82 / 4.31	2.66 / 3.99	2.55 / 3.76	2.47 / 3.59	2.40 / 3.45	2.35 / 3.35	2.30 / 3.26	2.26 / 3.18	2.23 / 3.12	2.18 / 3.02	2.13 / 2.94	2.07 / 2.83	2.03 / 2.75	1.98 / 2.67	1.93 / 2.58	1.91 / 2.53	1.87 / 2.46	1.84 / 2.42	1.81 / 2.37	1.80 / 2.33	1.78 / 2.31
23	4.28 / 7.88	3.42 / 5.66	3.03 / 4.76	2.80 / 4.26	2.64 / 3.94	2.53 / 3.71	2.45 / 3.54	2.38 / 3.41	2.32 / 3.30	2.28 / 3.21	2.24 / 3.14	2.20 / 3.07	2.14 / 2.97	2.10 / 2.89	2.04 / 2.78	2.00 / 2.70	1.96 / 2.62	1.91 / 2.53	1.88 / 2.48	1.84 / 2.41	1.82 / 2.37	1.79 / 2.32	1.77 / 2.28	1.76 / 2.26
24	4.26 / 7.82	3.40 / 5.61	3.01 / 4.72	2.78 / 4.22	2.62 / 3.90	2.51 / 3.67	2.43 / 3.50	2.36 / 3.36	2.30 / 3.25	2.26 / 3.17	2.22 / 3.09	2.18 / 3.03	2.13 / 2.93	2.09 / 2.85	2.02 / 2.74	1.98 / 2.66	1.94 / 2.58	1.89 / 2.49	1.86 / 2.44	1.82 / 2.36	1.80 / 2.33	1.76 / 2.27	1.74 / 2.23	1.73 / 2.21
25	4.24 / 7.77	3.38 / 5.57	2.99 / 4.68	2.76 / 4.18	2.60 / 3.86	2.49 / 3.63	2.41 / 3.46	2.34 / 3.32	2.28 / 3.21	2.24 / 3.13	2.20 / 3.05	2.16 / 2.99	2.11 / 2.89	2.06 / 2.81	2.00 / 2.70	1.96 / 2.62	1.92 / 2.54	1.87 / 2.45	1.84 / 2.40	1.80 / 2.32	1.77 / 2.29	1.74 / 2.23	1.72 / 2.19	1.71 / 2.17
26	4.22 / 7.72	3.37 / 5.53	2.98 / 4.64	2.74 / 4.14	2.59 / 3.82	2.47 / 3.59	2.39 / 3.42	2.32 / 3.29	2.27 / 3.17	2.22 / 3.09	2.18 / 3.02	2.15 / 2.96	2.10 / 2.86	2.05 / 2.77	1.99 / 2.66	1.95 / 2.58	1.90 / 2.50	1.85 / 2.41	1.82 / 2.36	1.78 / 2.28	1.76 / 2.25	1.72 / 2.19	1.70 / 2.15	1.69 / 2.13
27	4.21 / 7.68	3.35 / 5.49	2.96 / 4.60	2.73 / 4.11	2.57 / 3.79	2.46 / 3.56	2.37 / 3.39	2.30 / 3.26	2.25 / 3.14	2.20 / 3.06	2.16 / 2.98	2.13 / 2.93	2.08 / 2.83	2.03 / 2.74	1.97 / 2.63	1.93 / 2.55	1.88 / 2.47	1.84 / 2.38	1.80 / 2.33	1.76 / 2.25	1.74 / 2.21	1.71 / 2.16	1.68 / 2.12	1.67 / 2.10
28	4.20 / 7.64	3.34 / 5.45	2.95 / 4.57	2.71 / 4.07	2.56 / 3.76	2.44 / 3.53	2.36 / 3.36	2.29 / 3.23	2.24 / 3.11	2.19 / 3.03	2.15 / 2.95	2.12 / 2.90	2.06 / 2.80	2.02 / 2.71	1.95 / 2.60	1.91 / 2.52	1.87 / 2.44	1.81 / 2.35	1.78 / 2.30	1.75 / 2.22	1.72 / 2.18	1.69 / 2.13	1.67 / 2.09	1.65 / 2.06
29	4.18 / 7.60	3.33 / 5.42	2.93 / 4.54	2.70 / 4.04	2.54 / 3.73	2.43 / 3.50	2.35 / 3.32	2.28 / 3.20	2.22 / 3.08	2.18 / 3.00	2.14 / 2.92	2.10 / 2.87	2.05 / 2.77	2.00 / 2.68	1.94 / 2.57	1.90 / 2.49	1.85 / 2.41	1.80 / 2.32	1.77 / 2.27	1.73 / 2.19	1.71 / 2.15	1.68 / 2.10	1.65 / 2.06	1.64 / 2.03
30	4.17 / 7.56	3.32 / 5.39	2.92 / 4.51	2.69 / 4.02	2.53 / 3.70	2.42 / 3.47	2.34 / 3.30	2.27 / 3.17	2.21 / 3.06	2.16 / 2.98	2.12 / 2.90	2.09 / 2.84	2.04 / 2.74	1.99 / 2.66	1.93 / 2.55	1.89 / 2.47	1.84 / 2.38	1.79 / 2.29	1.76 / 2.24	1.72 / 2.16	1.69 / 2.13	1.66 / 2.07	1.64 / 2.03	1.62 / 2.01
32	4.15 / 7.50	3.30 / 5.34	2.90 / 4.46	2.67 / 3.97	2.51 / 3.66	2.40 / 3.42	2.32 / 3.25	2.25 / 3.12	2.19 / 3.01	2.14 / 2.94	2.10 / 2.86	2.07 / 2.80	2.02 / 2.70	1.97 / 2.62	1.91 / 2.51	1.86 / 2.42	1.82 / 2.34	1.76 / 2.25	1.74 / 2.20	1.69 / 2.12	1.67 / 2.08	1.64 / 2.02	1.61 / 1.98	1.59 / 1.96
34	4.13 / 7.44	3.28 / 5.29	2.88 / 4.42	2.65 / 3.93	2.49 / 3.61	2.38 / 3.38	2.30 / 3.21	2.23 / 3.08	2.17 / 2.97	2.12 / 2.89	2.08 / 2.82	2.05 / 2.76	2.00 / 2.66	1.95 / 2.58	1.89 / 2.47	1.84 / 2.38	1.80 / 2.30	1.74 / 2.21	1.71 / 2.15	1.67 / 2.08	1.64 / 2.04	1.61 / 1.98	1.59 / 1.94	1.57 / 1.91
36	4.11 / 7.39	3.26 / 5.25	2.86 / 4.38	2.63 / 3.89	2.48 / 3.58	2.36 / 3.35	2.28 / 3.18	2.21 / 3.04	2.15 / 2.94	2.10 / 2.86	2.06 / 2.78	2.03 / 2.72	1.98 / 2.62	1.93 / 2.54	1.87 / 2.43	1.82 / 2.35	1.78 / 2.26	1.72 / 2.17	1.69 / 2.12	1.65 / 2.04	1.62 / 2.00	1.59 / 1.94	1.56 / 1.90	1.55 / 1.87
38	4.10 / 7.35	3.25 / 5.21	2.85 / 4.34	2.62 / 3.86	2.46 / 3.54	2.35 / 3.32	2.26 / 3.15	2.19 / 3.02	2.14 / 2.91	2.09 / 2.82	2.05 / 2.75	2.02 / 2.69	1.96 / 2.59	1.92 / 2.51	1.85 / 2.40	1.80 / 2.32	1.76 / 2.22	1.71 / 2.14	1.67 / 2.08	1.63 / 2.00	1.60 / 1.97	1.57 / 1.90	1.54 / 1.86	1.53 / 1.84
40	4.08 / 7.31	3.23 / 5.18	2.84 / 4.31	2.61 / 3.83	2.45 / 3.51	2.34 / 3.29	2.25 / 3.12	2.18 / 2.99	2.12 / 2.88	2.07 / 2.80	2.04 / 2.73	2.00 / 2.66	1.95 / 2.56	1.90 / 2.49	1.84 / 2.37	1.79 / 2.29	1.74 / 2.20	1.69 / 2.11	1.66 / 2.05	1.61 / 1.97	1.59 / 1.94	1.55 / 1.88	1.53 / 1.84	1.51 / 1.81
42	4.07 / 7.27	3.22 / 5.15	2.83 / 4.29	2.59 / 3.80	2.44 / 3.49	2.32 / 3.26	2.24 / 3.10	2.17 / 2.96	2.11 / 2.86	2.06 / 2.77	2.02 / 2.70	1.99 / 2.64	1.94 / 2.54	1.89 / 2.46	1.82 / 2.35	1.78 / 2.26	1.73 / 2.17	1.68 / 2.08	1.64 / 2.02	1.60 / 1.94	1.57 / 1.91	1.54 / 1.85	1.51 / 1.80	1.49 / 1.78
44	4.06 / 7.24	3.21 / 5.12	2.82 / 4.26	2.58 / 3.78	2.43 / 3.46	2.31 / 3.24	2.23 / 3.07	2.16 / 2.94	2.10 / 2.84	2.05 / 2.75	2.01 / 2.68	1.98 / 2.62	1.92 / 2.52	1.88 / 2.44	1.81 / 2.32	1.76 / 2.24	1.72 / 2.15	1.66 / 2.06	1.63 / 2.01	1.58 / 1.92	1.56 / 1.88	1.52 / 1.82	1.50 / 1.78	1.48 / 1.75
46	4.05 / 7.21	3.20 / 5.10	2.81 / 4.24	2.57 / 3.76	2.42 / 3.44	2.30 / 3.22	2.22 / 3.05	2.14 / 2.92	2.09 / 2.82	2.04 / 2.73	2.00 / 2.66	1.97 / 2.60	1.91 / 2.50	1.87 / 2.42	1.80 / 2.30	1.75 / 2.22	1.71 / 2.13	1.65 / 2.04	1.62 / 1.98	1.57 / 1.90	1.54 / 1.86	1.51 / 1.80	1.48 / 1.76	1.46 / 1.72
48	4.04 / 7.19	3.19 / 5.08	2.80 / 4.22	2.56 / 3.74	2.41 / 3.42	2.30 / 3.20	2.21 / 3.04	2.14 / 2.90	2.08 / 2.80	2.03 / 2.71	1.99 / 2.64	1.96 / 2.58	1.90 / 2.48	1.86 / 2.40	1.79 / 2.28	1.74 / 2.20	1.70 / 2.11	1.64 / 2.02	1.61 / 1.96	1.56 / 1.88	1.53 / 1.84	1.50 / 1.78	1.47 / 1.73	1.45 / 1.70
50	4.03 / 7.17	3.18 / 5.06	2.79 / 4.20	2.56 / 3.72	2.40 / 3.41	2.29 / 3.18	2.20 / 3.02	2.13 / 2.88	2.07 / 2.78	2.02 / 2.70	1.98 / 2.62	1.95 / 2.56	1.90 / 2.46	1.85 / 2.39	1.78 / 2.26	1.74 / 2.18	1.69 / 2.10	1.63 / 2.00	1.60 / 1.94	1.55 / 1.86	1.52 / 1.82	1.48 / 1.76	1.46 / 1.71	1.44 / 1.68

DEGREES OF FREEDOM FOR DENOMINATOR

	1	2	3	4	5	6	7	8	9	10	11	12	14	16	20	24	30	40	50	75	100	200	500	∞
55	4.02 / 7.12	3.17 / 5.01	2.78 / 4.16	2.54 / 3.68	2.38 / 3.37	2.27 / 3.15	2.18 / 2.98	2.11 / 2.85	2.05 / 2.75	2.00 / 2.66	1.97 / 2.59	1.93 / 2.53	1.88 / 2.43	1.83 / 2.35	1.76 / 2.23	1.72 / 2.15	1.67 / 2.06	1.61 / 1.96	1.58 / 1.90	1.52 / 1.82	1.50 / 1.78	1.46 / 1.71	1.43 / 1.66	1.41 / 1.64
60	4.00 / 7.08	3.15 / 4.98	2.76 / 4.13	2.52 / 3.65	2.37 / 3.34	2.25 / 3.12	2.17 / 2.95	2.10 / 2.82	2.04 / 2.72	1.99 / 2.63	1.95 / 2.56	1.92 / 2.50	1.86 / 2.40	1.81 / 2.32	1.75 / 2.20	1.70 / 2.12	1.65 / 2.03	1.59 / 1.93	1.56 / 1.87	1.50 / 1.79	1.48 / 1.74	1.44 / 1.68	1.41 / 1.63	1.39 / 1.60
65	3.99 / 7.04	3.14 / 4.95	2.75 / 4.10	2.51 / 3.62	2.36 / 3.31	2.24 / 3.09	2.15 / 2.93	2.08 / 2.79	2.02 / 2.70	1.98 / 2.61	1.94 / 2.54	1.90 / 2.47	1.85 / 2.37	1.80 / 2.30	1.73 / 2.18	1.68 / 2.09	1.63 / 2.00	1.57 / 1.90	1.54 / 1.84	1.49 / 1.76	1.46 / 1.71	1.42 / 1.64	1.39 / 1.60	1.37 / 1.56
70	3.98 / 7.01	3.13 / 4.92	2.74 / 4.08	2.50 / 3.60	2.35 / 3.29	2.23 / 3.07	2.14 / 2.91	2.07 / 2.77	2.01 / 2.67	1.97 / 2.59	1.93 / 2.51	1.89 / 2.45	1.84 / 2.35	1.79 / 2.28	1.72 / 2.15	1.67 / 2.07	1.62 / 1.98	1.56 / 1.88	1.53 / 1.82	1.47 / 1.74	1.45 / 1.69	1.40 / 1.62	1.37 / 1.56	1.35 / 1.53
80	3.96 / 6.96	3.11 / 4.88	2.72 / 4.04	2.48 / 3.56	2.33 / 3.25	2.21 / 3.04	2.12 / 2.87	2.05 / 2.74	1.99 / 2.64	1.95 / 2.55	1.91 / 2.48	1.88 / 2.41	1.82 / 2.32	1.77 / 2.24	1.70 / 2.11	1.65 / 2.03	1.60 / 1.94	1.54 / 1.84	1.51 / 1.78	1.45 / 1.70	1.42 / 1.65	1.38 / 1.57	1.35 / 1.52	1.32 / 1.49
100	3.94 / 6.90	3.09 / 4.82	2.70 / 3.98	2.46 / 3.51	2.30 / 3.20	2.19 / 2.99	2.10 / 2.82	2.03 / 2.69	1.97 / 2.59	1.92 / 2.51	1.88 / 2.43	1.85 / 2.36	1.79 / 2.26	1.75 / 2.19	1.68 / 2.06	1.63 / 1.98	1.57 / 1.89	1.51 / 1.79	1.48 / 1.73	1.42 / 1.64	1.39 / 1.59	1.34 / 1.51	1.30 / 1.46	1.28 / 1.43
125	3.92 / 6.84	3.07 / 4.78	2.68 / 3.94	2.44 / 3.47	2.29 / 3.17	2.17 / 2.95	2.08 / 2.79	2.01 / 2.65	1.95 / 2.56	1.90 / 2.47	1.86 / 2.40	1.83 / 2.33	1.77 / 2.23	1.72 / 2.15	1.65 / 2.03	1.60 / 1.94	1.55 / 1.85	1.49 / 1.75	1.45 / 1.68	1.39 / 1.59	1.36 / 1.54	1.31 / 1.46	1.27 / 1.40	1.25 / 1.37
150	3.91 / 6.81	3.06 / 4.75	2.67 / 3.91	2.43 / 3.44	2.27 / 3.13	2.16 / 2.92	2.07 / 2.76	2.00 / 2.62	1.94 / 2.53	1.89 / 2.44	1.85 / 2.37	1.82 / 2.30	1.76 / 2.20	1.71 / 2.12	1.64 / 2.00	1.59 / 1.91	1.54 / 1.83	1.47 / 1.72	1.44 / 1.66	1.37 / 1.56	1.34 / 1.51	1.29 / 1.43	1.25 / 1.37	1.22 / 1.33
200	3.89 / 6.76	3.04 / 4.71	2.65 / 3.88	2.41 / 3.41	2.26 / 3.11	2.14 / 2.90	2.05 / 2.73	1.98 / 2.60	1.92 / 2.50	1.87 / 2.41	1.83 / 2.34	1.80 / 2.28	1.74 / 2.17	1.69 / 2.09	1.62 / 1.97	1.57 / 1.88	1.52 / 1.79	1.45 / 1.69	1.42 / 1.62	1.35 / 1.53	1.32 / 1.48	1.26 / 1.39	1.22 / 1.33	1.19 / 1.28
400	3.86 / 6.70	3.02 / 4.66	2.62 / 3.83	2.39 / 3.36	2.23 / 3.06	2.12 / 2.85	2.03 / 2.69	1.96 / 2.55	1.90 / 2.46	1.85 / 2.37	1.81 / 2.29	1.78 / 2.23	1.72 / 2.12	1.67 / 2.04	1.60 / 1.92	1.54 / 1.84	1.49 / 1.74	1.42 / 1.64	1.38 / 1.57	1.32 / 1.47	1.28 / 1.42	1.22 / 1.32	1.16 / 1.24	1.13 / 1.19
1,000	3.85 / 6.66	3.00 / 4.62	2.61 / 3.80	2.38 / 3.34	2.22 / 3.04	2.10 / 2.82	2.02 / 2.66	1.95 / 2.53	1.89 / 2.43	1.84 / 2.34	1.80 / 2.26	1.76 / 2.20	1.70 / 2.09	1.65 / 2.01	1.58 / 1.89	1.53 / 1.81	1.47 / 1.71	1.41 / 1.61	1.36 / 1.54	1.30 / 1.44	1.26 / 1.38	1.19 / 1.28	1.13 / 1.19	1.08 / 1.11
∞	3.84 / 6.64	2.99 / 4.60	2.60 / 3.78	2.37 / 3.32	2.21 / 3.02	2.09 / 2.80	2.01 / 2.64	1.94 / 2.51	1.88 / 2.41	1.83 / 2.32	1.79 / 2.24	1.75 / 2.18	1.69 / 2.07	1.64 / 1.99	1.57 / 1.87	1.52 / 1.79	1.46 / 1.69	1.40 / 1.59	1.35 / 1.52	1.28 / 1.41	1.24 / 1.36	1.17 / 1.25	1.11 / 1.15	1.00 / 1.00

DEGREES OF FREEDOM FOR DENOMINATOR

Studentized Range Values (*q*)

The following table presents Studentized range values (*q*) for overall alpha levels of .05 and .01 for a set of comparisons. The values in the column headed "df for denominator" represent df_{WITHIN} when a between-subjects design is used and df_{ERROR} when a repeated measures design is used. The values in the column headed "alpha" represent overall alpha levels of .05 and .01 for each number of degrees of freedom within or degrees of freedom error. The values at the top of the table under the heading "*k* = number of levels of the independent variable" represent the number of levels of the independent variable of interest. For example, the Studentized range value (*q*) for $df_{WITHIN} = 12$, an overall alpha level of .05, and *k* = 3 (that is, an independent variable that has three levels) is 3.77.

df for denominator	Alpha	k = Number of Levels of the Independent Variable									
		2	3	4	5	6	7	8	9	10	11
5	.05	3.64	4.60	5.22	5.67	6.03	6.33	6.58	6.80	6.99	7.17
	.01	5.70	6.98	7.80	8.42	8.91	9.32	9.67	9.97	10.24	10.48
6	.05	3.46	4.34	4.90	5.30	5.63	5.90	6.12	6.32	6.49	6.65
	.01	5.24	6.33	7.03	7.56	7.97	8.32	8.61	8.87	9.10	9.30
7	.05	3.34	4.16	4.68	5.06	5.36	5.61	5.82	6.00	6.16	6.30
	.01	4.95	5.92	6.54	7.01	7.37	7.68	7.94	8.17	8.37	8.55
8	.05	3.26	4.04	4.53	4.89	5.17	5.40	5.60	5.77	5.92	6.05
	.01	4.75	5.64	6.20	6.62	6.96	7.24	7.47	7.68	7.86	8.03
9	.05	3.20	3.95	4.41	4.76	5.02	5.24	5.43	5.59	5.74	5.87
	.01	4.60	5.43	5.96	6.35	6.66	6.91	7.13	7.33	7.49	7.65
10	.05	3.15	3.88	4.33	4.65	4.91	5.12	5.30	5.46	5.60	5.72
	.01	4.48	5.27	5.77	6.14	6.43	6.67	6.87	7.05	7.21	7.36
11	.05	3.11	3.82	4.26	4.57	4.82	5.03	5.20	5.35	5.49	5.61
	.01	4.39	5.15	5.62	5.97	6.25	6.48	6.67	6.84	6.99	7.13
12	.05	3.08	3.77	4.20	4.51	4.75	4.95	5.12	5.27	5.39	5.51
	.01	4.32	5.05	5.50	5.84	6.10	6.32	6.51	6.67	6.81	6.94
13	.05	3.06	3.73	4.15	4.45	4.69	4.88	5.05	5.19	5.32	5.43
	.01	4.26	4.96	5.40	5.73	5.98	6.19	6.37	6.53	6.67	6.79
14	.05	3.03	3.70	4.11	4.41	4.64	4.83	4.99	5.13	5.25	5.36
	.01	4.21	4.89	5.32	5.63	5.88	6.08	6.26	6.41	6.54	6.66
15	.05	3.01	3.67	4.08	4.37	4.59	4.78	4.94	5.08	5.20	5.31
	.01	4.17	4.84	5.25	5.56	5.80	5.99	6.16	6.31	6.44	6.55
16	.05	3.00	3.65	4.05	4.33	4.56	4.74	4.90	5.03	5.15	5.26
	.01	4.13	4.79	5.19	5.49	5.72	5.92	6.08	6.22	6.35	6.46
17	.05	2.98	3.63	4.02	4.30	4.52	4.70	4.86	4.99	5.11	5.21
	.01	4.10	4.74	5.14	5.43	5.66	5.85	6.01	6.15	6.27	6.38
18	.05	2.97	3.61	4.00	4.28	4.49	4.67	4.82	4.96	5.07	5.17
	.01	4.07	4.70	5.09	5.38	5.60	5.79	5.94	6.08	6.20	6.31
19	.05	2.96	3.59	3.98	4.25	4.47	4.65	4.79	4.92	5.04	5.14
	.01	4.05	4.67	5.05	5.33	5.55	5.73	5.89	6.02	6.14	6.25
20	.05	2.95	3.58	3.96	4.23	4.45	4.62	4.77	4.90	5.01	5.11
	.01	4.02	4.64	5.02	5.29	5.51	5.69	5.84	5.97	6.09	6.19
24	.05	2.92	3.53	3.90	4.17	4.37	4.54	4.68	4.81	4.92	5.01
	.01	3.96	4.55	4.91	5.17	5.37	5.54	5.69	5.81	5.92	6.02
30	.05	2.89	3.49	3.85	4.10	4.30	4.46	4.60	4.72	4.82	4.92
	.01	3.89	4.45	4.80	5.05	5.24	5.40	5.54	5.65	5.76	5.85
40	.05	2.86	3.44	3.79	4.04	4.23	4.39	4.52	4.63	4.73	4.82
	.01	3.82	4.37	4.70	4.93	5.11	5.26	5.39	5.50	5.60	5.69
60	.05	2.83	3.40	3.74	3.98	4.16	4.31	4.44	4.55	4.65	4.73
	.01	3.76	4.28	4.59	4.82	4.99	5.13	5.25	5.36	5.45	5.53
120	05	2.80	3.36	3.68	3.92	4.10	4.24	4.36	4.47	4.56	4.64
	.01	3.70	4.20	4.50	4.71	4.87	5.01	5.12	5.21	5.30	5.37
∞	.05	2.77	3.31	3.63	3.86	4.03	4.17	4.29	4.39	4.47	4.55
	.01	3.64	4.12	4.40	4.60	4.76	4.88	4.99	5.08	5.16	5.23

df for denominator	Alpha	\multicolumn{9}{c}{k = Number of Levels of the Independent Variable}
		12	13	14	15	16	17	18	19	20
5	.05	7.32	7.47	7.60	7.72	7.83	7.93	8.03	8.12	8.21
	.01	10.70	10.89	11.08	11.24	11.40	11.55	11.68	11.81	11.93
6	.05	6.79	6.92	7.03	7.14	7.24	7.34	7.43	7.51	7.59
	.01	9.48	9.65	9.81	9.95	10.08	10.21	10.32	10.43	10.54
7	.05	6.43	6.55	6.66	6.76	6.85	6.94	7.02	7.10	7.17
	.01	8.71	8.86	9.00	9.12	9.24	9.35	9.46	9.55	9.65
8	.05	6.18	6.29	6.39	6.48	6.57	6.65	6.73	6.80	6.87
	.01	8.18	8.31	8.44	8.55	8.66	8.76	8.85	8.94	9.03
9	.05	5.98	6.09	6.19	6.28	6.36	6.44	6.51	6.58	6.64
	.01	7.78	7.91	8.03	8.13	8.23	8.33	8.41	8.49	8.57
10	.05	5.83	5.93	6.03	6.11	6.19	6.27	6.34	6.40	6.47
	.01	7.49	7.60	7.71	7.81	7.91	7.99	8.08	8.15	8.23
11	.05	5.71	5.81	5.90	5.98	6.06	6.13	6.20	6.27	6.33
	.01	7.25	7.36	7.46	7.56	7.65	7.73	7.81	7.88	7.95
12	.05	5.61	5.71	5.80	5.88	5.95	6.02	6.09	6.15	6.21
	.01	7.06	7.17	7.26	7.36	7.44	7.52	7.59	7.66	7.73
13	.05	5.53	5.63	5.71	5.79	5.86	5.93	5.99	6.05	6.11
	.01	6.90	7.01	7.10	7.19	7.27	7.35	7.42	7.48	7.55
14	.05	5.46	5.55	5.64	5.71	5.79	5.85	5.91	5.97	6.03
	.01	6.77	6.87	6.96	7.05	7.13	7.20	7.27	7.33	7.39
15	.05	5.40	5.49	5.57	5.65	5.72	5.78	5.85	5.90	5.96
	.01	6.66	6.76	6.84	6.93	7.00	7.07	7.14	7.20	7.26
16	.05	5.35	5.44	5.52	5.59	5.66	5.73	5.79	5.84	5.90
	.01	6.56	6.66	6.74	6.82	6.90	6.97	7.03	7.09	7.15
17	.05	5.31	5.39	5.47	5.54	5.61	5.67	5.73	5.79	5.84
	.01	6.48	6.57	6.66	6.73	6.81	6.87	6.94	7.00	7.05
18	.05	5.27	5.35	5.43	5.50	5.57	5.63	5.69	5.74	5.79
	.01	6.41	6.50	6.58	6.65	6.73	6.79	6.85	6.91	6.97
19	.05	5.23	5.31	5.39	5.46	5.53	5.59	5.65	5.70	5.75
	.01	6.34	6.43	6.51	6.58	6.65	6.72	6.78	6.84	6.89
20	.05	5.20	5.28	5.36	5.43	5.49	5.55	5.61	5.66	5.71
	.01	6.28	6.37	6.45	6.52	6.59	6.65	6.71	6.77	6.82
24	.05	5.10	5.18	5.25	5.32	5.38	5.44	5.49	5.55	5.59
	.01	6.11	6.19	6.26	6.33	6.39	6.45	6.51	6.56	6.61
30	.05	5.00	5.08	5.15	5.21	5.27	5.33	5.38	5.43	5.47
	.01	5.93	6.01	6.08	6.14	6.20	6.26	6.31	6.36	6.41
40	.05	4.90	4.98	5.04	5.11	5.16	5.22	5.27	5.31	5.36
	.01	5.76	5.83	5.90	5.96	6.02	6.07	6.12	6.16	6.21
60	.05	4.81	4.88	4.94	5.00	5.06	5.11	5.15	5.20	5.24
	.01	5.60	5.67	5.73	5.78	5.84	5.89	5.93	5.97	6.01
120	.05	4.71	4.78	4.84	4.90	4.95	5.00	5.04	5.09	5.13
	.01	5.44	5.50	5.56	5.61	5.66	5.71	5.75	5.79	5.83
∞	.05	4.62	4.68	4.74	4.80	4.85	4.89	4.93	4.97	5.01
	.01	5.29	5.35	5.40	5.45	5.49	5.54	5.57	5.61	5.65

Critical Values for Pearson *r*

The following table presents critical values of Pearson *r* for directional (one-tailed) and nondirectional (two-tailed) tests of the null hypothesis $\rho = 0$. The column headed "df" lists the degrees of freedom associated with the distribution of interest. The values .05, .025, .01, and .005 at the top of the table under the heading "Level of Significance for Directional Test" represent alpha levels for directional tests. For example, the critical value of *r* for df = 25 and an alpha level of .05 for a directional test using the upper tail of the distribution (that is, for alternative hypotheses of the form $\rho > 0$) is +.323. Since the distribution is symmetrical, the critical value of *r* for a directional test using the lower tail of the distribution (that is, for null hypotheses of the form $\rho < 0$) is −.323. The values .10, .05, .02, and .01 under the heading "Level of Significance for Nondirectional Test" represent alpha levels for nondirectional tests. For example, the critical values of *r* for a nondirectional test for df = 25 and an alpha level of .05 are ±.381.

From *Statistical Analysis in Psychology and Education*, by G. A. Ferguson. Copyright ©1976 McGraw-Hill. Reprinted with permission.

df	Level of Significance for Directional Test			
	.05	.025	.01	.005
	Level of Significance for Nondirectional Test			
	.10	.05	.02	.01
1	.988	.997	.9995	.9999
2	.900	.950	.980	.990
3	.805	.878	.934	.959
4	.729	.811	.882	.917
5	.669	.754	.833	.874
6	.622	.707	.789	.834
7	.582	.666	.750	.798
8	.549	.632	.716	.765
9	.521	.602	.685	.735
10	.497	.576	.658	.708
11	.476	.553	.634	.684
12	.458	.532	.612	.661
13	.441	.514	.592	.641
14	.426	.497	.574	.623
15	.412	.482	.558	.606
16	.400	.468	.542	.590
17	.389	.456	.528	.575
18	.378	.444	.516	.561
19	.369	.433	.503	.549
20	.360	.423	.492	.537
21	.352	.413	.482	.526
22	.344	.404	.472	.515
23	.337	.396	.462	.505
24	.330	.388	.453	.496
25	.323	.381	.445	.487
26	.317	.374	.437	.479
27	.311	.367	.430	.471
28	.306	.361	.423	.463
29	.301	.355	.416	.456
30	.296	.349	.409	.449
35	.275	.325	.381	.418
40	.257	.304	.358	.393
45	.243	.288	.338	.372
50	.231	.273	.322	.354
60	.211	.250	.295	.325
70	.195	.232	.274	.303
80	.183	.217	.256	.283
90	.173	.205	.242	.267
100	.164	.195	.230	.254

Fisher's Transformation of Pearson r (r')

The following table presents Fisher's transformation of Pearson r for use in testing null hypotheses other than $\rho = 0$. The columns headed "r" list values of the Pearson correlation coefficient, and the adjacent columns headed "r'" list the corresponding transformed values. For example, the transformed value (r') corresponding to an r of .200 is .203. Transformed values for negative correlations are found by inserting a negative sign before the tabled values of r and r'.

From *Statistical Analysis in Psychology and Education,* by G. A. Ferguson. Copyright ©1976 McGraw-Hill. Reprinted with permission.

r	r'	r	r'	r	r'	r	r'	r	r'
.000	.000	.200	.203	.400	.424	.600	.693	.800	1.099
.005	.005	.205	.208	.405	.430	.605	.701	.805	1.113
.010	.010	.210	.213	.410	.436	.610	.709	.810	1.127
.015	.015	.215	.218	.415	.442	.615	.717	.815	1.142
.020	.020	.220	.224	.420	.448	.620	.725	.820	1.157
.025	.025	.225	.229	.425	.454	.625	.733	.825	1.172
.030	.030	.230	.234	.430	.460	.630	.741	.830	1.188
.035	.035	.235	.239	.435	.466	.635	.750	.835	1.204
.040	.040	.240	.245	.440	.472	.640	.758	.840	1.221
.045	.045	.245	.250	.445	.478	.645	.767	.845	1.238
.050	.050	.250	.255	.450	.485	.650	.775	.850	1.256
.055	.055	.255	.261	.455	.491	.655	.784	.855	1.274
.060	.060	.260	.266	.460	.497	.660	.793	.860	1.293
.065	.065	.265	.271	.465	.504	.665	.802	.865	1.313
.070	.070	.270	.277	.470	.510	.670	.811	.870	1.333
.075	.075	.275	.282	.475	.517	.675	.820	.875	1.354
.080	.080	.280	.288	.480	.523	.680	.829	.880	1.376
.085	.085	.285	.293	.485	.530	.685	.838	.885	1.398
.090	.090	.290	.299	.490	.536	.690	.848	.890	1.422
.095	.095	.295	.304	.495	.543	.695	.858	.895	1.447
.100	.100	.300	.310	.500	.549	.700	.867	.900	1.472
.105	.105	.305	.315	.505	.556	.705	.877	.905	1.499
.110	.110	.310	.321	.510	.563	.710	.887	.910	1.528
.115	.116	.315	.326	.515	.570	.715	.897	.915	1.557
.120	.121	.320	.332	.520	.576	.720	.908	.920	1.589
.125	.126	.325	.337	.525	.583	.725	.918	.925	1.623
.130	.131	.330	.343	.530	.590	.730	.929	.930	1.658
.135	.136	.335	.348	.535	.597	.735	.940	.935	1.697
.140	.141	.340	.354	.540	.604	.740	.950	.940	1.738
.145	.146	.345	.360	.545	.611	.745	.962	.945	1.783
.150	.151	.350	.365	.550	.618	.750	.973	.950	1.832
.155	.156	.355	.371	.555	.626	.755	.984	.955	1.886
.160	.161	.360	.377	.560	.633	.760	.996	.960	1.946
.165	.167	.365	.383	.565	.640	.765	1.008	.965	2.014
.170	.172	.370	.388	.570	.648	.770	1.020	.970	2.092
.175	.177	.375	.394	.575	.655	.775	1.033	.975	2.185
.180	.182	.380	.400	.580	.662	.780	1.045	.980	2.298
.185	.187	.385	.406	.585	.670	.785	1.058	.985	2.443
.190	.192	.390	.412	.590	.678	.790	1.071	.990	2.647
.195	.198	.395	.418	.595	.685	.795	1.085	.995	2.994

Critical Values for the Chi-Square Distribution

The following table presents critical values of chi-square. The column headed "df" lists the degrees of freedom associated with the chi-square distribution of interest. The values at the top of the table under the heading "Level of Significance" represent alpha levels. For our purposes, we are concerned only with the low alpha levels that appear at the right of the table. The high alpha levels are used in certain advanced statistical applications. As an example of the use of this table, the critical value of chi-square for df = 4 and an alpha level of .05 is 9.488.

From *Fundamentals of Behavioral Statistics,* by R. P. Runyon & A. Haber, Addison-Wesley Longman Publishing Co., 1976.

Level of Significance

df	.99	.98	.95	.90	.80	.70	.50	.30	.20	.10	.05	.02	.01
1	.000157	.00628	.00393	.0158	.0642	.148	.455	1.074	1.642	2.706	3.841	5.412	6.635
2	.0201	.0404	.103	.211	.446	.713	1.386	2.408	3.219	4.605	5.991	7.824	9.210
3	.115	.185	.352	.584	1.005	1.424	2.366	3.665	4.642	6.251	7.815	9.837	11.341
4	.297	.429	.711	1.064	1.649	2.195	3.357	4.878	5.989	7.779	9.488	11.668	13.277
5	.554	.752	1.145	1.610	2.343	3.000	4.351	6.064	7.289	9.236	11.070	13.388	15.086
6	.872	1.134	1.635	2.204	3.070	3.828	5.348	7.231	8.558	10.645	12.592	15.033	16.812
7	1.239	1.564	2.167	2.833	3.822	4.671	6.346	8.383	9.803	12.017	14.067	16.622	18.475
8	1.646	2.032	2.733	3.490	4.594	5.527	7.344	9.524	11.030	13.362	15.507	18.168	20.090
9	2.088	2.532	3.325	4.168	5.380	6.393	8.343	10.656	12.242	14.684	16.919	19.679	21.666
10	2.558	3.059	3.940	4.865	6.179	7.267	9.342	11.781	13.442	15.987	18.307	21.161	23.209
11	3.053	3.609	4.575	5.578	6.989	8.148	10.341	12.899	14.631	17.275	19.675	22.618	24.725
12	3.571	4.178	5.226	6.304	7.807	9.034	11.340	14.011	15.812	18.549	21.026	24.054	26.217
13	4.107	4.765	5.892	7.042	8.634	9.926	12.340	15.119	16.985	19.812	22.362	25.472	27.688
14	4.660	5.368	6.571	7.790	9.467	10.821	13.339	16.222	18.151	21.064	23.685	26.873	29.141
15	5.229	5.985	7.261	8.547	10.307	11.721	14.339	17.322	19.311	22.307	24.996	28.259	30.578
16	5.812	6.614	7.962	9.312	11.152	12.624	15.338	18.418	20.465	23.542	26.296	29.633	32.000
17	6.408	7.255	8.672	10.085	12.002	13.531	16.338	19.511	21.615	24.769	27.587	30.995	33.409
18	7.015	7.906	9.390	10.865	12.857	14.440	17.338	20.601	22.760	25.989	28.869	32.346	34.805
19	7.633	8.567	10.117	11.651	13.716	15.352	18.338	21.689	23.900	27.204	30.144	33.687	36.191
20	8.260	9.237	10.851	12.443	14.578	16.266	19.337	22.775	25.038	28.412	31.410	35.020	37.566
21	8.897	9.915	11.591	13.240	15.445	17.182	20.337	23.858	26.171	29.615	32.671	36.343	38.932
22	9.542	10.600	12.338	14.041	16.314	18.101	21.337	24.939	27.301	30.813	33.924	37.659	40.289
23	10.196	11.293	13.091	14.848	17.187	19.021	22.337	26.018	28.429	32.007	35.172	38.968	41.638
24	10.856	11.992	13.848	15.659	18.062	19.943	23.337	27.096	29.553	33.196	36.415	40.270	42.980
25	11.524	12.697	14.611	16.473	18.940	20.867	24.337	28.172	30.675	34.382	37.652	41.566	44.314
26	12.198	13.409	15.379	17.292	19.820	21.792	25.336	29.246	31.795	35.563	38.885	42.856	45.642
27	12.879	14.125	16.151	18.114	20.703	22.719	26.336	30.319	32.912	36.741	40.113	44.140	46.963
28	13.565	14.847	16.928	18.939	21.588	23.647	27.336	31.391	34.027	37.916	41.337	45.419	48.278
29	14.256	15.574	17.708	19.768	22.475	24.577	28.336	32.461	35.139	39.087	42.557	46.693	49.588
30	14.953	16.306	18.493	20.599	23.364	25.508	29.336	33.530	36.250	40.256	43.773	47.962	50.892

Critical Values for the Mann–Whitney *U* Test

The following tables present critical values of the *U* statistic for the Mann–Whitney *U* test for alpha levels of .05, .025, .01, and .005 for directional (one-tailed) tests, and alpha levels of .10, .05, .02, and .01 for nondirectional (two-tailed) tests. The critical values for directional alpha levels of .01 (roman type) and .005 (**boldface**) and nondirectional alpha levels of .02 (roman type) and .01 (**boldface**) are contained in the first table. The critical values for directional alpha levels of .05 (roman type) and .025 (**boldface**) and nondirectional alpha levels of .10 (roman type) and .05 (**boldface**) are contained in the second table.

The values at the top of each table represent sample sizes (n_1) for group 1, and the values at the left of each table represent sample sizes (n_2) for group 2. The critical value of *U* is defined by the point where the sample sizes for the two groups under study intersect. For example, the critical value of *U* for a nondirectional test for an alpha level of .05 and $n_1 = 10$ and $n_2 = 12$ is 29. The observed *U* is statistically significant if it is *equal to or less than* the critical *U*.

From *Elementary Statistics*, Second Edition, by R. E. Kirk, Brooks/Cole Publishing Co., 1984.

Critical Values of *U* for Alpha Levels of .01 (Roman Type) and .005 (**Boldface**) for Directional Tests, and Alpha Levels of .02 (Roman Type) and .01 (**Boldface**) for Nondirectional Tests.

n_2 \ n_1	1	2	3	4	5	6	7	8	9	10	11	12	13	14	15	16	17	18	19	20
1	—	—	—	—	—	—	—	—	—	—	—	—	—	—	—	—	—	—	—	—
	—	—	—	—	—	—	—	—	—	—	—	—	—	—	—	—	—	—	—	—
2	—	—	—	—	—	—	—	—	—	—	—	—	0	0	0	0	0	0	1	1
	—	—	—	—	—	—	—	—	—	—	—	—	—	—	—	—	—	—	**0**	**0**
3	—	—	—	—	—	—	0	0	1	1	1	2	2	2	3	3	4	4	4	5
	—	—	—	—	—	—	—	—	**0**	**0**	**0**	**1**	**1**	**1**	**2**	**2**	**2**	**2**	**3**	**3**
4	—	—	—	—	0	1	1	2	3	3	4	5	5	6	7	7	8	9	9	10
	—	—	—	—	—	**0**	**0**	**1**	**1**	**2**	**2**	**3**	**3**	**4**	**5**	**5**	**6**	**6**	**7**	**8**
5	—	—	—	0	1	2	3	4	5	6	7	8	9	10	11	12	13	14	15	16
	—	—	—	—	**0**	**1**	**1**	**2**	**3**	**4**	**5**	**6**	**7**	**7**	**8**	**9**	**10**	**11**	**12**	**13**
6	—	—	—	1	2	3	4	6	7	8	9	11	12	13	15	16	18	19	20	22
	—	—	—	**0**	**1**	**2**	**3**	**4**	**5**	**6**	**7**	**9**	**10**	**11**	**12**	**13**	**15**	**16**	**17**	**18**
7	—	—	0	1	3	4	6	7	9	11	12	14	16	17	19	21	23	24	26	28
	—	—	—	**0**	**1**	**3**	**4**	**6**	**7**	**9**	**10**	**12**	**13**	**15**	**16**	**18**	**19**	**21**	**22**	**24**
8	—	—	0	2	4	6	7	9	11	13	15	17	20	22	24	26	28	30	32	34
	—	—	—	**1**	**2**	**4**	**6**	**7**	**9**	**11**	**13**	**15**	**17**	**18**	**20**	**22**	**24**	**26**	**28**	**30**
9	—	—	1	3	5	7	9	11	14	16	18	21	23	26	28	31	33	36	38	40
	—	—	**0**	**1**	**3**	**5**	**7**	**9**	**11**	**13**	**16**	**18**	**20**	**22**	**24**	**27**	**29**	**31**	**33**	**36**
10	—	—	1	3	6	8	11	13	16	19	22	24	27	30	33	36	38	41	44	47
	—	—	**0**	**2**	**4**	**6**	**9**	**11**	**13**	**16**	**18**	**21**	**24**	**26**	**29**	**31**	**34**	**37**	**39**	**42**
11	—	—	1	4	7	9	12	15	18	22	25	28	31	34	37	41	44	47	50	53
	—	—	**0**	**2**	**5**	**7**	**10**	**13**	**16**	**18**	**21**	**24**	**27**	**30**	**33**	**36**	**39**	**42**	**45**	**48**
12	—	—	2	5	8	11	14	17	21	24	28	31	35	38	42	46	49	53	56	60
	—	—	**1**	**3**	**6**	**9**	**12**	**15**	**18**	**21**	**24**	**27**	**31**	**34**	**37**	**41**	**44**	**47**	**51**	**54**
13	—	0	2	5	9	12	16	20	23	27	31	35	39	43	47	51	55	59	63	67
	—	—	**1**	**3**	**7**	**10**	**13**	**17**	**20**	**24**	**27**	**31**	**34**	**38**	**42**	**45**	**49**	**53**	**56**	**60**
14	—	0	2	6	10	13	17	22	26	30	34	38	43	47	51	56	60	65	69	73
	—	—	**1**	**4**	**7**	**11**	**15**	**18**	**22**	**26**	**30**	**34**	**38**	**42**	**46**	**50**	**54**	**58**	**63**	**67**
15	—	0	3	7	11	15	19	24	28	33	37	42	47	51	56	61	66	70	75	80
	—	—	**2**	**5**	**8**	**12**	**16**	**20**	**24**	**29**	**33**	**37**	**42**	**46**	**51**	**55**	**60**	**64**	**69**	**73**
16	—	0	3	7	12	16	21	26	31	36	41	46	51	56	61	66	71	76	82	87
	—	—	**2**	**5**	**9**	**13**	**18**	**22**	**27**	**31**	**36**	**41**	**45**	**50**	**55**	**60**	**65**	**70**	**74**	**79**
17	—	0	4	8	13	18	23	28	33	38	44	49	55	60	66	71	77	82	88	93
	—	—	**2**	**6**	**10**	**15**	**19**	**24**	**29**	**34**	**39**	**44**	**49**	**54**	**60**	**65**	**70**	**75**	**81**	**86**
18	—	0	4	9	14	19	24	30	36	41	47	53	59	65	70	76	82	88	94	100
	—	—	**2**	**6**	**11**	**16**	**21**	**26**	**31**	**37**	**42**	**47**	**53**	**58**	**64**	**70**	**75**	**81**	**87**	**92**
19	—	1	4	9	15	20	26	32	38	44	50	56	63	69	75	82	88	94	101	107
	—	**0**	**3**	**7**	**12**	**17**	**22**	**28**	**33**	**39**	**45**	**51**	**56**	**63**	**69**	**74**	**81**	**87**	**93**	**99**
20	—	1	5	10	16	22	28	34	40	47	53	60	67	73	80	87	93	100	107	114
	—	**0**	**3**	**8**	**13**	**18**	**24**	**30**	**36**	**42**	**48**	**54**	**60**	**67**	**73**	**79**	**86**	**92**	**99**	**105**

Note: To be statistically significant for any given n_1 and n_2, the observed *U* must be *equal to or less than* the value shown in the table. Dashes in the body of the table indicate that no decision is possible at the relevant level of alpha.

Critical Values of *U* for Alpha levels of .05 (Roman Type) and **.025 (Boldface)** for Directional Tests, and Alpha Levels of .10 (Roman Type) and **.05 (Boldface)** for Nondirectional Tests.

n_2 \ n_1	1	2	3	4	5	6	7	8	9	10	11	12	13	14	15	16	17	18	19	20
1	—	—	—	—	—	—	—	—	—	—	—	—	—	—	—	—	—	—	0	0
	—	—	—	—	—	—	—	—	—	—	—	—	—	—	—	—	—	—	—	—
2	—	—	—	—	0	0	0	1	1	1	1	2	2	2	3	3	3	4	4	4
	—	—	—	—	—	—	—	**0**	**0**	**0**	**0**	**1**	**1**	**1**	**1**	**1**	**2**	**2**	**2**	**2**
3	—	—	0	0	1	2	2	3	3	4	5	5	6	7	7	8	9	9	10	11
	—	—	—	—	**0**	**1**	**1**	**2**	**2**	**3**	**3**	**4**	**4**	**5**	**5**	**6**	**6**	**7**	**7**	**8**
4	—	—	0	1	2	3	4	5	6	7	8	9	10	11	12	14	15	16	17	18
	—	—	—	**0**	**1**	**2**	**3**	**4**	**4**	**5**	**6**	**7**	**8**	**9**	**10**	**11**	**11**	**12**	**13**	**13**
5	—	0	1	2	4	5	6	8	9	11	12	13	15	16	18	19	20	22	23	25
	—	—	**0**	**1**	**2**	**3**	**5**	**6**	**7**	**8**	**9**	**11**	**12**	**13**	**14**	**15**	**17**	**18**	**19**	**20**
6	—	0	2	3	5	7	8	10	12	14	16	17	19	21	23	25	26	28	30	32
	—	—	**1**	**2**	**3**	**5**	**6**	**8**	**10**	**11**	**13**	**14**	**16**	**17**	**19**	**21**	**22**	**24**	**25**	**27**
7	—	0	2	4	6	8	11	13	15	17	19	21	24	26	28	30	33	35	37	39
	—	—	**1**	**3**	**5**	**6**	**8**	**10**	**12**	**14**	**16**	**18**	**20**	**22**	**24**	**26**	**28**	**30**	**32**	**34**
8	—	1	3	5	8	10	13	15	18	20	23	26	28	31	33	36	39	41	44	47
	—	**0**	**2**	**4**	**6**	**8**	**10**	**13**	**15**	**17**	**19**	**22**	**24**	**26**	**29**	**31**	**34**	**36**	**38**	**41**
9	—	1	3	6	9	12	15	18	21	24	27	30	33	36	39	42	45	48	51	54
	—	**0**	**2**	**4**	**7**	**10**	**12**	**15**	**17**	**20**	**23**	**26**	**28**	**31**	**34**	**37**	**39**	**42**	**45**	**48**
10	—	1	4	7	11	14	17	20	24	27	31	34	37	41	44	48	51	55	58	62
	—	**0**	**3**	**5**	**8**	**11**	**14**	**17**	**20**	**23**	**26**	**29**	**33**	**36**	**39**	**42**	**45**	**48**	**52**	**55**
11	—	1	5	8	12	16	19	23	27	31	34	38	42	46	50	54	57	61	65	69
	—	**0**	**3**	**6**	**9**	**13**	**16**	**19**	**23**	**26**	**30**	**33**	**37**	**40**	**44**	**47**	**51**	**55**	**58**	**62**
12	—	2	5	9	13	17	21	26	30	34	38	42	47	51	55	60	64	68	72	77
	—	**1**	**4**	**7**	**11**	**14**	**18**	**22**	**26**	**29**	**33**	**37**	**41**	**45**	**49**	**53**	**57**	**61**	**65**	**69**
13	—	2	6	10	15	19	24	28	33	37	42	47	51	56	61	65	70	75	80	84
	—	**1**	**4**	**8**	**12**	**16**	**20**	**24**	**28**	**33**	**37**	**41**	**45**	**50**	**54**	**59**	**63**	**67**	**72**	**76**
14	—	2	7	11	16	21	26	31	36	41	46	51	56	61	66	71	77	82	87	92
	—	**1**	**5**	**9**	**13**	**17**	**22**	**26**	**31**	**36**	**40**	**45**	**50**	**55**	**59**	**64**	**67**	**74**	**78**	**83**
15	—	3	7	12	18	23	28	33	39	44	50	55	61	66	72	77	83	88	94	100
	—	**1**	**5**	**10**	**14**	**19**	**24**	**29**	**34**	**39**	**44**	**49**	**54**	**59**	**64**	**70**	**75**	**80**	**85**	**90**
16	—	3	8	14	19	25	30	36	42	48	54	60	65	71	77	83	89	95	101	107
	—	**1**	**6**	**11**	**15**	**21**	**26**	**31**	**37**	**42**	**47**	**53**	**59**	**64**	**70**	**75**	**81**	**86**	**92**	**98**
17	—	3	9	15	20	26	33	39	45	51	57	64	70	77	83	89	96	102	109	115
	—	**2**	**6**	**11**	**17**	**22**	**28**	**34**	**39**	**45**	**51**	**57**	**63**	**67**	**75**	**81**	**87**	**93**	**99**	**105**
18	—	4	9	16	22	28	35	41	48	55	61	68	75	82	88	95	102	109	116	123
	—	**2**	**7**	**12**	**18**	**24**	**30**	**36**	**42**	**48**	**55**	**61**	**67**	**74**	**80**	**86**	**93**	**99**	**106**	**112**
19	0	4	10	17	23	30	37	44	51	58	65	72	80	87	94	101	109	116	123	130
	—	**2**	**7**	**13**	**19**	**25**	**32**	**38**	**45**	**52**	**58**	**65**	**72**	**78**	**85**	**92**	**99**	**106**	**113**	**119**
20	0	4	11	18	25	32	39	47	54	62	69	77	84	92	100	107	115	123	130	138
	—	**2**	**8**	**13**	**20**	**27**	**34**	**41**	**48**	**55**	**62**	**69**	**76**	**83**	**90**	**98**	**105**	**112**	**119**	**127**

Note: To be statistically significant for any given n_1 and n_2, the observed *U* must be *equal to or less than* the value shown in the table. Dashes in the body of the table indicate that no decision is possible at the relevant level of alpha.

Critical Values for the Wilcoxon Signed-Rank Test

The following table presents critical values of the T statistic for directional (one-tailed) and nondirectional (two-tailed) Wilcoxon signed-rank tests. The columns headed "N" list the sample size for the investigation of interest. The values .05, .025, .01, and .005 at the top of the table under the headings "Level of Significance for Directional Test" represent alpha levels for directional tests. The values .10, .05, .02, and .01 under the headings "Level of Significance for Nondirectional Test" represent alpha levels for nondirectional tests. For example, the critical value of T for a nondirectional test for $N = 35$ and an alpha level of .05 is 195. The observed T is statistically significant if it is *equal to or less than* the critical T.

From *Critical Values and Probability Levels of the Wilcoxon Rank Sum Test and the Wilcoxon Signed Rank Test,* by F. Wilcoxon, S. Katti, and R. A. Wilcox. Copyright ©1963 and reproduced with the permission of American Cyanamid Company.

	Level of Significance for Directional Test					Level of Significance for Directional Test			
	.05	.025	.01	.005		.05	.025	.01	.005
	Level of Significance for Nondirectional Test					Level of Significance for Nondirectional Test			
N	.10	.05	.02	.01	N	.10	.05	.02	.01
5	0	—	—	—	28	130	116	101	91
6	2	0	—	—	29	140	126	110	100
7	3	2	0	—	30	151	137	120	109
8	5	3	1	0	31	163	147	130	118
9	8	5	3	1	32	175	159	140	128
10	10	8	5	3	33	187	170	151	138
11	13	10	7	5	34	200	182	162	148
12	17	13	9	7	35	213	195	173	159
13	21	17	12	9	36	227	208	185	171
14	25	21	15	12	37	241	221	198	182
15	30	25	19	15	38	256	235	211	194
16	35	29	23	19	39	271	249	224	207
17	41	34	27	23	40	286	264	238	220
18	47	40	32	27	41	302	279	252	233
19	53	46	37	32	42	319	294	266	247
20	60	52	43	37	43	336	310	281	261
21	67	58	49	42	44	353	327	296	276
22	75	65	55	48	45	371	343	312	291
23	83	73	62	54	46	389	361	328	307
24	91	81	69	61	47	407	378	345	322
25	100	89	76	68	48	426	396	362	339
26	110	98	84	75	59	446	415	379	355
27	119	107	92	83	50	466	434	397	373

Note: To be statistically significant for any given N, the observed T must be *equal to or less than* the value shown in the table. Slight discrepancies will be found between the critical values appearing in the table above and in Table 2 of the 1964 revision of F. Wilcoxon and R. A. Wilcox, *Some Rapid Approximate Statistical Procedures*, New York, Lederle Laboratories, 1964. The disparity reflects the latter's policy of selecting the critical value nearest a given significance level, occasionally overstepping that level. For example, for N = 8, the probability of a T of 3 is .0390 (nondirectional) and the probability of a T of 4 is .0546 (nondirectional). Wilcoxon and Wilcox select a T of 4 as the critical value at the .05 level of significance (nondirectional), whereas this table reflects a more conservative policy by setting a T of 3 as the critical value at this level.

Critical Values for Spearman *r*

The following table presents critical values of Spearman *r* for directional (one-tailed) and nondirectional (two-tailed) tests of the null hypothesis $\rho_s = 0$. The column headed "*N*" lists the sample size for the investigation of interest. The values .05, .025, .01, and .005 at the top of the table under the heading "Level of Significance for Directional Test" represent alpha levels for directional tests. For example, the critical value of r_s for $N = 22$ and an alpha level of .05 for a directional test using the upper tail of the distribution (that is, for alternative hypotheses of the form $\rho_s > 0$) is +.359. Since the distribution is symmetrical, the critical value of r_s for a directional test using the lower tail of the distribution (that is, for null hypotheses of the form $\rho_s < 0$) is −.359. The values .10, .05, .02, and .01 under the heading "Level of Significance for Nondirectional Test" represent alpha levels for nondirectional tests. For example, the critical values of r_s for a nondirectional test for $N = 22$ and an alpha level of .05 are ±.428.

From *Statistical Analysis in Psychology and Education*, by G. A. Ferguson. Copyright ©1976 McGraw-Hill. Reprinted with permission.

	Level of Significance for Directional Test			
	.05	.025	.01	.005
	Level of Significance for Nondirectional Test			
N	.10	.05	.02	.01
5	.900	1.000	1.000	—
6	.829	.886	.943	1.000
7	.714	.786	.893	.929
8	.643	.738	.833	.881
9	.600	.683	.783	.833
10	.564	.648	.746	.794
12	.506	.591	.712	.777
14	.456	.544	.645	.715
16	.425	.506	.601	.665
18	.399	.475	.564	.625
20	.377	.450	.534	.591
22	.359	.428	.508	.562
24	.343	.409	.485	.537
26	.329	.392	.465	.515
28	.317	.377	.448	.496
30	.306	.364	.432	.478

Glossary of Major Symbols

Numbers in parentheses indicate the sections where the symbols are first discussed.

A	Factor A (17.4)
a	Intercept of a line (5.2)/Number of levels of factor A (17.4)
$A \times B$	Interaction of factor A and factor B (17.4)
B	Factor B (17.4)
b	Slope of a line (5.2)/Number of levels of factor B (17.4)
C	Number of pairs of mean ranks or rank sums to be tested using the Dunn procedure (16.6)
c	Number of levels of the column variable in a contingency table (15.4)
$_nC_r$	Number of combinations of n things taken r at a time (6.7)
CD	Critical difference for the Tukey HSD test (12.5)
cf	Cumulative frequency (2.1)
CI	Confidence interval (8.10)
CMF_j	Column marginal frequency associated with cell j (15.4)
crf	Cumulative relative frequency (2.1)
D	Difference between raw (11.2) or rank scores (16.8) for an individual
d	Signed deviation of an individual's score from the group mean (12.2)
$\bar{D}$	Mean difference score in a sample (11.2)
df	Degrees of freedom (*Note:* If df is subscripted, the subscript indicates the source of variability to which the degrees of freedom apply.) (7.4)
E	Expected value of the sum of ranks for the Wilcoxon signed-rank test (16.5)
E_j	Expected frequency for cell j (15.4) or category j (15.14)/Expected rank sum for group j for the Wilcoxon rank sum test (16.4)
E_R^2	Epsilon-squared (16.6)
eta^2	Eta-squared statistic (10.3)
F	F ratio (12.2)
f	Frequency of a score (2.1)
G	Grand mean (10.3)
H	H statistic in the Kruskal–Wallis test (16.6)
H_0	Null hypothesis (8.2)
H_1	Alternative hypothesis (8.2)

i	Size of the interval of a numerical category (3.1)
IQR	Interquartile range (3.2)
k	Number of levels of a variable (12.2)
L	Lower real limit of a numerical category (3.1)/Number of levels of the variable that has the fewer values in a contingency table (15.6)
$\log_e$	Calculation of the natural logarithm of a number (Appendix 14.1)
$\underline{M}$	Sample mean in American Psychological Association format (3.9)
Mdn	Median (3.1)
MS	Mean square (*Note:* If MS is subscripted, the subscript indicates the source of variability to which the mean square applies.) (7.4)
N	Total sample size (1.9)
n	Number of scores in a subgroup (3.9)/Number of trials in the binomial expression (6.8)
n'	Adjusted per-cell sample size necessary to achieve a desired level of power for two-way between-subjects analysis of variance (17.10)
n_L	Number of individuals with scores less than a specified value (3.1)
n_T	Tabled per-cell sample size necessary to achieve a desired level of power for two-way between-subjects analysis of variance (17.10)
n_W	Number of individuals with scores within a numerical category (3.1)
O_j	Observed frequency for cell j (15.4) or category j (15.14)
P	Percentile (4.1)
p	Probability (2.8)/Probability of a "success" in the binomial expression (6.8)/Significance or probability level (8.11)/Sample proportion (Appendix 15.1)
$_nP_r$	Number of permutations of n things taken r at a time (6.7)
$p(A)$	Probability of event A (1.8)
$p(A|B)$	Conditional probability of event A, given event B (6.2)
$p(A, B)$	Joint probability of *both* event A *and* event B (6.3)
$p(A \text{ or } B)$	Probability of *at least one* of event A and event B (6.4)
PR_X	Percentile rank of the score X (4.1)
q	Probability of a "failure" in the binomial expression (6.8)/Studentized range value (12.5)
r	Sample Pearson correlation coefficient (5.3)/Number of "successes" in the binomial expression (6.8)/Number of levels of the row variable in a contingency table (15.4)
R^2	Multiple correlation coefficient (18.6)
r^2	Coefficient of determination (14.3)
r' (r prime)	Fisher's logarithmic transformation of r (Appendix 14.1)
r_c	Matched-pairs rank biserial correlation coefficient (16.5)
r_g	Glass rank biserial correlation coefficient (16.4)
r_i	Number of mutually exclusive and exhaustive events that can occur on trial i (6.7)

R_j	R statistic in the Wilcoxon rank sum test (16.4)/Sum of the ranks in condition j for the Kruskal–Wallis test (16.6) and Friedman analysis of variance by ranks (16.7)
R_n	Sum of the ranks of the negative differences for the Wilcoxon signed-rank test (16.5)
R_p	Sum of the ranks of the positive differences for the Wilcoxon signed-rank test (16.5)
r_s	Sample Spearman rank-order correlation coefficient (16.8)
rf	Relative frequency (2.1)
rf_j	Relative frequency in the population for category j (15.14)
RMF_j	Row marginal frequency associated with cell j (15.4)
s	Sample standard deviation (3.2)
s^2	Sample variance (3.2)
$\hat{s}$ (s-hat)	Standard deviation estimate (7.3)
$\hat{s}^2$	Variance estimate (7.3)
$\hat{s}_D$	Estimated standard deviation of population difference scores (11.2)
$\hat{s}_{\overline{D}}$	Estimated standard error of the mean of difference scores (11.2)
s_i^2	Square of the sum of the scores of individual i (13.2)
$\hat{s}^2_{\text{pooled}}$	Pooled variance estimate (10.2)
$\hat{s}_{\overline{X}}$	Estimated standard error of the mean (7.5)
$\hat{s}_{\overline{X}_1 - \overline{X}_2}$	Estimated standard error of the difference (10.2)
s_{YX}	Sample standard error of estimate (5.6)
$\hat{s}_{YX}$	Estimated standard error of estimate (14.8)
SCP	Sum of cross-products (5.3)
<u>SD</u>	Standard deviation estimate in American Psychological Association format (3.9)
SS	Sum of squares (*Note:* If SS is subscripted, the subscript indicates the source of variability to which the sum of squares applies.) (3.2)
T	T score (transformed standard score in a distribution that has a mean of 50.00 and a standard deviation of 10.00) (4.5)/Treatment effect (10.3)/T statistic in the Wilcoxon signed-rank test (16.5)
t	t statistic (8.9)/Number of scores tied at a particular rank for the Wilcoxon rank sum test and the Kruskal–Wallis test (Appendix 16.1)
T_j^2	Square of the sum of the scores in condition j (12.2)
$T_{A_i}^2$	Square of the sum of the scores at level i of factor A (17.4)
$T_{A_i B_j}^2$	Square of the sum of the scores in cell $A_i B_j$ (17.4)
$T_{B_j}^2$	Square of the sum of the scores at level j of factor B (17.4)
U	Mann–Whitney U statistic (16.4)
u	Uniqueness of a score (Applications to the Analysis of a Social Problem, Chapter 4)
V	Fourfold point correlation coefficient/Cramer's statistic (15.6)

W	Concordance coefficient (16.7)
X	General name for a variable (1.9)/A predictor variable (14.8)
$\overline{X}$	Sample mean for variable X (3.1)
$\overline{X}_i$	Mean X score for individual i across conditions (11.3)
X_n	Nullified score on variable X (10.3)
X_P	Score value defining the Pth percentile (4.1)
Y	A criterion variable (14.8)
$\hat{Y}$ (predicted Y)	Predicted score on variable Y (5.6)
z	Standard score in a normal distribution (4.3)/z statistic (8.2)
α (alpha)	Probability of a Type I error (8.5)/Intercept of a line for the population (14.1)
β (beta)	Probability of a Type II error (8.5)/Slope of a line for the population (14.1)/ Regression coefficient (18.6)
χ^2 (chi-square)	Chi-square statistic (15.4)
χ_r^2	Test statistic for Friedman analysis of variance by ranks (16.7)
ε (epsilon)	Error score for an individual in the population (14.1)
μ (mu)	Mean of a population (3.8) and of a sampling distribution of the mean (7.5)
μ_D	Mean difference score in a population and in a sampling distribution of the mean of difference scores (11.2)
π (pi)	Population proportion (Appendix 15.1)
ρ (rho)	Population Pearson correlation coefficient (14.1)
ρ'	Log-transformed value of ρ (Appendix 14.1)
ρ_s	Population Spearman rank-order correlation coefficient (16.8)
Σ (sigma)	Summation sign (1.9)
σ (sigma)	Population standard deviation (3.8)
σ^2	Population variance (3.8)
$\sigma_{\overline{D}}$	Standard error of the mean of difference scores (11.2)
σ_R	Standard deviation of the sampling distribution of R for the Wilcoxon rank sum test (16.4)
$\sigma_{r'}$	Standard error of r' (Appendix 14.1)
σ_T	Standard deviation of the sampling distribution of T for the Wilcoxon signed-rank test (16.5)
$\sigma_{\overline{X}}$	Population standard error of the mean (7.5)
$\sigma_{\overline{X}_1-\overline{X}_2}$	Population standard error of the difference (10.2)
! (factorial)	Factorial of a number (6.7)
$1-\beta$	Power of a statistical test (8.5)
. . .	Indication to include all relevant values that fall between the written values in the algebraic operation (6.7)

References

Ainsworth, M. S., Blehar, M. C., Waters, E., & Wall, S. (1978). *Patterns of attachment: A psychological study of the stranger situation.* Hillsdale, NJ: Erlbaum.

Amabile, T. M., & Gitomer, J. (1984). Children's artistic creativity: Effects of choice in task materials. *Personality and Social Psychology Bulletin, 10,* 209–215.

American Psychological Association. (1994). *Publication manual of the American Psychological Association* (4th ed.). Washington, DC: Author.

Anderson, C. A., & Ford, C. M. (1986). Affect of the game player: Short-term effects of highly and mildly aggressive video games. *Personality and Social Psychology Bulletin, 12,* 390–402.

Anderson, N. H. (1968). Likeableness ratings of 555 personality-trait words. *Journal of Personality and Social Psychology, 9,* 272–279.

Anderson, N. H. (1970). Functional measurement and psychophysical judgment. *Psychological Review, 77,* 153–170.

Bandura, A., Ross, D., & Ross, S. A. (1963). Imitation of film-mediated aggressive models. *Journal of Abnormal and Social Psychology, 66,* 3–11.

Barcus, F. E. (1971). *Saturday children's television: A report of TV programming and advertising on Boston commercial television.* Newton, MA: Action for Children's Television.

Barron, F. (1965). The psychology of creativity. In T. Newcomb (Ed.), *New directions in psychology: 11* (pp. 1–113). New York: Holt, Rinehart & Winston.

Becker, M. A., & Gaeddert, W. P. (1988). *Sex-role orientation and perceived sexual attractiveness.* Unpublished manuscript, The Pennsylvania State University at Harrisburg, Division of Behavioral Science and Education.

Becker, M. A., & Suls, J. (1982). Test performance as a function of the hard driving and speed components of the Type A coronary prone behavior pattern. *Journal of Psychosomatic Research, 26,* 435–440.

Bennett, E. L., Krech, D., & Rosenzweig, M. R. (1964). Reliability and regional specificity of cerebral effects of environmental complexity and training. *Journal of Comparative and Physiological Psychology, 57,* 440–441.

Bohrnstedt, G. W., & Carter, T. M. (1971). Robustness in regression analysis. *Sociological Methodology, 12,* 118–146.

Boneau, C. A. (1960). The effects of violations of assumptions underlying the t test. *Psychological Bulletin, 57,* 49–64.

Borden, R. (1978). *Environmental attitudes and beliefs in technology.* Unpublished manuscript, Purdue University, Department of Psychological Sciences, West Lafayette, IN.

Brehm, J. (1956). Post-decision changes in desirability of alternatives. *Journal of Abnormal and Social Psychology, 52,* 384–389.

Burger, J. M., & Smith, N. G. (1985). Desire for control and gambling behavior among problem gamblers. *Personality and Social Psychology Bulletin, 11,* 145–152.

Bushman, B. J. (1988). The effects of apparel on compliance: A field experiment with a female authority figure. *Personality and Social Psychology Bulletin, 14,* 459–467.

Camilli, G., & Hopkins, K. D. (1978). Applicability of chi-square to 2×2 contingency tables with small expected cell frequencies. *Psychological Bulletin, 85,* 163–167.

Carrol, R. M., & Nordholm, L. A. (1975). Sampling characteristics of Kelley's η^2 and Hays' ω^2. *Educational and Psychological Measurement, 35,* 541–554.

Casler, L. (1964). The effects of hypnosis on ESP. *Journal of Parapsychology, 28,* 126–134.

Cohen, J. (1967). An alternative to Marascuilo's large sample multiple comparisons for proportions. *Psychological Bulletin, 67,* 199–201.

Cohen, J. (1977). *Statistical power analysis for the behavioral sciences* (2nd ed.). New York: Academic Press.

Cohen, J. (1988). *Statistical power analysis for the behavioral sciences.* Hillsdale, NJ: Erlbaum.

Cohen, J. (1994). The earth is round ($p < .05$). *American Psychologist, 49,* 997–1003.

Conover, W. J. (1974a). Rejoinder. *Journal of the American Statistical Association, 69,* 382.

Conover, W. J. (1974b). Some reasons for not using the Yates continuity correction on 2×2 contingency tables. *Journal of the American Statistical Association, 69,* 374–376.

Conover, W. J. (1980). *Practical nonparametric statistics.* New York: Wiley.

Conover, W. J., & Iman, R. L. (1981). Rank transformations as a bridge between parametric and non-parametric statistics. *American Statistician, 35,* 124–129.

Cox, C. (1926). *Genetic studies of genius.* Stanford, CA: Stanford University Press.

Cox, M., & Key, C. (1993). Post hoc pair-wise comparisons for the chi square test of homogeneity of proportions. *Educational and Psychological Measurement, 53,* 951–962.

Crusco, A., & Wetzel, C. G. (1984) The Midas touch: The effects of interpersonal touch on restaurant tipping. *Personality and Social Psychology Bulletin, 4,* 512–517.

Davison, M. & Sharma, A. (1988). Parametric statistics and levels of measurement. *Psychological Bulletin, 104,* 137–144.

Deutsch, F. M., & Mackesy, M. E. (1985). Friendship and the development of self-schemas: The effects of talking about others. *Personality and Social Psychology Bulletin, 11,* 399–408.

Dunn, O. J. (1964). Multiple comparisons using rank sums. *Technometrics, 6,* 241–252.

Eron, L. D. (1963). Relationship of TV viewing habits and aggressive behavior in children. *Journal of Abnormal and Social Psychology, 67,* 193–196.

Fancher, R. E. (1985). *The intelligence men: Makers of the IQ controversy.* New York: Norton.

Feldman-Summers, S., & Ashworth, C. D. (1980). *Factors related to intentions to report a rape.* Unpublished manuscript, University of Washington, Department of Psychology, Seattle.

Ferguson, G. A. (1976). *Statistical analysis in psychology and education.* New York: McGraw-Hill.

Festinger, L., & Carlsmith, J. M. (1959). Cognitive consequences of forced compliance. *Journal of Abnormal and Social Psychology, 58,* 203–210.

Fiedler, F. (1967). *A theory of leadership effectiveness.* New York: McGraw-Hill.

Fisher, R. (1950). *Statistical methods for research workers.* New York: Hafner.

Fleishman, A. I. (1980). Confidence intervals for correlation ratios. *Educational and Psychological Measurement, 40,* 659–670.

Fox, J., & Long, J. S. (1990). *Modern methods of data analysis.* Newbury Park, CA: Sage.

Freund, L. (1962). *Mathematical statistics.* Englewood Cliffs, NJ: Prentice-Hall.

Friedman, H. (1972). *Introduction to statistics.* New York: Random House.

Frieze, I. H., Parsons, J. E., Johnson, P. B., Ruble, D. N., & Zellman, G. L. (1978). *Women and sex roles: A social psychological perspective.* New York: Norton.

Gabrielcik, A., & Fazio, R. H. (1984). Priming and frequency estimates: A strict test of the availability heuristic. *Personality and Social Psychology Bulletin, 10,* 85–89.

Gallup, G. (1976). *The sophisticated poll watcher's guide*. Princeton, NJ: Princeton Opinion Press.

Glass, G. V., & Hakstian, A. R. (1969). Measures of association in comparative experiments: Their development and interpretation. *American Educational Research Journal, 6*, 403–414.

Glass, G., & Hopkins, K. D. (1984). *Statistical methods in education and psychology*. Englewood Cliffs, NJ: Prentice-Hall.

Glass, G., & Stanley, J. C. (1970). *Statistical methods in education and psychology*. Englewood Cliffs, NJ: Prentice-Hall.

Goldberg, P. (1968). Are women prejudiced against women? *Transaction, 5*, 28–30.

Goodman, L. A. (1964). Simultaneous confidence intervals for contrasts among multinomial populations. *Annals of Mathematical Statistics, 35*, 716–725.

Greenberg, J., Williams, K. D., & O'Brien, M. K. (1986). Considering the harshest verdict first: Biasing effects on mock jury verdicts. *Personality and Social Psychology Bulletin, 12*, 41–50.

Greenwald, A. (1975). Consequences of prejudice against the null hypothesis. *Psychological Bulletin, 82*, 1–20.

Guilford, J. P. (1965). *Fundamental statistics in psychology and education*. New York: McGraw-Hill.

Gulliksen, H. (1960). *Test theory*. New York: McGraw-Hill.

Haggard, E. A. (1958). *Intraclass correlation and analysis of variance*. New York: Dryden Press.

Hamilton, L. C. (1992). *Regression with graphics*. Pacific Grove, CA: Brooks/Cole.

Harvath, J. (1943). Problem solving performance and music. *American Journal of Psychiatry, 22*, 211–212.

Harwell, M. R., Rubinstein, E. N., & Hayes, W. (1992). Summarizing Monte Carlo results in methodological research: The one and two factor fixed effects ANOVA cases. *Journal of Educational Statistics, 17*, 315–339.

Havlicek, L. L., & Peterson, N. L. (1977). Effect of the violation of assumptions upon significance levels of the Pearson *r*. *Psychological Bulletin, 84*, 373–377.

Hays, W. L. (1981). *Statistics* (3rd ed.). New York: Holt, Rinehart & Winston.

Hayter, A. J. (1986). The maximum familywise error rate of Fisher's least significant difference test. *Journal of the American Statistical Association, 81*, 1000–1004.

Heslin, R., & Boss, D. (1980). Nonverbal intimacy in airport arrival and departure. *Personality and Social Psychology Bulletin, 6*, 248–252.

Hoaglin, D. C., Mosteller, F., & Tukey, J. W. (1983). *Understanding robust and exploratory data analysis*. New York: Wiley.

Holland, B. S., & Copenhaver, M. (1988) Improved Bonferroni-type multiple testing procedures. *Psychological Bulletin, 104*, 145–149.

Howell, D. C. (1985). *Fundamental statistics for the behavioral sciences*. Boston: Duxbury Press.

Howson, C., & Urbach, P. (1989). *Scientific reasoning: The Bayesian approach*. LaSalle, IL: Open Court Press.

Hsu, T. C., & Feldt, I. S. (1969). The effect of limitations on the number of criterion score values on the significance level of the *F*-test. *American Educational Research Journal, 6*, 515–527.

Huck, S. W., & Sandler, H. M. (1979). *Rival hypotheses: Alternative interpretations of data based conclusions*. New York: Harper & Row.

Huff, D. (1954). *How to lie with statistics*. New York: Norton.

Hurlock, E. (1925). An evaluation of certain incentives used in schoolwork. *Journal of Educational Psychology, 16*, 145–159.

Jaccard, J. (1980). *Factors affecting the acceptance of male oral contraceptives.* Unpublished manuscript, State University of New York at Albany, Department of Psychology.

Jaccard, J., Becker, M. A., & Wood, G. (1984). Pairwise multiple comparison procedures: A review. *Psychological Bulletin, 96,* 589–596.

Jensen, A. R. (1973). *Educability and group differences.* New York: Basic Books.

Johnson, M. K., & Liebert, R. M. (1977). *Statistics.* Englewood Cliffs, NJ: Prentice-Hall.

Johnson, T. (1976). *Luck versus ability: A replication.* Unpublished manuscript, Purdue University, Department of Psychological Sciences, West Lafayette, IN.

Joreskog, K., & Sorbom, D. (1993). *SIMPLIS.* Chicago: Scientific Software.

Judge, G. G., Griffiths, W., Hill, R. C., Lutkephol, H., & Lee, T. C. (1985). *The theory and practice of econometrics.* New York: Wiley.

Kelman, H. C., & Hovland, C. I. (1953). Reinstatement of the communicator in delayed measurement of opinion change. *Journal of Abnormal and Social Psychology, 48,* 327–335.

Kennedy, J. J. (1970). The eta coefficient in complex ANOVA designs. *Educational and Psychological Measurement, 30,* 885–889.

Kerlinger, F. N. (1973). *Foundations of behavioral research.* New York: Holt, Rinehart & Winston.

Kesselman, H. J. (1975). A Monte Carlo investigation of three estimates of treatment magnitude: Epsilon squared, eta squared, and omega squared. *Canadian Psychological Review, 16,* 44–48.

Kesselman, H. J., Rogan, J. C., Mendoza, J. L., & Breen, L. J. (1980). Testing the validity of the repeated measures F tests. *Psychological Bulletin, 87,* 479–481.

Kirk, R. E. (1968). *Experimental design: Procedures for the behavioral sciences.* Pacific Grove, CA: Brooks/Cole.

Kirk, R. E. (1972). *Statistical issues: A reader for the behavioral sciences.* Pacific Grove, CA: Brooks/Cole.

Kirk, R. E. (1978). *Introductory statistics.* Pacific Grove, CA: Brooks/Cole.

Kirk, R. E. (1995). *Experimental design: Procedures for the behavioral sciences* (3rd ed.), Pacific Grove, CA: Brooks/Cole.

Kleinke, C. L., Meeker, F. B., & Staneski, R. A. (1986). Preferences for opening lines: Comparing ratings by men and women. *Sex Roles, 15,* 585–600.

Koch, G., & Edwards, S. (1988). Clinical efficacy trials with categorical data. In C. E. Pearce (Ed.), *Biopharmaceutical statistics for drug development.* New York: Marcel Dekker Press.

Krantz, D. H., Luce, R. D., Suppes, P., & Tversky, A. (1971). *Foundations of measurement.* New York: Academic Press.

Kruskal, W. H., & Wallis, W. A. (1952). Use of ranks in one criterion variance analysis. *Journal of the American Statistical Association, 47,* 583–621.

Larkin, J. E. (1987). Are good teachers perceived as high self-monitors? *Personality and Social Psychology Bulletin, 13,* 64–72.

Levy, K. J. (1977). Pairwise comparisons involving unequal sample sizes associated with correlations, proportions, and variances. *British Journal of Mathematical and Statistical Psychology, 30,* 137–139.

Lord, F. M. (1953). On the statistical treatment of football numbers. *American Psychologist, 8,* 750–751.

Lunney, G. H. (1970). Using analysis of variance with a dichotomous dependent variable: An empirical study. *Journal of Educational Measurement, 7,* 263–269.

Mantel, N. (1974). Comment and suggestion. *Journal of the American Statistical Association, 69,* 378–380.

Marascuilo, L. A., & McSweeney, M. (1977). *Nonparametric and distribution-free methods for the social sciences.* Pacific Grove, CA: Brooks/Cole.

Maxwell, S., & Delaney, H. (1990). *Designing experiments and analyzing data.* Belmont, CA: Wadsworth.

McArthur, L. Z., & Eisen, S. V. (1976). Television and sex role stereotyping. *Journal of Applied Social Psychology, 6,* 329–351.

McCall, R. B. (1980). *Fundamental statistics for psychology* (3rd ed.). New York: Harcourt Brace Jovanovich.

McConnell, J. V. (1966). New evidence for the transfer of training effect in planaria. In *The biological basis of memory traces.* Symposium conducted at the International Congress of Psychology, Moscow.

McNemar, Q. (1962). *Psychological statistics.* New York: Wiley.

Mehta, C., & Patel, N. (1992). *StatXact-Turbo: Statistical software for exact nonparametric inference.* Cambridge, MA: Cytel.

Miller, R. G. (1966). *Simultaneous statistical inference.* New York: McGraw-Hill.

Milligan, G. W., Wong, D. S., & Thompson, P. A. (1987). Robustness properties of nonorthogonal analysis of variance. *Psychological Bulletin, 101,* 464–470.

Minium, E. (1978). *Statistical reasoning in psychology and education.* New York: Wiley.

Moore, D. S., & McCabe, G. P. (1993). *Introduction to the practice of statistics.* New York: Freeman.

Morrow, F., & Davidson, D. (1976). *Race and family size decisions.* Unpublished manuscript, Purdue University, Department of Psychological Sciences, West Lafayette, IN.

Nezlek, J. (1978) *Social behavior and diaries.* Unpublished manuscript, College of William and Mary, Department of Psychology, Williamsburg, VA.

Norusis, M. J. (1992). *SPSS for Windows base system users guide.* Chicago: SPSS Systems.

Parrish, M., Lundy, R. M., & Leibowitz, H. W. (1969). Effect of hypnotic age regression on the magnitude of the Ponzo and Poggendorff illusions. *Journal of Abnormal Psychology, 74,* 693–698.

Pearson, E. S., & Please, N. W. (1978). Relations between the shape of the population distribution of four simple test statistics. *Biometrika, 62,* 223–241.

Pedhazur, E. J. (1982). *Multiple regression in behavioral research* (2nd ed.). New York: Holt, Rinehart & Winston.

Petty, R. E., Wells, G. L., Heesacker, M., Brock, T. C., & Cacioppo, J. T. (1983). The effects of recipient posture on persuasion: A cognitive response analysis. *Personality and Social Psychology Bulletin, 9,* 209–222.

Posten, I. (1978). The robustness of the two-sample *t* test over the Pearson system. *Journal of Statistical Computation and Simulation, 6,* 295–311.

Posten, I. (1979). The robustness of the one-sample *t* test over the Pearson system. *Journal of Statistical Computation and Simulation, 9,* 133–149.

Robbins, R. A. (1988). *Objective and subjective factors in estimating life expectancy.* Unpublished manuscript, The Pennsylvania State University at Harrisburg, Division of Behavioral Science and Education.

Rosenthal, R. (1995). Methodology. In A. Tesser (Ed.), *Advanced social psychology* (pp. 17–50). New York: McGraw-Hill.

Rubovits, P. C., & Maehr, M. L. (1973). Pygmalion black and white. *Journal of Personality and Social Psychology, 25,* 210–218.

Ryan, T. A. (1959). Multiple comparisons in psychological research. *Psychological Bulletin, 56,* 26–47.

Sawilosky, S., & Hillman, S. (1992). A more realistic look at the robustness and Type II error properties of the *t* test to departures from population normality. *Psychological Bulletin, 111,* 352–360.

Sears, D. O. (1969). Political behavior. In G. Lindzey and E. Aronson (Eds.), *The handbook of social psychology* (pp. 315–458). Reading, MA: Addison-Wesley.

Siegel, S., & Castellan, N. J. (1988). *Nonparametric statistics for the behavioral sciences.* New York: McGraw-Hill.

Simpson, J. A., Campbell, B., & Berscheid, E. (1986). The association between romantic love and marriage: Kephart (1967) twice revisited. *Personality and Social Psychology Bulletin, 12,* 363–372.

Smith, W. L., Phillipus, M. J., & Guard, H. L. (1968). Psychometric study of children with learning problems and 14-6 positive spike EEG patterns, treated with ethosuximide (zarontin) and placebo. *Archives of Disease in Childhood, 43,* 616–619.

Sroufe, L. A., & Waters, E. (1977). Attachment as an organizational construct. *Child Development, 48,* 1184–1199.

Steel, R. G. (1960). A rank sum test for comparing all pairs of treatments. *Technometrics, 2,* 197–207.

Steiner, I. D. (1972). *Group process and productivity.* New York: Academic Press.

Stephen, G. (1975). *Psychology and law.* New York: McGraw-Hill.

Stevens, S. S. (1951). Mathematics, measurement, and psychophysics. In S. S. Stevens (Ed.), *Handbook of experimental psychology* (pp. 28–42). New York: Wiley.

Thorndike, R. (1942). Regression fallacies in the matched groups experiment. *Psychometrika, 7,* 85–102.

Touhey, J. C. (1974). Effects of additional women professionals on ratings of occupational prestige and desirability. *Journal of Personality and Social Psychology, 29,* 86–89.

Tukey, J. W. (1953). *The problem of multiple comparisons.* Unpublished manuscript, Princeton University, Department of Statistics, Princeton, NJ.

Van den Brink, W. (1988). The robustness of the *t* test of the correlation coefficient and the need for simulation studies. *British Journal of Mathematical and Statistical Psychology, 41,* 251–256.

Warkov, S., & Greeley, A. (1966). Parochial school origins and education achievement. *American Sociological Review, 31,* 406–414.

Wechsler, D. (1958). *The measurement and appraisal of adult intelligence.* Baltimore: Williams & Wilkins.

Weil, A. T., Zinberg, N. E., & Nelson, J. (1968). Clinical and psychological effects of marijuana in man. *Science, 162,* 1234–1242.

Wike, E. L. (1971). *Data analysis: A statistical primer for psychology students.* Chicago: Aldine.

Wilcox, R., Charlin, V., & Thompson, K. (1986). New Monte Carlo results on the robustness of the ANOVA *F*, *W*, and *F** statistics. *Communications in Statistics—Simulation and Computation, 15,* 933–943.

Wilcoxon, F. (1949). *Some rapid approximate statistical procedures.* New York: American Cyanamid.

Willingham, W. W. (1974). Predicting success in graduate education. *Science, 183,* 275–278.

Winer, B. J. (1971). *Statistical principles in experimental design.* New York: McGraw-Hill.

Witte, R. S. (1980). *Statistics.* New York: Holt, Rinehart & Winston.

Wyer, R., & Goldberg, L. (1970). A probabilistic analysis of the relationship between beliefs and attitudes. *Psychological Review, 77,* 100–120.

Zeisel, H., & Kalven, H., Jr. (1978). Parking tickets and missing women: Statistics and the law. In J. M. Tanur, F. Mosteller, W. H. Kruskal, R. F. Link, R. S. Pieters, G. R. Rising & E. L. Lehmann (Eds.), *Statistics: A guide to the unknown* (2nd ed., pp. 139–149). San Francisco: Holden-Day.

Zimmerman, D. W., & Zumbo, B. D. (1993). The relative power of parametric and non-parametric statistical methods. In G. Keren & C. Lewis (Eds.), *A handbook for data analysis in the behavioral sciences: Methodological issues* (pp. 481–517). Hillsdale, NJ: Erlbaum.

Answers to Selected Exercises

Chapter 1

2. a. a constant because there are always 24 hours in a day
   b. a variable because different people have different attitudes toward abortion
   c. a constant because all presidents of the United States must be born in the United States
   d. a constant because a number divided by itself always equals 1.00
   e. a variable because different total numbers of points are scored in different football games
   f. a variable because different months have different numbers of days

3. a. quantitative   d. quantitative
   b. qualitative   e. qualitative
   c. quantitative   f. qualitative

4. The independent variable is children's preference for aggressive television shows. The dependent variable is the peer ratings of aggression. Both are quantitative.

5. The independent variable is the type of occupation, which is qualitative. The dependent variable is perceptions of occupational prestige, which is quantitative.

10. a. discrete   c. discrete
    b. continuous   d. continuous

11. a. 21,384.105 and 21,384.115   c. 12.5 and 13.5
    b. .6885 and .6895   d. 12.95 and 13.05
                                        e. 12.995 and 13.005

13. The newspaper's sample is probably not a representative sample of the community in general because it represents only those people who read the newspaper and who were willing to take the time and trouble to send in a ballot. This casts doubt on the validity of the newspaper's conclusion about the election.

16. You probably found that only a very small number of individuals were selected to participate in both samples. This indicates that every member of a population has an equal chance of being selected when random sampling is used and, thus,

that random sampling will tend to yield representative samples.

19. .05

21. a. ΣX e. $(\Sigma Y)^2$
    b. $\Sigma_{i=3}^{5} X_i^2$ f. $\Sigma_{i=1}^{3} Y_i$
    c. $\Sigma (X - 5)$ g. $\Sigma (X - 1)(Y - 6)$
    d. $\Sigma X^2/N$ h. ΣXY

22. a. 36   d. 9
    b. 36   e. 9
    c. $\Sigma Xk = k \Sigma X$ f. $\Sigma (X/k) = (\Sigma X)/k$

23. a. 4.893   h. .396
    b. 8.975   i. 1.000
    c. 1.415   j. 3.667
    d. 4.145   k. 12.254
    e. 6.245   l. 9.724
    f. 2.616   m. 1.995
    g. 6.316   n. 2.005

25. Calculations for the original scores:
    a. 19.96   c. 398.34
    b. 3.99   d. 80.06
    Calculations for the rounded scores:
    a. 19.98   c. 399.20
    b. 4.00   d. 80.24
    The difference between the two sets of results is because the original scores were to four decimal places and the rounded scores were to two decimal places. The answers based on the original scores are thus more precise.

27. b   34. a
29. c   36. d
32. a

Chapter 2

1.–2. Sick days	f	rf	%	cf	crf
8	2	.100	10.0	20	1.000
7	4	.200	20.0	18	.900
6	10	.500	50.0	14	.700
5	0	.000	0.0	4	.200
4	2	.100	10.0	4	.200
3	2	.100	10.0	2	.100

3. .100; 1.000; .900
4. .200; .300; .100
6. .100; .600; .900

7.
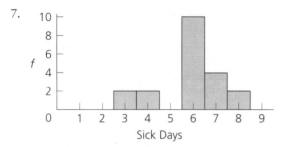
Sick Days

10.
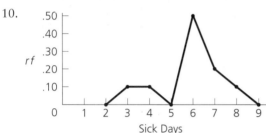
Sick Days

The shape of this graph is identical to the shape of the frequency polygon from Exercise 8.

12. In an ungrouped frequency distribution, each different score value is listed. In a grouped frequency distribution, scores are grouped together into intervals.

16.

Intelligence score	f
120–129	5
110–119	10
100–109	20
90–99	10
80–89	5

18. .700; .300; .300

20.

Intelligence Score

22.
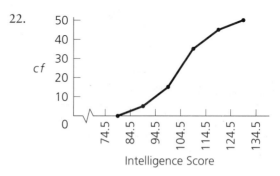
Intelligence Score

The similarity in shapes is because the cumulative frequency line will always remain level or increase as it moves from left to right.

23. 10; 80
24. 5; 80
30. The ordinate of a frequency graph should be presented such that its height at the demarcation for the highest frequency is approximately three-fourths to two-thirds the length of the abscissa. This helps to ensure the uniform, clearly interpretable presentation of graphed results.
32. A probability distribution is a distribution that represents the probabilities associated with all possible score values for a variable. The nature of probability distributions for qualitative and discrete variables is different from that for continuous variables because in the former case it is possible to list possible values of the variable and their corresponding probabilities. Since the number of possible values a continuous variable can have is, in principle, infinite, this is not possible for continuous variables. Instead, probabilities for a continuous variable are conceptualized as the areas under corresponding intervals of the density curve.

34.

Attitude	f	rf	cf	crf
5	370	.200	1,850	1.000
4	555	.300	1,480	.800
3	185	.100	925	.500
2	407	.220	740	.400
1	333	.180	333	.180

35. b   43. a
37. a   45. c
41. c
42. c

Chapter 3

3. 4.33
6. Mode = 0, median = 0.00, mean = 0.00
8. The mean for the first set of scores is 12.00; the mean for the second set is 15.00; the mean for the third set is 2.00. If a constant, k, is added to each score (X) in a set, the mean of the new set of scores will equal $\overline{X} + k$. If a constant is subtracted from each score in a set, the mean of the new set of scores will equal $\overline{X} - k$.
10. The mean for the first set of scores is 30.00; the mean for the second set is 90.00; the mean for the third set is 3.00. If each score (X) in a set is multiplied by a constant, k, the mean of the new set of scores will equal $k\overline{X}$. If each score in a set is divided by a constant, the mean of the new set of scores will equal $\overline{X}/k$.
11. The median for the first set of scores in Exercise 8 is 12; the median for the second set is 15; the median for the third set is 2. Adding a constant to or subtracting a constant from each score in a set has the same effects on the median as on the mean.

    The median for the first set of scores in Exercise 10 is 30; the median for the second set is 90; the median for the third set is 3.00. Multiplying or dividing each score in a set by a constant has the same effects on the median as on the mean.

    The above operations would have the same effects on the mode as on the mean and the median.
13. The mean is a poorer descriptor of central tendency for Set I. This is because the extreme score of 300 in Set I substantially adjusts (increases) the mean and thus leads to a distorted picture of the central tendency of the data.
15. The range is a misleading index of variability when there is an extreme score in a set of scores that are otherwise similar to one another.
18. 10.00
20. All three measures of variability must equal 0 because there is no variability when all of the scores are the same.
21. The standard deviation is more "interpretable" than the variance because it represents an average deviation from the mean *in the original unit of measurement*. In contrast, the variance is in terms of squared deviation units.
23. The sum of squares equals 104.00 using both approaches. Typically, it is more efficient to apply the computational formula than the defining formula because the computational formula requires fewer steps.
24. Range = 10, SS = 144.95, s^2 = 6.90, s = 2.63
26. The variance for each set of scores is 2.00, and the standard deviation for each set is 1.41. Adding a constant to or subtracting a constant from each score in a set does not affect the variance or the standard deviation.
28. For the first set of scores, the variance is 2.00 and the standard deviation is 1.41; for the second set, the variance is 18.00 and the standard deviation is 4.24; for the third set, the variance is .50 and the standard deviation is .71. If each score in a set is multiplied by a constant, k, the variance of the new set of scores will equal the variance of the old set of scores multiplied by k^2, and the standard deviation will equal the standard deviation of the old set of scores multiplied by k. If each score in a set is divided by a constant, the variance of the new set of scores will equal the variance of the old set of scores divided by k^2, and the standard deviation will equal the standard deviation of the old set of scores divided by k.
29. One example of such scores is:

Set I	Set II
0	9
5	10
10	10
15	10
20	11

Although the mean for both sets of scores is 10.00, the standard deviation is 7.07 for Set I and .63 for Set II.
31. Considering the mean score across all participants, the speed estimates are quite accurate (28.67 compared with the actual speed of 30). However, the relatively large standard deviation (7.74) indicates that the *individual* estimates are not all that accurate.
33. The consultant should be fired: Standard deviations cannot be negative.
36. The first set of scores is positively skewed, as the three measures of central tendency all take on different values and the mean is greater than the median. The second set of scores is negatively skewed, as the three measures of central tendency all take on different values and the mean is

smaller than the median. The third set of scores is not skewed, as equal numbers of scores occur above and below the mean, as indicated by the fact that the mean is also the median.

39. b   46. b
41. a   48. b
43. d

Chapter 4

2. a. 2.99   d. 5.52
    b. 3.75   e. 7.67
    c. 4.25   f. 4.00
4. a. 94.40   c. 20.20
    b. 74.90   d. 5.20
9. a. .62   d. .03
    b. 1.00   e. −1.46
    c. −.36   f. 1.43
11. −1.23
13. A positive standard score indicates that the original score is greater than the mean. A negative standard score indicates that the original score is less than the mean.
14. 0; 1.00
16. John's performance on the English exam was 1.00 standard deviation above the mean, and his performance on the math exam was 6.67 standard deviations above the mean. Hence, John's performance was better (relative to his classmates') on the math exam.
18. One situation is determining the relative abilities on a given task of people of different ages. For instance, the bowling averages of a child and an adult can be compared in this manner.
22. a. .9913   e. .4798
    b. .1210   f. .3550
    c. .1210   g. .3550
    d. .4798   h. .7850
23. a. .1587   d. .9836
    b. .1587   e. .5000
    c. .7865   f. .7257
25. The standard score corresponding to a galvanic skin response score of 61.40 is 4.00. Since only .003% of the standard scores in a normal distribution are 4.00 or greater, the individual displayed an extreme galvanic skin response when asked the critical question. The implication is that he is lying.
27. a. 128.60   d. 85.00
    b. 75.60   e. 115.90
    c. 100.00   f. 107.50

28. a. 102.20   d. 109.80
    b. 108.22   e. 90.20
    c. 100.00
30. a. 58.7   d. 34.4
    b. 70.0   e. 50
    c. 65.6   f. 90.4
31. b   40. d
35. b   41. c
37. d   44. a
38. c   45. c

Chapter 5

2. The slope indicates the number of units variable Y changes when variable X changes by 1 unit.
5. 3.00 units; 6.00 units; 21.00 units
6. The magnitude of a correlation coefficient indicates the degree to which two variables approximate a linear relationship.
8. a. +.37   d. +.26
    b. −.37   e. −.44
    c. −.76   f. +.61
9. One example of such a scatterplot is:

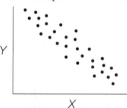

11. Two variables that are probably positively correlated are calorie consumption and weight gain.
13. .18
16. Two variables that are probably (positively) correlated but not causally related are the number of automobiles in a household and the number of television sets in a household. The correlation between these two variables is probably attributable to such factors as the number of occupants and their income levels.
17. The general form of the regression equation is $\hat{Y} = a + bX$. This differs from the linear model in that the values yielded by $a + bX$ are *predicted Y* scores rather than actual Y scores.
18. These are defined such that the sum of the squared discrepancies between individuals' actual Y scores and their predicted Y scores based on the regression equation is minimized.
19. $\hat{Y} = 5.00 + .20X$
20. 5.60; 6.20; 5.40
24. 2.41

26. If two variables are related in a curvilinear fashion, restricting the range of variable X tends to increase the magnitude of the observed correlation coefficient. If two variables are linearly related, the effect of restricting the range of one of them is often to reduce the magnitude of the observed correlation coefficient.

30. b
32. a
34. b
35. c

37. a
39. c
41. a

Chapter 6

2. 180
4. 350
5. .514; .486; .500; .500
6. .343; .686; .657; .314
7. Being a man (event A) and favoring the ERA (event B) are not independent because the probability of being a man $[p(A) = .50]$ is not the same as the conditional probability of being a man, given that one favors the ERA $[p(A|B) = .333]$.
8. .171; .157
9. Four joint probabilities can be computed: (1) being a man who favors the ERA, (2) being a man who opposes the ERA, (3) being a woman who favors the ERA, and (4) being a woman who opposes the ERA.
10. .843; .829
17. .486
20. .516
23. a. 60   d. 120
    b. 720   e. 24
    c. 12
24. 5! equals 120, which is the same answer as for **d** of Exercise 23. 4! equals 24, which is the same answer as for **e** of Exercise 23. From this, we can generalize that $_nP_n = n!$.
26. a. 10   d. 1
    b. 6   e. 1
    c. 6
28. 216
30. a. .010   d. .010
    b. .044   e. .044
    c. .172
33. The correspondence between the binomial and normal distributions is influenced by the number of trials (n) and the probability of success (p). The correspondence improves as n increases and as p becomes closer to .50.

34. $\mu = 90.00$, $\sigma = 6.00$
36. A score of 30 translates into a z score of 2.50. From Appendix B, the probability of obtaining a z score of 2.50 or greater is .0062. Since .0062 is less than the criterion value of .05, the researcher should conclude that the psychotherapeutic approach is more effective than no treatment in helping individuals recover from their symptoms.

38. c   47. a
40. c   48. a
42. a
44. b

Chapter 7

2. Sampling error refers to the fact that values of sample statistics are likely to differ from values of their corresponding population parameters because they are based on only a portion of the overall population. The amount of sampling error can be represented as the difference between the value of a sample statistic and the value of the corresponding population parameter.

3. An unbiased estimator is a statistic whose average (mean) over all possible random samples of a given size equals the value of the parameter. A biased estimator is a statistic whose average over all possible random samples of a given size does not equal the value of the parameter.

6. The variance and standard deviation estimates (2.67 and 1.63, respectively) are larger than the variance and standard deviation for the sample (2.50 and 1.58, respectively).

8. Degrees of freedom are the number of pieces of statistical information that are independent of one another. There are $N - 1$ degrees of freedom associated with a sum of squares around a sample mean because, given all but one deviation score for a distribution of scores, the last deviation score is determined by the other $N - 1$ deviation scores.

10. A sampling distribution of the mean is a theoretical distribution consisting of the mean scores for all possible samples of a given size that can be drawn from a population. A frequency distribution, in contrast, is concerned with the frequency with which score values occur within a set of scores.

11. The central limit theorem addresses the mean, the standard deviation, and the shape of a sampling distribution of the mean.

13.

Sample	Sample mean
2, 2	2.00
2, 4	3.00
2, 6	4.00
4, 2	3.00
4, 4	4.00
4, 6	5.00
6, 2	4.00
6, 4	5.00
6, 6	6.00

Sum = 36.00

$$\text{Mean} = \frac{36.00}{9} = 4.00$$

The mean across the sample means is equal to the population mean because $\mu = (2 + 4 + 6)/3 = 4.00$. This illustrates that, as stated in the central limit theorem, the mean of a sampling distribution of the mean is always equal to the population mean.

15. A standard deviation of a set of raw scores represents an average deviation from the mean of the distribution. A standard error of the mean is the standard deviation of a sampling distribution of the mean and represents an average deviation of the sample means from the population mean.

17. A standard error of the mean of 0 indicates that the means of all samples drawn from the relevant population are equal to the population mean.

18. A standard error of the mean of 0 indicates that there is no variability in the scores in the population—that is, that all of the scores are the same and σ therefore equals 0.

19. The sample mean for a random sample of size 30 drawn from population A is probably a better estimator of its population mean than is the sample mean for a random sample of size 30 drawn from population B because, as indexed by its smaller standard deviation, there is less variability in population A than in population B. Consequently, the standard error of the mean for population A (.91) is smaller than the standard error of the mean for population B (1.28), thus indicating that, on the average, means for samples of a given size drawn from population A will be closer to the true population mean than will means for samples of the same size drawn from population B.

21. $\bar{X} = 5.00, \hat{s}_{\bar{X}} = .41$

26. The mean is usually preferred to the mode and the median because, given the same population, the sampling distribution of the mean will show less variability (that is, it will have a smaller standard error) than either the sampling distribution of the median or the sampling distribution of the mode.

27. b	36. d
30. d	38. b
31. b	39. b
34. a	

Chapter 8

2. Assuming the null hypothesis is true allows us to compare the observed result of a statistical test with an expected result and, thus, to make inferences about population values from sample values.

4. This is because, due to sampling error, we can never unambiguously conclude that the true population mean is equal to any one specific value based on sample data.

5. $H_0: \mu = \$100$
   $H_1: \mu \neq \$100$

6. $z = 1.79$

7. Since the observed z of 1.79 is neither less than the negative critical value of -1.96 nor greater than the positive critical value of $+1.96$, we fail to reject the null hypothesis. We cannot confidently conclude that the actual value of μ differs from 4.

11. The probability of a Type I error is equal to alpha because if the null hypothesis is true, the null hypothesis will be incorrectly rejected any time the observed value of the test statistic falls in the rejection region, and the probability of this occurring is equal to alpha.

12. Power is equal to 1 minus the probability of a Type II error because if the probability of making a Type II error by failing to reject a false null hypothesis is equal to β, the probability of making a correct decision by rejecting the null hypothesis under this circumstance (power) must be equal to $1 - \beta$.

13. Power decreases as alpha is set at a lower (more conservative) level because the lower the alpha level, the lower is the probability of rejecting the null hypothesis. Thus, when the null hypothesis is false, the probability of detecting a difference between the hypothesized population mean and the actual population mean (power) decreases as does alpha.

15. The phrases "statistically significant" and "statistically nonsignificant" emphasize the *statistical* nature of a conclusion—that is, whether or not the null hypothesis was rejected. A statistically significant result may or may not have important practical implications.

18. a. ±2.093   d. +2.262
    b. +1.729   e. +1.833
    c. −1.729   f. −1.833

19. a. The critical values of t for an alpha level of .05 and 9,999 degrees of freedom are approximately ±1.960. Since the observed t value of 10.00 is greater than +1.960, we reject the null hypothesis.
    b. The critical values of t for an alpha level of .05 and 99 degrees of freedom are approximately ±1.987, as determined through interpolation. Since the observed t value of 1.00 is neither less than −1.987 nor greater than +1.987, we fail to reject the null hypothesis.
    c. The observed t value of 5.00 is greater than the positive critical value of approximately +1.987 for an alpha level of .05 and 99 degrees of freedom, so we reject the null hypothesis.
        The null hypothesis is rejected in part **a** but not in part **b** because the t test in part **a** is based on a larger number of cases. The null hypothesis is rejected in part **c** but not in part **b** because the t test in part **c** is based on a smaller standard deviation estimate.

23. The 95% confidence interval is 72.44 to 76.36. The 99% confidence interval is 71.82 to 76.98.

24. The 95% confidence interval is 72.60 to 76.20. The 99% confidence interval is 72.03 to 76.77. The effect of increasing N is to decrease the width of the intervals.

26. The 95% confidence interval is 116.98 to 125.02. The 99% confidence interval is 115.58 to 126.42.

30. c   37. d
31. c   39. c
33. a   41. a
36. a
42.   Results
    The mean number of children in the sample ($\underline{M}$ = 2.96, $\underline{SD}$ = 1.81) was compared against a hypothesized fertility rate of 2.11 using a one-sample $\underline{t}$ test. The alpha level was .05. This test was found to be statistically significant, $\underline{t}$(24) = 2.34, $\underline{p}$ < .05,

indicating that Catholics are reproducing at a rate above zero population growth.

Chapter 9

2. Observational strategy

3. Experimental strategy

6. A control group is a group in an experiment that is not exposed to the independent variable. The advantage of including a control group is that it provides a baseline for evaluating the effects of the experimental manipulation.

9. One limitation of random assignment is that it is not applicable when an observational research strategy is used. A second limitation is that it does not *guarantee* that the research groups will not differ beforehand on the dependent variable.

10. Sampling error can be reduced by increasing the sample sizes for the various groups or by defining the groups such that the variances of scores in the populations will be relatively small (that is, holding potential disturbance variables constant).

12. Disturbance variables are variables that are unrelated to the independent variable but that affect the dependent variable. They can be controlled by holding a variable constant.

13. One advantage of a within-subjects design is that it is more economical in terms of participants. A second advantage is that it offers better control of confounding variables related to individual differences. A potential problem with within-subjects designs is that carry-over effects can occur.

14. The independent variables in the studies described in Exercises 2 (race), 3 (noise level), and 5 (gender) are all between-subjects in nature. The independent variable in the study described in Exercise 4 (observer status) is within-subjects in nature.

16. Robustness is the extent to which conclusions drawn on the basis of a statistical test are unaffected by violations of the assumptions underlying the test. Robustness is important because when a test is robust to violation of an assumption, it can be applied even when that assumption is violated.

19. Three factors that influence the robustness of a statistical test are sample size, the degree of violation of distributional assumptions, and the form of the violation of distributional assumptions.

23. a

26. b	31. a
27. c	33. b
29. b	35. a

Chapter 10

2. The mean of a sampling distribution of the difference between two independent means will always equal the difference between the relevant population means.
3. The independent groups t test assumes that the two population variances are homogeneous. Thus, our goal is to estimate σ^2, the variance of both populations. By pooling the variance estimates from the two samples, we increase the degrees of freedom on which the estimate of σ^2 is based and thereby obtain a better estimate.
4. $\hat{s}^2_{pooled} = 5.48$, $\hat{s}_{\overline{X}_1 - \overline{X}_2} = .98$
5. a. .21
   b. .23
   c. .31
   d. This is because there is more variability in a sampling distribution of the difference between two independent means than in the corresponding sampling distributions of the mean. For instance, if the smallest sample mean in each of two sampling distributions of the mean were 2.00 and the largest sample mean were 7.00, the range in each case would be 7.00 − 2.00 = 5.00. However, the smallest mean difference in the sampling distribution of the difference between two independent means based on the two distributions of sample means would be 2.00 − 7.00 = −5.00, and the largest mean difference would be 7.00 − 2.00 = 5.00, a range of 5.00 − (−5.00) = 10.00. Since the estimated standard error of the difference and the estimated standard errors of the mean also reflect variability within the corresponding sampling distributions, it follows that, as with the ranges, the former measure will always be larger than the latter measures.
8. a. ±2.101   d. +1.701
   b. +1.734   e. ±2.021
   c. ±2.048   f. −1.684
10. The critical values of t for an alpha level of .05 and 8 degrees of freedom are ±2.306. Since the observed t value of 6.71 is greater than +2.306, we reject the null hypothesis and conclude that there is a relationship between gender and discriminatory attitudes.

11. $SS_{TOTAL} = 26.50$
12. $T_M = 1.50$, $T_F = -1.50$

13. Gender	X_n
Male	6.50
Male	5.50
Male	5.50
Male	5.50
Male	4.50
Female	6.50
Female	5.50
Female	5.50
Female	5.50
Female	4.50

14. $SS_{ERROR} = 4.00$, $SS_{EXPLAINED} = 22.50$
15. The value of eta-squared is .85 using both Equation 10.10 and Equation 10.11. This represents a strong effect.
16. Men ($\overline{X} = 7.00$) are significantly more discriminatory than women ($\overline{X} = 4.00$).
24. The sum of squares total is equal to the sum of squares explained plus the sum of squares error. In other words, the total variability in the dependent variable, as represented by SS_{TOTAL}, can be partitioned into two components, one ($SS_{EXPLAINED}$) reflecting the influence of the independent variable and one (SS_{ERROR}) reflecting the influence of disturbance variables.
26. It is inappropriate because eta-squared is a biased estimator. Specifically, eta-squared tends to slightly overestimate the strength of the relationship in the population across random samples.
29. $n = 87$

33. c	40. b
36. d	41. a
37. c	42. a
38. d	45. b
47.	

Results

An independent groups t test with an alpha level of .05 indicated that the cortex weight of rats will be greater when they are raised in an enriched environment ($M = 660.00$ mg, $SD = 24.15$ mg) than when they are raised in an isolated environment ($M = 626.00$ mg, $SD = 23.15$ mg), $t(12) = 2.69$, $p < .02$. As indexed by eta^2, the strength of the relationship between the type of environment and cortex weight was .38.

Chapter 11

3. a. ±2.056
   b. +1.706
   c. ±2.160
   d. −1.771
6. The rationale is that potential confounding variables will be evenly distributed across conditions and, thus, turned into disturbance variables.
7. This is because variability due to individual differences is extracted from the dependent variable as part of the correlated groups t test procedure. Because the degrees of freedom for the correlated groups test $(N - 1)$ are less than the degrees of freedom for the independent groups test $(n_1 + n_2 - 2)$, a correlated groups t test will not be more powerful than a corresponding independent groups t test when the correlation between scores in the two conditions is so close to 0 that the magnitude of the estimated standard errors is comparable for the two tests. This reflects the fact that the t distribution requires more extreme values of t in order to reject the null hypothesis as the degrees of freedom become smaller.
8. The critical values of t for an alpha level of .05 and 4 degrees of freedom are ±2.776. Since the observed t value of −3.77 is less than −2.776, we reject the null hypothesis and conclude that there is a relationship between the amount of noise one is exposed to and learning scores.
9. The value of eta-squared is .78. This represents a strong effect.
10. Learning scores are significantly higher under quiet conditions ($\overline{X} = 13.00$) than under noisy conditions ($\overline{X} = 8.60$).
11. Analyzing the data as if the independent variable were between-subjects in nature, the observed value of t, $t(8) = -1.18$, is neither less than the negative critical value of −2.306 nor greater than the positive critical value of +2.306, so we fail to reject the null hypothesis. The value of eta-squared in this instance is .15. This represents a moderate effect. The fact that eta-squared was .78 in the correlated groups case but only .15 in the independent groups case indicates that individual differences had a sizeable effect on the dependent variable. The fact that the null hypothesis was rejected when a correlated groups t test was applied but not when an independent groups t test was applied illustrates the increased power of the statistical analysis when variability due to individual differences is extracted from the dependent variable and, thus, the advantage of within-subjects research designs.
15. See table at bottom of page. The respective mean values are the same because we have extracted the effects of individual differences in their role as disturbance variables. Since disturbance variables are unrelated to the independent variable, the means will not be affected.
16. The observed value of t is −3.16. This is the same value obtained in Exercise 12.
17. Independent groups t test
18. Correlated groups t test
23. .60   31. c
26. b   34. b
27. a   36. a
30. b   37. b
38.   Results

    The mean desirability ratings for the unchosen alternative before versus after choosing between the two products were compared using a correlated groups t test. The alpha level was .05. This showed that the alternative in question was rated as significantly more desirable on the first occasion ($\underline{M} = 6.00$, $\underline{SD} = 1.49$) than

Individual	X for time 1	X for time 2	$\overline{X}_i$	Nullified X for time 1	Nullified X for time 2
1	10	12	11.00	12.00	14.00
2	11	13	12.00	12.00	14.00
3	12	14	13.00	12.00	14.00
4	13	17	15.00	11.00	15.00
5	14	14	14.00	13.00	13.00
Mean =	12.00	14.00	13.00	12.00	14.00

on the second ($\underline{M}$ = 4.00, $\underline{SD}$ = 1.15), $\underline{t}$(9) = 3.00, $\underline{p}$ < .02. The strength of the relationship between the time of assessment and product desirability was .50, as indexed by eta^2.

Chapter 12

2. The alternative hypothesis states that the population means in question are not all equal to one another. It cannot be summarized in a single mathematical statement because there are a number of ways in which three or more population means can pattern themselves so that they are not all equal.

3. Between-group variability concerns the differences between the mean scores in the various groups under study. Within-group variability concerns the variability of scores *within* each of the groups.

6. The F ratio, over the long run, will approach 1.00 when the null hypothesis is true. The F ratio, over the long run, will be greater than 1.00 when the null hypothesis is not true.

8. The value of the sum of squares within must be 0 because all scores within a given group are the same.

9. The mean square between is the sum of squares between divided by the degrees of freedom between. The mean square within is the sum of squares within divided by the degrees of freedom within.

11. a. 3.55   c. 3.35
    b. 3.24   d. 2.50

13.

Source	SS	df	MS	F
Between	30.00	2	15.00	5.62
Within	152.00	57	2.67	
Total	182.00	59		

16.

Null hypothesis tested	Absolute difference between sample means	Value of CD	Null hypothesis rejected?
$\mu_S = \mu_M$	$\lvert 6.00 - 8.00 \rvert = 2.00$	2.48	No
$\mu_S = \mu_D$	$\lvert 6.00 - 10.00 \rvert = 4.00$	2.48	Yes
$\mu_M = \mu_D$	$\lvert 8.00 - 10.00 \rvert = 2.00$	2.48	No

Divorced individuals ($\overline{X}_D$ = 10.00) have more positive attitudes than single individuals ($\overline{X}_S$ =

6.00). However, we cannot confidently conclude that either divorced or single individuals differ in their divorce attitudes from married individuals ($\overline{X}_M$ = 8.00).

17.

Source	SS	df	MS	F
Between	40.00	2	20.00	10.00
Within	24.00	12	2.00	
Total	64.00	14		

The critical value of F for an alpha level of .05 and 2 and 12 degrees of freedom is 3.88. Since the observed F value of 10.00 is greater than 3.88, we reject the null hypothesis and conclude that there is a relationship between the type of car and repair records.

18. The value of eta-squared is .62 using both Equations 12.15 and 12.16. This represents a strong effect.

19.

Null hypothesis tested	Absolute difference between sample means	Value of CD	Null hypothesis rejected?
$\mu_X = \mu_Y$	$\lvert 2.00 - 4.00 \rvert = 2.00$	2.38	No
$\mu_X = \mu_Z$	$\lvert 2.00 - 6.00 \rvert = 4.00$	2.38	Yes
$\mu_Y = \mu_Z$	$\lvert 4.00 - 6.00 \rvert = 2.00$	2.38	No

Car X ($\overline{X}_X$ = 2.00) performs better than car Z ($\overline{X}_Z$ = 6.00). However, we cannot confidently conclude that either car X or car Z differs in performance from car Y ($\overline{X}_Y$ = 4.00).

23.

Source	SS	df	MS	F
Between	90.00	1	90.00	36.00
Within	20.00	8	2.50	
Total	110.00	9		

The critical value of F for an alpha level of .05 and 1 and 8 degrees of freedom is 5.32. Since the observed F value of 36.00 is greater than 5.32, we reject the null hypothesis and conclude that a relationship exists between the independent and dependent variables.

24. The observed value of t is −6.00. The square of this value is 36.00, which is equal to the observed value of F from Exercise 23. The critical values of t for an alpha level of .05 and 8 degrees of freedom are ±2.306. The square of these values is

5.32, which is equal to the critical value of F from Exercise 23. This indicates that one-way between-subjects analysis of variance bears a mathematical relationship to the independent groups t test in the two-group case such that $F = t^2$.

25. One-way between-subjects analysis of variance
28. $n = 37$
30. d   38. a
34. b   39. a
35. c   41. b
36. b   43. d

46.

Source	SS	df	MS	F
Between	140.00	2	70.00	7.38
Within	256.00	27	9.48	
Total	396.00	29		

Null hypothesis tested	Absolute difference between sample means	Value of CD	Null hypothesis rejected?
$\mu_W = \mu_A$	$|3.00 - 8.00| = 5.00$	3.42	Yes
$\mu_W = \mu_H$	$|3.00 - 7.00| = 4.00$	3.42	Yes
$\mu_A = \mu_H$	$|8.00 - 7.00| = 1.00$	3.42	No

Results

For an alpha level of .05, a one-way analysis of variance of judgments of the probability of guilt as a function of the defendant's race was found to be statistically significant, $F(2, 27) = 7.38$, $p < .01$. The strength of the relationship was .35, as indexed by eta^2. A Tukey HSD test revealed that the mean guilt-probability judgment for the white defendant ($M = 3.00$, $SD = 2.79$) was significantly lower than the mean guilt-probability judgment for both the African-American ($M = 8.00$, $SD = 3.02$) and the Hispanic ($M = 7.00$, $SD = 3.40$) defendants. The mean guilt-probability judgments for the African-American and the Hispanic defendants did not significantly differ.

Chapter 13

3. This is because variability due to individual differences is removed from the dependent variable

in the form of the sum of squares across subjects as part of the repeated measures analysis of variance procedure.

4.

Source	SS	df	MS	F
IV	90.00	4	5.00	1.67
Error	132.00	44	3.00	
Across subjects	46.00	11		
Total	198.00	59		

6. $df_{IV} = 4$, $df_{ACROSS\ SUBJECTS} = 20$, $df_{ERROR} = 80$, $df_{TOTAL} = 104$

7. a. 3.25   c. 3.34
   b. 2.76   d. 2.58

9. The Huynh–Feldt epsilon and the Greenhouse–Geisser epsilon increase the robustness by adjusting the degrees of freedom for the F test so that they are less than or equal to the degrees of freedom IV and the degrees of freedom error usually used in assessing the significance of the observed F ratio. The use of these adjusted degrees of freedom serves to decrease the Type I error rate and, thus, to increase the robustness of the statistical test.

11. A one-way repeated measures analysis of variance might be less powerful than a corresponding one-way between-subjects analysis of variance when the F ratio for the repeated measures test is similar to the F ratio for the between-subjects test because individual differences have only a minimal influence on the dependent variable. This reflects that the degrees of freedom for the denominator of the F test will always be less in the repeated measures case than in the between-subjects case and that the value of F required to reject the null hypothesis becomes more extreme as the degrees of freedom become smaller.

13.

Source	SS	df	MS	F
IV	14.53	2	7.26	4.78
Error	12.14	8	1.52	
Across subjects	24.26	4		
Total	50.93	14		

The critical value of F for an alpha level of .05 and 2 and 8 degrees of freedom is 4.46. Since the observed F value of 4.78 is greater than 4.46, we reject the null hypothesis and conclude that there is a relationship between the time of assessment and anxiety.

14. The value of eta-squared is .54 using both Equations 13.14 and 13.15. This represents a strong effect.

15.

Null hypothesis tested	Absolute difference between sample means	Value of CD	Null hypothesis rejected?
$\mu_1 = \mu_2$	$|3.80 - 5.20| = 1.40$	2.23	No
$\mu_1 = \mu_3$	$|3.80 - 6.20| = 2.40$	2.23	Yes
$\mu_2 = \mu_3$	$|5.20 - 6.20| = 1.00$	2.23	No

Anxiety was greater at time 3 ($\overline{X}_3 = 6.20$) than at time 1 ($\overline{X}_1 = 3.80$). However, we cannot confidently conclude that anxiety at either time 1 or time 3 differs from anxiety at time 2 ($\overline{X}_2 = 5.20$).

16.

Source	SS	df	MS	F
Between	14.53	2	7.26	2.40
Within	36.40	12	3.03	
Total	50.93	14		

Analyzing the data as if the independent variable were between-subjects in nature, the observed value of F does not exceed the critical value of 3.88, so we fail to reject the null hypothesis. The value of eta-squared in this instance is .29. This represents a strong effect. The fact that eta-squared was .54 in the repeated measures case but only .29 in the between-subjects case indicates that individual differences had a sizeable effect on the dependent variable. The fact that the null hypothesis was rejected when a one-way repeated measures analysis of variance was applied but not when a one-way between-subjects analysis of variance was applied illustrates the increased power of the statistical analysis when variability due to individual differences is identified and removed from the dependent variable.

20. One-way between-subjects analysis of variance
21. Correlated groups t test
25. .50   35. c
26. b   38. b
27. d   40. c
30. c   41. d
32. c   42. b

44.

Source	SS	df	MS	F
IV	220.00	3	73.33	36.66
Error	24.00	12	2.00	
Across subjects	16.00	4		
Total	260.00	19		

Null hypothesis tested	Absolute difference between sample means	Value of CD	Null hypothesis rejected?
$\mu_H = \mu_E$	$|18.00 - 14.00| = 4.00$	2.66	Yes
$\mu_H = \mu_C$	$|18.00 - 10.00| = 8.00$	2.66	Yes
$\mu_H = \mu_V$	$|18.00 - 10.00| = 8.00$	2.66	Yes
$\mu_E = \mu_C$	$|14.00 - 10.00| = 4.00$	2.66	Yes
$\mu_E = \mu_V$	$|14.00 - 10.00| = 4.00$	2.66	Yes
$\mu_C = \mu_V$	$|10.00 - 10.00| = 0.00$	2.66	No

Results

A one-way repeated measures analysis of variance compared the mean ratings of how important the four factors (health risks, effectiveness, cost, and convenience) were considered to be in deciding whether to use a male oral contraceptive. The alpha level was .05. This test was found to be statistically significant, $F(3, 12) = 36.66$, $p < .01$. The strength of the relationship, as indexed by eta^2, was .90. A Tukey HSD test indicated that the health risks factor ($M = 18.00$, $SD = 1.58$) was rated as significantly more important than the other three factors and that the effectiveness factor ($M = 14.00$, $SD = 1.58$) was rated as significantly more important than both the cost ($M = 10.00$, $SD = 1.58$) and the convenience ($M = 10.00$, $SD = 1.58$) factors. The means for these last two factors did not significantly differ.

Chapter 14

2. The procedures are equivalent because the critical values of r using the latter approach are the values of r that correspond to the critical values of t using the former approach.

3. a. ±.349   b. ±.423

c. +.360   e. ±.241
d. −.441

6. The critical values of r for an alpha level of .05 and 8 degrees of freedom are ±.632. Since the observed r value of +.74 is greater than +.632, we reject the null hypothesis and conclude that there is some degree of a linear relationship between the two variables.

7. The value of r^2 is .55. This represents a strong effect.

8. The two variables approximate a direct linear relationship.

12. Independent groups t test

13. One-way repeated measures analysis of variance

14. Correlated groups t test

17. $N = 54$

20. The estimated standard error of estimate estimates the average error that will be made across the population when predicting scores on Y from the regression equation.

21. $\hat{Y} = 6.87 + .46X$; 8.25; 10.09; 11.93

22. 1.71

25. b   34. c
27. d   35. b
30. b   37. a
31. a   40. b
33. d   41. c
44.

```
                    Results
     A Pearson correlation with an alpha
level of .05 addressed the relation-
ship between the leader's LPC score
(M = 67.30, SD = 4.79) and minutes
until solution of the problem (M =
10.90, SD = 4.04). This was found to
be statistically significant, r(8) =
.67, p < .05, indicating that these
two variables are positively related.
This suggests that group problem-
solving performance deteriorates as
the leader's LPC score increases. The
regression equation for predicting
minutes until solution from the
leader's LPC score was found to be
Ŷ = -27.39 + .57X, and the estimated
standard error of estimate was found
to be 3.18.
```

47. The critical values of r for an alpha level of .05 and 8 degrees of freedom are ±.632. Since the observed r value of .75 is greater than +.632, we reject the null hypothesis and conclude that some degree of linear relationship exists between GRE-A scores and graduate GPA. The regression equation for predicting graduate GPA from GRE-A scores is $\hat{Y} = .52 + .005X$. Based on this equation, the predicted 2-year grade point averages of applicants 1, 3, 5, and 7 are all greater than 3.00, so these applicants are viable candidates for admission.

Chapter 15

4. Observed frequencies are the frequencies that are actually observed in an investigation. Expected frequencies are the frequencies that we would expect to observe if the two variables under study were unrelated in the population.

5. 110

7. 255

8.

Cell	E
Romance-infrequent	10.59
Romance-moderate	15.00
Romance-frequent	19.41
Comedy-infrequent	21.18
Comedy-moderate	30.00
Comedy-frequent	38.82
Drama-infrequent	28.24
Drama-moderate	40.00
Drama-frequent	51.76

9. a. 3.841   c. 5.991
   b. 9.488   d. 16.919

11. The disadvantage is that Yates' correction tends to reduce statistical power while adding little control over Type I errors.

13. The critical value of chi-square for an alpha level of .05 and 1 degree of freedom is 3.841. Since the observed chi-square value of 7.11 is greater than 3.841, we reject the null hypothesis and conclude that there is a relationship between smoking behavior and cause of death.

14. .22

15. Examination of the $(O - E)^2/E$ and $O - E$ values suggests that smokers are more likely than expected to die of cancer and less likely than expected to die of other causes, whereas the reverse is true of nonsmokers.

16. The observed chi-square using Equation 15.4 is 7.11. This is the same value as was obtained in Exercise 13 using Equation 15.2.

20. The advantage of analyzing quantitative variables with parametric tests rather than the chi-square test is that parametric tests tend to be more powerful. An advantage of the chi-square approach is that a quantitative variable need be measured on only an ordinal level as opposed to the approximately interval level required for parametric tests.

21. Pearson correlation

22. Chi-square test

27. .85

28. The critical value of chi-square for an alpha level of .05 and 3 degrees of freedom is 7.815. Since the observed chi-square value of 11.37 is greater than 7.815, we reject the null hypothesis and conclude that the distribution of evening news program preference for the population of college students is not the same as national ratings.

29. Examination of the $(O - E)^2/E$ and $O - E$ values suggests that college students are more likely than expected to prefer the CBS and NBC evening news programs, and less likely than expected to prefer the ABC evening news program or to have no evening news program preference.

34. b

36. a

37. d

40. b   43. a

41. b   44. c

46.   Results

    The relationship between the gender of the central character and the role portrayed by that character was analyzed using a chi-square test. The alpha level was .05. This test was found to be statistically significant, $\chi^2(1, \underline{N} = 315) = 10.68$, $\underline{p} < .01$. The observed frequencies for the four cells can be found in Table 1.

    The strength of the relationship, as indexed by the fourfold point correlation coefficient, was .18. This reflects primarily the fact that female central characters are less likely than expected to be portrayed as authorities and more likely than expected to be portrayed as users.

    (*Note:* Table 1 would be similar to the table in the exercise, but it would also include the marginal frequencies.)

Chapter 16

3.

Set I	Set II	Set III	Set IV
5	1.5	1.5	4
2	1.5	6	2
6	6	4.5	2
1	5	1.5	2
3.5	3	4.5	5
3.5	4	3	6

5. The rank transformation approach to nonparametric analysis involves converting a set of scores on a variable to ranks and then analyzing these rank scores using the traditional parametric formulas.

7. The critical values of z for an alpha level of .05 are ±1.96. Since the observed z value of .39 is neither less than −1.96 nor greater than +1.96, we fail to reject the null hypothesis of no relationship between car ownership and performance in school.

8. .09

9. The critical value of U for an alpha level of .05 and $n_1 = n_2 = 8$ is 13. Since the observed U value of 25 is not equal to or less than 13, we fail to reject the null hypothesis of no relationship between gender and attitudes toward the pill.

10. .22

15. The critical value of χ_r^2 for an alpha level of .05 and 2 degrees of freedom is 5.991. Since the observed χ_r^2 value of 7.20 is greater than 5.991, we reject the null hypothesis and conclude that there is a relationship between the brand of picture tube and quality ratings. (*Note:* The nature of the relationship can be determined using the Dunn procedure.)

16. .36

17. The critical values of r_s for an alpha level of .05 and $N = 18$ are ±.475. Since the observed r_s value of +.48 is greater than +.475, we reject the null hypothesis and conclude that there is a relationship between crime rates in cities and the size of a city's police force. The strength of the relationship, as indicated by the magnitude of the correlation coefficient, is .48. The nature of the relationship, as indicated by the sign of the correlation coefficient, is that the two variables approximate a direct linear relationship.

18. d

21. a

22. b
24. a
25. a
30. a
31. a
32. d
36. d
37. b
38. b
42.

                          Results
        A Wilcoxon signed-rank test with an
    alpha level of .05 compared individu-
    als' self-esteem before versus after
    participating in the encounter ses-
    sion. The rank sums were found to not
    significantly differ, $\underline{N}$ = 10, $\underline{T}$ = 22,
    $\underline{p}$ > .10.

43.

                          Results
        A Kruskal-Wallis test was applied
    to the ranked data relating the af-
    fective content of words to memory.
    The alpha level was .05. The result-
    ing value of $\underline{H}$ was found to be sta-
    tistically nonsignificant, $\underline{H}$(2, $\underline{N}$ =
    15) = 2.78, $\underline{p}$ > .30.

Chapter 17

2. 2; 2; 3
5. The terminology "main effect" has two related meanings in the context of two-way factorial designs. First, a main effect refers to the comparison of the means for the levels of one independent variable collapsed across the second independent variable. Second, a main effect is said to be present if the null hypothesis concerning that effect is rejected. An interaction effect refers to the comparison of the cell means in terms of whether the nature of the relationship between one of the independent variables and the dependent variable differs as a function of the other independent variable. If the null hypothesis of no interaction effect is rejected, an interaction is said to be present.

6.

	Main effect of factor A?	Main effect of factor B?	Interaction?
a.	No	Yes	No
b.	Yes	No	No
c.	Yes	Yes	No

8.

	B_1	B_2	B_3
A_1	6.00	5.00	4.00
A_2	1.00	2.00	3.00

10. Nonparallel lines in a graph of population means indicate that an interaction is present. Nonparallel lines in a graph of sample means indicate that an interaction *might* be present. The difference in the two situations is due to the fact that, because of the role of sampling error, nonparallel lines in a graph of sample means do not necessarily indicate that an interaction exists in the population.

13.

Source	SS	df	MS	F
A	20.00	2	10.00	5.00
B	45.00	3	15.00	7.50
$A \times B$	60.00	6	10.00	5.00
Within	216.00	108	2.00	
Total	341.00	119		

15. a. 4.00; 3.15; 3.15
    b. 4.04; 2.80; 2.80
    c. 3.10; 3.10; 2.47
    d. 2.76; 3.15; 2.25
17. 2; 3; 6
18. 54; 9
19. 130.00
20. Hypotheses for the main effect of factor A:
    H_0: $\mu_{A_1} = \mu_{A_2}$
    H_1: $\mu_{A_1} \neq \mu_{A_2}$
    Hypotheses for the main effect of factor B:
    H_0: $\mu_{B_1} = \mu_{B_2} = \mu_{B_3}$
    H_1: The three population means are not all equal
    Hypotheses for the interaction effect:
    H_0: The mean difference on the dependent variable as a function of factor A is the same at each level of factor B
    H_1: The mean difference on the dependent variable as a function of factor A is not the same at each level of factor B
21. The critical value of F for an alpha level of .05 and 1 and 48 degrees of freedom is 4.04. Since the observed F value of 10.00 for the main effect of factor A is greater than 4.04, we reject the null hypothesis and conclude that there is a relationship between this factor and the dependent variable.

The critical value of F for an alpha level of .05 and 2 and 48 degrees of freedom is 3.19. Since the observed F value of 4.00 for the main effect of factor B is greater than 3.19, we reject the null hypothesis and conclude that a relationship exists between this factor and the dependent variable.

The observed F value of 4.00 for the interaction effect is greater than the critical value of 3.19 for an alpha level of .05 and 2 and 48 degrees of freedom, so we reject the null hypothesis and conclude that there is an interaction.

22. The value of eta-squared for the main effect of factor A is .14. This represents a moderate effect. The value of eta-squared for both the main effect of factor B and the interaction effect is .11. These also represent moderate effects.

23. 6; 10; 9

24. The critical value of F for an alpha level of .05 and 1 and 60 degrees of freedom is 4.00. Since the observed F value of 24.00 for the main effect of factor A is greater than 4.00, we reject the null hypothesis and conclude that a relationship exists between this factor and the dependent variable. Scores in condition A_2 ($\overline{X}_{A_2} = 19.33$) are significantly higher than scores in condition A_1 ($\overline{X}_{A_1} = 15.00$).

25. The critical value of F for an alpha level of .05 and 2 and 60 degrees of freedom is 3.15. Since the observed F value of 45.50 for the main effect of factor B is greater than 3.15, we reject the null hypothesis and conclude that there is a relationship between this factor and the dependent variable.

The HSD test can be applied as follows:

Null hypothesis tested	Absolute difference between sample means	Value of CD	Null hypothesis rejected?
$\mu_{B_1} = \mu_{B_2}$	$|12.00 - 17.50| = 5.50$	2.29	Yes
$\mu_{B_1} = \mu_{B_3}$	$|12.00 - 22.00| = 10.00$	2.29	Yes
$\mu_{B_2} = \mu_{B_3}$	$|17.50 - 22.00| = 4.50$	2.29	Yes

Scores in condition B_3 ($\overline{X}_{B_3} = 22.00$) are significantly higher than scores in either condition B_2 ($\overline{X}_{B_2} = 17.50$) or condition B_1 ($\overline{X}_{B_1} = 12.00$). In turn, scores in condition B_2 are significantly higher than scores in condition B_1.

26. The critical value of F for an alpha level of .05 and 2 and 60 degrees of freedom is 3.15. Since the observed F value of .50 for the interaction effect is not greater than 3.15, we fail to reject the null hypothesis of no interaction between factor A and factor B.

35. Unequal cell sizes are problematic because they introduce a relationship between the two independent variables into the sample. The introduction of such a relationship creates a number of statistical and conceptual issues in testing the two main effects and the $A \times B$ interaction.

36. $n = 38$
39. c
41. b
43. b
44. b
47. b

49.

Source	SS	df	MS	F
A (supposed time)	18.87	2	9.43	11.79
B (weight)	22.53	1	22.53	28.17
A × B	11.27	2	5.63	7.04
Within	19.20	24	.80	
Total	71.87	29		

HSD test for the main effect of supposed time:

Null hypothesis tested	Absolute difference between sample means	Value of CD	Null hypothesis rejected?
$\mu_E = \mu_T$	$|6.00 - 7.30| = 1.30$	1.00	Yes
$\mu_E = \mu_O$	$|6.00 - 7.90| = 1.90$	1.00	Yes
$\mu_T = \mu_O$	$|7.30 - 7.90| = .60$	1.00	No

interaction comparisons [*Note*: When the F ratios for two interaction comparisons are the same, the determination of which is tested against the smaller "critical alpha" should be based on theoretical considerations, when relevant. When theory does not suggest an ordering, as in our example, both F ratios can be tested against the average of the two "critical alphas" that would otherwise be used. Thus, the "critical alpha" is $(.05/3 + .05/2)/2 = (.017 + .025)/2 = .021$ for the first two interaction comparisons.]:

Interaction comparison	F ratio	p value	Critical alpha	Null hypothesis rejected?
A × B$_{(1)}$	10.56	.003	.021	Yes
A × B$_{(2)}$	10.56	.003	.021	Yes
A × B$_{(3)}$	0	1.000	.05/1 = .050	No

Results

Hunger scores were subjected to a two-way analysis of variance having three levels of supposed time (11:00, 12:00, and 1:00) and two levels of weight (overweight versus normal). All effects were found to be statistically significant for an alpha level of .05. The main effect of weight was such that overweight individuals (M = 7.93, SD = 1.67) reported significantly greater hunger than normal-weight individuals (M = 6.20, SD = .86), F (1, 24) = 28.17, p < .01. The strength of the relationship, as indexed by eta^2, was .31. The nature of the main effect of supposed time, F (2, 24) = 11.79, p < .01, was determined using a Tukey HSD test. This indicated that hunger was significantly greater at both 12:00 (M = 7.30, SD = 1.64) and 1:00 (M = 7.90, SD = 1.66) than at 11:00 (M = 6.00, SD = .67). The 12:00 and 1:00 means did not significantly differ. As indexed by eta^2, the strength of the relationship was .26.

The interaction effect, F (2, 24) = 7.04, p < .01, was analyzed using interaction comparisons in conjunction with a modified Bonferroni procedure (Holland & Copenhaver, 1988) based on an overall alpha level of .05. The relevant means and standard deviations can be found in Table 1. The 2 x 2 subtables for 11:00 versus 12:00 and 11:00 versus 1:00 yielded statistically significant interaction effects. Overweight individuals reported greater hunger at 12:00 than at 11:00 (8.60 - 6.00 = 2.60) relative to normal-weight individuals (6.00 - 6.00 = .00). The former individuals also reported greater hunger at 1:00 than at 11:00 (9.20 − 6.00 = 3.20) relative to the latter (6.60 − 6.00 = .60). The change in hunger between 1:00 and 12:00 did not differ for overweight (9.20 − 8.60 = .60) versus normal-weight (6.60 − 6.00 = .60) participants. The strength of

the overall interaction effect, as indexed by eta^2, was .16.

Table 1
Mean Hunger As a Function of Supposed Time and Weight

	Supposed time		
Weight	11:00	12:00	1:00
Overweight	6.00 (.71)	8.60 (1.14)	9.20 (.84)
Normal	6.00 (.71)	6.00 (.71)	6.60 (1.14)

Note: Standard deviations are in parentheses. For each cell, n = 5.

Chapter 18

1. The chi-square test.

4. The most common method of analysis for two between-subjects variables that are measured on a level that at least approximates interval characteristics is Pearson correlation. This technique is appropriate when the researcher wants to evaluate a linear relationship. If the expected relationship is nonlinear, procedures for nonlinear relationships can be applied. When one or both variables are measured on an ordinal level that seriously departs from interval level characteristics, Spearman rank-order correlation or some other nonparametric correlational index is the test of choice. The Case II statistics given in Table 18.2 might be applicable when the independent variable is within-subjects in nature or has fewer than five or so values associated with it. If both variables have only two or three values, the chi-square test can be used.

5. The independent variable is the time of year. This variable is qualitative in nature. The dependent variable is scores on the depression scale. This variable is quantitative in nature. Hence, this is a Case II situation. Since the independent variable is within-subjects and has two levels, the appropriate parametric and nonparametric tests for analyzing the relationship between the variables are the correlated groups t test and the Wilcoxon signed-rank test, respectively.

6. The independent variable is the color of the ice cream. This variable is qualitative in nature. The dependent variable is the taste ratings. This

variable is quantitative in nature. Hence, this is a Case II situation. Since the independent variable is within-subjects and has three levels, the appropriate parametric and nonparametric tests for analyzing the relationship between the variables are one-way repeated measures analysis of variance and Friedman analysis of variance by ranks, respectively.

7. The independent variable is the noise level. This variable is quantitative in nature. The dependent variable is the amount of growth. This variable is also quantitative in nature. Hence, this is a Case IV situation. Since the independent variable is between-subjects and has only two levels, the appropriate parametric and nonparametric tests for analyzing the relationship between the variables are the independent groups t test and the Wilcoxon rank sum test/Mann–Whitney U test, respectively.

8. The independent variable is parents' marital status. This variable is qualitative in nature. The dependent variable is the presence or absence of an imaginary friend. This variable is also qualitative in nature. Hence, this is a Case I situation. Since both variables are between-subjects, the appropriate test for analyzing the relationship between the variables is the chi-square test.

9. The independent variable is social class. This variable is quantitative in nature. The dependent variable is dogmatism. This variable is also quantitative in nature. Hence, this is a Case IV situation. The appropriate parametric and nonparametric tests for analyzing the relationship between the variables are Pearson correlation (assuming the expected relationship is linear) and Spearman rank-order correlation, respectively.

10. The independent variable is the number of close friends people have. This variable is quantitative in nature. The dependent variable is whether or not people feel these friends would, in general, "be there" for them in an emergency. This variable is qualitative in nature. Hence, this is a Case III situation. The appropriate test for analyzing the relationship between the variables is logistic regression or polychotomous logistic regression.

11. The independent variable is the sipping rate of the experimental assistant. This variable is quantitative in nature. The dependent variable is participants' consumption time. This variable is also quantitative in nature. Hence, this is a Case IV situation. Since the independent variable is be-

tween-subjects and has only three levels, the appropriate parametric and nonparametric tests for analyzing the relationship between the variables are one-way between-subjects analysis of variance and the Kruskal–Wallis test, respectively.

12. The independent variable is whether or not hypnosis was administered. This variable is qualitative in nature. The dependent variable is the change in hand temperature. This variable is quantitative in nature. Hence, this is a Case II situation. Since the independent variable is between-subjects and has two levels, the appropriate parametric and nonparametric tests for analyzing the relationship between the variables are the independent groups t test and the Wilcoxon rank sum test/Mann–Whitney U test, respectively.

13. The independent variable is the time before taking the exam. This variable is quantitative in nature. The dependent variable is scores on the Palmer Sweat Index. This variable is also quantitative in nature. Hence, this is a Case IV situation. Since the independent variable is within-subjects and has four levels, the appropriate parametric and nonparametric tests for analyzing the relationship between the variables are one-way repeated measures analysis of variance and Friedman analysis of variance by ranks, respectively.

23. The defining characteristic of multivariate statistics is that they analyze the relationship between three or more variables.

25. Conducting multiple analyses of variance when analyzing two or more dependent variables is problematic because it increases the probability of making a Type I error for at least one of the tests beyond the probability specified by the alpha level for any single analysis.

27. A regression coefficient (β) represents the number of units the criterion variable in multiple regression is predicted to change for each unit change in a given predictor variable when the effects of the other predictor variables are held constant. A squared multiple correlation coefficient (R^2) is an index of the strength of the relationship between the criterion variable and the set of predictor variables. Specifically, it indicates the proportion of variability in the criterion variable that is associated with the predictor variables considered simultaneously.

28. The goal of factor analysis is to determine whether the correlations among a set of variables

can be accounted for by one or more underlying
dimensions. This is accomplished through the
analysis of the patterning of correlations (or covariances) between all pairs of variables in the
data set in terms of the underlying factor
structure.

32. c
33. a
36. a
37. b
39. a

Index

Abscissa, 43
Alpha level, 171, 210
    for Dunn procedure, 464, 467
    effect on Type I errors, 213–215
    effect on Type II errors/power, 214–215, 282
    relationship to significance level, 230
    for Tukey HSD test, 340
Alternative hypothesis, 171, 207, 266–267. *See also spe-cific tests*
    directional versus nondirectional, 216–218
American Psychological Association style. *See* Method of presentation
Analysis of variance
    multivariate, 538–539
    one-way between-subjects. *See* One-way between-subjects analysis of variance
    one-way repeated measures. *See* One-way repeated measures analysis of variance
    two-way between-subjects. *See* Two-way between-subjects analysis of variance
    two-way between-within, 538
    two-way repeated measures, 537–538
Assumptions. *See also specific tests*
    assessment of violations, 253
    bivariate normal distribution, 396
    effect on test selection, 251–254, 451–452, 532–537
    expected frequencies, 428
    homogeneity of variance, 252–253, 264
    normality, 222, 252–253, 535
    sphericity, 371–372

Bayesian statistics, 541
Bayes' theorem, 164, 541
Beta, 212
Biased estimator, 184
Binomial distribution, 171–173
Binomial expression, 169–173
Bivariate normal distribution assumption, 396
Bivariate relationships, 240

Carry-over effects, 250
    control by counterbalancing, 306–307, 375–376
    control by random ordering of conditions, 375–376, 382
Causation, inference of, 138–139, 248, 406, 432
Cell, 158, 422, 486
Central Limit Theorem, 189
Central tendency
    comparison of measures, 75–77, 80, 90, 197
    defined, 67–78
    for discrete variables, 77

Central tendency *(continued)*
    graphs, 86–89
        bar graph, 86–87
        box and whisker plot, 87–89
        boxplot, 87–89
        line plot, 87
    mean. *See* Mean
    median, 68–72, 75–77, 90
    mode, 68, 75–77
    relationship to variability, 84–86, 95
Chi-square distribution, 426–427
Chi-square goodness-of-fit test, 439–441
Chi-square statistic, 425–426
Chi-square test of independence/homogeneity
    assumptions, 428
    computer output, 442–443
    conditions for use, 421
    degrees of freedom, 426
    expected frequencies, 423–426, 428
    formula, 425
    hypotheses, 424
    inference of a relationship, 426
    method of presentation, 436
    nature of the relationship, 431
        intuitive approach, 431–432
        modified Bonferroni procedure, 444–445
    observed frequencies, 422–426, 428
    power, 435
    use of quantitative variables, 434–435
    random versus fixed variables, 423
    relationship to log-linear analysis, 540
    strength of the relationship (fourfold point correlation coefficient/Cramer's statistic), 430–431
    for 2×2 contingency tables, 429–430
Coefficient of determination, 397
Cohort effects, 255
Combinations, 167–168
Computer packages, 26, 60
Concordance coefficient, 467
Confidence intervals
    computer output, 232
    for eta-squared, 276–277
    for the mean, 225–228
Confidence limits, 225
Confounding variables
    control of, 244–250, 306–307, 375–376
    defined, 244
    relationship to causation, 248
Constant, 5
Contingency table, 158, 422
Continuous variables, 14–15

Control group, 241
Correlated groups *t* test
    assumptions, 303–304
    computer output, 314–316
    conditions for use, 298
    degrees of freedom, 301
    formula, 301
    hypotheses, 299–300
    inference of a relationship, 301–302
    method of presentation, 312–313
    nature of the relationship, 306
    nonparametric counterpart, 459–462
    nullified score approach, 316–317
    power, 308–309, 311–312
    strength of the relationship (eta-squared), 304–306
Correlation matrix, 151
Correlation. *See* Pearson correlation; Spearman rank-order
        correlation
Counterbalancing, 306–307
Counting rules, 166–169
Covariance, 135
Cramer's statistic, 430–431
Criterion variable, 401
Critical difference
    for one-way between-subjects analysis of variance, 341
    for one-way repeated measures analysis of variance, 373
    for two-way between-subjects analysis of variance,
        502–503
Critical values, 209–210
Cross-sectional design, 254–255
Crosstabulation table, 158, 422
Curvilinear regression, 145

Degrees of freedom, 187–188. *See also specific tests*
Density curve, 55
Dependent variables, 5–6, 241–242
Descriptive statistics, 18–19
    computer output, 93–94
Deviation scores, 72–74
Directional tests, 216–218
Discrete variables, 14–15
Distribution-free statistics. *See* Nonparametric statistics
Distributions
    chi-square, 426–427
    effect of standardization, 113
    empirical, 56–57
    F, 331–332, 338
    frequency. *See* Frequency distributions
    normal. *See* Normal distribution
    probability, 53–56, 159
    t, 219–220, 338
    theoretical, 56–57
Disturbance variables
    control of, 245–246, 299, 307, 375–376, 505
    defined, 245
    relationship to unexplained variance, 274
Dunn procedure, 464, 467

Epsilon-squared, 464
Error variance, 274
Estimated standard error. *See* Standard error

Eta-squared, 271–277. *See also specific tests*
Expected frequencies, 423–426
    assumption, 428
Experimental research strategy, 241
Explained variance, 274

Factor analysis, 540
Factorial, 167–168
Factorial designs, 485–486
Factor structure, 540
F distribution, 331–332
    relationship to *t* distribution, 338
Fisher's exact test, 428, 430
Fisher's transformation of Pearson *r*, 412–413
Fixed effects analysis of variance, 324
Fourfold point correlation coefficient, 430–431
F ratio, 325–326, 328–330
    for interaction comparisons, 505
    multivariate, 538–539
    for one-way between-subjects analysis of variance,
        325–326, 328–330
    for one-way repeated measures analysis of variance,
        364, 367–369, 374–375
    sampling distribution, 330–332
    for two-way between-subjects analysis of variance, 496,
        500–501
Frequencies
    absolute, 35
    cumulative, 37
    cumulative relative, 37
    computer output, 60–62
    expected, 423–426, 428
    observed, 422–426
    relative, 36–37
        comparison with probability, 53
Frequency distributions, 34–42
    defined, 35
    for qualitative variables, 41–42
    for quantitative variables
        grouped scores, 38–40
        ungrouped scores, 34–38
Frequency graphs
    bar graph, 50–51
    of cumulative frequencies, 52–53
    of cumulative relative frequencies, 53
    frequency histogram, 43–48
        for multiple groups, 46–47
    frequency polygon, 44–46
        for multiple groups, 45–46
    for grouped scores, 46–48
    line plot, 45–47
    misleading graphs, 51–52
    for qualitative variables, 50–51
    of relative frequencies, 52
    stem and leaf plot, 48–50
Frequency table, 158, 422
Friedman analysis of variance by ranks, 465–468

Glass rank biserial correlation coefficient, 458
Goodness-of-fit test, 439–441
Greenhouse–Geisser epsilon, 371

Homogeneity of variance assumption, 252–253, 264
Homogeneity of variance, tests of, 253, 287–289, 352, 522
Hotelling T^2 test, 539
HSD test. *See* Tukey HSD test
H statistic, 462–464
Huynh–Feldt epsilon, 371, 384
Hypothesis, 4
Hypothesis testing, 170–171, 205–210. *See also specific tests*
    alpha level, 171, 210
    alternative hypothesis, 171, 207, 266–267
    alternatives to, 540–541
    directional versus nondirectional tests, 216–218
    errors in, 212–214
    null hypothesis, 171, 206–207, 209, 266–267
        failing to reject versus accepting, 211–212
    relationship to confidence intervals, 228
    steps for independent groups t test, 266–269
    steps for one-sample z test, 209–210

Independent groups t test
    assumptions, 270
    computer output, 287–290
    conditions for use, 259
    degrees of freedom, 267
    formula, 268
    hypotheses, 261, 266–267
    inference of a relationship, 268–269
    method of presentation, 283–284
    nature of the relationship, 277–278
    nonparametric counterparts, 455–459
    power, 281–283, 308–309
    strength of the relationship (eta-squared), 271–276
Independent variables, 5–6, 241, 485–486
    relationship to confounding variables, 244
Inference of a relationship. *See specific tests*
Inferential statistics, 18–19
Interaction comparisons, 503–505, 511–513
Interaction effects, 488–493, 495, 501, 503–505, 511–513
Intercept, 129, 393
    of regression line, 141–142
Interpolation, 221
Interquartile range, 78–79, 103, 118
Interval estimation, 228, 541. *See also* Confidence intervals
Interval measurement, 7–10

Kruskal–Wallis test, 462–465, 473–474
Kurtosis, 90–91

Least squares analysis of variance, 515, 523
Least squares criterion, 141–142
Levene test, 287–289, 352
Linear equation, 130
Linear model, 126–130, 393, 401, 539
Log-linear analysis, 438, 540
Longitudinal design, 254–255
"Lower-bound" epsilon, 384

Main effects, 487–493, 500–503
Mann–Whitney U test, 455–459
Marginal frequencies, 158, 422–423
    column, 425
    row, 425
Margin of error, 192
Matched-pairs rank biserial correlation coefficient, 461
Matching, 246
Mauchly test, 371–372, 382–384
Mean, 73–77, 186
    defined, 73
    for dichotomous variables, 94
    estimated, 182–184
    formula, 73
    grand, 272, 305
    in normal distributions, 75, 90
    sampling distribution of, 188–197
    in skewed distributions, 89–90
Mean square(s)
    defined, 188
    for interaction comparisons, 505
    for one-way between-subjects analysis of variance, 328–330
    for one-way repeated measures analysis of variance, 364, 367–368
    for two-way between-subjects analysis of variance, 496, 499
Measurement, 6–13, 291
Median 68–72, 75–77, 90, 101–102
Median of absolute deviations (MAD), 118
M estimators, 408–409
Method of presentation, 57–59. *See also specific tests*
    directional versus nondirectional tests, 230
    frequency information, 51–52, 58–59
    graphs, 51–52, 59, 517
    measures of central tendency, 91–92
    measures of variability, 91–92
    referencing statistical procedures, 465
    research report sections, 229
    results of statistical tests, 229–230
    scores on educational and psychological tests, 113–115
    statistical symbols, 59
    tables, 58–59, 436, 516–518
Mode, 68
Modified Bonferroni procedure
    for chi-square test, 444–445
    for one-way repeated measures analysis of variance, 373, 385–386
    for two-way between-subjects analysis of variance, 505, 512–513
Monte Carlo studies, 253
Multiple comparison procedures, 339–340. *See also* Tukey HSD test
Multiple regression, 407, 539–540
Multivariate analysis of variance, 538–539
Multivariate statistics, 537–540

Nature of the relationship. *See specific tests*
Nominal measurement, 7–10
Nondirectional tests, 216–218
Nonparametric statistics, 251–252, 451–452

Nonparametric statistics *(continued)*
    outliers, 454
    rank transformation approach, 455
Normal distribution, 56–57
    characteristics, 56, 109–110
    defined, 56
    formula, 121
    relationship to binomial distribution, 171–173
    relationship to *t* distribution, 220
    standard scores in, 109–113
    values of mean, median, and mode, 75
Normality assumption, 222, 252–253, 535
Null hypothesis, 171, 206–207, 209, 266–267. *See also specific tests*
    failing to reject versus accepting, 211–212
Nullified scores
    for correlated groups *t* test, 305–306
    for independent groups *t* test, 273–274
    for one-way repeated measures analysis of variance, 362–363

Observational research strategy, 241
Observed frequencies, 422–426
Odds ratios, 541
Omega-squared, 276
One-sample *t* test, 218–223
    assumptions, 222–223, 253
    conditions for use, 222
    degrees of freedom, 221
    formula, 221
    hypotheses, 206–207, 209
    method of presentation, 229–230
One-sample *z* test, 207–210
    conditions for use, 207
    formula, 209
    hypotheses, 206–207, 209
One-tailed tests, 216–218
One-way between-subjects analysis of variance
    assumptions, 337–338
    computer output, 352–353
    conceptualization, 325–330
    conditions for use, 323
    degrees of freedom, 329, 354, 375
    formulas, 329–330, 333–335
    F ratio, 325–326, 328–330, 375
    hypotheses, 324–325
    inference of a relationship, 332
    mean squares, 328–330
    method of presentation, 347–348
    nature of the relationship, 339–343
        Tukey HSD test, 340–343
    nonparametric counterpart, 462–465
    power, 347
    strength of the relationship (eta-squared), 338–339
    summary table, 331–332
    sums of squares, 327–328, 363–364, 375
        formulas, 333–335
One-way repeated measures analysis of variance
    assumptions, 371–372
    computer output, 382–384
    conditions for use, 361

One-way repeated measures analysis of variance *(continued)*
    degrees of freedom, 367, 371, 375
    formulas, 364–368
    F ratio, 364, 368, 374–375
    hypotheses, 363
    inference of a relationship, 368–369
    mean squares, 364, 367–368
    method of presentation, 380
    nature of the relationship
        when sphericity is met (Tukey HSD test), 373
        when sphericity is violated (modified Bonferroni procedure), 373, 385–386
    nonparametric counterpart, 465–468
    power, 379–380
    strength of the relationship (eta-squared), 372
    summary table, 368
    sums of squares, 363–367, 374–375
Ordinal measurement, 7–10
Ordinate, 43
Outliers, 42–43, 76–77, 88, 149–150, 454

Parameters, 18, 270–271
Parametric statistics, 251–252, 451–452
Pearson correlation
    assumptions, 396
    coefficient
        defined, 130–131
        formulas, 134–136
    computer output, 409–410
    conditions for use, 125, 392
    degrees of freedom, 394
    hypotheses, 393–394
    inference of a relationship, 395–396
    issues
        nonlinear relationships, 145
        outliers, 149–150
        restricted range, 147–148
    method of presentation, 398–399
    nature of the relationship, 131, 397
    nonparametric counterpart, 468–471
    power, 398
    relationship to linear model, 126–130, 393
    relationship to regression, 131
    strength of the relationship (r^2), 397
    testing null hypotheses other than $\rho = 0$, 412–413
Percentage, 37, 101
    cumulative, 37
Percentile, 101–103
Percentile rank, 101, 104–105
Permutations, 167–168
Point-biserial correlation coefficient, 275
Point estimation, 277
Polynomial regression, 145
Pooled variance estimate, 264–265
    relationship to mean square within, 330
Population
    defined, 16
    finite versus infinite, 181–182
Power, 281–283. *See also specific tests*
    as affected by alpha, 214–215, 281–283

Power *(continued)*
    as affected by sample size, 215, 281–283
    as affected by the strength of the relationship, 281–283
    of correlated groups versus independent groups *t* tests, 308–309
    defined, 213
    of directional versus nondirectional tests, 217–218
    of one-way between-subjects versus one-way repeated measures analysis of variance, 364, 374–375
    of parametric versus nonparametric statistics, 252, 434–435
Predictor variable, 401
Presentation of statistical results. *See* Method of presentation
Probability
    adding probabilities, 162, 164
    as affected by sampling with versus without replacement, 166
    binomial expression, 169–173
    comparison with relative frequency, 53
    conceptualizations of, 19–20, 157–158
    conditional, 160, 162–164, 514
    counting rules, 166–169
    independence of events, 160
    joint, 161–164
    simple events, 19–20, 157–159, 162–164
    subjective, 165
Probability density function, 55–56
Probability distributions, 53–56, 159
Probability level. *See* Significance level
p value. *See* Significance level

Qualitative variables
    defined, 11
    frequency distributions, 41–42
    measures of central tendency and variability, 89
Quantitative variables
    defined, 11
    frequency distributions, 34–40

Random assignment, 242, 244, 246
Random number table, 18
Random ordering of conditions, 375–376
Random sampling, 16–18, 181–182, 191–192
    area sampling, 28, 191
    defined, 16
Range, 78
Rank scores, 452–454
    corrections for ties, 474
Rank transformation approach, 455
Ratio measurement, 8–10
Real limits, 15–16
Regression
    coefficient, 539
    curvilinear (polynomial), 145
    equation, 140–141, 401
        for standardized scores, 148–149
    issues
        equation for standardized scores, 148–149
        predicting *X* from *Y*, 146–147
        restricted range, 148

Regression *(continued)*
    least squares criterion, 141–142
    line, 140–142
    method of presentation, 407
    multiple, 407, 539–540
    prediction, 141–142, 401–402
    relationship to Pearson correlation, 131
    standard error of estimate, 143–144, 402–403
Regression coefficient, 539
Regression equation, 140–141, 401
Regression line, 140–142
Rejection region, 209–210
Representative sample, 16
Research design, 240–251
    between-subjects designs, 248–250
    control of variables. *See* Confounding variables; Disturbance variables
    correlated groups designs, 248–250
    cross-sectional design, 254–255
    experimental strategy, 241
    factorial designs, 485–486
    independent groups designs, 248–250
    inference of causation, 248
    longitudinal design, 254–255
    matched-subjects designs, 250, 259
    nonexperimental strategy, 241
    observational strategy, 241
    reduction of sampling error, 242–244
    repeated measures designs, 248–250
    within-subjects designs, 248–250
Research process, 4–5
Robustness, 223, 252–254, 451
Rounding, 25–26
R statistic, 457

Sample, 16
Sampling
    area, 28, 191
    random, 16–18, 181–182, 191–192
    without replacement, 164–166
    with replacement, 164–166, 182
Sampling distribution, 197
    of the chi-square statistic, 426–427
    of the correlation coefficient, 394
    of the difference between two independent means, 262–264
    of the *F* ratio, 330–332
    of the mean, 188–190, 192–197
        mean, 189
        shape, 194–196
        standard deviation, 189–190, 192–197, 207–208
    of the mean of difference scores, 300–301
    of a proportion, 198–199
Sampling error, 183–184
    reduction of, 242–244
    relationship to between- and within-group variability, 325–326
    relationship to sampling distribution of the mean, 188–190, 192, 207–210
Scatterplot, 126
Sensitivity of statistical tests, 299

Significance, statistical versus real-world, 215–216
Significance level
  defined, 230
  reporting of, 229–230
Simple main effects analysis, 503
Skewness, 89–90
Slope, 126–129, 393
  of regression line, 141–142
Smoothing techniques, 410–411
Spearman rank-order correlation, 468–471
Sphericity assumption, 371–372
Squared multiple correlation coefficient, 540
Standard deviation, 81–83, 186
Standard deviation estimate, 185–186, 274
Standard error
  of the correlation coefficient, 394
  of the difference, 263–266
  of estimate, 143–144, 402–403
  of the mean, 189–190, 192–197, 207–208, 218–219
  of the mean of difference scores, 301
Standard scores, 105–113
  defined, 106
  formulas, 106, 110–111
  in normal distributions, 109–113
  relationship to T scores, 114–115
  shape of distribution, 113
Statistics, 18
  descriptive versus inferential, 18–19
Strength of the relationship, 271–277. *See also specific tests*
  following a nonsignificant test, 277
Studentized range values
  for one-way between-subjects analysis of variance, 341
  for one-way repeated-measures analysis of variance, 373
  for two-way between-subjects analysis of variance, 503
Summary table
  for one-way between-subjects analysis of variance, 331–332
  for one-way repeated measures analysis of variance, 368
  for two-way between-subjects analysis of variance, 500
Summation notation, 20–24
Sum of cross-products, 134–136
Sum(s) of squares
  calculation, 79–80, 83–84
  comparison with variance, 80–81
  defined, 79–80
  degrees of freedom, 187
  for interaction comparisons, 504
  for one-way between-subjects analysis of variance, 327–328, 333–335, 363–364, 375
  for one-way repeated-measures analysis of variance, 363–367, 374–375
  for two-way between-subjects analysis of variance, 495–498

Tau, 470
t distribution, 219–220
  relationship to F distribution, 338
  relationship to normal distribution, 220
Test selection, 254–255, 531–537
Test statistic, 209
Treatment effects, 272–273

T scores, 114–115
T statistic, 460
t test
  correlated groups. *See* Correlated groups t test
  independent groups. *See* Independent groups t test
  one-sample. *See* One-sample t test
Tukey HSD test, 339–340
  for one-way between-subjects analysis of variance, 340–343
  for one-way repeated-measures analysis of variance, 373
  for two-way between-subjects analysis of variance, 502–503
Two-tailed tests, 216–218
Two-way between-subjects analysis of variance
  assumptions, 501
  computer output, 522–523
  conditions for use, 486–487
  degrees of freedom, 498–499
    for interaction comparisons, 505
  formulas, 496–500
  F ratios, 496, 500–501,
    for interaction comparisons, 505
  hypotheses
    for interaction effects, 488–489, 495
    for main effects, 487–488
  inference of a relationship, 500–501
  interaction effects, 488–489, 496
  main effects, 487–488, 496
  mean squares, 495–496, 499–501
  method of presentation, 516–518
  nature of the relationship, 502–505
    interaction comparisons, 503–505, 511–513
    modified Bonferroni procedure, 505, 512–513
    simple main effects analysis, 503
    Tukey HSD test, 502–503
  power, 515–516
  strength of the relationship (eta-squared), 502
  summary table, 500
  sums of squares, 495–498
  unequal sample sizes, 513–515
Type I errors, 212–214
  control of, 282, 340. *See also* Modified Bonferroni procedure; Tukey HSD test
Type II errors, 212–214, 282–283

Unbiased estimator, 183–184
Unexplained variance, 274
Uniqueness, 117–118
U statistic, 457–458

Variability
  between-group, 325–326, 496
  comparison of measures, 82–83, 186
  defined, 67–68, 78
  graphs of, 86–89
  interquartile range, 78–79, 103, 118
  partitioning of
    for independent groups t test, 271–274
    for one-way between-subjects analysis of variance, 326–328

Variability *(continued)*
    for one-way repeated measures analysis of variance,
        363–364
    for two-way between-subjects analysis of variance,
        495–496
    range, 78
    relationship to central tendency, 84–86, 95
    standard deviation, 81–82
    standard deviation estimate, 185–186, 274
    sum of squares. *See* Sum(s) of squares
    variance, 80–81
    variance estimate, 184–186, 274
        pooled, 264–265, 330
    within-group, 326
Variables
    between-subjects, 249
    confounding. *See* Confounding variables
    continuous, 14–15
    criterion, 401
    defined, 5
    dependent, 5–6, 241–242
    discrete, 14–15
    disturbance. *See* Disturbance variables
    holding constant, 245–246
    independent, 5–6, 241–242, 485–486
    predictor, 401
    qualitative. *See* Qualitative variables

Variables *(continued)*
    quantitative. *See* Quantitative variables
    within-subjects, 249
Variance, 80–83, 186
Variance estimate, 184–186, 274
    pooled, 264–265, 330
    relationship to mean square between, 329–330
Variance extraction
    for correlated groups *t* test, 304–306
    for independent groups *t* test, 272–274
Variance ratio. *See* F ratio

Wilcoxon rank sum test, 455–459, 474
Wilcoxon signed-rank test, 459–462
Within-subjects designs, 249–250

Yates' correction for continuity, 429

z scores, 109–113
    defined, 110
z test
    one-sample, 206–210
    for proportions, 444–445

Credits

TO THE OWNER OF THIS BOOK:

We hope that you have found *Statistics for the Behavioral Sciences*, Third Edition, useful. So that this book can be improved in a future edition, would you take the time to complete this sheet and return it? Thank you.

School and address: _____

Department: _____

Instructor's name: _____

1. What I like most about this book is: _____

_____

_____

2. What I like least about this book is: _____

_____

_____

3. My general reaction to this book is: _____

_____

4. The name of the course in which I used this book is: _____

_____

5. Were all of the chapters of the book assigned for you to read? _____

   If not, which ones weren't? _____

6. In the space below, or on a separate sheet of paper, please write specific suggestions for improving this book and anything else you'd care to share about your experience in using the book.

_____

_____

_____

_____

_____

Optional:

Your name: _____ Date: _____

May Brooks/Cole quote you, either in promotion for *Statistics for the Behavioral Sciences,* Third Edition, or in future publishing ventures?

Yes: _____ No: _____

Sincerely,

James Jaccard
Michael A. Becker

- -
FOLD HERE

- -
FOLD HERE

Statistical Tests for Analyzing Relationships Between Variables

Parametric Test

Independent Groups *t* Test

Inference of a relationship:	*t* statistic
Strength of the relationship:	Eta-squared
Nature of the relationship:	Inspection of group means

Correlated Groups *t* Test

Inference of a relationship:	*t* statistic
Strength of the relationship:	Eta-squared
Nature of the relationship:	Inspection of group means

One-Way Between-Subjects Analysis of Variance

Inference of a relationship:	*F* ratio
Strength of the relationship:	Eta-squared
Nature of the relationship:	Tukey HSD test

One-Way Repeated Measures Analysis of Variance

Inference of a relationship:	*F* ratio
Strength of the relationship:	Eta-squared
Nature of the relationship:	Tukey HSD test (when sphericity assumption is met)
	Modified Bonferroni procedure (when sphericity assumption is violated)

Pearson Correlation

Inference of a relationship:	*r* statistic
Strength of the relationship:	r^2
Nature of the relationship:	Sign of *r*

Two-Way Between-Subjects Analysis of Variance

Inference of a relationship:	*F* ratio
Strength of the relationship:	Eta-squared
Nature of the relationship:	
Main effects	Inspection of group means (if the main effect has two levels)
	Tukey HSD test (if the main effect has more than two levels)
Interaction effect	Inspection of cell means preceded by interaction comparisons and modified Bonferroni procedure for other than 2×2 designs